Karl Heinrich Lieser

Nuclear and Radiochemistry

Fundamentals and Applications

Second, Revised Edition

 WILEY-VCH

Karl Heinrich Lieser

Nuclear and Radiochemistry

Fundamentals and Applications

Second, Revised Edition

 **WILEY-VCH**

Berlin · Weinheim · New York · Chichester
Brisbane · Singapore · Toronto

Chemistry Library

Prof. Dr. Karl Heinrich Lieser
Fachbereich Chemie
TU Darmstadt
Eduard-Zintl-Institut
Hochschulstraße 4
D-64289 Darmstadt

1st edition, 1997
2nd edition, 2001

Library of Congress Card No.: applied for.

A catalogue record for this book is available from the British Library.

Die Deutsche Bibliothek – CIP Cataloguing-in-Publication-Data

A catalogue record for this publication is available from Die Deutsche Bibliothek

ISBN 3-527-30317-0

Composition: Asco Typesetters, Hong Kong.
Printing: Strauss Offsetdruck GmbH, 69503 Mörlenbach. Bookbinding: J. Schäffer & Co. KG, 67269 Grünstadt.
Printed in the Federal Republic of Germany. _ac*_

Preface

This textbook gives a complete and concise description of the up-to-date knowledge of nuclear and radiochemistry and applications in the various fields of science. It is based on teaching courses and on research spanning over 40 years.

The book is mainly addressed to chemists desiring sound information about this branch of chemistry dealing with the properties of radioactive matter. Students and scientists working in other branches of chemistry, in enviromental science, physics, geology, mineralogy, biology, medicine, technology and other fields will also find valuable information about the principles and applications of nuclear and radio-chemistry.

Nuclear science comprises three overlapping fields, nuclear physics, nuclear and radiochemistry, and nuclear technology. Whereas nuclear physics deals with the physical properties of the atomic nucleus and the energetic aspects of nuclear reactions, in nuclear and radiochemistry the chemical aspects of atomic nuclei and of nuclear reactions (nuclear chemistry) and the chemical properties, preparation and handling of radioactive substances (radiochemistry) are considered. The concern of nuclear technology, on the other hand, is the use of nuclear energy, in particular the production of nuclear fuel and the operation of nuclear reactors and reprocessing plants. A well-founded knowledge of nuclear reactions and of nuclear and radio-chemistry is needed in nuclear technology. Another related field, radiation chem-istry, deals with the chemical effects of radiation, in particular nuclear radiation, and is more closely related to physical chemistry.

Research in nuclear and radiochemistry comprises: Study of radioactive matter in nature, investigation of radioactive transmutations and of nuclear reactions by chem-ical methods, hot atom chemistry (chemical effects of nuclear reactions) and influ-ence of chemical bonding on nuclear properties, production of radionuclides and labelled compounds, and the chemistry of radioelements – which represent more than a quarter of all chemical elements.

Applications include the use of radionuclides in geo- and cosmochemistry, dating by nuclear methods, radioanalysis, the use of radiotracers in chemical research, Mössbauer spectrometry and related methods, the use of radionuclides in the life sciences, in particular in medicine, technical and industrial applications and inves-tigations of the behaviour of natural and man-made radionuclides, in particular actinides and fission products, in the environment (geosphere and biosphere). Dosi-metry and radiation protection are considered in the last chapter of the book.

Fundamentals and principles are presented first, before progressing into more complex aspects and into the various fields of application. With regard to the fact that radioactivity is a property of matter, chemical and phenomenological points of view are presented first, before more theoretical aspects are discussed. Physical prop-erties of the atomic nucleus are considered insofar as they are important for nuclear and radiochemists.

Endeavours are made to present the subjects in clear and comprehensible form and to arrange them in a logical sequence. All the technical terms used are defined when they are first introduced, and applied consistently. A glossary can be found at the end of the text. In order to restrict the volume of the book, detailed derivations of equations are avoided and relevant information is compiled in tables, as far as possible. More complex relations are preferably elucidated by examples rather than by giving lengthy explanations.

For further reading, relevant literature is listed abundantly at the end of each chapter. Generally, it is arranged in chronological order, beginning with literature of historical relevance and subdivided according the subject matter, into general and more special aspects.

I am indebted to many colleagues for valuable suggestions, and I wish to thank Mrs. Boatman for reading the manuscript.

Darmstadt, April 1996   K. H. Lieser

Preface to the second edition

After concept and structure of the book proved to be useful, they have not been changed in the second edition. However, new developments and results have been considered and the text has been revised taking into account new data.

In preparing this edition, I enjoyed the assistance of my son Joachim Lieser, who gave me many valuable hints.

I acknowledge the readiness of the publishers to supplement the text and to make the corrections necessary to bring this book up to date.

Darmstadt, April 2000   K. H. Lieser

Contents

1 Radioactivity in Nature

1.1 Discovery of Radioactivity

Radioactivity was discovered in 1896 in Paris by Henri Becquerel, who investigated the radiation emitted by uranium minerals. He found that photographic plates were blackened in the absence of light, if they were in contact with the minerals. Two years later (1898) similar properties were discovered for thorium by Marie Curie in France and by G. C. Schmidt in Germany. That radioactivity had not been discovered earlier is due to fact that human beings, like animals, do not have sense organs for radioactive radiation. Marie Curie found differences in the radioactivity of uranium and uranium minerals and concluded that the minerals must contain still other radioactive elements. Together with her husband, Pierre Curie, she discovered polonium in 1898, and radium later in the same year.

Radioactivity is a property of matter and for the detection of radioactive substances detectors are needed, e.g. Geiger–Müller counters or photographic emulsions. It was found that these detectors also indicate the presence of radiation in the absence of radioactive substances. If they are shielded by thick walls of lead or other materials, the counting rate decreases appreciably. On the other hand, if the detectors are brought up to greater heights in the atmosphere, the counting rate increases to values that are higher by a factor of about 12 at a height of 9000 m above ground. This proves the presence of another kind of radiation that enters the atmosphere from outside. It is called cosmic radiation to distinguish it from the terrestrial radiation that is due to the radioactive matter on the earth. By cascades of interactions with the gas molecules in the atmosphere, cosmic radiation produces a variety of elementary particles (protons, neutrons, photons, electrons, positrons, mesons) and of radioactive atoms.

1.2 Radioactive Substances in Nature

Radioactive substances are widely distributed on the earth. Some are found in the atmosphere, but the major part is present in the lithosphere. The most important ones are the ores of uranium and thorium, and potassium salts, including the radioactive decay products of uranium and thorium. Uranium and thorium are common elements in nature. Their concentrations in granite are about 4 and 13 mg/kg, respectively, and the concentration of uranium in seawater is about 3 µg/l. Some uranium and thorium minerals are listed in Table 1.1. The most important uranium mineral is pitchblende (U_3O_8). Uranium is also found in mica. The most important thorium mineral is monazite, which contains between about 0.1 and 15% Th.

The measurement of natural radioactivity is an important tool for dating, e.g. for the determination of the age of minerals (see section 16.1).

Table 1.1. Uranium and thorium minerals.

Mineral	Composition	Conc of U [%]	Conc of Th [%]	Deposits
Pitchblende	U_3O_8	60–90		Bohemia, Congo, Colorado (USA)
Becquerelite	$2UO_3 \cdot 3H_2O$	74		Bavaria, Congo
Uraninite		65–75	0.5–10	Japan, USA, Canada
Broeggerite	$UO_2 \cdot UO_3$	48–75	6–12	Norway
Cleveite		48–66	3.5–4.5	Norway, Japan, Texas
Carnotite	$K(UO_2)(VO_4) \cdot nH_2O$	45		USA, Congo, Russia, Australia
Casolite	$PbO \cdot UO_3 \cdot SiO_2 \cdot H_2O$	40		Congo
Liebigite	Carbonates of U and Ca	30		Austria, Russia
Thorianite	$(Th, U)O_2$	4–28	60–90	Ceylon, Madagascar
Thorite	$ThSiO_4 \cdot H_2O$	1–19	40–70	Norway, USA
Monazite	Phosphates of Th and Rare Earths		0.1–15	Brazil, India, Russia, Norway, Madagascar

The radioactive atoms with half-lives >1 d that are found in nature are listed in Table 1.2. The table shows that radioactivity is mainly observed with heavier elements and seldom with light ones (e.g. ^{40}K and ^{87}Rb). ^{14}C, ^{10}Be, ^7Be, and ^3H (tritium) are produced in the atmosphere by cosmic radiation. The production of ^{14}C is about $22 \cdot 10^3$ atoms of ^{14}C per second per square metre of the earth's surface and that of ^3H about $2.5 \cdot 10^3$ atoms per second per square metre of the earth's surface. Taking into account the radioactive decay and the residence times in the atmosphere, this means a global equilibrium inventory of about 63 tons of ^{14}C and of about 3.5 kg of ^3H.

The measurement of the natural radioactivity of ^{14}C or of ^3H is also used for dating. However, interferences have to be taken into account due to the production of these radionuclides in nuclear reactors and by nuclear explosions.

The energy produced by decay of natural radioelements in the earth is assumed to contribute considerably to the temperature of the earth. In particular, the relatively high temperature gradient of about 30 °C per 1 km depth observed up to several kilometres below the surface is explained by radioactive decay taking place in the minerals, e.g. in granite.

Table 1.2. Naturally occurring radioactive species (radionuclides) with half-lives >1 d (decay modes are explained in chapter 5).

Radioactive species (radionuclides)	Half-life	Decay mode	Isotopic abundance [%]	Remarks
^{238}U	$4.468 \cdot 10^9$ y	α, γ, e^- (sf)	99.276	
^{234}U	$2.455 \cdot 10^5$ y	α, γ, e^- (sf)	0.0055	
^{234}Th	24.1 d	β^-, γ, e^-		
^{230}Th (Ionium)	$7.54 \cdot 10^4$ y	α, γ (sf)		Uranium family $A = 4n + 2$
^{226}Ra	1600 y	α, γ		
^{222}Rn	3.825 d	α, γ		
^{210}Po	138.38 d	α, γ		
^{210}Bi	5.013 d	$\beta^-, \gamma(\alpha)$		
^{210}Pb	22.3 y	$\beta^-, \gamma, e^-(\alpha)$		
^{235}U	$7.038 \cdot 10^8$ y	α, γ (sf)	0.720	
^{231}Th	25.5 h	β^-, γ		
^{231}Pa	$3.276 \cdot 10^4$ y	α, γ		Actinium family $A = 4n + 3$
^{227}Th	18.72 d	α, γ, e^-		
^{227}Ac	21.773 y	β^-, γ, e^- (α)		
^{223}Ra	11.43 d	α, γ		
^{232}Th	$1.405 \cdot 10^{10}$ y	α, γ, e^-(sf)	100	
^{228}Th	1.913 y	α, γ, e^-		Thorium family $A = 4n$
^{228}Ra	5.75 y	β^-, γ, e^-		
^{224}Ra	3.66 d	α, γ		
^{190}Pt	$6.5 \cdot 10^{11}$ y	α	0.013	
^{186}Os	$2.0 \cdot 10^{15}$ y	α	1.58	
^{187}Re	$5.0 \cdot 10^{10}$ y	β^-	62.60	
^{174}Hf	$2.0 \cdot 10^{15}$ y	α	0.16	
^{176}Lu	$3.8 \cdot 10^{10}$ y	β^-, γ, e^-	2.60	
^{152}Gd	$1.1 \cdot 10^{14}$ y	α	0.20	
^{147}Sm	$1.06 \cdot 10^{11}$ y	α	15.0	
^{148}Sm	$7 \cdot 10^{15}$ y	α	11.3	
^{144}Nd	$2.29 \cdot 10^{15}$ y	α	23.80	
^{138}La	$1.05 \cdot 10^{11}$ y	$\varepsilon, \beta^-, \gamma$	0.09	
^{123}Te	$1.24 \cdot 10^{13}$ y	ε	0.908	
^{115}In	$4.4 \cdot 10^{14}$ y	β^-	95.7	
^{113}Cd	$9.3 \cdot 10^{15}$ y	β^-	12.22	
^{87}Rb	$4.80 \cdot 10^{10}$ y	β^-	27.83	
^{40}K	$1.28 \cdot 10^9$ y	$\beta^-, \varepsilon, \beta^+, \gamma$	0.0117	
^{14}C	5730 y	β^-		
^{10}Be	$1.6 \cdot 10^6$ y	β^-		Produced in the atmosphere by cosmic radiation
^{7}Be	53.3 d	ε, γ		
^{3}H	12.323 y	β^-		

Literature

General and Historical

H. Becquerel, Sur les Radiations Invisibles Émises par les Corps Phosphorescents, C. R. Acad. Sci. Paris, *122*, 501 **(1896)**

M. Curie, Rayons Émis par les Composés de l'Uranium et du Thorium, C. R. Acad. Sci. Paris, *126*, 1102 **(1898)**

P. Curie, M. Slodowska-Curie, Sur une Nouvelle Substance Radio-active Contenu dans la Pechblende, C. R. Acad. Sci. Paris, *127*, 175 **(1898)**

G. C. Schmidt, Über die von Thorium und den Thorverbindungen ausgehende Strahlung, Verh. Phys. Ges. Berlin, *17*, 4 **(1898)**

P. Curie, Traité de Radioactivité, Gauthier-Villard, Paris, **1910**; Radioactivité, Hermann, Paris, **1935**

G. Hevesy, F. Paneth, Lehrbuch der Radioaktivität, 2nd ed., Akad. Verlagsges., Leipzig, **1931**

E. Cotton, Les Curies, Seghers, Paris, **1963**

A. Romer, The Discovery of Radioactivity and Transmutation, Dover, New York, **1964**

A. Ivimey, Marie Curie, Pioneer of the Atomic Age, Praeger, New York, **1980**

C. Ronneau, Radioactivity: A Natural Phenomenon, Chem. Educ. *66*, 736 **(1990)**

M. Genet, The Discovery of Uranic Rays: A Short Step for Henri Becquerel but a Giant Step for Science, Radiochim. Acta *70/71*, 3 **(1995)**

J. P. Adloff, H. J. MacCordick, The Dawn of Radiochemistry, Radiochim. Acta *70/71*, 13 **(1995)**

More Special

S. Flügge (Ed.), Kosmische Strahlung, Handbuch der Physik, Vol. XLV, 1/1, Springer, Berlin, **1961**

Textbooks and Handbooks on Inorganic Chemistry

Gmelins Handbook of Inorganic Chemistry, 8th ed., Uranium, Supplement Vol. Al, Springer, Berlin, **1979**

Gmelins Handbook of Inorganic Chemistry, 8th ed., Thorium, Supplement Vol. Al, Springer, Berlin, **1990**

A. Zeman, P. Benes, St. Joachimsthal Mines and their Importance in the Early History of Radioactivity, Radiochim. Acta *70/71*, 23 **(1995)**

2 Radioelements, Isotopes and Radionuclides

2.1 Periodic Table of the Elements

The Periodic Table of the elements was set up in 1869 by Lothar Meyer and independently by D. Mendeleyev, in order to arrange the elements according to their chemical properties and to make clear the relationships between the elements. This table allowed valuable predictions to be made about unknown elements. With respect to the order of the elements according to their atomic numbers Moseley's rule proved to be very useful:

$$\sqrt{v} = a(Z - b)$$

(v is the frequency of a certain series of X rays, Z is the atomic number, and a and b are constants).

The Periodic Table initiated the discovery of new elements which can be divided into three phases, overlapping chronologically:

(a) Discovery of stable elements: The last of this group were hafnium (discovered in 1922) and rhenium (discovered in 1925). With these, the group of stable elements increased to 81 (atomic numbers 1 (hydrogen) to 83 (bismuth) with the exception of the atomic numbers 43 and 61). In addition, the unstable elements 90 (thorium) and 92 (uranium) were known.

(b) Discovery of naturally occurring unstable elements: Uranium had already been discovered in 1789 (Klaproth) and thorium in 1828 (Berzelius). The investigation of the radioactive decay of these elements, mainly by Marie and Pierre Curie, led to the discovery of the elements with the atomic numbers 84 (Po = polonium), 86 (Rn = radon), 87 (Fr = francium), 88 (Ra = radium), 89 (Ac = actinium), and 91 (Pa = protactinium).

(c) Discovery of artificial elements: The missing elements 43 (Tc = technetium) and 61 (Pm = promethium) have been made artificially by nuclear reactions. Element 85 (At = astatine) was also first produced by nuclear reaction, and later it was found in the decay products of uranium and thorium.

Of special interest was the discovery of the transuranium elements, because this meant an extension of the Periodic Table of the elements. At present, 23 transuranium elements are known, beginning with elements 93 (Np = neptunium), 94 (Pu = plutonium), 95 (Am = Americium) .... and ending with elements 112, 114, 116, 118. The first transuranium elements were discovered at Berkeley, California, by G. T. Seaborg and his group, first reports about elements 104 to 106 came from Dubna, Russia, synthesis of elements 107 to 112 was first accomplished at Darmstadt, that of element 114 at Dubna, and that of elements 116 and 118 at Berkeley. With increasing atomic number the stability decreases appreciably to values of the order of milliseconds, and the question whether an "island" of higher stability may be reached at atomic numbers of about 114 (or 120 or 126) is still open.

The radioactive elements mentioned under b) and c) are called radioelements. They exist only in unstable forms and comprise the elements with the atomic numbers 43, 61, and all the elements with atomic numbers ≥ 84.

Some artificial elements (group c) have probably been produced in the course of the genesis of the elements and were present on the earth at the time of its formation. The age of the earth is estimated to be about $4.5 \cdot 10^9$ y. During this time elements of shorter half-life disappeared by nuclear transformations. After waiting for about 10^6 y, most of the artificial elements would have decayed again, and after a much longer time (about 10^{12} y) the radioelements U and Th would also not exist any more on the earth in measurable amounts, with the consequence that the Periodic Table of the elements would end with element 83 ($Bi = bismuth$).

2.2 Isotopes and the Chart of the Nuclides

The investigation of the natural radioelements (group b) led to the realization that the elements must exist in various forms differing from each other by their mass and their nuclear properties. In fact, about 40 kinds of atoms with different half-lives were found, for which only 12 places in the Periodic Table of the elements were available on the basis of their chemical properties. The problem was solved in 1913 by Soddy, who proposed putting several kinds of atoms in the same place in the Periodic Table. This led to the term isotope which means "in the same place". Isotopes differ by their mass, but their chemical properties are the same, if the relatively small influence of the mass on the chemical properties is neglected. Immediately after Soddy's proposal, the existence of isotopes of stable elements was proved by Thompson (1913) using the method for analysis of positive rays, and with appreciably higher precision by Aston (1919), who developed the method of mass spectrography.

For many elements a great number of isotopes is known, e.g. tin has 10 stable and 18 unstable isotopes. Some elements have only one stable isotope, e.g. Be, F, Na, Al, P, I. Cs.

The various kinds of atoms differing from each other by their atomic number or by their mass are called nuclides. The correct name of unstable (radioactive) nuclides is radionuclides, and the terms radioelements for unstable elements and radionuclides for unstable nuclides are analogous. For identification, the symbol (or the atomic number) and the mass number are used. For example, $^{14}_{6}C$ is carbon with the mass number 14 and the atomic number 6. The atomic number can be omitted (^{14}C), because it is known by the symbol. ^{14}C can also be written as C-14. For complete information, the kind and the energy of transmutation and the half-life may also be indicated:

$$^{14}C \xrightarrow{\ \beta^-\,(0.156\,\text{MeV})\ } {}^{14}N$$

It is evident that the Periodic Table of the elements does not have room to include information about all the isotopes of the elements. For that purpose the chart of the nuclides has been designed, which is based on the proton–neutron model of atomic nuclei. In this model, protons and neutrons are considered to be the elementary particles building up the nuclei of atoms. They are therefore called nucleons. The number of nucleons in the nucleus is equal to the mass number, and the number of protons is equal to the atomic number. By combination of various numbers of protons and neutrons the atomic nuclei are obtained, as shown in Table 2.1 for light nuclei. Stable nuclei result from the combination of about equal numbers of protons and neutrons. The transfer of this information into a diagram in which the number of protons is plotted as ordinate and the number of protons as abscissa gives the chart

Table 2.1. Proton–neutron model of the nuclides (P = number of protons; N = number of neutrons).

P	N	Nuclide	Nuclide mass [u]	Natural abundance [%]	Atomic mass [u]	Remarks
1	0	^{1}H	1.007825	99.985	} 1.00797	Stable
1	1	^{2}H (D)	2.014102	0.0155		Stable
1	2	^{3}H (T)	3.016049			Unstable
2	1	^{3}He	3.016030	0.000137	} 4.00260	Stable
2	2	^{4}He	4.002603	99.999863		Stable
2	3	^{5}He				Unstable
2	4	^{6}He	6.018891			Unstable
3	2	^{5}Li				Unstable
3	3	^{6}Li	6.015123	7.5	} 6.940	Stable
3	4	^{7}Li	7.016004	92.5		Stable
3	5	^{8}Li	8.022487			Unstable
3	6	^{9}Li	9.026790			Unstable
4	3	^{7}Be	7.016930			Unstable
4	4	^{8}Be	8.005305			Unstable
4	5	^{9}Be	9.012183	100.00	9.01218	Stable
4	6	^{10}Be	10.013535			Unstable
4	7	^{11}Be	11.021660			Unstable
5	3	^{8}B	8.024608			Unstable
5	4	^{9}B				Unstable
5	5	^{10}B	10.012938	19.9	} 10.811	Stable
5	6	^{11}B	11.009305	80.1		Stable
5	7	^{12}B	12.014353			Unstable
5	8	^{13}B	13.017780			Unstable
6	4	^{10}C	10.016858			Unstable
6	5	^{11}C	11.011433			Unstable
6	6	^{12}C	12.000000	98.892	} 12.0112	Stable
6	7	^{13}C	13.003354	1.108		Stable
6	8	^{14}C	14.003242			Unstable
6	9	^{15}C	15.010599			Unstable
6	10	^{16}C	16.014700			Unstable
7	5	^{12}N	12.018613			Unstable
7	6	^{13}N	13.005739			Unstable
7	7	^{14}N	14.003074	99.635	} 14.0067	Stable
7	8	^{15}N	15.000108	0.365		Stable
7	9	^{16}N	16.006099			Unstable
7	10	^{17}N	17.008449			Unstable

of the nuclides, the first part of which is shown in Fig. 2.1. The atomic number Z is equal to the number of protons P $(Z = P)$, and the mass number A is equal to the number of protons P plus the number of neutrons N $(A = P + N)$. Therefore, $N = A - Z$.

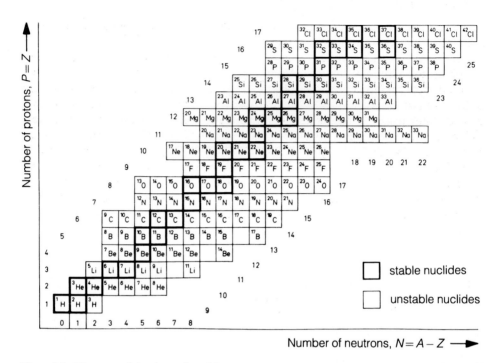

Figure 2.1. First part of the chart of nuclides.

About 2800 nuclides are known. About 340 of these are found in nature and may be subdivided into four groups: (1) 258 are indisputably stable. (2) For 25 nuclides with atomic numbers $Z < 80$ radioactive decay has been reported, but not confirmed for 7 of these. Many exhibit extremely long half-lives (9 nuclides $>10^{16}$ y and 4 nuclides $>10^{20}$ y), and radioactivity has not been proved unambiguously. Some have later be reported to be stable, and the 15 nuclides with half-lives $>10^{15}$ y may be considered to be quasistable. (3) Main sources of natural radioactivity comprising 46 nuclides are ^{238}U, ^{235}U and ^{232}Th and their radioactive decay products. (4) Several radionuclides are continuously produced by the impact of cosmic radiation, and the main representatives of this group are ^{14}C, ^{10}Be, ^{7}Be and ^{3}H. Radionuclides present in nature in extremely low concentrations, such as ^{244}Pu and its decay products or products of spontaneous fission of U and Th, are not considered in this list. Radionuclides existing from the beginning, i.e. since the genesis of the elements, are called primordial radionuclides. They comprise the radionuclides of group (2) and ^{238}U, ^{235}U, ^{232}Th and ^{244}Pu.

The following groups of nuclides can be distinguished:

– Isotopes: $Z = P$ equal
– Isotones: $N = A - Z$ equal
– Isobars: $A = N + Z$ equal
– Isodiaspheres: $A - 2Z = N - Z$ equal

The positions of these groups of nuclides in the chart of the nuclides is shown in Fig. 2.2.

For certain nuclides, different physical properties (half-lives, mode of decay) are observed. They are due to different energetic states, the ground state and one or more metastable excited states of the same nuclide. These different states are called isomers or nuclear isomers. Because the transition from the metastable excited states to the ground states is "forbidden", they have their own half-lives, which vary between some milliseconds and many years. The excited states (isomers) either change to the ground state by emission of a γ-ray photon (isomeric transition; IT) or transmutation to other nuclides by emission of α or β particles. Metastable excited states (isomers) are characterized by the suffix m behind the mass number A, for instance ^{60m}Co and ^{60}Co. Sometimes the ground state is indicated by the suffix g. About 400 nuclides are known to exist in metastable states.

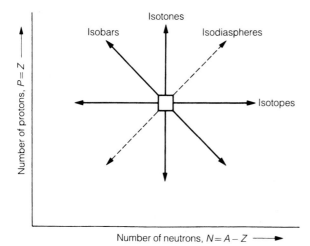

Figure 2.2. Isotopes, isotones, isobars (and isodiaspheres) in the chart of nuclides.

By comparison of the number of protons P and the number of neutrons N in stable nuclei, it is found that for light elements (small Z) $N \approx P$. With increasing atomic number Z, however, an increasing excess of neutrons is necessary in order to give stable nuclei. $A - 2Z$ is a measure of the neutron excess. For ^{4}He the neutron excess is zero. It is 3 for ^{45}Sc, 11 for ^{89}Y, 25 for ^{139}La, and 43 for ^{209}Bi. Thus, if in the chart of the nuclides the stable nuclides are connected by a mean line, which starts from the origin with a slope of 1 and is bent smoothly towards the abscissa. This mean line is called the line of β stability (Fig. 2.3).

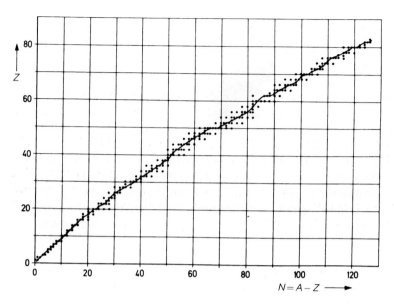

Figure 2.3. Stable nuclides and the line of β stability.

2.3 Stability and Transmutation of Nuclides

On the basis of the proton–neutron model of atomic nuclei the following combinations may be distinguished:

P even, N even (even–even nuclei)	Very common, 158 nuclei
P even, N odd (even–odd nuclei)	Common, 53 nuclei
P odd, N even (odd–even nuclei)	Common, 50 nuclei
P odd, N odd (odd–odd nuclei)	Rare, only 6 nuclei (^{2}H, ^{6}Li, ^{10}B, ^{14}N, ^{50}V, ^{180}Ta)

This unequal distribution does not correspond to statistics. The high abundance of even–even nuclei indicates the high stability of this combination. On the other hand, odd–odd nuclei seem to be exceptions. Four of the stable odd–odd nuclei are very light.

Alpha activity is preferably found for heavier elements $Z = P > 83$ (Bi). Elements with even atomic numbers exhibit mainly β activity or electron capture. In the case of β decay or electron capture, the mass number A remains constant. Either a neutron is changed into a proton or a proton into a neutron. Thus, odd–odd nuclei are transformed into even–even nuclei – for instance, ^{40}K into ^{40}Ar or into ^{40}Ca.

In finding nuclides of natural radioactivity, the Mattauch rule has proved to be very helpful. It states that stable neighbouring isobars do not exist (exceptions: $A = 50, 180$). For instance, in the following sequences of isobars, the middle one is radioactive:

^{40}Ar	^{40}K	^{40}Ca
^{138}Ba	^{138}La	^{138}Ce
^{176}Y	^{176}Lu	^{176}Hf

Detailed study of the chart of nuclides makes evident that for certain values of P and N a relatively large number of stable nuclides exist. These numbers are 2, 8, 20, 28, 50, 82 (126, only for N). The preference of these "magic numbers" is explained by the shell structure of the atomic nuclei (shell model). It is assumed that in the nuclei the energy levels of protons and of neutrons are arranged into shells, similar to the energy levels of electrons in the atoms. Magic proton numbers correspond to filled proton shells and magic neutron numbers to filled neutron shells. Because in the shell model each nucleon is considered to be an independent particle, this model is often called the independent particle model.

2.4 Binding Energies of Nuclei

The high stability of closed shells (magic numbers) is also evident from the binding energies of the nucleons. Just below each magic number the binding energy of an additional proton or neutron is exceptionally high, and just above each magic number it is exceptionally low, similarly to the binding energies of an additional electron by a halogen atom or a noble gas atom, respectively.

Not all properties of the nuclei can be explained by the shell model. For calculation of binding energies and the description of nuclear reactions, in particular nuclear fission, the drop model of the nucleus has proved to be very useful. In this model it is assumed that the nucleus behaves like a drop of a liquid, in which the nucleons correspond to the molecules. Characteristic properties of such a drop are cohesive forces, surface tension, and the tendency to split if the drop becomes too big.

In order to calculate the binding energy (E_B) of the nuclei, Weizsäcker developed a semi-empirical formula based on the drop model:

$$E_B = E_V + E_C + E_F + E_S + E_g \tag{2.1}$$

E_B is the total binding energy of all nucleons. The most important contribution is the volume energy

$$E_V = +a_V A \tag{2.2}$$

where a_V is a constant and A is the mass number. The mutual repulsion of the protons is taken into account by the Coulomb term E_C:

$$E_C = -a_C \frac{Z(Z-1)}{A^{1/3}} \tag{2.3}$$

where a_C is a constant and Z is the atomic number. $A^{1/3}$ is a measure of the radius of the nucleus and therefore also of the distance between the protons. With increasing surface energy a drop of water becomes more and more unstable. Accordingly, in the drop model of the nucleus a surface energy term E_F is subtracted:

$$E_F = -a_F A^{2/3} \tag{2.4}$$

where a_F is again a constant and $A^{2/3}$ is a measure for the surface. Neutrons are necessary to build up stable nuclei. But the excess of neutrons diminishes the total energy of the nucleus. This contribution is called the symmetry energy E_S:

$$E_S = -a_S \frac{(A - 2Z)^2}{A} \tag{2.5}$$

Finally, the relatively high stability of even–even nuclei is taken into account by a positive contribution to the total binding energy E_B of the nucleus, and the relatively low stability of odd–odd nuclei by a negative contribution. The following values are taken for this odd–even energy E_g:

$$E_g = \begin{cases} +\delta(A, Z) & \text{for even–even nuclei} \\ 0 & \text{for even–odd and odd–even neclei} \\ -\delta(A, Z) & \text{for odd–odd nuclei} \end{cases} \tag{2.6}$$

The value of δ is given approximately by $\delta \approx a_g/A$, where a_g is a constant. From the masses of the nuclides the following values have been calculated for the constants: $a_V \approx 14.1$ MeV, $a_C \approx 0.585$ MeV, $a_F \approx 13.1$ MeV, $a_S \approx 19.4$ MeV, $a_g \approx 33$ MeV.

As can be seen from the various terms, E_B plotted as a function of Z will give parabolas, one parabola for odd mass numbers A ($E_g = 0$) and two parabolas for even mass numbers A ($E_g = \pm\delta$). These parabolas give the energetics for a certain row of isobars. Two examples are given in Figs. 2.4 and 2.5. The transmutation into stable nuclei proceeds by β decay or electron capture. As increasing binding energies are drawn towards the bottom, the stable nuclei are at the bottom of the curves. The unstable isobars are transformed stepwise, either by β^- decay from lower to higher atomic numbers or by β^+ decay (alternatively by electron capture, EC, symbol ε) from higher to lower atomic numbers into stable nuclides. Nuclides at the bottom of the parabolas for odd–odd nuclei have two possibilities, β^- decay and β^+ decay (or electron capture). The continuation of these parabolas through the whole chart of the nuclides gives the "valley of stable nuclei" and the "line of β stability".

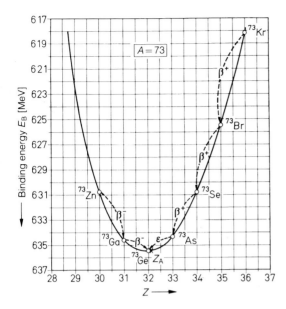

Figure 2.4. Binding energy and transmutation of nuclides with odd mass numbers.

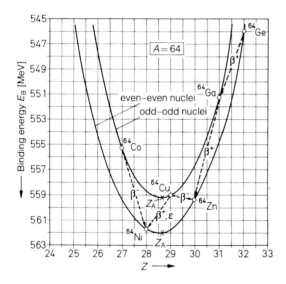

Figure 2.5. Binding energy and transmutation of nuclides with even mass numbers.

2.5 Nuclide Masses

The mass number A is equal to the number of nucleons, $A = P + N$, and is always an integer. The nuclide mass M, on the other hand, is the exact mass of the nuclides in universal atomic mass units u, and the atomic mass is the mean of the nuclide masses of the stable nuclides in their natural abundance.

The basis of the atomic mass unit u is the mass of the carbon isotope ^{12}C: $M(^{12}\text{C}) = 12.000000$. Nuclide masses and atomic masses include the mass of the electrons of the neutral atom: M = mass of the nucleus $+Zm_{\text{e}}$, where Z is the atomic number and m_{e} the mass of one electron in atomic mass units u. One atomic mass unit is equal to $(1.660566 \pm 0.000009) \cdot 10^{-24}$ g. The most accurate determinations are made by mass spectrometry with an error varying between 10^{-8} and 10^{-5} u.

The mass m of particles travelling with very high velocities increases as the velocity approaches the velocity of light c:

$$m = \frac{m_0}{\sqrt{1 - (v/c)^2}} \tag{2.7}$$

where m_0 is the mass of the particle at rest and v its velocity. Eq. (2.7) was derived by Einstein in his special theory of relativity. Another result of this theory is the equivalence of mass and energy:

$$E = mc^2 \tag{2.8}$$

The conversion factor is given by the square of the velocity of light c. Eq. (2.8) is of fundamental importance for all branches of nuclear science. For instance, it allows calculation of the energy that can be gained by conversion of matter into energy in nuclear reactions like nuclear fission or fusion. Since $1\,\text{u} = 1.660566 \cdot 10^{-24}$ g and $c = 2.997925 \cdot 10^8\,\text{m s}^{-1}$, $1\,\text{u}$ is equivalent to $1.49244 \cdot 10^{-10}$ J. The energy units mainly used in nuclear science are eV (energy gained by an electron passing in vacuo a potential of 1 V; $1\,\text{eV} = 1.60219 \cdot 10^{-19}$ J), keV and MeV. By application of these units it follows that

$$1\,\text{u} = 931.5\,\text{MeV} \tag{2.9}$$

On the basis of the proton–neutron model of atomic nuclei, the following equation can be written for the mass of a nuclide:

$$M = ZM_{\text{H}} + NM_{\text{n}} - \delta M \tag{2.10}$$

where M_{H} is the nuclide mass of ^{1}H and comprises the mass of one proton as well as that of one electron. M_{n} is the mass of the neutron in atomic mass units, and δM is called the mass defect. It is due to the fact that the binding energy E_{B} of the nucleons according to eq. (2.8) results in a decrease in the mass compared with the sum of the masses of the individual particles. The effect of the binding energy of the electrons is very small with respect to the binding energy of the nucleons and can be neglected.

Application of eq. (2.8) gives

$$\delta M = \frac{E_B}{c^2} = ZM_H + NM_n - M \qquad (2.11)$$

If E_B is divided by the mass number, the mean binding energy per nucleon is obtained, which is a measure of the stability of the nucleus:

$$\frac{E_B}{A} = \frac{c^2}{A}(ZM_H + NM_n - M) \qquad (2.12)$$

The mean binding energy per nucleon is plotted in Fig. 2.6 as a function of the mass number A. The figure shows that the elements with atomic numbers around that of iron have the highest mean binding energies per nucleon. Above $Z \approx 90$ the mean binding energy decreases continuously. Thus, it can be deduced immediately from Fig. 2.6 that the fission of heavy nuclei into two smaller ones leads to a gain in energy. For instance, the difference in the mean binding energy for uranium atoms and two nuclides with half the mass number of uranium is about 1 MeV. As uranium contains about 200 nucleons, about 200 MeV should be gained by the fission of one uranium atom into two smaller atoms. This is the energy set free by nuclear fission in nuclear reactors.

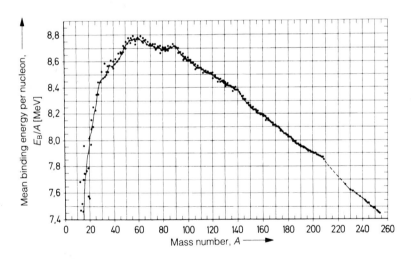

Figure 2.6. Mean binding energy per nucleon.

On the other hand, in the range of light atoms, the even–even nuclei ^{4}He, ^{12}C and ^{16}O have particularly high mean binding energies of the nucleons. Values for light nuclei are plotted separately in Fig. 2.7. It is obvious from this figure that ^{4}He is an extremely stable combination of nucleons, and very high energies may be obtained by fusion of hydrogen atoms (^{1}H, ^{2}H or ^{3}H) to ^{4}He. This is the aim of the development of fusion reactors.

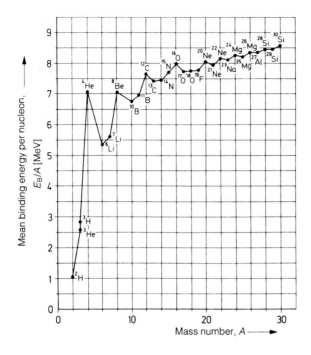

Figure 2.7. Mean binding energy per nucleon for light nuclides.

As the increase of the curve in Fig. 2.6 in the range of light nuclides is much steeper than the decrease in the range of heavy nuclides, the energy gained per mass unit of "fuel" is much higher for fusion than for fission. In the sun and in the stars the energy is produced mainly by nuclear fusion.

Literature

Isotopes

J. Biegeleisen, Isotopes, Annu. Rev. Nucl. Sci. *3*, 221 **(1953)**

G. H. Clewlett, Chemical Separation of Stable Isotopes, Annu. Rev. Nucl. Sci. *4*, 293 **(1954)**

J. Kistemacher, J. Biegeleisen, A. O. C. Nier (Eds.), Proc. Int. Symp. Isotope Separation, North Holland, Amsterdam, **1958**

L. Melander, Isotope Effects on Reaction Rates, Ronald Press, New York, **1960**

H. London, Separation of Isotopes, G. Newnes, London, **1961**

A. E. Brodsky, Isotopenchemie, Akademie Verlag, Berlin, **1961**

S. S. Roginski, Theoretische Grundlagen der Isotopenchemie, VEB Deutscher Verlag der Wissenschaften, Berlin, **1962**

F. A. White, Mass Spectrometry in Science and Technology, Wiley, New York, **1968**

A. Romer, Radiochemistry and the Discovery of Isotopes, Dover, New York, **1970**

K. H. Lieser, Einführung in die Kernchemie, 3rd ed., VCH, Weinheim, **1991**

Nuclide Charts and Tables

Nuclear Data Sheets, Section A, 1 **(1965)** et seq., Section B, 1 **(1966)** et seq., Academic Press, New York

A. H. Wapstra, K. Bos, The **1977** Atomic Mass Evaluation, Atomic Data and Nuclear Data Tables, Vol. 19, Academic Press, New York, **1977**

G. Pfennig, H. Klewe-Nebenius, W. Seelmann-Eggebert, Chart of the Nuclides (Karlsruher Nuklidkarte), 6th ed., revised, Forschungszentrum Karlsruhe, **1998**

E. Browne, R. B. Firestone, Table of Radioactive Isotopes (Ed.: V. S. Shirley), Wiley, New York, **1986**

Nuclides and Isotopes, 14th ed., General Electric Company, Nuclear Energy Operations, 175 Curtner Ave, San Jose, CA, **1989**

M. S. Antony, Chart of the Nuclides – Strasbourg **1992**, Jan-Claude Padrines, AGECOM, Séléstat, **1992**

3 Physical Properties of Atomic Nuclei and Elementary Particles

3.1 Properties of Nuclei

Whereas the diameters of atoms vary between about $0.8 \cdot 10^{-10}$ and $3.0 \cdot 10^{-10}$ m, the diameters of nuclei are in the range of about $0.3 \cdot 10^{-14}$ to $1.6 \cdot 10^{-14}$ m.

The first concepts of nuclear forces and nuclear radii were developed by Rutherford in 1911 on the basis of the scattering of α particles in metal foils. The experiments showed that the positive charge of the atoms is concentrated in a very small part of the atom, the nucleus. The scattering of the α particles could be explained by the Coulomb interaction with the nuclei, whereas the electrons did not influence the path of the α particles. The radius of an atomic nucleus can be described by the formula

$$r_{\mathrm{N}} = r_0 A^{1/3} \tag{3.1}$$

where $r_0 = (1.28 \pm 0.05)$ fm $(1\ \mathrm{fm} = 10^{-15}\ \mathrm{m})$ is a constant and A the mass number. The charge distribution (distribution of the protons) is practically constant in the interior of the nucleus and decreases near its surface, as shown in Fig. 3.1.

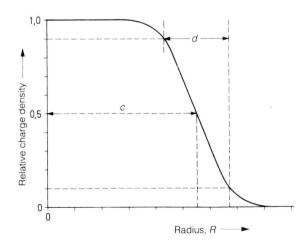

Figure 3.1. Charge distribution in nuclei ($c =$ half-density radius; $d =$ skin thickness).

The layer of decreasing density is about 2.5 fm, independently of the atomic number. The distribution of the neutrons is assumed to be approximately the same as that of the protons. Then the mass distribution in the nucleus is also the same as the charge distribution, and it follows from eq. (3.1) that the density of nuclear matter in the interior of the nuclei is given by

$$\rho = \frac{A}{\frac{4}{3}\pi r_N^3 N_{Av}} = \frac{1}{\frac{4}{3}\pi r_0^3 N_{Av}} \approx 2 \cdot 10^{14}\,\text{g/cm}^3 \tag{3.2}$$

where N_{Av} is Avogadro's constant $(6.022094 \pm 0.000006) \cdot 10^{-23}\,\text{mol}^{-1}$.

If the nuclear forces between two nucleons are plotted as a function of their distance from each other, the curve shown in Fig. 3.2 is obtained. The nucleon–nucleon interaction becomes effective only at distances less than 2.4 fm. The interaction itself is very strong, resulting in a high negative potential of about 50 MeV and a very small equilibrium distance of about 0.6 fm. The shape of the curve in Fig. 3.2 is very similar to the shape of the potential for covalent chemical bonds (Morse function), but the distances are different by five orders of magnitude and the energies by more than six orders of magnitude. The approximately constant value of about $(8.2 \pm 0.8)\,\text{MeV}$ for the mean binding energy per nucleon for mass numbers $A > 12$ in Fig. 2.6 indicates that each nucleon interacts only with a limited number of other nucleons, showing also an analogy to chemical bonds between atoms. Nuclear forces tend towards saturation, much as chemical bonds do.

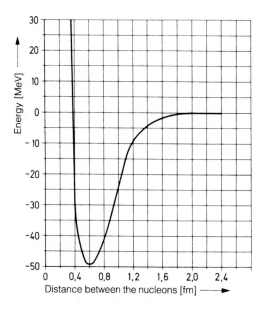

Figure 3.2. Nucleon–nucleon interaction as a function of the distance between the nucleons. (The influence of charges is not taken into account).

The Coulomb repulsion energy E_C between two protons is given by

$$E_C = \frac{e^2}{4\pi\varepsilon_0 r} \tag{3.3}$$

where e is the electric charge of a proton, ε_0 the electric field constant, and r their distance apart within the nucleus. Since $r \approx 3\,\text{fm}$, E_C is about 0.5 MeV. This repulsion energy is small compared with the mean binding energy of about 8 MeV. For a greater number of protons the total repulsion energy increases according to the formula

$$E_c = \frac{3}{5} Z(Z-1) \frac{e^2}{4\pi\varepsilon_0 r} \tag{3.4}$$

where Z is the atomic number and r the effective distance between the protons, which can be set equal to the radius of the nucleus. Whereas the nuclear forces strive for saturation, the Coulomb repulsion energy between protons increases continuously with the atomic number Z, causing the instability of heavy nuclei with high atomic numbers.

Nuclear forces are due to the strong interaction between nucleons. Besides the strong interaction, weak interaction and electromagnetic interaction are important for nuclei and elementary particles. Weak interaction also has a limited range, of the order of some femtometres. It is responsible for β-decay processes. Electromagnetic interaction is observed for all particles carrying an electromagnetic field (charged particles such as protons and neutral particles with a magnetic momentum such as neutrons). Electromagnetic interaction is also responsible for chemical bonding. As weak and electromagnetic interactions have some common features, they are assumed to have a common origin. The fourth kind of interaction is gravitation, the range of which is extremely large. As the name indicates, gravitation is responsible for gravity and the motion of the planets. The four fundamental types of interaction are summarized in Table 3.1. According to quantum theory, virtual mediating particles are responsible for the interactions, e.g. exchange of gluons for strong interaction and exchange of photons for electromagnetic interaction.

Table 3.1. The four fundamental types of interaction.

Type of interaction	Mediating particle	Relative force constant
Strong	Gluon	1
Electromagnetic	Photon	10^{-2}
Weak	Bosons (Z, W$^-$, W$^+$)	10^{-5}
Gravitation	Graviton	10^{-40}

The hyperfine structure of atomic spectra that is observed under the influence of an external magnetic field is due to the interaction of electrons and nuclei. This hyperfine structure may be caused (a) by different masses of the atoms (if the element contains two or more isotopes) and/or (b) by the interaction of the magnetic momenta of the electrons and the nuclei (if the latter have an angular momentum). (a) represents an isotope effect and (b) is proof of the existence of a nuclear angular momentum. The nuclear angular momentum is measured in units of $h/2\pi$, as well as the angular momentum of an electron, a proton or a neutron, which is $\frac{1}{2}h/2\pi$, for each of these particles. It is a vector of magnitude $\sqrt{I(I+1)}h/2\pi$, where I is the quantum number of the nuclear angular momentum, called the nuclear spin. Nuclei with even mass numbers A have integral nuclear spins, $I = 0, 1, 2, \ldots$, whereas nuclei with odd mass numbers have half-numbered nuclear spins, $I =$

$\frac{1}{2}, \frac{3}{2}, \frac{5}{2}, \ldots$ Even–even nuclei in the ground state always have $I = 0$. Odd–odd nuclei have an integral spin, in most cases $I = 0$; and even–odd and odd–even nuclei have half-numbered spins varying between $I = \frac{1}{2}$ and $I = \frac{11}{2}$. It is assumed that protons and neutrons compensate their spins in pairs. The main contribution to the nuclear spin comes from the last unpaired nucleon.

The nuclear angular momentum originates from the individual angular momenta of the nucleons, which, on their part, have two contributions, spin angular momenta and orbital angular momenta, which are due to the spin and orbital motions, respectively, of the nuclei. The spin angular momenta $\vec{s_i}$ of the nucleons as well as their orbital angular momenta $\vec{l_i}$ are vectors. With respect to the interaction of particles in a system, two cases may be distinguished:

(a) The interaction of the individual $\vec{s_i}$ and $\vec{l_i}$ of each particle is strong compared with the interaction between the particles; in other words, the spin–orbital coupling is strong. In this case the resulting angular momentum $\vec{j_i}$ of each particle is calculated according to the rules of vector addition, $\vec{j_i} = \vec{s_i} + \vec{l_i}$, and the angular momentum of the system is given by $\vec{I} = \Sigma \vec{j_i}$. This kind of coupling is called *jj* coupling.
(b) The interaction of the individual $\vec{s_i}$ and $\vec{l_i}$ of each particle is weak compared with the interaction between the particles; in other words, the spin–orbital coupling is weak. Then the resultant spin angular momentum $\vec{S} = \Sigma \vec{s_i}$ and the resultant orbital angular momentum $\vec{L} = \Sigma \vec{l_i}$ are calculated first, and the angular momentum of the system is given by $\vec{I} = \vec{S} + \vec{L}$. This kind of coupling is called *LS* or Russell–Saunders coupling.

jj coupling holds for the nucleons in nuclei and for the electrons of heavy atoms, *LS* coupling for the electrons of light and medium–heavy atoms. Strictly speaking, the term "nuclear spin" is correct for the spin momentum of a single nucleon, but it is commonly used for the quantum number for the resultant angular momentum of a nucleus consisting of two or more nucleons.

The law of conservation of momentum, known from mechanics, is also valid for nuclear angular momenta: In all changes in a given system (including nuclear reactions) the total angular momentum is conserved.

Rotation of a charged particle causes a magnetic momentum (dipole momentum). According to theory, the magnetic momentum of an electron is

$$\mu_B = \frac{\mu_0 e h}{4\pi m_e} = 1.1653 \cdot 10^{-29} \, \text{V s m} \tag{3.5}$$

where μ_0 is the magnetic field constant, e the electrical elementary unit, and m_e the mass of an electron. μ_B is called the Bohr magneton. The experimental value is in very good agreement with theory. The magnetic momentum of the nucleus is much smaller; according to theory,

$$\mu_N = \frac{\mu_0 e h}{4\pi m_p} = 6.3466 \cdot 10^{-33} \, \text{V s m} \tag{3.6}$$

where m_p is the mass of a proton and μ_N is called the nuclear magneton. The magnetic momentum of the proton is much greater than the calculated value ($+2.7926\,\mu_B$, parallel to the spin). Surprisingly, the neutron has also a magnetic momentum ($-1.9135\,\mu_N$, antiparallel to the spin). These values are explained by the inner structure of the proton and the neutron. The magnetic momentum of a nucleus μ_I is also a vector and is written

$$\vec{\mu_I} = g_I \vec{I} \mu_N \tag{3.7}$$

where g_I is called the nuclear g factor. As can be seen from this formula, all nuclei with nuclear spin $I = 0$ (for instance, even–even nuclei) have no magnetic momentum.

If the magnetic momentum of a nucleus is not zero, the nucleus performs a precession with frequency ν_0 (Larmor frequency) under the influence of an outer magnetic field:

$$\nu_0 = \frac{g_I \mu_N}{h} B_0 \tag{3.8}$$

where h is Planck's constant and B_0 is the magnetic flux density. For $B_0 = 1$ tesla, ν_0 is $42.6 \cdot 10^6\,\mathrm{s}^{-1}$, which is in the region of radiofrequencies. The nucleus may adopt $2I + 1$ energy levels differing from each other by

$$\Delta E = h\nu = g_I \mu_N B_0 \tag{3.9}$$

By absorption or emission of photons of frequency

$$\nu = \frac{g_I \mu_N}{h} B_0$$

which is identical to the Larmor frequency, the nucleus can pass from a certain energy level to a neighbouring level. This process is known as nuclear magnetic resonance (NMR) and is an important tool in the study of chemical bonds.

Many nuclei also have an electrical quadrupole momentum, which is a measure of the deviation of charge distribution from spherical symmetry. The electrical quadrupole momentum is given by

$$Q = \frac{2}{5} Z(a^2 - b^2) = \frac{4}{5} Z \bar{r}^2 \frac{\Delta r}{\bar{r}} \tag{3.10}$$

Z is the atomic number, a and b are the radii of an ellipsoid of revolution along the axis of symmetry and perpendicular to it, respectively, $\bar{r}$ is the mean radius, $\Delta r = a - b$, and $\Delta r / \bar{r}$ is a measure of the deformation. Q may be positive ($a > b$) or negative ($a < b$). Nuclides with $I = 0$ or $\frac{1}{2}$ do not have an electrical quadropole momentum; that means their nuclei have spherical symmetry.

The properties of groups of photons or electrons, protons, neutrons or atomic nuclei cannot be described by classical statistics. By application of quantum mechanics two kinds of statistics are derived, Fermi–Dirac statistics and Bose–Einstein statistics. In Fermi–Dirac statistics, the wave function changes its sign if all the coordinates of two identical particles are exchanged (antisymmetric wave function). This means that each quantum state can be occupied by only one particle, and the Pauli principle is valid. Fermi–Dirac statistics is obeyed by electrons, protons, neutrons and nuclei with odd mass numbers. In Bose–Einstein statistics the sign of the wave function does not change if the coordinates of two identical particles are exchanged (symmetric wave function), two or more particles may be in the same quantum state, and the Pauli principle is not valid. Bose–Einstein statistics is obeyed by photons and nuclei with even mass numbers.

Parity is also related to the symmetry properties of nuclei. The wave function ψ of a particle is a function of the coordinates x, y, z and of the spin quantum number s. The probability of finding a particle at a certain position (x, y, z), and with a certain spin s, is given by $|\psi|^2$. This probability must not change if the coordinates are inverted $(-x, -y, -z, s)$, whereas ψ may or may not change its sign. Parity is said to be even if $\psi(-x, -y, -z) = \psi(x, y, z)$ and is indicated by "+" behind the nuclear spin (e.g. 0+), and it is said to be odd if $\psi(-x, -y, -z) = -\psi(x, y, z)$ and is marked by "−" at the nuclear spin (e.g. $\frac{3}{2}$−). For parity a law of conservation is also valid: in case of strong and electromagnetic interaction the parity of the system does not change, whereas it may change in case of weak interaction.

Further properties of nuclei are discussed in nuclear physics, for instance isospin and strangeness. However, as these properties are of less importance in nuclear and radiochemistry and for understanding of decay schemata, they are not explained here in detail.

3.2 Elementary Particles and Quarks

Several decades ago the number of elementary particles known was limited, and the system of elementary particles seemed to be comprehensible. Electrons had been known since 1858 as cathode rays, although the name electron was not used until 1881. Protons had been known since 1886 in the form of channel rays and since 1914 as constituents of hydrogen atoms. The discovery of the neutron in 1932 by Chadwick initiated intensive development in the field of nuclear science. In the same year positrons were discovered, which have the same mass as electrons, but positive charge. All these particles are stable with the exception of the neutron, which decays in the free state with a half-life of 10.25 min into a proton and an electron. In the following years a series of very unstable particles were discovered: the mesons, the muons, and the hyperons. Research in this field was stimulated by theoretical considerations, mainly by the theory of nuclear forces put forward by Yukawa in 1935. The half-lives of mesons and muons are in the range up to 10^{-6} s, the half-lives of hyperons in the order of up to 10^{-10} s. They are observed in reactions of high-energy particles.

Open questions with respect to β^- decay (e.g. conservation of energy and spin) led to Fermi's hypothesis of the existence of another elementary particle, the neutrino, in 1934. This particle should be neutral and have a mass of approximately zero. Due to these properties its detection was very difficult. The first successful experiment was completed in 1956. The neutrino is a stable particle, but its interaction with matter is extremely small. To obtain more detailed information about the properties of neutrinos, further experiments have been started, for instance the "Gallex" experiment in the Gran Sasso National Laboratory (Italy). In this experiment, some of the solar neutrinos that pass through a tank containing 30 tons of gallium metal react with ^{71}Ga to form ^{71}Ge, which is converted to gaseous GeH_4 and counted in a proportional counter.

The great number of particles that were found mainly by the application of new accelerators led to the question whether all these particles are really elementary particles or whether heavier particles might be built up from more fundamental particles. Particles are referred to as "fundamental", if they exhibit no inner structure. In 1964, Gell-Mann proposed the existence of those fundamental particles, which he called quarks. In the standard model, three families of quarks are distinguished, "up" and "down", "charm" and "strange", "top" and "bottom". The quarks and their properties are listed in Table 3.2. The existence of quarks has been proved by application of modern accelerators and storage rings. For example, pairs of quarks and antiquarks are observed if high-energy electrons and positrons collide. Generally, quarks are firmly included in neutrons and protons. Only under extreme conditions, they are able to exist in the form of a quark-gluon plasma.

Table 3.2. Quarks (according to the standard model).

Name	Symbol	Rest mass [u]	Electric charge [units]	Corresponding antiparticle[a]
Up	u	0.33	$+\frac{2}{3}$	$\bar{u}$
Down	d	0.33	$-\frac{1}{3}$	$\bar{d}$
Charm	c	1.6	$+\frac{2}{3}$	$\bar{c}$
Strange	s	0.54	$-\frac{1}{3}$	$\bar{s}$
Top	t	24.2	$+\frac{2}{3}$	$\bar{t}$
Bottom	b	5.3	$-\frac{1}{3}$	$\bar{b}$

[a] Electric charge and quantum numbers are opposite to those of the corresponding particles.

Another group of fundamental particles are the leptons (light particles), comprising also three families, electron and electron neutrino, muon and muon neutrino, tau particle and tau neutrino. Properties of the leptons are summarized in Table 3.3. The most important particles of this group are the electron and the electron neutrino, which are both stable.

It is now under investigation whether fundamental particles, such as quarks and electrons, are really point-like or structured.

The other particles, with the collective name hadrons, are divided into two groups, the mesons and the baryons. Mesons are composed of two quarks and baryons of three quarks. Some mesons and baryons are listed in Table 3.4. All mesons are rather unstable, with lifetimes up to about 10^{-8} s. The baryons are also very unstable, with the exception of the neutron (lifetime 890 ± 10 s) and the proton, which is considered to be stable whereas theoretical considerations predict a certain instability (lifetime $>10^{40}$ s).

Table 3.3. Leptons.

Name	Symbol	Rest mass [u]	Mean lifetime [s]	Electric charge [units]	Corresponding antiparticle[a]
Electron	e^-	0.0005486	stable	−1	Positron, e^+
Electron neutrino	ν_e	0 to $2 \cdot 10^{-7}$	stable	0	Electron antineutrino, $\bar{\nu}_e$
Muon	μ^-	0.1144	$2 \cdot 10^{-6}$	−1	Positive muon, μ^+
Muon neutrino	ν_μ	0 to $2 \cdot 10^{-7}$	stable	0	Muon antineutrino, $\bar{\nu}_\mu$
Heavy lepton	τ^-	1.915	$3 \cdot 10^{-12}$	−1	Heavy antilepton, τ^+
Tau neutrino	ν_τ	0.18	?	0	Tau antineutrino, $\bar{\nu}_\tau$

[a] Electric charge and quantum numbers are opposite to those of the corresponding particles.

Table 3.4. Some hadrons.

Symbol	Quark composition	Rest mass [u]	Mean lifetime [s]
Mesons			
π^+	$\bar{d}$ u	0.150	$\approx 2 \cdot 10^{-8}$
π^0	$\bar{u}$ u or $\bar{d}$ d	0.145	$\approx 1 \cdot 10^{-16}$
π^-	$\bar{u}$ d	0.150	$\approx 2 \cdot 10^{-8}$
ρ^+	$\bar{d}$ u	0.833	
ρ^0	u $\bar{u}$ or d $\bar{d}$	0.833	
ρ^-	$\bar{u}$ d	0.833	
K^-	$\bar{u}$ s	0.530	$\approx 1 \cdot 10^{-8}$
B^-	$\bar{u}$ b	1.32	
B^0	$\bar{d}$ b	1.32	
Baryons			
n } Nucleons	u d d	1.0087	890 ± 10
p }	u u d	1.0073	Stable
Λ	u d s	1.198	$\approx 2 \cdot 10^{-10}$
Σ^+	u u s	1.277	$\approx 1 \cdot 10^{-10}$
Σ^0	u d s	1.280	$\approx 1 \cdot 10^{-14}$
Ξ^-	d s s	1.419	$\approx 2 \cdot 10^{-10}$
Ω^-	s s s	1.795	$\approx 1 \cdot 10^{-10}$

According to quantum mechanics, antiparticles have to be assigned to all particles. Particle and antiparticle differ from each other in the sign of electric charge. Antiparticles interact strongly with the corresponding particles forming other particles with lower or zero rest masses. The most frequent case of such annihilation processes is the transformation of an electron and a positron into two photons.

A positron, having given off its energy by interaction with matter, may coexist with an electron for a short time in the form of a positronium atom (e^+e^-) before annihilation occurs. Absorption of other short-lived elementary particles such as muons, pions, kaons or sigma particles may lead to substitution of protons or electrons, respectively, in atoms or molecules, with the result of formation of so-called exotic atoms or molecules. Although the lifetime of these species is very short

($\leq$ about 10^{-6} s), information may be obtained about their properties, and the unstable particles may serve as probes to study the properties of nuclei. The special features of short-lived elementary particles in atoms or molecules will be discussed in more detail in section 6.6.

Antimatter was first produced at the European research centre CERN, near Geneva, in 1995 by interaction of antiprotons with a beam of Xe atoms. The anti-hydrogen atoms $\bar{p}^- e^+$ disappeared quickly after a short lifetime of about 30 ns by annihilation and liberation of large amounts of energy.

Literature

General

R. D. Evans, The Atomic Nucleus, McGraw-Hill, New York, **1955**

E. Segré, Nuclei and Particles, Benjamin, New York, **1964**

I. Kaplan, Nuclear Physics, 2nd ed., Addison–Wesley, Reading, MA, **1964**

E. B. Paul, Nuclear and Particle Physics, North-Holland, Amsterdam, **1969**

M. G. Bowler, Nuclear Physics, Pergamon, Oxford, **1973**

G. Friedlander, J. W. Kennedy, E. S. Macias, J. M. Miller, Nuclear and Radiochemistry, 3rd ed., Wiley, New York, **1981**

L. Valentin, Subatomic Physics. Nuclei and Particles, Vols. 1 and 2, Hermann, Paris, **1982** (in French)

A. Vertes, I. Kiss, Nuclear Chemistry, Elsevier, Amsterdam, **1987**

K. S. Krane, Introductory Nuclear Physics, Wiley, New York, **1988**

More Special

G. Gamow, C. L. Critchfield, The Theory of Atomic Nucleus and Nuclear Energy Sources, Clarendon Press, Oxford, **1949**

J. M. Blatt, V. F. Weisskopf, Theoretical Nuclear Physics, Wiley, New York, **1952**

M. G. Mayer, J. H. D. Jensen, Elementary Theory of Nuclear Shell Structure, Wiley, New York, **1955**

H. A. Bethe, P. Morrison, Elementary Nuclear Theory, 2nd ed., Wiley, New York, **1956**

F. Ajzenberg-Selove (Ed.), Nuclear Spectroscopy, Part B, Academic Press, New York, **1960**

M. A. Preston, Physics of the Nucleus, Addison–Wesley, Reading, MA, **1962**

L. C. L. Yuan, C. S. Wu (Eds.), Methods of Experimental Physics, Vol. 5B, Nuclear Physics, Academic Press, New York, **1963**

W. D. Myers, W. J. Swiatecki, Nuclear Masses and Deformations, Nucl. Phys. *81*, 1 **(1966)**

E. K. Hyde, Nuclear Models, Chemistry *40*, 12 **(1967)**

P. Marmier, E. Sheldon, Physics of Nuclei and Particles, McGraw-Hill, New York, **1971**

N. Y. Kim, Mesic Atoms and Nuclear Structure, North-Holland, Amsterdam, **1971**

M. Eisenberg, W. Greiner, Microscopic Theory of the Nucleus, North-Holland, Amsterdam, **1972**

H. Frauenfelder, E. Henley, Subatomic Physics, Prentice-Hall, Englewood Cliffs, NJ, **1974**

J. A. Rasmussen, Models of Heavy Nuclei, in: Nuclear Spectroscopy and Reactions, Part C (Ed. J. Cerny), Academic Press, New York, **1974**

A. Bohr, B. R. Mottelson, Nuclear Structure, 2 Vols., Benjamin, New York, **1969** and **1975**

M. A. Preston, R. K. Bhaduri, Structure of the Nucleus, Addison–Wesley, London, **1975**

Th. J. Trenn (Ed.), Radioactivity and Atomic Theory, Taylor and Francis, London, **1975**

R. C. Barrett, D. F. Jackson, Nuclear Size and Structure, Clarendon Press, New York, **1977**

E. Segré, Nuclei and Particles, 2nd ed., Benjamin, Reading, MA, **1978**

F. Halzen, A. D. Martin, Quarks and Leptons, Wiley, New York, **1984**

F. Close, M. Marten, C. Sutton, The Particle Explosion, Oxford University Press, Oxford, **1987**

R. L. Hahn, The Physics and (Radio) chemistry of Solar Neutrino Experiments, Radiochim. Acta *70/71*, 177 **(1995)**

Tables

E. Browne, R. B. Firestone (Ed. V. S. Shirley), Table of Radioactive Isotopes, Wiley, New York, **1986**

A. H. Wapstra, G. J. Nijgh, R. van Lieshout, Nuclear Spectroscopy Tables, North Holland, Amsterdam, **1959**

H. Behrens, J. Jänecke, Numerical Tables for Beta Decay and Electron Capture, Landolt-Börnstein, New Series, Vol. I/4, Springer, Berlin, **1964**

4 Radioactive Decay

4.1 Decay Series

The terms "radioactive transmutation" and "radioactive decay" are synonymous. Generally, the term "decay" is preferred in the English literature. As already mentioned in section 2.1, many radionuclides were found after the discovery of radioactivity in 1896. These radionuclides were named UX_1, UX_2, ...; or mesothorium 1, mesothorium 2, ...; or actinouranium, ..., in order to indicate their genesis. Their atomic and mass numbers were determined later, after the concept of isotopes had been established.

The great variety of radionuclides present in thorium and uranium ores are listed in Tables 4.1, 4.2 and 4.3. Whereas thorium has only one isotope with a very long half-life (^{232}Th), uranium has two (^{238}U and ^{235}U), giving rise to one decay series for Th and two for U. In order to distinguish the two decay series of U, they were named after long-lived members of practical importance: the uranium–radium series and the actinium series. The uranium–radium series includes the most important radium isotope (^{226}Ra) and the actinium series the most important actinium isotope (^{227}Ac).

Table 4.1. Thorium decay series (thorium family): $A = 4n$.

Nuclide	Half-life	Decay mode	Maximum energy of the radiation [MeV]
^{232}Th	$1.405 \cdot 10^{10}$ y	α	4.01
^{228}Ra (MsTh$_1$)	5.75 y	β^-	0.04
^{228}Ac (MsTh$_2$)	6.13 h	β^-	2.11
^{228}Th (RdTh)	1.913 y	α	5.42
^{224}Ra (ThX)	3.66 d	α	5.69
^{220}Rn (Tn)	55.6 s	α	6.29
^{216}Po (ThA)	0.15 s	α	6.78
^{212}Pb (ThB)	10.64 h	β^-	0.57
^{212}Bi (ThC)	60.6 min	α, β^-	α : 6.09; β : 2.25
^{212}Po (ThC′)⌐	$3.0 \cdot 10^{-7}$ s	α	8.79
^{208}Tl (ThC″)⊣	3.053 min	β^-	1.80
^{208}Pb (ThD) ⤶		Stable	

In all of these decay series, only α and β^- decay are observed. With emission of an α particle (^{4}He) the mass number decreases by 4 units, and the atomic number by 2 units ($A' = A - 4$; $Z' = Z - 2$). With emission of a β^- particle the mass number does not change, whereas the atomic number increases by 1 unit ($A' = A$; $Z' = Z + 1$). These are the first and second displacement laws formulated by Soddy and Fajans in 1913. By application of the displacement laws it can easily be deduced

Table 4.2. Uranium–radium decay series (uranium family): $A = 4n + 2$.

Nuclide	Half-life	Decay mode	Maximum energy of the radiation [MeV]
^{238}U(UI)	4.468·10^9 y	α	4.20
^{234}Th(UX$_1$)	24.1 d	β^-	0.199
^{234m}Pa(UX$_2$)	1.17 min	β^-	2.28
^{234}Pa(UZ)	6.7 h	β^-	1.24
^{234}U(UII)	2.455·10^5 y	α	4.78
^{230}Th(Io)	7.54·10^4 y	α	4.69
^{226}Ra	1600 y	α	4.78
^{222}Rn	3.825 d	α	5.49
^{218}Po(RaA)	3.05 min	$\alpha, \beta^{-\,(a)}$	α : 6.00
^{214}Pb(RaB)	26.8 min	β^-	1.02
^{218}At	1.6 s	$\alpha, \beta^{-\,(a)}$	α : 6.75
^{218}Rn	0.035 s	α	7.13
^{214}Bi(RaC)	19.9 min	$\alpha^{(a)}, \beta^-$	α : 5.51; β^- : 3.27
^{214}Po(RaC′)	1.64·10^{-4} s	α	7.69
^{210}Tl(RaC″)	1.3 min	β^-	2.34
^{210}Pb(RaD)	22.3 y	$\alpha^{(a)}, \beta^-$	α : 3.72; β^- : 0.061
^{206}Hg	8.15 min	β^-	1.31
^{210}Bi(RaE)	5.013 d	$\alpha^{(a)}, \beta^-$	α : 4.69; β^- : 1.16
^{206}Tl(RaE″)	4.2 min	β^-	1.53
^{210}Po(RaF)	138.38 d	α	5.31
^{206}Pb(RaG)		Stable	

(a) <0.1%.

Table 4.3. Actinium decay series (actinium family): $A = 4n + 3$.

Nuclide	Half-life	Decay mode	Maximum energy of the radiation [MeV]
^{235}U(AcU)	7.038 · 10^8 y	α	4.40
^{231}Th(UY)	25.5 h	β^-	0.31
^{231}Pa	3.28 · 10^4 y	α	5.03
^{227}Ac	21.77 y	$\alpha^{(a)}, \beta^-$	α : 4.95; β^- : 0.046
^{227}Th(RdAc)	18.72 d	α	6.04
^{223}Fr(AcK)	21.8 min	$\alpha^{(a)}, \beta^-$	α : 5.34; β^- : 1.15
^{223}Ra(AcX)	11.43 d.	α	5.72
^{219}At	54 s	$\alpha, \beta^{-\,(a)}$	α : 6.27
^{219}Rn(An)	3.96 s	α	6.82
^{215}Bi	7.6 min	β^-	2.2
^{215}Po(AcA)	1.78 · 10^{-3} s	$\alpha, \beta^{-\,(a)}$	α : 7.39
^{211}Pb(AcB)	36.1 min	β^-	1.38
^{215}At	$\approx 10^{-4}$ s	α	8.02
^{211}Bi(AcC)	2.17 min	$\alpha, \beta^{-\,(a)}$	α : 6.62; β^- : 0.29
^{211m}Po	25.2 s	α	8.88
^{211}Po(AcC′)	0.516 s	α	7.45
^{207}Tl(AcC″)	4.77 min	β^-	1.44
^{207}Pb(AcD)		Stable	

(a) <5%.

that all members of a certain decay series may differ from each other in their mass numbers only by multiples of 4 units. The mass number of ^{232}Th is 232, which can be written $4n$ ($n = 58$). By variation of n, all possible mass numbers of the members of the decay series of ^{232}Th, also called the thorium family, are obtained. Thus, $A = 4n$ is a common label for the thorium family. For the uranium–radium family the label is $A = 4n + 2$, and for the actinium family $A = 4n + 3$. By comparing these labels, it must be concluded that one radioactive decay series with $A = 4n + 1$ is missing in nature. Members of this family have been produced artificially by nuclear reactions. Since the longest half-life in this family is exhibited by ^{237}Np, it is called the neptunium family, and the decay series is called the neptunium series. The decay series of neptunium is listed in Table 4.4. It was probably present in nature for some millions of years after the genesis of the elements, but decayed due to the relatively short half-life of ^{237}Np, compared with the time elapsed since the genesis of the elements and the age of the earth (about $5 \cdot 10^9$ y).

Table 4.4. Neptunium decay series (neptunium family): $A = 4n + 1$.

Nuclide	Half-life	Decay mode	Maximum energy of the radiation [MeV]
^{237}Np	$2.144 \cdot 10^6$ y	α	4.79
^{233}Pa	27.0 d	β^-	0.57
^{233}U	$1.59 \cdot 10^5$ y	α	4.82
^{229}Th	$7.88 \cdot 10^3$ y	α	4.90
^{225}Ra	14.8 d	β^-	0.32
^{225}Ac	10.0 d	α	5.83
^{221}Fr	4.9 min	α	6.34
^{217}At	0.032 s	α	7.07
^{213}Bi	45.6 min	$\alpha^{(a)}, \beta^-$	$\alpha : 5.87; \beta^- : 1.42$
^{213}Po	$4.2 \cdot 10^{-6}$ s	α	8.38
^{209}Tl	2.16 min	β^-	1.83
^{209}Pb	3.25 h	β^-	0.64
^{209}Bi		Stable	

(a) 2.2%.

The genetic correlations of the radionuclides within the families are often characterized by the terms "mother" and "daughter". Thus, ^{238}U is the mother nuclide of all members of the uranium family, ^{226}Ra is the mother nuclide of ^{222}Rn, and so forth.

The final members of the decay series are stable nuclides: ^{208}Pb at the end of the thorium family, ^{206}Pb at the end of the uranium–radium family, ^{207}Pb at the end of the actinium family, and ^{209}Bi at the end of the neptunium family. In all four decay series one or more branchings are observed. For instance, ^{212}Bi decays with a certain probability by emission of an α particle into ^{208}Tl, and with another probability by emission of an electron into ^{212}Po. ^{208}Tl decays by emission of an electron into ^{208}Pb, and ^{212}Po by emission of an α particle into the same nuclide (Table 4.1), thus closing the branching. In both branches the sequence of decay alternates: either α decay is followed by β^- decay or β^- decay is followed by α decay.

4.2 Law and Energy of Radioactive Decay

Radioactive decay follows the laws of statistics. If a sufficiently great number of radioactive atoms are observed for a sufficiently long time, the law of radioactive decay is found to be

$$-\frac{dN}{dt} = \lambda N \qquad (4.1)$$

where N is the number of atoms of a certain radionuclide, $-dN/dt$ is the disintegration rate, and λ is the disintegration or decay constant (dimension s^{-1}). It is a measure of the probability of radioactive decay. The law of radioactive decay describes the kinetics of the reaction

$$A \rightarrow B + x + \Delta E \qquad (4.2)$$

where A denotes the radioactive mother nuclide, B the daughter nuclide, x the particle emitted and ΔE the energy set free by the decay process, which is also called the Q-value. Eq. (4.2) represents a first-order reaction and is in the present case a mononuclear reaction.

Radioactive decay is only possible if $\Delta E > 0$. ΔE can be determined by comparison of the masses. According to the relation found by Einstein (eq. (2.8)),

$$\Delta E = \Delta M \, c^2 = [M_A - (M_B + M_x)]c^2 \qquad (4.3)$$

By calculation of ΔE it can be decided whether a decay process is possible or not.

Even if $\Delta E > 0$, the question of the probability of a radioactive decay process is still open. It can only be answered if the energy barrier is known. The energetics of radioactive decay are plotted schematically in Fig. 4.1. The energies of the mother nuclide and the products of the mononuclear reaction differ by ΔE. But the nuclide A has to surmount an energy barrier with the threshold energy E_S. The nuclide may occupy discrete energy levels above ground level. However, only if its excitation

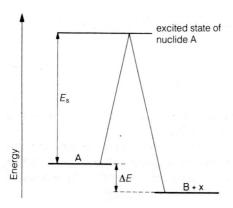

Figure 4.1. Energy barrier of radioactive decay.

energy is high enough can decay occur. The energy barrier must either be surmounted or crossed by quantum mechanical tunnelling.

The law governing radioactive decay (eq. (4.1)) is analoguous to that of first-order
chemical kinetics. The excited state on top of the energy barrier corresponds to the
activated complex, and E_S is equivalent to the activation energy.

Integration of eq. (4.1) gives

$$N = N_0 e^{-\lambda t} \tag{4.4}$$

where N_0 is the number of radioactive atoms at the time $t = 0$. Instead of the decay
constant λ, the half-life $t_{1/2}$ is frequently used. This is the time after which half the
radioactive atoms have decayed: $N = N_0/2$. Introducing the half-life $t_{1/2}$ in eq. (4.4),
it follows that

$$t_{1/2} = \frac{\ln 2}{\lambda} = \frac{0.693}{\lambda} \tag{4.5}$$

and

$$N = N_0 \left(\frac{1}{2}\right)^{t/t_{1/2}} \tag{4.6}$$

From this equation it is seen immediately that the number of radioactive atoms has
decreased to one-half after one half-life, to 1/128 (less than 1%) after 7 half-lives, and
to 1/1024 (about 0.1%) after 10 half-lives. If the time t is small compared with the
half-life of the radionuclide ($t \ll t_{1/2}$), the following approximation formula may be
used:

$$e^{-\lambda t} = 1 - \lambda t + \frac{(\lambda t)^2}{2} - \cdots$$

$$= 1 - (\ln 2)\left(\frac{t}{t_{1/2}}\right) + \frac{(\ln 2)^2}{2}\left(\frac{t}{t_{1/2}}\right)^2 - \cdots \tag{4.7}$$

The average lifetime τ is obtained by the usual calculation of an average value

$$\tau = \frac{1}{N_0}\int_0^\infty N\,dt = \int_0^\infty e^{-\lambda t}\,dt = \frac{1}{\lambda} \tag{4.8}$$

From eq. (4.4) it follows that after the average lifetime τ the number of radioactive
atoms has decreased from N_0 to N_0/e ($\tau = t_{1/2}/(\ln 2)$).

Generally, the half-life of a radionuclide does not depend on pressure, temperature, state of matter or chemical bonding. However, in some special cases in which
low-energy transitions occur, these parameters have been found to have a small
influence (section 10.2).

The activity A of a radionuclide is given by its disintegration rate:

$$A = -\frac{dN}{dt} = \lambda N = \frac{\ln 2}{t_{1/2}} N \tag{4.9}$$

The dimension is s^{-1}, and the unit is called becquerel (Bq): $1\,Bq = 1\,s^{-1}$. An older unit is the curie (Ci). It is still used sometimes, related to the activity of $1\,g$ of ^{226}Ra, and defined as $1\,Ci = 3.700 \cdot 10^{10}\,s^{-1} = 37\,GBq$. Smaller units are 1 milli-curie (mCi) $= 37\,MBq$, 1 microcurie (μCi) $= 37\,kBq$, 1 nanocurie (nCi) $= 37\,Bq$, and 1 picocurie (pCi) $= 0.37\,Bq$. 1 Ci is a rather high activity which cannot be handled directly but needs special installations, such as hot cells. Activities of the order of several mCi are applied in medicine for diagnostic purposes, activities of the order of 1 μCi are usually sufficient for the investigation of the behaviour of radionuclides, and activities of the order of 1 nCi are measurable without special efforts.

As the activity A is proportional to the number N of radioactive atoms, the exponential law, eq. (4.4), holds also for the activity:

$$A = A_0 e^{-\lambda t} \tag{4.10}$$

The mass m of the radioactive atoms can be calculated from their number N and their activity A:

$$m = \frac{N \cdot M}{N_{Av}} = \frac{A \cdot M}{N_{Av}\lambda} = \frac{A \cdot M}{N_{Av}\ln 2} t_{1/2} \tag{4.11}$$

where M is the nuclide mass and N_{Av} is Avogadro's number ($6.022 \cdot 10^{23}$).

In laboratory experiments with radionuclides, knowledge of the mass of the radioactive substances is very important. For example, the mass of 1 MBq of ^{32}P ($t_{1/2} = 14.3\,d$) is only about 10^{-10} g, and that of 1 MBq of ^{99m}Tc ($t_{1/2} = 6.0\,h$) is only about $5 \cdot 10^{-12}$ g. If there is no carrier present in the form of a large excess of inactive atoms of the same element in the same chemical state, these small amounts of radio-nuclides may easily be lost, for instance by adsorption on the walls. Whereas in the case of radioisotopes of stable elements the condition of the presence of carriers is often fulfilled due to the ubiquity of most stable elements, it is not fulfilled in case of short-lived isotopes of radioelements, and extraordinary behaviour may be observed (section 13.3).

The ratio of the activity to the total mass m of the element (the sum of radioactive and stable isotopes) is called the specific activity A_s:

$$A_s = \frac{A}{m} [Bq/g] \tag{4.12}$$

Sometimes high or well-defined specific activities are required, for instance in the case of the application of radionuclides or labelled compounds in medicine, or as tracers in other fields of research.

4.3 Radioactive Equilibria

Genetic relations between radionuclides, as in the decay series, can be written in the form

$$\text{nuclide } 1 \rightarrow \text{nuclide } 2 \rightarrow \text{nuclide } 3 \tag{4.13}$$

In words: nuclide 1 is transformed by radioactive decay into nuclide 2, and the latter into nuclide 3. Nuclide 1 is the mother nuclide of nuclide 2, and nuclide 2 the daughter nuclide of nuclide 1. At any instant, the net production rate of nuclide 2 is given by the decay rate of nuclide 1 diminished by the decay rate of nuclide 2:

$$\frac{dN_2}{dt} = -\frac{dN_1}{dt} - \lambda_2 N_2 = \lambda_1 N_1 - \lambda_2 N_2 \tag{4.14}$$

With the decay rate of nuclide 1 it follows:

$$\frac{dN_2}{dt} + \lambda_2 N_2 - \lambda_1 N_1^0 e^{-\lambda_1 t} = 0 \tag{4.15}$$

where N_1^0 is the number of atoms of nuclide 1 at time 0. The solution of the first-order differential equation (4.15) is

$$N_2 = \frac{\lambda_1}{\lambda_2 - \lambda_1} N_1^0 (e^{-\lambda_1 t} - e^{-\lambda_2 t}) + N_2^0 e^{-\lambda_2 t} \tag{4.16}$$

N_2^0 is the number of atoms of nuclide 2 present at $t = 0$. If nuclides 1 and 2 are separated quantitatively at $t = 0$, the situation becomes simpler and two fractions are obtained. In the fraction containing nuclide 2, this nuclide is not produced any more by decay of nuclide 1, and for the fraction containing nuclide 1 it follows with $N_2^0 = 0$:

$$N_2 = \frac{\lambda_1}{\lambda_2 - \lambda_1} N_1^0 (e^{-\lambda_1 t} - e^{-\lambda_2 t}) \tag{4.17}$$

Rearrangement gives:

$$N_2 = \frac{\lambda_1}{\lambda_2 - \lambda_1} N_1 [1 - e^{-(\lambda_2 - \lambda_1)t}] \tag{4.18}$$

or, after substitution of the decay constants λ by the half-lives $t_{1/2}$:

$$N_2 = \frac{t_{1/2}(2)/t_{1/2}(1)}{1 - t_{1/2}(2)/t_{1/2}(1)} N_1 \left[1 - \left(\frac{1}{2}\right)^{t_{1/2}(2) - t_{1/2}(1)}\right] \tag{4.19}$$

The term in the exponent of $\frac{1}{2}$ in eq. (4.19) may be rewritten to show the influence of the ratio of the half-lives $t_{1/2}(1)/t_{1/2}(2)$:

$$\frac{t}{t_{1/2}(2)} - \frac{t}{t_{1/2}(1)} = \left[1 - \frac{t_{1/2}(2)}{t_{1/2}(1)}\right]\frac{t}{t_{1/2}(2)} \tag{4.20}$$

The time necessary to attain radioactive equilibrium depends on the half-life of the daughter nuclide as well as on the ratio of the half-lives. This is seen in Fig. 4.2. After a sufficiently long time, the exponential function in eq. (4.18) becomes zero and radioactive equilibrium is established:

$$N_2 = \frac{\lambda_1}{\lambda_2 - \lambda_1}N_1 = \frac{t_{1/2}(2)/t_{1/2}(1)}{1 - t_{1/2}(2)/t_{1/2}(1)}N_1 \tag{4.21}$$

In the radioactive equilibrium, the ratio N_2/N_1, the ratio of the masses and the ratio of the activities are constant. It should be mentioned that this is not an equilibrium in the sense used in thermodynamics and chemical kinetics, because it is not reversible, and, in general, it does not represent a stationary state.

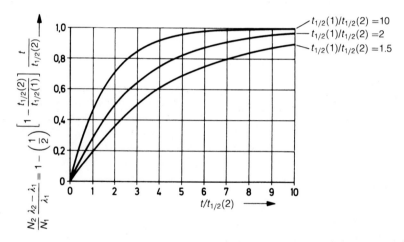

Figure 4.2. Attainment of radioactive equilibrium as a function of $t/t_{1/2}(2)$ for different ratios of the half-lives of the mother and daughter nuclides.

Four cases can be distinguished:

(a) The half-life of the mother nuclide is much longer than that of the daughter nuclide, $t_{1/2}(1) \gg t_{1/2}(2)$.
(b) The half-life of the mother nuclide is longer than that of the daughter nuclide, but the decay of the mother nuclide cannot be neglected, $t_{1/2}(1) > t_{1/2}(2)$.
(c) The half-life of the mother nuclide is shorter than that of the daughter nuclide: $t_{1/2}(1) < t_{1/2}(2)$.
(d) The half-lives of the mother nuclide and the daughter nuclide are similar: $t_{1/2}(1) \approx t_{1/2}(2)$.

These four cases are considered in the following sections in more detail, because they are of practical importance in radiochemistry.

4.4 Secular Radioactive Equilibrium

In secular radioactive equilibrium $(t_{1/2}(1) \gg t_{1/2}(2))$, eq. (4.18) reduces to

$$N_2 = \frac{\lambda_1}{\lambda_2} N_1 (1 - e^{-\lambda_2 t}) \tag{4.22}$$

Assuming that mother and daughter nuclide are separated from each other at time $t = 0$, the growth of the daughter nuclide in the fraction of the mother nuclide and the decay of the daughter nuclide in the separated fraction are plotted in Fig. 4.3. The logarithms of the activities are plotted in Fig. 4.4. The solid curves can be

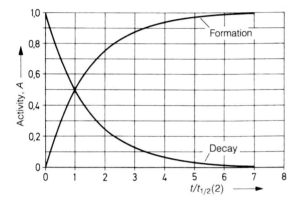

Figure 4.3. Decay of the daughter nuclide and its formation from the mother nuclide in the case of secular equilibrium as a function of $t/t_{1/2}(2)$.

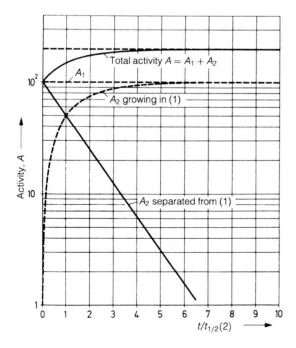

Figure 4.4. Secular equilibrium: activities of mother and daughter nuclide as a function of $t/t_{1/2}(2)$.

measured directly in the two fractions after chemical separation, whereas the broken curves are found by extrapolation or by subtraction, respectively.

After $t \gg t_{1/2}(2)$ (in practice, after about 10 half-lives of nuclide 2), radioactive equilibrium is established and the following relations hold:

$$\frac{N_2}{N_1} = \frac{\lambda_1}{\lambda_2} = \frac{t_{1/2}(2)}{t_{1/2}(1)} \tag{4.23}$$

$$A_1 = A_2 \tag{4.24}$$

The activities of the mother nuclide and of all the nuclides emerging from it by nuclear transformation or a sequence of nuclear transformations are the same, provided that secular radioactive equilibrium is established.

Secular radioactive equilibrium has several practical applications:

(a) Determination of the long half-life of a mother nuclide by measuring the mass ratio of daughter and mother nuclides, provided that the half-life of the daughter nuclide is known. Examples are the determinations of the half-lives of ^{226}Ra and ^{238}U which cannot be obtained directly by measuring their radioactive decay, because of the long half-lives. The half-life of ^{226}Ra is obtained by measuring the absolute activity of the daughter nuclide ^{222}Rn in radioactive equilibrium with ^{226}Ra, and its half-life. From the activity and the half-life, the number of radioactive atoms of ^{222}Rn is calculated by use of eq. (4.9), and the half-life of ^{226}Ra is obtained from eq. (4.23). The half-life of ^{238}U is determined by measuring the mass ratio of ^{226}Ra and ^{238}U in a uranium mineral. With the known half-life of ^{236}Ra, that of ^{238}U is calculated by application of eq. (4.23).

(b) Calculation of the mass ratios of radionuclides that are in secular radioactive equilibrium. From the half-lives, the masses of all radionuclides of the natural decay series in radioactive equilibrium with the long-lived mother nuclides can be calculated by use of eq. (4.11).

(c) Calculation of the mass of a mother nuclide from the measured activity of a daughter nuclide. For example, the amount of ^{238}U in a sample can be determined by measuring the activity of ^{234}Th or of ^{234m}Pa. The latter emits high-energy β^- radiation and can therefore be measured easily. The mass of ^{238}U is obtained by application of eq. (4.11) with $A_1 = A_2$:

$$m_1 = \frac{M_1}{N_{Av}} \frac{A_2}{\ln 2} t_{1/2}(1) \tag{4.25}$$

where m_1 is the mass and M_1 the nuclide mass of the long-lived mother nuclide, and N_{Av} is Avogadro's number.

(d) Finally, the previous application can be reversed inasmuch as a sample of U or U_3O_8 can be weighed to provide a source of known activity of ^{234m}Pa. The α radiation of ^{238}U is filtered from the high-energy β^- radiation of ^{234m}Pa by covering the sample with thin aluminium foil. From eq. (4.11) it follows with $A_1 = A_2$ that 1 mg of ^{238}U is a radiation source emitting 740 β^- particles from ^{234m}Pa per minute. Such a sample may be used as a β^- standard.

4.5 Transient Radioactive Equilibrium

The attainment of a transient radioactive equilibrium is plotted in Fig. 4.5 for $t_{1/2}(1)/t_{1/2}(2) = 5$. Now $t_{1/2}(2)$ alone does not regulate the attainment of the radioactive equilibrium; its influence is modified by a factor containing the ratio $t_{1/2}(1)/t_{1/2}(2)$, as already explained in section 4.3. Again, as in Fig. 4.4, the solid curves can be measured experimentally, and the broken curves are obtained by extrapolation or by subtraction, respectively.

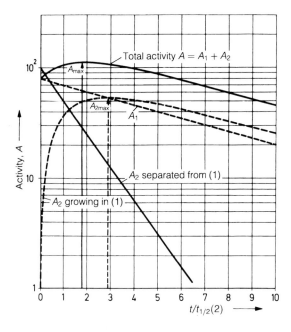

Figure 4.5. Transient equilibrium: activities of mother and daughter nuclide as a function of $t/t_{1/2}(2)$ $(t_{1/2}(1)/t_{1/2}(2) = 5)$.

After attainment of radioactive equilibrium, eq. (4.19) is valid. Introducing the half-lives, this equation becomes

$$\frac{N_2}{N_1} = \frac{t_{1/2}(2)}{t_{1/2}(1) - t_{1/2}(2)} \tag{4.26}$$

Whereas in secular radioactive equilibrium the activities of the mother and the daughter nuclide are the same, in transient radioactive equilibrium the daughter activity is always higher:

$$\frac{A_1}{A_2} = \frac{\lambda_1 N_1}{\lambda_2 N_2} = 1 - \frac{\lambda_1}{\lambda_2} = 1 - \frac{t_{1/2}(2)}{t_{1/2}(1)} \tag{4.27}$$

The possibilities of application of transient radioactive equilibrium are similar to those explained for secular radioactive equilibrium. Instead of eq. (4.25), the following equation holds:

$$m_1 = \frac{M_1}{N_{Av}} \frac{A_2}{\ln 2} [t_{1/2}(1) - t_{1/2}(2)] \tag{4.28}$$

4.6 Half-life of Mother Nuclide Shorter than Half-life of Daughter Nuclide

In this case the mother nuclide decays faster than the daughter nuclide, and the ratio between the two changes continuously, until the mother nuclide has disappeared and only the daughter nuclide is left. The situation is plotted in Fig. 4.6. No radioactive equilibrium is attained.

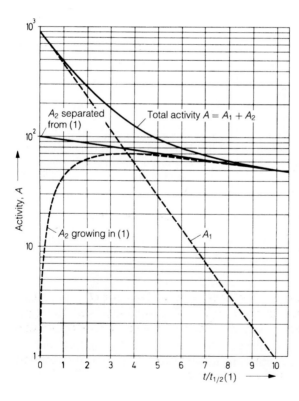

Figure 4.6. Half-life of mother nuclide shorter than that of daughter nuclide – no radioactive equilibrium $(t_{1/2}(1)/t_{1/2}(2) = 0.1)$.

4.7 Similar Half-lives

As the difference between the half-lives of mother nuclide and daughter nuclide becomes smaller and smaller, the attainment of radioactive equilibrium is more and more delayed, as can be seen from eq. (4.20) and from Fig. 4.2, provided that $t_{1/2}(1) > t_{1/2}(2)$. In this situation, the following questions are of practical interest:

(a) How much time must elapse before the decay curve of the longer-lived radio-nuclide can be observed?
(b) At which time after separation of mother and daughter nuclide does the daughter nuclide reach maximum activity?

In answering question (a), it is assumed that an error ε can be accepted in measuring the decay curve:

$$\varepsilon = e^{-(\lambda_2 - \lambda_1)t} \quad \text{if } \lambda_1 < \lambda_2 \tag{4.29a}$$

or

$$\varepsilon = e^{-(\lambda_1 - \lambda_2)t} \quad \text{if } \lambda_1 > \lambda_2 \tag{4.29b}$$

Introducing the half-lives, the following equations are calculated for the time after which the decay curve of the longer-lived radionuclide is observed with an error ε:

$$t \geq \frac{\log(1/\varepsilon)}{\log 2} \frac{t_{1/2}(1)t_{1/2}(2)}{t_{1/2}(1) - t_{1/2}(2)} \quad \text{if } t_{1/2}(1) > t_{1/2}(2) \tag{4.30a}$$

or

$$t \geq \frac{\log(1/\varepsilon)}{\log 2} \frac{t_{1/2}(1)t_{1/2}(2)}{t_{1/2}(2) - t_{1/2}(1)} \quad \text{if } t_{1/2}(1) < t_{1/2}(2) \tag{4.30b}$$

Application of this formula to the case of the sequence of radionuclides

$$^{135}\text{I} \xrightarrow[6.6\,\text{h}]{\beta^-} {}^{135}\text{Xe} \xrightarrow[9.1\,\text{h}]{\beta^-} {}^{135}\text{Cs}$$

gives the result that 160 h elapse before the half-life of the longer-lived ^{135}Xe can be observed in the decay curve with an error of 1%. This is a very long time compared with the half-lives, and the activity of ^{135}Xe will have decreased by 5 orders of magnitude.

In order to answer question (b), eq. (4.17) is differentiated with respect to time and dN_2/dt is set equal to zero. The resulting equation is

$$t_{\max}(2) = \frac{1}{\lambda_2 - \lambda_1} \ln \frac{\lambda_2}{\lambda_1} \tag{4.31}$$

In the sequence $^{135}\text{I} \rightarrow {}^{135}\text{Xe} \rightarrow {}^{135}\text{Cs}$, the maximum activity of ^{135}Xe is reached after 11.1 h.

4.8 Branching Decay

Branching decay is often observed for odd–odd nuclei on the line of β stability. For example, ^{40}K, which is responsible for the natural radioactivity of potassium, decays into ^{40}Ca with a probability of 89.3% by emission of β^- particles and into ^{40}Ar with a probability of 10.7% by electron capture. Branching decay is also observed in the decay series, as already mentioned in section 4.1.

For a certain radionuclide A showing branching decay into a nuclide B and a nuclide C,

the two probabilities of decay, given by the decay constants, may be denoted by λ_b and λ_c, respectively. As these two probabilities are independent of each other, the decay constant λ_A of the radionuclide A is given by the sum of of λ_b and λ_c, and the decay rate of A is

$$-\frac{dN_A}{dt} = \lambda_b N_A + \lambda_c N_A = \lambda_A N_A \tag{4.32}$$

Integration of this equation gives

$$N_A = N_A^0 e^{-(\lambda_b + \lambda_c)t} \tag{4.33}$$

The rates of production of the nuclides B and C are

$$\frac{dN_B}{dt} = \lambda_b N_A \quad \text{and} \quad \frac{dN_C}{dt} = \lambda_c N_A \tag{4.34}$$

and the decay rates of these nuclides are

$$-\frac{dN_B}{dt} = \lambda_B N_B \quad \text{and} \quad -\frac{dN_C}{dt} = \lambda_c N_C \tag{4.35}$$

The net rate of production of B is

$$\frac{dN_B}{dt} = \lambda_b N_A - \lambda_B N_B \tag{4.36}$$

or, introducing eq. (4.33),

$$\frac{dN_B}{dt} + \lambda_B N_B - \lambda_b N_A^0 e^{-(\lambda_b + \lambda_c)t} = 0 \tag{4.37}$$

Integration of this equation with $N_B = 0$ at $t = 0$ gives

$$N_B = \frac{\lambda_b}{\lambda_B - (\lambda_b + \lambda_c)} N_A^0 [e^{-(\lambda_b + \lambda_c)t} - e^{-\lambda_B t}] \tag{4.38}$$

A similar relation holds for nuclide C. In the case of secular equilibrium $(\lambda_b + \lambda_c \ll \lambda_B)$, it follows that

$$\frac{N_B}{N_A} = \frac{\lambda_b}{\lambda_B} \quad \text{and} \quad \frac{N_C}{N_A} = \frac{\lambda_c}{\lambda_C} \tag{4.39}$$

Whereas there are two probabilities in branching decay, λ_b and λ_c, there is only one half-life:

$$t_{1/2}(A) = \frac{\ln 2}{\lambda_A} = \frac{\ln 2}{\lambda_b + \lambda_c} \tag{4.40}$$

In the case of secular equilibrium, two partial half-lives may be distinguished formally:

$$t_{1/2}(A)_b = \frac{\ln 2}{\lambda_b} \quad \text{and} \quad t_{1/2}(A)_c = \frac{\ln 2}{\lambda_c} \tag{4.41}$$

Introduction of these partial half-lives leads to the relations

$$\frac{N_B}{N_A} = \frac{t_{1/2}(B)}{t_{1/2}(A)_b} \quad \text{and} \quad \frac{N_C}{N_A} = \frac{t_{1/2}(C)}{t_{1/2}(A)_c} \tag{4.42}$$

which are analogous to eq. (4.23).

If the daughter nuclides are longer-lived or even stable (as in the case of the decay of ^{40}K), the following equations are valid:

$$N_B = \frac{\lambda_b}{\lambda_b + \lambda_c} N_A [e^{(\lambda_b + \lambda_c)t} - 1] \tag{4.43a}$$

$$N_C = \frac{\lambda_c}{\lambda_b + \lambda_c} N_A [e^{(\lambda_b + \lambda_c)t} - 1] \tag{4.43b}$$

and

$$\frac{N_B}{N_C} = \frac{\lambda_b}{\lambda_c} \tag{4.44}$$

If the time t is small compared with the half-life of the mother nuclide A $(t \ll t_{1/2}(A))$, it follows from eqs. (4.43) that

$$\frac{N_B}{N_A} = \lambda_b t \quad \text{and} \quad \frac{N_C}{N_A} = \lambda_c t \tag{4.45}$$

4.9 Successive Transformations

In the previous sections the radioactive equilibrium between a mother nuclide and a daughter nuclide according to eq. (4.13) has been considered. This can be extended to a longer sequence of successive transformations:

$$(1) \rightarrow (2) \rightarrow (3) \rightarrow (4) \cdots \rightarrow (n) \cdots \tag{4.46}$$

For such a sequence, eq. (4.14) can be written in a more general form

$$\frac{dN_n}{dt} = \lambda_{n-1}N_{n-1} - \lambda_n N_n \tag{4.47}$$

Solution of the series of differential equations with n = 1, 2, 3, 4, ... n, for the initial conditions $N_1 = N_1^0$, $N_2 = N_3 = \cdots = N_n = 0$, gives for the number of atoms $N_n(t)$ of nuclide number n in the series at the time *t*:

$$N_n = c_1 e^{-\lambda_1 t} + c_2 e^{-\lambda_2 t} + \cdots + c_n e^{-\lambda_n t} \tag{4.48}$$

The coefficients in this equation are

$$c_1 = \frac{\lambda_1 \lambda_2 \dots \lambda_{n-1}}{(\lambda_2 - \lambda_1)(\lambda_3 - \lambda_1) \dots (\lambda_n - \lambda_1)} N_1^0$$

$$c_2 = \frac{\lambda_1 \lambda_2 \dots \lambda_{n-1}}{(\lambda_1 - \lambda_2)(\lambda_3 - \lambda_2) \dots (\lambda_n - \lambda_2)} N_1^0 \tag{4.49}$$

$$\dots$$

$$c_n = \frac{\lambda_1 \lambda_2 \dots \lambda_{n-1}}{(\lambda_1 - \lambda_n)(\lambda_2 - \lambda_n) \dots (\lambda_{n-1} - \lambda_n)} N_1^0$$

By use of these equations the number of atoms in any series of successive transformations can be calculated. For the daughter nuclide 2, eq. (4.17) is obtained.

In some practical cases the equations for n = 3 are useful:

$$N_3 = \lambda_1 \lambda_2 N_1^0 \left[\frac{e^{-\lambda_1 t}}{(\lambda_2 - \lambda_1)(\lambda_3 - \lambda_1)} + \frac{e^{-\lambda_2 t}}{(\lambda_1 - \lambda_2)(\lambda_3 - \lambda_2)} + \frac{e^{-\lambda_3 t}}{(\lambda_2 - \lambda_3)(\lambda_1 - \lambda_3)} \right] \tag{4.50}$$

If nuclide 3 is stable ($\lambda_3 = 0$), the increase of N_3 is given by

$$N_3 = N_1^0 \left[1 - \frac{\lambda_2}{\lambda_2 - \lambda_1} e^{-\lambda_1 t} - \frac{\lambda_1}{\lambda_1 - \lambda_2} e^{-\lambda_2 t} \right] \tag{4.51}$$

With the decay law for nuclide 1 and eq. (4.17) it follows that:

$$N_3 = N_1^0 - N_1 - N_2 \tag{4.52}$$

i.e. the number of atoms of the stable end product is given by the number of atoms of the mother nuclide 1 at the beginning, diminished by the number of atoms 1 that are still present and the number of atoms of the intermediate 2.

If the half-life of the mother nuclide is much longer than those of the succeeding radionuclides (secular equilibrium), eq. (4.48) becomes much simpler, provided that radioactive equilibrium is established. As in this case $\lambda_1 \ll \lambda_2, \lambda_3 \ldots \lambda_n$, all terms are small compared with the first one, giving

$$N_n = c_1 e^{-\lambda_1 t} \tag{4.53}$$

and

$$c_1 = \frac{\lambda_1}{\lambda_n} N_1^0 \tag{4.54}$$

Furthermore, under these conditions the following relations are valid:

$$\frac{N_n}{N_1} = \frac{\lambda_1}{\lambda_n} \quad \text{or} \quad \frac{N_n}{N_1} = \frac{t_{1/2}(n)}{t_{1/2}(1)} \tag{4.55}$$

and

$$A_n = A_1 \tag{4.56}$$

These equations are the same as those derived for radioactive equilibrium between mother and daughter nuclide (eqs. (4.23) and (4.24)); i.e. in secular equilibrium the relations in section 4.4 are not only valid for the directly succeeding daughter nuclide, but also for all following radionuclides of the decay series. This has already been applied in the examples given in section 4.4.

If secular equilibrium is not established, the activities of succeeding radionuclides can also be calculated by use of the equations given in this section. An example is the decay of the naturally occurring ^{218}Po (^{218}Po $\xrightarrow{\alpha}$ ^{214}Pb $\xrightarrow{\beta^-}$ ^{214}Bi $\xrightarrow{\alpha}$ ^{210}Pb...). The activities of ^{218}Po and its first decay products are plotted in Fig. 4.7 as a function of time.

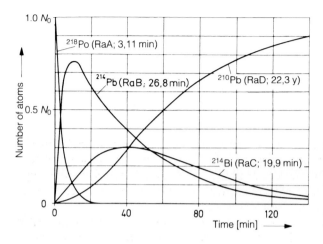

Figure 4.7. Several successive transformations: decay of ^{218}Po (according to E. Rutherford, J. Chadwick, C. D. Ellis, *Radiation from Radioactive Substances*, Cambridge University Press, **1930**).

Literature

General

E. Rutherford, J. Chadwick, C. D. Ellis, Radiation from Radioactive Substances, Cambridge University Press, Cambridge, **1930**

R. D. Evans, The Atomic Nucleus, McGraw-Hill, New York, **1955**

E. Segrè (Ed.), Radioactive Decay, in: Experimental Nuclear Physics, Vol. III, Wiley, New York, **1959**

I. Kaplan, Nuclear Physics, 2nd ed., Addison–Wesley, Reading, MA, **1964**

G. Friedlander, J. W. Kennedy, E. S. Macias, J. M. Miller, Nuclear and Radiochemistry, 3rd ed., Wiley, New York, **1981**

E. Browne, R. B. Firestone (Ed. V. S. Shirley), Table of Radioactive Isotopes, Wiley, New York, **1986**

K. H. Lieser, Einführung in die Kernchemie, 3rd ed., VCH, Weinheim, **1991**

More Special

H. Bateman, The Solution of a System of Differential Equations Occurring in the Theory of Radioactive Transformations, Proc. Cambridge Philos. Soc. *15*, 423 **(1910)**

G. W. A. Newton, History of the Unraveling of the Natural Decay Series, Radiochim. Acta *70/71*, 31 **(1995)**

5 Decay Modes

5.1 Survey

The various decay modes are listed in Table 5.1. Unstable, radioactive nuclei may be transformed by emission of nucleons (α decay and, very rarely, emission of protons or neutrons) or by emission of electrons or positrons (β^- and β^+ decay, respectively). Alternatively to the emission of a positron, the unstable nucleus may capture an electron of the electron shell of the atom (symbol ε).

In most cases the emission of nucleons, electrons or positrons leads to an excited state of the new nucleus, which gives off its excitation energy in the form of one or several photons (γ rays). This de-excitation occurs most frequently within about 10^{-13} s after the preceding α or β decay, but in some cases the transition to the ground state is "forbidden" resulting in a metastable isomeric state that decays independently of the way it was formed.

Alpha decay is observed for heavy nuclei with atomic numbers $Z > 83$ and for some groups of nuclei far away from the line of β stability. Radionuclides with very long half-lives are mainly α emitters. Proton emission has been found for nuclei with a high excess of protons far away from the line of β stability and more frequently as a two-stage process after β^+ decay (β delayed proton emission).

With increasing atomic numbers spontaneous fission begins to compete with α decay and prevails for some radionuclides with $Z \geq 96$. However, due to high fission barriers, α decay is still the dominating mode of decay for many heavy nuclides with $Z > 105$.

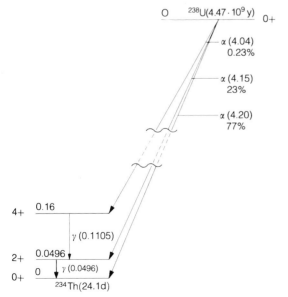

Figure 5.1. Decay scheme of ^{238}U (energies of excited states, α decay and γ transitions in MeV, 0 for ground states; nuclear spin and parity are indicated).

Table 5.1. Decay modes.

Decay mode	Symbol	Radiation emitted	Decay process and example (in short form)	Remarks
α decay	α	Helium nuclei $^4_2\mathrm{He}^{2+}$	$^A Z \to {}^{A-4}(Z-2) + {}^4_2\mathrm{He}^{2+}$ $^{238}\mathrm{U}(\alpha)^{234}\mathrm{Th}$	Preferably heavy nuclei with $Z > 83$
β decay	β^-	Electrons $^0_{-1}\mathrm{e}^-$	$^1_0\mathrm{n} \to {}^1_1\mathrm{p} + {}^0_{-1}\mathrm{e}^- + {}^0_0\bar{\nu}_\mathrm{e}$ (in the nucleus) $^A Z \to {}^A(Z+1)$ $^{14}\mathrm{C}(\beta^-)^{14}\mathrm{N}$	Below the line of β stability
	β^+	Positrons $^0_1\mathrm{e}^+$	$^1_1\mathrm{p} \to {}^1_0\mathrm{n} + {}^0_1\mathrm{e}^+ + {}^0_0\nu_\mathrm{e}$ (in the nucleus) $^A Z \to {}^A(Z-1)$ $^{11}\mathrm{C}(\beta^+)^{11}\mathrm{B}$	Above the line of β stability
Electron capture (EC)	ε	Characteristic X rays of the daughter nuclide	$^1_1\mathrm{p}$ (nucleus) $+ {}^0_{-1}\mathrm{e}^-$ (electron shell) $\to {}^1_0\mathrm{n} + {}^0_0\nu_\mathrm{e}$ $^A Z \to {}^A(Z-1)$ $^{37}\mathrm{Ar}(\varepsilon)^{37}\mathrm{Cl}$	
γ transition	γ	Photons (hν)	Emission of excitation energy	In most cases about 10^{-16} to 10^{-13} s after preceding α or β decay
Isomeric transition (IT)	I_γ	Photons (hν)	Delayed emission of excitation energy: $^{Am}Z \to {}^A Z$	Metastable excited states; preferably below magic numbers
Internal conversion (IC)	e^-	Conversion electrons and characteristic X rays	Transfer of excitation energy to an electron in the shell $^{58m}\mathrm{Co}(\mathrm{e}^-)^{58}\mathrm{Co}$	Preferably at low excitation energies (<0.2 MeV)
Proton decay	p	Protons $^1_1\mathrm{p}$	$^A Z \to {}^{A-1}(Z-1) + {}^1_1\mathrm{p}$ $^{147}\mathrm{Tm}(\mathrm{p})^{146}\mathrm{Er}$	Far away from the line of β stability
Spontaneous fission	sf	Fission products and neutrons	$^A Z \to {}^{A'}Z' + {}^{A-A'-v}(Z - Z') + vn$ $^{254}\mathrm{Cf}(\mathrm{sf})\dots$	Preferably at mass numbers $A > 245$

Details of the decay of radionuclides are recorded in the form of decay schemes, in which the energy levels are plotted and the half-lives, the nuclear spins, the parity and the transitions are indicated. Nuclei with higher atomic numbers are put to the right, and energies are given in MeV. As an example, the decay scheme of ^{238}U is plotted in Fig. 5.1.

5.2 Alpha Decay

As indicated in Table 5.1, ^4_2He nuclei are emitted by α decay, and so the atomic number decreases by two units and the mass number by four units (first displacement law of Soddy and Fajans).

The energy ΔE of α decay can be calculated by means of the Einstein formula $\Delta E = \Delta m\,c^2$:

$$\Delta E = (m_1 - m_2 - m_\alpha)c^2 \tag{5.1}$$

where m_1, m_2 and m_α are the masses of the mother nucleus, the daughter nucleus and the α particle, respectively. Introducing the masses of the nuclides (nucleus plus electrons), $M = m + Zm_e$, gives

$$\Delta E = (M_A - M_B - M_\alpha)c^2 \tag{5.2}$$

where M_A, M_B and M_α are the nuclide masses of the mother nuclide, the daughter nuclide and the α particle, respectively. By application of eq. (5.2) it is found that all nuclides with mass numbers $A > 140$ are unstable with respect to α decay. The reason that the binding energy of an α particle in the nucleus is relatively small is the high binding energy of the four nucleons in the α particle. However, as long as ΔE is small, α decay is not observed due to the energy barrier which has to be surmounted by the α particle. Therefore, nuclides with $A > 140$ are energetically unstable, but kinetically more or less stable with respect to α decay.

All α particles originating from a certain decay process are monoenergetic, i.e. they have the same energy. The energy of the decay process is split into two parts, the kinetic energy of the α particle, E_α, and the kinetic energy of the recoiling nucleus, E_N:

$$\Delta E = E_\alpha + E_N \tag{5.3}$$

From the law of conservation of momentum it follows that

$$m_\alpha v_\alpha = m_N v_N \tag{5.4}$$

where m_α and m_N are the masses and v_α and v_N are the velocities of the α particle and the nucleus, respectively, and eq. (5.3) becomes

$$\Delta E = E_\alpha\left(1 + \frac{m_\alpha}{m_N}\right) \tag{5.5}$$

Because the mass of heavy nuclei is appreciably higher than that of an α particle ($m_N \gg m_\alpha$), E_α is only about 2% smaller than ΔE.

Geiger and Nuttall in 1911 found that the decay constants of the α emitters in the natural decay series and the ranges R of the α particles in air are correlated for a certain decay series by equations of the form

$$\log \lambda = a \log R + b \tag{5.6}$$

where a and b are constants. These relations are called Geiger–Nuttall rules and are plotted in Fig. 5.2. As the range of α particles in air is a function of their energy E_α, eq. (5.6) may also be written

$$\log \lambda = a' \log E_\alpha + b' \tag{5.7}$$

Equations of the same kind have been found for other α particles.

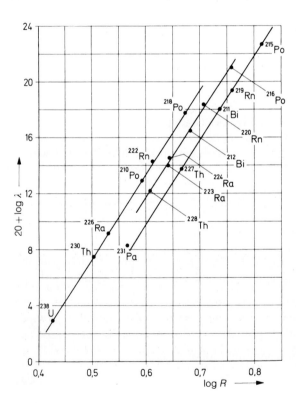

Figure 5.2. Relation between the range of α particles in air and the decay constant λ (Geiger–Nuttall rules) (according to: H. Geiger: Z. Physik *8*, 45 (1921)).

Substitution of λ by the half-life $t_{1/2}$ gives

$$\log t_{1/2} = \bar{b} - \bar{a} \log E_\alpha \tag{5.8}$$

where $\bar{b}$ and $\bar{a}$ are new constants. This relation is shown for a greater number of even–even nuclei in Fig. 5.3.

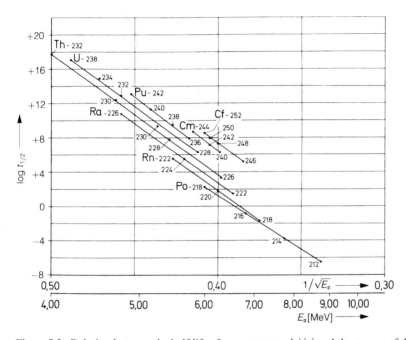

Figure 5.3. Relation between the half-life of even–even nuclei [s] and the energy of the α particles.

For some time, the theoretical interpretation of α decay encountered a fundamental problem: scattering experiments showed that the energy barrier for α particles entering the nucleus is relatively high (>9 MeV), whereas α particles leaving the nucleus have energies of only about 4 MeV. This problem was solved by the concept of quantum mechanical tunnelling (Gamow, 1928), according to which there is a certain probability that the α particles are able to tunnel through the energy barrier instead of passing over it. Theoretical calculations led to the following approximation formula for this probability, given by the decay constant λ:

$$\log \lambda \approx a - b \frac{Z-2}{\sqrt{2E_\alpha}} \tag{5.9}$$

This equation shows some similarity to eq. (5.7).

Several types of α spectra are distinguished in practice:

(a) Spectra showing only one line, which indicates only one group of α particles, all of the same energy. All transmutations go to the ground state of the daughter nuclide. Examples are ^{218}Po, ^{210}Po (Fig. 5.4).

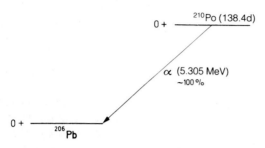

Figure 5.4. Decay scheme of ^{210}Po.

(b) Spectra with two or more lines or groups with similar energies. In this case, α decay leads to different excited states besides the ground state of the daughter nuclide. Examples are ^{212}Bi (Fig. 5.5), ^{223}Ra, ^{224}Ra, ^{227}Th, ^{231}Pa.

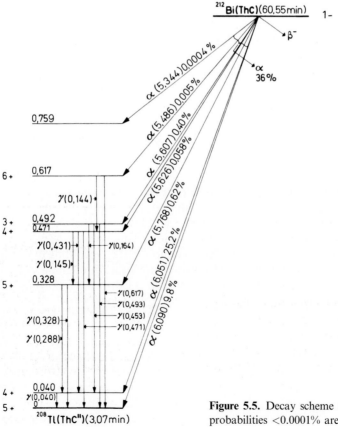

Figure 5.5. Decay scheme of ^{212}Bi. (Transitions with probabilities <0.0001% are not taken into account.)

(c) Spectra with one main line or group and other groups of much higher energy but much less intensity (about 10^{-2} to 10^{-5}% of the main group). Whereas the main group originates from the transmutation of the ground state of the mother nuclide to the ground state of the daughter nuclide, the other groups of higher intensity emerge from the transmutation of excited states of the mother nuclide to the ground state of the daughter nuclide. Examples are ^{214}Po, ^{212}Po (Fig. 5.6). The excited states of these nuclides are populated by preceding β^- decay of ^{214}Bi and ^{212}Bi, respectively.

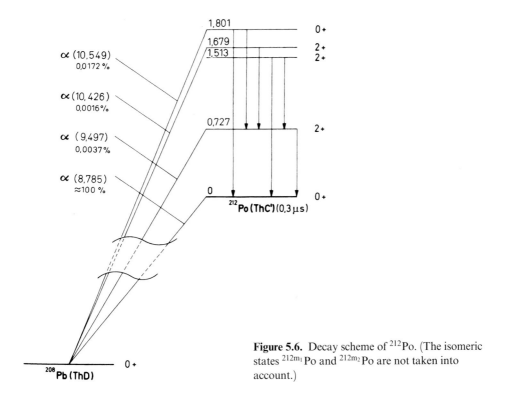

Figure 5.6. Decay scheme of ^{212}Po. (The isomeric states ^{212m_1}Po and ^{212m_2}Po are not taken into account.)

5.3 Beta Decay

Nuclides with an excess of neutrons experience β^- decay. In the nucleus a neutron is converted into a proton, an electron and an electron antineutrino, as indicated in Table 5.1. The atomic number increases by one unit, whereas the mass number does not change (second displacement law of Soddy and Fajans). The energy of the decay process can again be calculated by comparison of the masses according to Einstein:

$$\Delta E = (m_1 - m_2 - m_e)c^2 \tag{5.10}$$

where m_1, m_2 and m_e are the masses of the mother nucleus, the daughter nucleus and the electron, respectively. The mass of the antineutrino is neglected, because it is extremely small ($<2 \cdot 10^{-7}$ u, Table 3.3). Inserting the masses of the nuclides (nucleus plus electrons), $M = m + Zm_e$, gives

$$\Delta E = [M_1 - Z_1 m_e - M_2 + (Z_1 + 1)m_e - m_e]c^2$$

$$= (M_1 - M_2)c^2 \tag{5.11}$$

Nuclides with an excess of protons exhibit β^+ decay. A proton in the nucleus is converted into a neutron, a positron and an electron neutrino, as indicated in Table 5.1. The atomic number decreases by one unit, and the mass number remains unchanged. As in the case of β^- decay, the energy of the decay process is obtained by eq. (5.10). But because now $Z_2 = Z_1 - 1$, it follows that:

$$\Delta E = [M_1 - Z_1 m_e - M_2 + (Z_1 - 1)m_e - m_e]c^2$$

$$= (M_1 - M_2 - 2m_e)c^2 \tag{5.12}$$

This means that β^+ decay can occur only if M_1 is at least two electron masses higher than M_2:

$$M_1 > M_2 + 2m_e \tag{5.13}$$

An alternative to β^+ decay is electron capture (EC, symbol ε). Electrons are available from the electron shell of the nuclide, and the transformation of an excess proton into a neutron can also proceed by the taking up of an electron. As the K electrons have the highest probability of being close to the nucleus, they have also the highest probability of being captured. The result is the same as in the case of β^+ decay. But instead of positrons, characteristic X rays are emitted, because the empty position in the K shell of the atom is filled up by electrons of higher shells, a process that is associated with the emission of characteristic X rays. These X rays from the K shell are monoenergetic, and therefore radionuclides exhibiting electron capture are of practical interest as X-ray sources. To a smaller extent, X rays from the L shell may also be emitted. However, if the decay energy is smaller than the binding energies of the K electrons, only electrons from outer orbits (L, M, ...) can be captured.

In contrast to α particles, β particles do not have a distinct energy, but they show a continuous energy distribution (Fig. 5.7). The energy of the emitted electrons varies between zero and the maximum energy E_{max}, whereas the mean energy of the electrons is only about one-third of E_{max}. This seemed to be a contradiction to the law of conservation of energy, until Fermi postulated in 1934 that in addition to the electron another particle, the neutrino (correctly speaking an electron antineutrino), is emitted (neutrino hypothesis), which carries away the missing energy

$$E_{max} = E_e + E_v \tag{5.14}$$

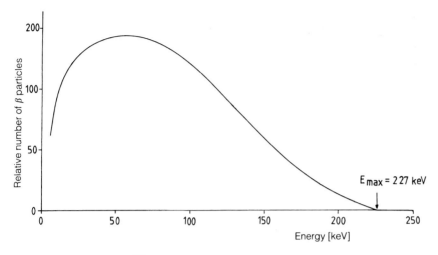

Figure 5.7. β spectrum of ^{147}Pm.

where E_e is the energy of the electron and E_v the energy of the electron neutrino or antineutrino, respectively. The neutrino has no charge, an extremely small mass ($m_v < 1/1000\, m_e$), the spin $\frac{1}{2}h/2\pi$ and it obeys the Fermi–Dirac statistics. These properties were postulated in order to fulfil the conservation laws.

Taking into account the formation of electron neutrinos in addition to the emission of electrons and positrons, respectively, the following equations are valid for β^- and β^+ decay.

$$\beta^-: \quad {}^1_0 n(\text{nucleus}) \rightarrow {}^1_1 p(\text{nucleus}) + {}^0_{-1}e^- + {}^0_0\bar{v}_e \tag{5.15}$$

$$\beta^+: \quad {}^1_1 p(\text{nucleus}) \rightarrow {}^1_0 n(\text{nucleus}) + {}^0_1 e^+ + {}^0_0 v_e \tag{5.16}$$

Electron capture is described by the following equation:

$$\varepsilon: \quad {}^1_1 p(\text{nucleus}) + {}^0_{-1}e^-(\text{shell}) \rightarrow {}^1_0 n(\text{nucleus}) + {}^0_0 v_e \tag{5.17}$$

The energy ΔE, given by eq. (5.10), is split up:

$$\Delta E = E_e + E_v + E_N \tag{5.18}$$

where E_e is the energy of the electron or positron, respectively, E_v that of the electron antineutrino or the electron neutrino, respectively, and E_N the recoil energy of the nucleus. As the mass of the electron is very small compared with the mass of a nucleus, it follows that:

$$\Delta E \approx E_{max} = E_e + E_v \tag{5.19}$$

For nuclei with $m_N > 5\,\text{u}$, the difference between ΔE and E_{max} is <0.01%. In the case of electron capture, E_e in eq. (5.18) is given by the binding energy of the electron in the electron shell which is very small compared with ΔE, and the neutrino receives the whole energy of the decay process:

$$\Delta E \approx E_\nu \tag{5.20}$$

As already mentioned in section 3.2, the interaction of neutrinos with matter is extremely small, and the first proof of their existence was only possible in 1956 on the basis of their reaction with the protons in a large tank of water:

$$^0_0\bar{\nu}_e + p \rightarrow \, ^1_0 n + \, ^0_1 e^+ \tag{5.21}$$

The simultaneous formation of a neutron and a positron was determined by means of large scintillation counters and cadmium. The annihilation of the positrons (by reaction with electrons) resulted in the production of two γ ray photons, whereas the neutrons lost their energy by collisions with protons and reacted with the cadmium to give another γ-ray photon with a delay of several microseconds.

Because neutrinos play an important role in astro- and cosmophysics, there is great interest in learning more about their properties and reactions.

Similarly to α decay, empirical relations between the decay constant λ of β emitters and the maximum energy $E_{\max}$, which is practically the same as the energy ΔE of the decay process (eq. (5.19)), were also found for β decay (Sargent, 1933):

$$\log \lambda = a + b \log E_{\max} \tag{5.22}$$

Different values of a and b were obtained for light, medium-weight and heavy nuclei (Sargent diagrams).

Application of quantum mechanics by Fermi led to the following theoretical formula for the probability of β decay:

$$P(E_k)\,dE = G^2|M|^2 F(Z, E_k)(E_k + m_0 c^2)(E_k^2 + 2m_0 c^2 E_k)^{1/2} \cdot (\Delta E - E_k)^2 \, dE_k \tag{5.23}$$

P is the fraction of nuclei decaying per unit time and emitting β particles with kinetic energy E_k. $G^2|M|^2$ is the relative probability of β decay. By the function F, the influence of the Coulomb forces of the nuclei is taken into account: electrons are slowed down by the positive charge of the nuclei, whereas positrons are accelerated. This causes a shift of the spectrum of electrons to the low-energy range, compared with the spectrum of positrons, as shown in Fig. 5.8. The other terms on the right-hand side of eq. (5.23) are statistical factors indicating the fractions of the decay energy ΔE transferred to the electrons, and m_0 is the rest mass of the electron. Introducing the momentum p_e of the electrons instead of their kinetic energy gives, under relativistic conditions, $(E_k + 2m_0 c^2 E_k)^{1/2} = p_e c$ and

$$P(E_k)\,dE = P(p_e)\,dp_e = cG^2|M|^2 F(Z, p_e)p_e^2(\Delta E - E_k)^2 \, dp_e \tag{5.24}$$

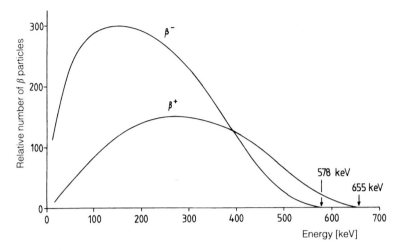

Figure 5.8. Spectra of the β^- and β^+ particles emitted by ^{64}Cu.

(c is the velocity of light). This relation can be tested by a "Fermi" or "Curie plot": M is assumed to be constant and

$$\left(\frac{P(p_e)}{p_e^2 F(Z, p_e)}\right)^{1/2} \sim \Delta E - E_k$$

is plotted as a function of E_k. As an example, the "Curie plot" for tritium is shown in Fig. 5.9. The extrapolation of the straight line gives relatively exact values of ΔE and E_{max}. For all "allowed" β transmutations the agreement with theory is good, whereas in the case of "forbidden" β transmutations relatively long half-lives are found.

Theoretical values of the decay constant λ are obtained by integrating eq. (5.23):

$$\lambda = \int_0^E P(E_k)\, dE_k \tag{5.25}$$

In case of "allowed" β decay, relatively high decay energy ΔE and low atomic numbers Z, the term $G^2|M|^2 F(Z, E_k)$ can be substituted by a constant G'^2 giving

$$\lambda \approx a \cdot \Delta E^5 \quad \text{or} \quad \log \lambda \approx \log a + 5 \log \Delta E \tag{5.26}$$

where a is a constant. The slope $+5$ agrees rather well with the slope of the curves in the Sargent diagrams (eq. (5.22)). For higher atomic numbers Z, the function $F(Z, E_k)$ can no longer be approximated by a constant. Putting the constant G and the matrix element M before the integral and abbreviating the integral by f gives the relation

$$\lambda = G^2|M|^2 f \tag{5.27}$$

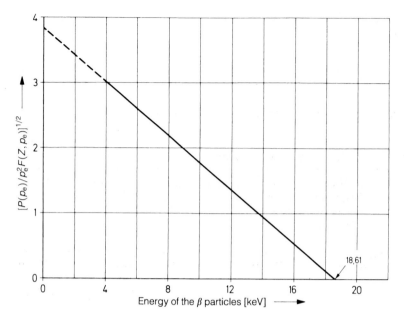

Figure 5.9. Curie plot for tritium (according to: F. T. Porter, *Phys. Rev.* **1959**, *115*, 450).

With $\lambda = (\ln 2)/t_{1/2}$ it follows

$$f \cdot t_{1/2} = \frac{\ln 2}{G^2 |M|^2} \tag{5.28}$$

The value of $f \cdot t_{1/2}$, known in short as the "*ft* value", should be similar for nuclei with similar matrix element M. Relatively low *ft* values are obtained for "allowed" transmutations and relatively high *ft* values for "forbidden" transmutations. Evaluation of the integral (5.25) gives the following approximate values for f:

$$\beta^- \text{decay: } \log f(\beta^-) = 4.0 \log \Delta E + 0.78 + 0.02\, Z - 0.005(Z - 1) \log \Delta E \tag{5.29}$$

$$\beta^+ \text{decay: } \log f(\beta^+) = 4.0 \log \Delta E + 0.79 + 0.007\, Z - 0.009(Z+1) \left(\log \frac{\Delta E}{3} \right)^2 \tag{5.30}$$

$$\text{Electron capture: } \log f(\varepsilon) = 2.0 \log \Delta E - 5.6 + 3.5(Z + 1) \tag{5.31}$$

ΔE is the decay energy in MeV and Z the atomic number of the nuclide. $f(\varepsilon)/f(\beta^+)$ gives the approximate ratio of electron capture to β^+ decay. It increases with Z and with decreasing ΔE.

Classification of β transmutations and selection rules are listed in Table 5.2. The lowest *ft* values are observed for nuclides that are transformed into a "mirror nuclide", i.e. a nuclide in which the numbers of protons and neutrons are inter-changed, for instance by the transmutation

$$^{17}_{9}F \rightarrow \, ^{17}_{8}O + \, ^{0}_{1}e^{+} + \, ^{0}_{0}\nu_{e} \tag{5.32}$$

Such transformations are said to be "favoured" (Table 5.2).

Table 5.2. Classification of β transmutations and selection rules.

Classification	Change of the quantum number of orbital spin, ΔL	Change of the nuclear spin, ΔI	Change of the parity	$\log ft$	Examples
Allowed (favoured)	0	0	No	2.7–3.7	n, ^{3}H, ^{6}He($\Delta I = 1$!), ^{11}C, ^{13}N, ^{15}O, ^{17}F, ^{19}Ne, ^{21}Na, ^{23}Mg, ^{25}Al, ^{27}Si, ^{29}P, ^{31}S, ^{33}Cl, ^{35}Ar, ^{37}K, ^{39}Ca, ^{41}Sc, ^{43}Ti
Allowed (normal)	0	0 or 1	No	4–7	^{12}B, ^{12}N, ^{35}S, ^{64}Cu, ^{69}Zn, ^{114}In
Allowed (l-forbidden)	2	1	No	6–9	^{14}C, ^{32}P
First forbidden	1	0 or 1	Yes	6–10	^{111}Ag, ^{143}Ce, ^{115}Cd, ^{187}W
First forbidden (special cases)	1	2	Yes	7–10	^{38}Cl, ^{90}Sr, ^{97}Zr, ^{140}Ba
Second forbidden	2	2	No	11–14	^{36}Cl, ^{99}Tc, ^{135}Cs, ^{137}Cs
Second forbidden (special cases)	2	3	No	≈ 14	^{10}Be, ^{22}Na
Third forbidden	3	3	Yes	17–19	^{87}Rb
Third forbidden (special cases)	3	4	Yes	18	^{40}K
Fourth forbidden	4	4	No	≈ 23	^{115}In

The β^{-} decay of ^{46}Sc (Fig. 5.10) may be taken as an example of the application of the selection rules in Table 5.2. The half-life of ^{46}Sc is 83.8 d $= 7.24 \cdot 10^{6}$ s, and nearly 100% of ^{46}Sc are converted into the second excited state of ^{46}Ti. The ft value is calculated from eq. (5.29) to be $\log ft = 6.3$, in good agreement with the value for allowed (normal) transmutations in Table 5.2 (no change of spin, $\Delta I = 0$, no change of parity). Transition to the first excited state occurs very seldom. The half-life with respect to this transition is higher by a factor of $100/0.004$, which gives $\log ft = 13.1$. This value is also in rather good agreement with the theoretical value for twofold forbidden transmutations in Table 5.2. (change of spin, $\Delta I = 2$, no change of parity).

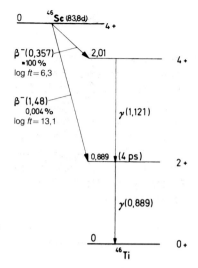

Figure 5.10. Decay scheme of ^{46}Sc.

Direct transition to the ground state of ^{46}Ti is not observed. This would be a fourfold forbidden transmutation (change of spin, $\Delta I = 4$, no change of parity) with $\log ft \approx 23$ according to Table 5.2, corresponding to a probability of about $2 \cdot 10^{-12}\%$, which is too low to be measured.

The decay scheme of ^{64}Cu is plotted in Fig. 5.11: 39.6% of ^{64}Cu show β^- decay to the ground state of ^{64}Zn, 19.3% β^+ decay to the ground state of ^{64}Ni, 40.5% electron capture (ε) and transition to the ground state of ^{64}Ni and 0.6% electron capture and transition to the excited state of ^{64}Ni.

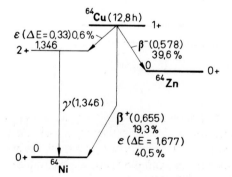

Figure 5.11. Decay scheme of ^{64}Cu.

5.4 Gamma Transitions

If a nucleus changes from an excited state to the ground state or another excited state of lower energy, γ-ray photons are emitted. As an example, the decay scheme of ^{198}Au is plotted in Fig. 5.12. With 98.7% probability, ^{198}Au changes into the first excited state of the daughter nuclide ^{198}Hg (0.412 MeV above ground level), with 1.3% probability into the second excited state (1.087 MeV above ground level), and with 0.025% to the ground level of ^{198}Hg. Accordingly, three γ transitions are observed: the second excited state changes with 20% probability directly to the ground state and with 80% probability to the first excited state at 0.412 MeV, resulting in the following intensities relative to the total β activity: γ (1.087), $1.3 \cdot 0.2 = 0.26\%$; $\gamma(0.676)$, $1.3 \cdot 0.8 = 1.04\%$; $\gamma(0.412)$, $98.7 + 1.3 \cdot 0.8 = 99.74\%$.

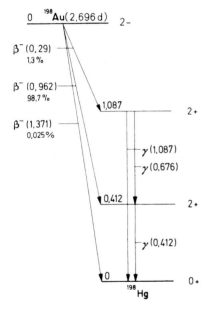

Figure 5.12. Decay scheme of ^{198}Au.

All γ rays emitted by a certain nucleus are monoenergetic, i.e. they have well-defined energies. Because the recoil energies transmitted to the nuclei by emission of the γ-ray photons are very small compared with the energies of the γ rays, the latter are practically equal to the excitation energies or the differences in the excitation energies of the nuclei:

$$E_\gamma = \Delta E \tag{5.33}$$

Gamma spectrometry is therefore the most important tool for studying the properties of atomic nuclei.

Generally, the lifetime of the excited states is very small, of the order of 10^{-16} to 10^{-13} s; the γ radiation is emitted immediately after a preceding α or β decay. However, if immediate γ transitions are "forbidden", because of high differences of the nuclear spins of the excited state and the ground state in combination with the laws of conservation of nuclear momentum and of parity, a metastable state or nuclear isomer results which decays with its own half-life (section 2.2). The transition from the metastable isomeric state into the ground state is called isomeric transition (IT). Isomeric transition is free from accompanying α or β radiation, and some nuclear isomers are of great practical importance as pure γ emitters. For example, ^{99m}Tc has found broad application in nuclear medicine.

Instead of emitting a γ-ray photon, the excited nucleus may transmit its excitation energy to an electron of the atomic shell, preferably a K electron, a process called internal conversion (IC). The probability of this alternative increases with increasing atomic number and with decreasing excitation energy. The conversion electron (symbol e^-) is emitted instead of a γ-ray photon and its energy is

$$E_e = E_\gamma - E_B \tag{5.34}$$

where E_γ is the energy of the γ-ray photon and E_B the binding energy of the electron. In contrast to β particles, conversion electrons are monoenergetic. Internal conversion is followed by emission of characteristic X rays, as in the case of electron capture (section 5.3).

The ratio of conversion electrons and γ-ray photons emitted per unit time is called the conversion coefficient:

$$\alpha = \frac{\lambda_{e^-}}{\lambda_\gamma} \tag{5.35}$$

λ_e^- and λ_γ are the partial decay constants (probabilities) of conversion (electron emission) and γ-ray emission. The conversion coefficient α is the sum of the partial conversion coefficients of the K shell, L shell, ...

$$\alpha = \alpha_k + \alpha_L + \cdots = \frac{\lambda_{e^-}(K)}{\lambda_\gamma} + \frac{\lambda_{e^-}(L)}{\lambda_\gamma} + \cdots \tag{5.36}$$

A very rare alternative to γ-ray emission is the simultaneous emission of an electron and a positron. This possibility only exists if the excitation energy is greater than the energy necessary for the generation of an electron and a positron (pair formation), which amounts to 1.02 MeV.

The theory of γ-ray emission is based on the model of an electromagnetic multipole. Such a multipole may change its electric or magnetic momentum by emission of electromagnetic radiation. Consequently, electric multipole radiation (E) and magnetic multipole radiation (M) are distinguished. The quantum number L of the nuclear angular momentum may change by one or several units, and because of the conservation of momentum, the γ-ray photon carries with it the corresponding angular momentum $L \cdot h/2\pi$. L is an integer ($L = 1, 2, \ldots; L \neq 0$) and characterizes the multipole radiation: $L = 1$ is the dipole radiation, $L = 2$ the quadrupole radiation, $L = 3$ the octupole radiation etc. Electric and magnetic multipole radiation have different parity. Electric multipole radiation has even parity if L is even, and magnetic multipole radiation has even parity if L is odd. Because emission of electromagnetic radiation is an electromagnetic interaction, the law of conservation of parity is valid.

The probability of γ-ray emission is given by the sum of the probabilities for the emission of the individual multipole radiations, which decrease drastically with increasing L. Furthermore, for a certain multipole, the probability of the emission of electric multipole radiation is about two orders of magnitude higher than that of the emission of magnetic multipole radiation.

On the basis of the shell model of the nuclei, Weisskopf derived the following equations for the probabilities of γ-ray emission, given by the decay constants λ_E for electric multipole radiation and λ_M for magnetic multipole radiation:

$$\lambda_E = 2.4 \; S(r_0 A^{1/3})^{2L} \left(\frac{E}{197}\right)^{2L+1} \cdot 10^{21} \; \text{s}^{-1} \tag{5.37}$$

$$\lambda_M = 0.55 \; SA^{-2/3}(r_0 A^{1/3})^{2L} \left(\frac{E}{197}\right)^{2L+1} \cdot 10^{21} \; \text{s}^{-1} \tag{5.38}$$

where $r_0 A^{1/3}$ is the radius of the nucleus in fm ($r_0 = 1.28$ fm, section 3.1), A the mass number, E the energy of the γ-ray photons, and S is given by

$$S = \frac{2(L+1)}{L[1 \cdot 3 \cdot 5 \cdot \ldots (2L+1)]^2} \left(\frac{3}{L+3}\right)^2 \tag{5.39}$$

Values calculated by means of these formulas for the half-life of excited states are listed in Table 5.3. Because not all properties of nuclei are considered in the model, the table gives only approximate values.

Table 5.3. Half-lives of γ transitions calculated by application of the model of multipole radiation.

Type of radiation	Change of the orbital spin quantum number, ΔL	Change of the parity	Half-life (s) at energies of		
			1 MeV	0.2 MeV	0.05 MeV
E_1	1	Yes	$2 \cdot 10^{-16}$	$3 \cdot 10^{-14}$	$2 \cdot 10^{-12}$
M_1	1	No	$2 \cdot 10^{-14}$	$2 \cdot 10^{-12}$	$2 \cdot 10^{-10}$
E_2	2	No	$1 \cdot 10^{-11}$	$3 \cdot 10^{-8}$	$3 \cdot 10^{-5}$
M_2	2	Yes	$9 \cdot 10^{-10}$	$3 \cdot 10^{-6}$	$3 \cdot 10^{-3}$
E_3	3	Yes	$7 \cdot 10^{-7}$	$6 \cdot 10^{-2}$	$9 \cdot 10^{2}$
M_3	3	No	$7 \cdot 10^{-5}$	5	$8 \cdot 10^{4}$
E_4	4	No	$8 \cdot 10^{-2}$	$2 \cdot 10^{5}$	$4 \cdot 10^{10}$
M_4	4	Yes	7	$1 \cdot 10^{7}$	$4 \cdot 10^{12}$

The selection rules for γ-ray emission are summarized in the equation

$$I_i + I_e \geq L \geq |I_i - I_e| \tag{5.40}$$

where I_i and I_e are the nuclear spins before and after γ-ray emission, respectively. According to the law of conservation of parity, electric multipole radiation with even L and magnetic multipole radiation with odd L are allowed, if the initial and the final state have the same parity; on the other hand, electric multipole radiation with odd L and magnetic multipole radiation with even L are allowed, if the initial and the final state have different parity. Because photons have the spin quantum number 1, transitions from $I_i = 0$ to $I_e = 0$ are not possible by emission of γ-ray photons. If in such a transition the parity does not change, conversion electrons may be emitted instead of γ-ray photons (example ^{72}Ge), or, if the excitation energy is high enough ($\Delta E > 1.02$ MeV), an electron and a positron may be emitted (example ^{214}Po).

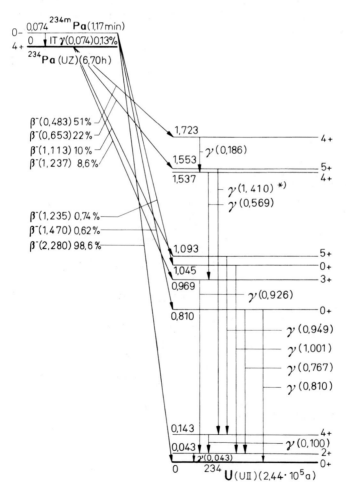

Figure 5.13. Decay scheme of ^{234}Pa and ^{234m}Pa. (*) Transition in several steps.)

The values calculated by means of eqs. (5.37) and (5.38) are in rather good agreement with experimental values, which is also a proof of the applicability of the shell model of the nuclei. This model leads to the expectation that excited states of low excitation energy and with nuclear spins differing appreciably from the nuclear spins of the ground states are to be found most frequently for nuclei with atomic numbers Z or neutron numbers N just below the magic numbers Z or $N = 50$, 82 and 126, respectively. According to Table 5.3, these excited states should exhibit long half-lifes. Actually, in these regions of the chart of nuclides "islands of isomeric nuclei" are observed, i.e. many metastable excited states with measurable half-lives are found in these regions.

The first case of a nuclear isomer was found in 1921 by Hahn, who proved by chemical methods the existence of two isomeric states of ^{234}Pa which were called UX$_2$ and UZ. The decay scheme of ^{234}Pa is plotted in Fig. 5.13. Both nuclear isomers are produced by decay of ^{234}Th. ^{234m}Pa $(t_{1/2} = 1.17\,\text{m})$ changes at nearly 100% directly into ^{234}U. Later, the production of artificial radionuclides by nuclear reactions led to the discovery of a great number of nuclear isomers. In the case of ^{80}Br, for instance, two isomeric states were found (Fig. 5.14), and chemical separation of ^{80m}Br and ^{80}Br is also possible. From the change of nuclear spin and of parity half-lives can be assessed by application of the selection rules (eq. (5.40)) and of eqs. (5.37) and (5.38). The half-lives of nuclear isomers may vary between seconds and many years.

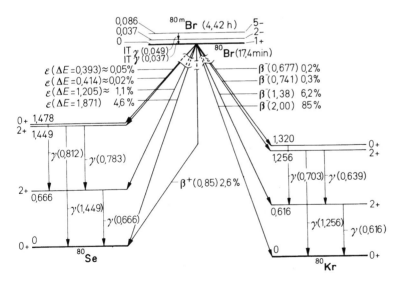

Figure 5.14. Decay scheme of ^{80}Br and ^{80m}Br.

5.5 Proton Decay and Other Rare Decay Modes

With an increasing excess of protons, on the left-hand side of the line of β stability, the binding energy of the last proton decreases markedly, and a region is expected in which this binding energy approaches zero and proton emission from the ground state becomes energetically possible. However, as in the case of α decay, the protons leaving the nucleus have to pass an energy barrier by tunnelling; this gives these nuclei that are unstable with respect to proton decay a certain lifetime.

Proton activity was observed for the first time for ^{147}Tm ($t_{1/2} = 0.56$ s) and ^{151}Lu ($t_{1/2} = 90$ ms), both produced at the UNILAC accelerator of GSI (Darmstadt, Germany) in 1982. Both nuclides emit monoenergetic protons of 1.05 and 1.23 MeV, respectively, by transmutation of the ground state of the mother nuclide into the ground state of the daughter nuclide:

$$^{147}\text{Tm} \rightarrow \,^{146}\text{Er} + \,^{1}\text{p} \tag{5.41}$$

$$^{151}\text{Lu} \rightarrow \,^{150}\text{Yb} + \,^{1}\text{p} \tag{5.42}$$

Thus, besides α decay, β decay and γ transition, a fourth type of decay is known since 1982. In the meantime, further examples of proton decay have been discovered, all on the extreme proton-rich side of the chart of nuclides. In this region, proton emission (p decay) competes with emission of positrons (β^+ decay), and because in most cases β^+ decay is favoured, p decay is observed relatively seldom.

More frequently, p emission occurs after β^+ decay in a two-stage process: β^+ decay leads to an excited state of the daughter nuclide, and from this excited state the proton can easily surmount the energy barrier. This two-stage process is called β^+-delayed proton emission. It is observed for several β^+ emitters from ^{9}C to ^{41}Ti with $N = Z - 3$, with half-lives in the range of 1 ms to 0.5 s. Simultaneous emission of two protons has been observed for a few proton-rich nuclides, e.g. ^{16}Ne ($t_{1/2} \approx 10^{-20}$ s). Proton decay from the isomeric state is observed in case of ^{53m}Co (probability $\approx 1.5\%$, $t_{1/2} \approx 0.25$ s).

The situation on the right-hand (neutron-rich) side of the line of β stability is different from that on the left-hand side, because the binding energy of additional neutrons is higher than that of additional protons and the binding energy of the last neutron approaches zero only at great distances from the line of β stability. This leads to an extended region of nuclides that are energetically stable to neutron emission from the ground state, and up to now neutron emission from the ground state has not been observed. All neutron-rich nuclides in this region exhibit sequences of β^- transmutations leading to nuclides of increasing atomic number Z.

Neutron emission immediately following β^- transmutation (β^--delayed neutron emission) is observed for many neutron-rich nuclides, such as ^{87}Br and many fission products. Delayed neutron emission is very important for the operation of nuclear reactors (chapter 10).

Spontaneous emission of particles heavier than α particles is called cluster radioactivity. Theoretical calculations, published since 1980, showed that spontaneous fragmentation of nuclei with atomic numbers $Z > 40$ by emission of cluster nuclei, such as $^{5-7}$Li, $^{7-9}$Be, $^{11-14}$C, $^{14-16}$N, $^{19-22}$F and $^{20-25}$Ne is energetically possible with half-lives ranging from about 10^{22} to 10^{92} y. Thus, compared with other modes of decay, cluster radioactivity is a very rare event.

Emission of ^{14}C by ^{223}Ra was observed in 1984 by Rose and Jones with a branching ratio $\lambda_c : \lambda_\alpha = (8.5 \pm 2.5) \cdot 10^{-10}$ (subscript c for cluster). The daughter nucleus is ^{209}Pb. In the meantime, more than 14 cluster emitters have been found with branching ratios $\lambda_c : \lambda_\alpha$ between 10^{-9} and 10^{-16}. The clusters have atomic numbers between $Z = 6$ and $Z = 14$ (isotopes of C, O, F, Ne, Mg, Si) which are preferably even, and the daughter nuclei are near to the magic nucleus ^{208}Pb.

Stable nuclei may become unstable if their electron shell is stripped off; then they show a special kind of β^- transmutation in which the electron set free in the nucleus by transformation of a neutron into a proton occupies a free place in the empty electron shell of the atom. This has been observed with nuclei of ^{163}Dy in the storage ring ESR at GSI; these are converted with a half-life of about 47 d by β^- decay into ^{163}Ho if there is no electron shell present. Thus, under extreme conditions, such as they exist in stars, additional kinds of transmutation are possible.

5.6 Spontaneous Fission

Spontaneous fission (symbol sf) was found in 1940 by Flerov and Petrzhak at Dubna, after fission by neutrons had been discovered in 1938 by Hahn and Strassmann in Berlin. Spontaneous fission is another mode of radioactive decay, which is observed only for high mass numbers A. For ^{238}U the ratio of the probability of spontaneous fission to that of α decay is about $1 : 10^6$. It increases with the atomic number Z and the number of neutrons in the nucleus. For ^{256}Fm the probability of spontaneous fission relative to the total probability of decay is already 92%.

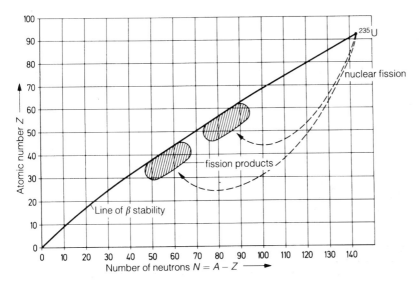

Figure 5.15. Neutron excess of the fission products due to the neutron excess of heavy nuclei.

Spontaneous fission can be described by the equation

$$^{A}Z \rightarrow \underset{(1)}{^{A'}Z'} + \underset{(2)}{^{A-A'-\nu}(Z - Z')} + \nu n + \Delta E \tag{5.43}$$

as already indicated in Table 5.1. ν is the number of neutrons and ΔE the energy set free by the fission process. The resulting nuclei (1) and (2) have, in general, different mass numbers A and atomic numbers Z. Because of the high neutron excess of heavy fissioning nuclei, the fission products (1) and (2) are found in the chart of nuclides on the neutron-rich side of the line of β stability, as illustrated in Fig. 5.15. Several excess neutrons ($\nu = 2$ to 4) are emitted promptly (prompt neutrons). Other neutrons may be emitted by the primary fission products and are called delayed neutrons.

The energy ΔE in eq. (5.43) can be calculated by comparing the masses of the mother nuclide and the products. It is found that for mass numbers $A \approx 200$, ΔE amounts to about 200 MeV. Comparison of the masses shows that ΔE is already positive for $A > 100$, indicating that all nuclides above $Z \approx 46$ (Pd) are unstable with respect to fission. The fact that these nuclides do not exhibit spontaneous fission is due to the high energy barrier.

Spontaneous fission proceeds by tunnelling, similarly to α decay. The steps of the fission process are illustrated in Fig. 5.16:

(a) The nucleus oscillates between a more spherical and a more ellipsoidal shape. By further distortion and constriction near the centre of the ellipsoid the nucleus

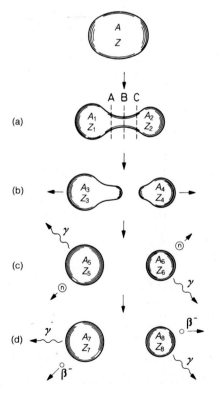

Figure 5.16. The steps of spontaneous fission.

attains the shape of a dumbbell, in which at least one part has magic numbers of protons and neutrons.

(b) The nucleus splits into two parts. If this splitting takes place at A, two parts of nearly equal mass, but different excitation energy, are formed (symmetric fission). Most probable is the fission at B, by which two parts of different mass, but similar excitation energy, are formed (asymmetric fission). Fission at C leads to two parts of very different mass and different excitation energy. The Coulomb repulsion energy, which has a much greater range than the nuclear forces, drives both products apart, and the fission products attain high kinetic energies.

(c) The highly excited fission products emit neutrons (prompt neutrons) and photons (prompt photons), and sometimes also charged particles. Up to this stage, the processes take place within about 10^{-15} s.

(d) The fission products change by one or several β^- transformations and emission of γ-ray photons into stable products. In the case of high excitation energies, emission of further neutrons (delayed neutrons) may be observed.

Table 5.4. Partial half-lives of spontaneous fission (values from R. Vandenbosch, J. R. Huizenga, Nuclear Fission, Academic Press, New York, 1973).

Nuclide	Partial half-life of spontaneous fission	Average number ν of neutrons set free	Nuclide	Partial half-life of spontaneous fission	Average number ν of neutrons set free
^{230}Th	$\geq 1.5 \cdot 10^{17}$ y		^{249}Cf	$6.5 \cdot 10^{10}$ y	
^{232}Th	$> 10^{21}$ y		^{250}Cf	$1.7 \cdot 10^{4}$ y	3.53 ± 0.09
^{232}U	$\approx 8 \cdot 10^{13}$ y		^{252}Cf	85 y	3.764
^{233}U	$1.2 \cdot 10^{17}$ y		^{254}Cf	60 d	3.88 ± 0.14
^{234}U	$1.6 \cdot 10^{16}$ y		^{253}Es	$6.4 \cdot 10^{5}$ y	
^{235}U	$3.5 \cdot 10^{17}$ y		^{254}Es	$> 2.5 \cdot 10^{7}$ y	
^{236}U	$2 \cdot 10^{16}$ y		^{255}Es	2440 y	
^{238}U	$9 \cdot 10^{15}$ y	2.00 ± 0.08	^{244}Fm	≥ 3.3 ms	
^{237}Np	$> 10^{18}$ y		^{246}Fm	≈ 20 s	
^{236}Pu	$3.5 \cdot 10^{9}$ y	2.22 ± 0.2	^{248}Fm	≈ 60 h	
^{238}Pu	$5 \cdot 10^{10}$ y	2.28 ± 0.08	^{250}Fm	≈ 10 y	
^{239}Pu	$5.5 \cdot 10^{15}$ y		^{252}Fm	115 y	
^{240}Pu	$1.4 \cdot 10^{11}$ y	2.16 ± 0.02	^{254}Fm	246 d	3.99 ± 0.20
^{242}Pu	$7 \cdot 10^{10}$ y	2.15 ± 0.02	^{255}Fm	$1.2 \cdot 10^{4}$ y	
^{244}Pu	$6.6 \cdot 10^{10}$ y	2.30 ± 0.19	^{256}Fm	2.63 h	3.83 ± 0.18
^{241}Am	$2.3 \cdot 10^{14}$ y		^{257}Fm	120 y	4.02 ± 0.13
^{242m_1}Am	$9.5 \cdot 10^{11}$ y		^{258}Fm	380 μs	
^{243}Am	$3.3 \cdot 10^{13}$ y		^{257}Md	≥ 30 h	
^{240}Cm	$1.9 \cdot 10^{6}$ y		^{252}No	≈ 7.5 s	
^{242}Cm	$6.5 \cdot 10^{6}$ y	2.59 ± 0.09	^{254}No	$\geq 9 \cdot 10^{4}$ s	
^{244}Cm	$1.3 \cdot 10^{7}$ y	2.76 ± 0.07	^{256}No	≈ 1500 s	
^{246}Cm	$1.8 \cdot 10^{7}$ y	3.00 ± 0.20	^{258}No	1.2 ms	
^{248}Cm	$4.2 \cdot 10^{6}$ y	3.15 ± 0.06	^{256}Lr	$> 10^{5}$ s	
^{250}Cm	$1.4 \cdot 10^{4}$ y	3.31 ± 0.08	^{257}Lr	$> 10^{5}$ s	
^{249}Bk	$1.7 \cdot 10^{9}$ y	3.64 ± 0.16	^{258}Lr	≥ 20 s	
^{246}Cf	$2.0 \cdot 10^{3}$ y	2.85 ± 0.19	^{261}104	≥ 650 s	
^{248}Cf	$3.2 \cdot 10^{4}$ y		^{261}105	8 s	

Partial half-lives of spontaneous fission and the number of neutrons set free are listed in Table 5.4. The partial half-lives are calculated by use of eq. (4.41) in section 4.8 (branching decay). They vary between the order of nanoseconds and about 10^{17} y. Although the fission barrier for nuclides such as ^{238}U (≈ 6 MeV) is small compared with the total binding energy of the nucleons (≈ 1800 MeV), spontaneous fission of ^{238}U has a low probability compared with α decay.

The main part of the energy ΔE released by spontaneous fission according to eq. (5.43) is set free promptly (about 89%) in the form of kinetic energy of the fission products (about 82%), kinetic energy of the neutrons (about 3%) and energy of the γ rays (about 4%). The rest appears as excitation energy of the fission products and is given off with some delay in the form of energy of β^- particles (about 3%), energy of neutrinos (about 5%) and energy of γ-ray photons (about 3%).

Due to the production of neutrons, spontaneously fissioning nuclides are of practical interest as neutron sources. An example is ^{252}Cf ($t_{1/2} = 2.64$ y), which is used for neutron activation.

As already mentioned, asymmetric fission prevails strongly. This is illustrated for ^{242}Cm in Fig. 5.17: the fission yields are in the range of several percent for mass numbers A between about 95 to 110 and about 130 to 145, and below 0.1% for symmetric fission ($A \approx 120$). The fission yields are the average numbers of nuclei with a certain mass number A produced per fission. Because two nuclei are generated, the sum of the fission yields amounts to 200%.

The drop model of nuclei (section 2.4) proved to be useful to explain fission (Bohr and Wheeler, 1939): due to the surface tension of a liquid, a droplet assumes a spherical shape. If energy is supplied, the droplet begins to oscillate between spherical and elongated shapes. With increasing distortion, elongation passes a threshold and the droplet splits into two parts. In nuclei, the repulsive Coulomb forces, which

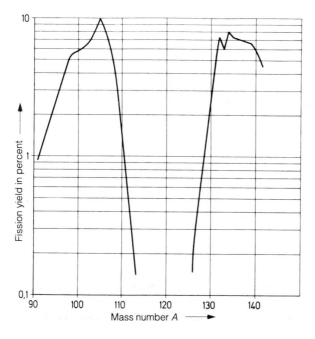

Figure 5.17. Fission yield as a function of the mass number for the spontaneous fission of ^{242}Cm (according to E. P. Steinberg, L. E. Glendenin, *Phys. Rev.* **1954**, *95*, 437).

are proportional to $Z(Z-1)/A^{1/3}$ (eq. (2.3) in section 2.4), tend to distort the nuclei, whereas the surface tension, which is proportional to $A^{2/3}$ (eq. (2.4) in section 2.4), tries to keep them in a spherical form. As long as the influence of the surface energy exceeds that of the Coulomb repulsion, there is a net restoring force and the nucleus returns to its spherical form. However, if at a certain deformation the influence of the Coulomb repulsion prevails, the nucleus becomes unstable toward splitting into two parts.

The ratio of the two opposing influences is taken as a measure of the instability x:

$$x \sim \frac{Z(Z-1)}{A^{1/3}A^{2/3}} \approx \frac{Z^2}{A} \qquad (5.44)$$

Z^2/A is called the fissionability parameter. The logarithm of the (partial) half-life of spontaneous fission $t_{1/2}(\mathrm{sf})$ of even–even nuclei is plotted as a function of the fissionability parameter in Fig. 5.18. The general trend is obvious from this figure. However, for each element a maximum of $t_{1/2}(\mathrm{sf})$ is observed. Furthermore, for even–odd, odd–even and odd–odd nuclei the (partial) half-lives for spontaneous fission are several orders of magnitude longer than those expected by interpolation of the values for even–even nuclei. Obviously, application of the simple drop model of nuclei does not lead to results that are in quantitative agreement with experimental data, and spontaneous fission is hindered, if odd numbers of protons or

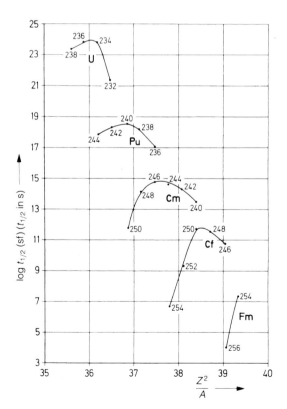

Figure 5.18. Half-life of spontaneous fission of even–even nuclei as a function of the fissionability parameter Z^2/A.

neutrons are present causing an increase of the fission barrier. This effect is similar to that observed for α decay (section 5.2).

Significant advances with respect to the quantitative theoretical description of spontaneous fission were achieved by the so-called shell-correction approach (Strutinsky, 1967) in which single-particle effects are combined with liquid-drop properties. This approach led to the prediction of a double potential barrier (Fig. 5.19) for some regions of Z and A. From the distorted state II the nuclei may pass much more easily over the fission barrier than from the ground state I.

The double-humped fission barrier in Fig. 5.19 also makes it possible to explain the very short half-lives, of the order of nano- to microseconds, observed for some spontaneously fissioning nuclear isomers (e.g. ^{242m}Am, ^{244m}Am, ^{246m}Am). By measuring the minimum energies needed to produce the ground and the isomeric states by nuclear reactions, the energy difference E' between these states can be determined. The values obtained are in good agreement with the results of the shell-correction theory.

The excitation energy set free by the fission process is not distributed evenly amongst the primary fission fragments. In a small fraction of spontaneous fission events, which is called cold fission, the excitation energy is so small that no neutrons are emitted (neutron-less fission). This fraction is about 0.2% in case of ^{252}Cf and about 3% in case of ^{235}U.

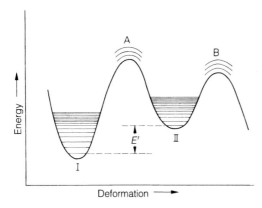

Figure 5.19. Double potential barrier of spontaneous fission.

Literature

General

R. D. Evans, The Atomic Nucleus, McGraw-Hill, New York, **1955**

E. Segrè (Ed.), Radioactive Decay, in: Experimental Nuclear Physics, Vol. III, Wiley, New York, **1959**

I. Kaplan, Nuclear physics, 2nd ed., Addison-Wesley, Reading, MA, **1964**

G. Friedlander, J. W. Kennedy, E. S. Macias, J. M. Miller, Nuclear and Radiochemistry, 3rd. ed., Wiley, New York, **1981**

D. N. Poenaru and M. Ivascu (Eds.), Particle Emission from Nuclei, 2 Vols., CRC Press, Boca Raton, FL, **1989**

D. N. Poenaru (Ed.), Handbook of Decay Modes, CRC Press, Boca Raton, FL, **1993**

More Special

I. Perlman and J. O. Rasmussen, Alpha Radioactivity, in: Handbuch der Physik, Vol. 42, Springer, Berlin, **1957**

K. Siegbahn (Ed.), Alpha-, Beta- and Gamma-Ray Spectroscopy, 2 Vols., North-Holland, Amsterdam, **1966**

C. S. Wu and S. A. Moszkowski, Beta Decay, Interscience, New York, **1966**

E. J. Konopinsky, The Theory of Beta Radioactivity, Oxford University Press, Oxford, **1966**

R. Vandenbosch and J. R. Huizenga, Nuclear Fission, Academic Press, New York, **1973**

E. Roeckl, Alpha Radioactivity, Radiochim. Acta *70/71*, 107 **(1995)**

S. Hofmann, Proton Radioactivity, Radiochim. Acta *70/71*, 93 **(1995)**

G. Ardisson and M. Hussonnois, Radiochemical Investigations of Cluster Radioactivities, Radiochim. Acta *70/71*, 123 **(1995)**

D. C. Hoffman and M. R. Lane, Spontaneous Fission, Radiochim. Acta *70/71*, 135 **(1995)**

Tables

E. Browne, R. B. Firestone (Ed. V. S. Shirley), Table of Radioactice Isotopes, Wiley, New York, **1986**

G. Erdtmann and G. Soyka, The Gamma Rays of the Radionuclides (Ed. K. H. Lieser), Verlag Chemie, Weinheim, **1979**

Nuclear Data Sheets, Section A, 1 **(1965)** et seq., Section B, 1 **(1966)** et seq., Academic Press, New York

6 Nuclear Radiation

6.1 General Properties

Knowledge of the properties of nuclear radiation is needed for the measurement and identification of radionuclides and in the field of radiation protection. The most important aspect is the interaction of radiation with matter.

Charged high-energy particles or photons, such as α particles, protons, electrons, positrons, γ-ray or X-ray photons, set off ionization processes in gases, liquids or solids:

$$M \rightsquigarrow M^+ + e^- \tag{6.1}$$

where M is an atom or a molecule. Any kind of radiation that is able to produce ions according to eq. (6.1) is called ionizing radiation. The arrow $\rightsquigarrow$ indicates a reaction induced by ionizing radiation. Excited atoms or molecules M^* may also be produced:

$$M \rightsquigarrow M^* \tag{6.2}$$

The chemical reactions induced by ionizing radiation in gases, liquids and solids are the field of radiation chemistry, whereas the concern of photochemistry is the chemical reactions induced by light (visible and UV).

The minimum energy needed for ionization or excitation of atoms or molecules is of the order of several eV, depending on the nature of the atoms or molecules. The photons of visible light have energies varying between about 1 eV ($\lambda = 1240$ nm) and 10 eV ($\lambda = 124$ nm). If their energy is high enough, as in the case of UV radiation, they lose it by one ionization process. Particles with energies of the order of 0.1 to 10 MeV, however, produce a great number of ions and electrons and of excited atoms or molecules. The products of the reactions (6.1) and (6.2) are accumulated in the track of the high-energy particle or photon. Heavy particles, such as α particles or protons, lead to a high density of the reaction products in the track, whereas their density is low in the track of electrons or γ-ray photons.

The ions M^+ and the excited atoms or molecules M^* produced in the primary reactions (6.1) and (6.2), respectively, give rise to further (secondary) reactions:

$$M^+ \rightarrow R^+ + R \cdot \qquad \text{(dissociation)} \tag{6.3}$$

$$M^+ + e^- \rightarrow M^* \qquad \text{(recombination)} \tag{6.4}$$

$$M^+ + X \rightarrow Y^+ \qquad \text{(chemical reaction)} \tag{6.5}$$

$$M^+ + X \rightarrow M + X^+ \qquad \text{(charge transfer)} \tag{6.6}$$

$$M^+ \rightarrow M^{n+} + (n-1)e^- \quad \text{(emission of Auger electrons)} \tag{6.7}$$

$$M^* \rightarrow M + h\nu \quad \text{(fluorescence)} \tag{6.8}$$

$$M^* \rightarrow 2R\cdot \quad \text{(dissociation into radicals)} \tag{6.9}$$

$$M^* \rightarrow R^+ + R^- \quad \text{(dissociation into ions)} \tag{6.10}$$

$$M^* + X \rightarrow Y \quad \text{(chemical reaction)} \tag{6.11}$$

$$M^* + X \rightarrow M + X^* \quad \text{(transfer of excitation energy)} \tag{6.12}$$

Many of these secondary reactions are very fast and occur within 10^{-10} to 10^{-7} s. Reactions (6.4), (6.8) and (6.9) are relatively frequent. Recombination is favoured in liquids and solids.

The concentration of the reaction products in the track is proportional to the energy lost by the ionizing particles per unit distance travelled along their path, which is called linear energy transfer (LET). For example, in water the LET value of 1 MeV α particles is 190 eV/nm and for 1 MeV electrons it is 0.2 eV/nm. This means that the concentration of reaction products in the track is higher by a factor of about 10^3 for α particles. In air, the ionizing radiation loses 34.0 eV (electrons) to 35.1 eV (α particles) per ion pair produced according to eq. (6.1). Because this value is about twice the ionization energy of N_2 (15.6 eV) and O_2 (12.1 eV), it follows that about half of the energy given off by the particles is used up by production of excited atoms and molecules, respectively, according to eq. (6.2).

If particles, in particular α particles, protons or photons, have sufficiently high energy, they may also give rise to nuclear reactions. Electrons entering the force field of nuclei give off a part of their energy in the form of photons (bremsstrahlung). If their energy is of the order of 1 MeV (e.g β^- radiation), these photons have the energy of X rays (X-ray bremsstrahlung), and at energies >10 MeV the photons have the energies of γ rays.

Neutrons may lose their energy in steps by collision with other particles or they may induce nuclear reactions. In contrast to particles, photons mostly give off their energy in one step.

The behaviour of various kinds of radiation in a magnetic field is shown in Fig. 6.1: γ radiation is not deflected, β^+ and β^- radiation are deflected in different directions, and the influence of the magnetic field on α particles is much smaller because the deflection depends on e/m (e is the charge and m the mass).

Whereas α radiation is easily absorbed (e.g. by one sheet of paper), for quantitative absorption of β radiation materials of several millimetres or centimetres thickness (depending on the energy) and for absorption of γ radiation either lead walls or thick walls of concrete are needed. At the same energy, the ratio of the absorption coefficients for α, β and γ radiation is about $10^4 : 10^2 : 1$. Furthermore, it has to be taken into account that α and β particles can be absorbed quantitatively, whereas the absorption of γ-ray photons is governed by an exponential law, and therefore only a certain fraction can be held back.

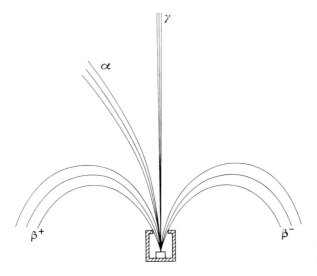

Figure 6.1. Behaviour of various kinds of radiation in a magnetic field.

6.2 Alpha Radiation

The limited range of α radiation can be seen in Fig. 6.2. The range depends on the energy of the α particles and amounts to several centimetres in air. Their course is practically not influenced by the collisions with electrons. Rarely an α particle collides with a nucleus and is strongly deflected, or it is captured by a nucleus and induces a nuclear reaction.

Figure 6.2. α rays in a cloud chamber.

The specific ionization of α particles in air is shown in Fig. 6.3: the number of ion pairs produced per millimetre of air increases strongly with the distance and falls off rather sharply near the end of the range of the α particles. The increase is due to the decreasing velocity of the α particles. As the energy decreases by about 35 eV per ion pair generated, an α particle with an initial energy of 3.5 MeV produces about 10^5 ion pairs. At the end of its path it forms a neutral He atom.

The relative number of α particles is plotted in Fig. 6.4 as a function of the distance from the source. The variation of the range is caused by the statistical variation of the number of collisions. Exact values of the range in air are obtained by extrapolating or by differentiating the curve in Fig. 6.4 (extrapolated range or average range, respectively). A simple device for the determination of the range of α particles

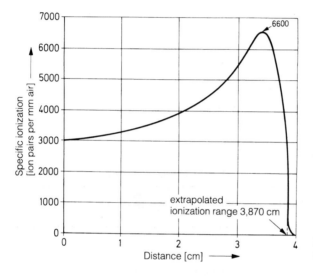

Figure 6.3. Specific ionization of the α particles of ^{210}Po in air.

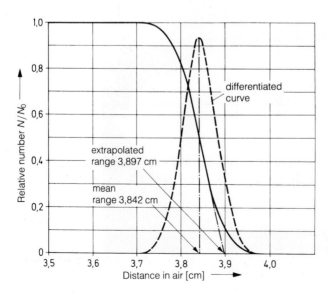

Figure 6.4. Relative number of the α particles from ^{210}Po as a function of the distance.

in air is shown in Fig. 6.5: ZnS emits light as long as it is being hit by α particles. At a certain distance between the ZnS screen and the α source the emission of light decreases very fast, indicating the range of the α particles in air. By application of this method, curves of the kind shown in Fig. 6.4 can be obtained.

The range of α particles in various substances is listed in Table 6.1. Multiplication of the range in cm by the density of the substance (g/cm^3) gives the range in g/cm^2. Table 6.1 shows that the ranges in mg/cm^2 are similar for very different substances; they increase markedly at higher atomic numbers.

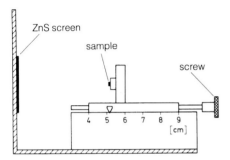

Figure 6.5. Device for the determination of the range of α particles in air.

Table 6.1. Range of the α particles of ^{214}Po $(E = 7.69\ \mathrm{MeV})$ in various substances.

Substance	Extrapolated range in [cm]	Density [g/cm^3]	Range in [mg/cm^2]
Air	6.95	0.001226	8.5
Mica	0.0036	2.8	10.1
Lithium	0.01291	0.534	6.9
Aluminium	0.00406	2.702	11.0
Zinc	0.00228	7.14	16.3
Iron	0.00187	7.86	14.7
Copper	0.00183	8.92	16.3
Silver	0.00192	10.50	20.2
Gold	0.00140	19.32	27.0
Lead	0.00241	11.34	27.3

As a measure of the absorption properties of a substance, the stopping power is used; it is defined as the energy lost per unit distance travelled by the particle:

$$B(E) = -\frac{dE}{dX} \tag{6.13}$$

The stopping power depends on the energy of the particle, just as the specific ionization does. The range of the particles is given by

$$R = \int_0^{E_0} \frac{dE}{B(E)} \tag{6.14}$$

where E_0 is the initial energy.

In Fig. 6.6 the range of α particles in air is plotted as a function of their initial energy. The curve can be used to determine the energy of α particles. Most α particles have ranges between about 3 and 7 cm in air, and for these the approximate relation R (cm) $= 0.318 \, E^{3/2}$ (E in MeV) is valid.

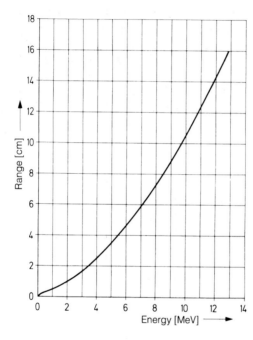

Figure 6.6. Range of α particles as a function of their initial energy (according to W. P. Jesse, J. Sadanskis, Physic. Rev. **78** (1950) 1).

For determination of the energy of α particles and for identification of α-emitting radionuclides, α spectrometers are usually applied, having been calibrated by use of α emitters of known energy. Relatively exact determination of the energy of α particles, protons and deuterons can be made by means of a magnetic spectrometer. The relation between the velocity v of the particles, the magnetic flux density B and the radius r of the particles is

$$v = B \, r \frac{Ze}{m} \tag{6.15}$$

where Ze is the charge and m the mass of the particles.

6.3 Beta Radiation

As already mentioned in section 6.1, the interaction of β radiation with matter is much weaker than that of α radiation. Whereas a 3 MeV α particle has a range of about 1.7 cm in air and produces several thousand ion pairs per millimetre, a β particle of the same energy covers a distance of about 10 m in air and produces only about 4 ion pairs per millimetre. On the other hand, the electrons are markedly deflected by collisions with other electrons, in contrast to the heavy α particles, and they therefore exhibit a zigzag course.

The absorption of the β particles of ^{32}P by aluminium is plotted in Fig. 6.7. The form of the absorption curve is due to the continuous energy distribution of the β particles and the scattering of the β radiation in the absorber. At the end of the absorption curve a nearly constant intensity of bremsstrahlung (X rays) is observed. By extrapolation of the absorption curve the maximum range R_{max} of the β particles can be found. In practice, this extrapolation is carried out by subtraction of the bremsstrahlung and extension of the curve to $10^{-4}I_0$ (Fig. 6.7).

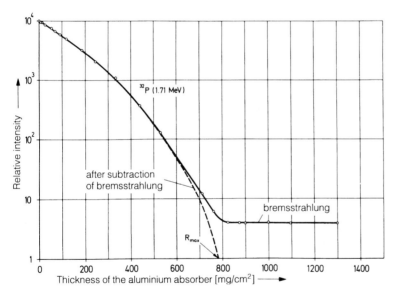

Figure 6.7. Absorption of the β particles of ^{32}P in aluminium.

Absorption curves of β^- and β^+ radiation are very similar. By use of the calibration curve in Fig. 6.8 the maximum energy E_{max} of β particles can be determined from their maximum range R_{max}.

Conversion electrons are monoenergetic and exhibit a nearly linear absorption curve (Fig. 6.9), if their energy is >0.2 MeV. At energies <0.2 MeV the absorption curve deviates more or less from linearity. To obtain the effective range of conversion electrons, the linear part of the absorption curve is extrapolated to the intensity $I = 0$.

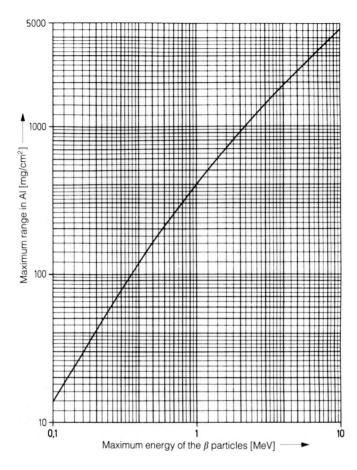

Figure 6.8. Maximum range R_{max} of β particles as a function of their maximum energy E_{max} (according to L. Katz, A. S. Penfold; Rev. mod. Physics **24** (1952) 28).

Beta radiation interacts with matter in three different ways:

(a) Interaction with electrons leads to excitation of the electron shell and ionization. The important parameter for this interaction is the electron density in the absorber, i.e. the number of electrons per mass unit given by Z/A. This is shown in Table 6.2 for three different energies E_{max} of β particles and different absorbers.

Table 6.2. Maximum range of β particles of three different energies in various substances.

Maximum energy [MeV]	Substance	Z/A	Maximum range [mg/cm^2]
0.156 (^{14}C)	Water	$8/18 = 0.44$	34
	Aluminium	$13/27 = 0.48$	28
1.71 (^{32}P)	Water	0.44	810
	Aluminium	0.48	800
1.0	Aluminium	0.48	400
	Gold	$79/197 = 0.40$	500

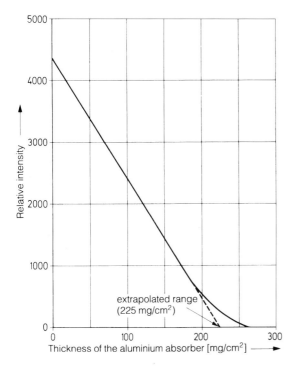

Figure 6.9. Absorption of the conversion electrons of ^{137m}Ba.

(b) Interaction with atomic nuclei increases with the energy of the β radiation. In the electric field of a nucleus high-energy electrons emit X rays of continuous energy distribution (bremsstrahlung) and lose their energy in steps. The ratio of energy loss by emission of bremsstrahlung and energy loss by ionization is approximately given by

$$\frac{E(\text{Bremsstr.})}{E(\text{Ionis.})} = \frac{E_{\max} \cdot Z}{800} \qquad (6.16)$$

where $E_{\max}$ is the maximum energy of the β radiation in MeV and Z the atomic number of the absorber.

(c) Backscattering can be measured by the method shown in Fig. 6.10. It also depends on the energy $E_{\max}$ of the β radiation and the atomic number Z of the absorber, as shown in Fig. 6.11 for three different energies $E_{\max}$ as a function of Z.

In the case of absorption of β^+ radiation, emission of γ-ray photons is observed: positrons are the antiparticles of electrons (section 3.2). After having given off their energy by the interactions (a) to (c), they react with electrons by annihilation and emission of predominantly two γ-ray photons with an energy of 0.51 MeV each in opposite directions (conservation of momentum). The energy of $2 \times 0.51 = 1.02$ MeV is equivalent to $2m_e$, the sum of the masses of the electron and the positron. This annihilation radiation allows identification and measurement of β^+ radiation.

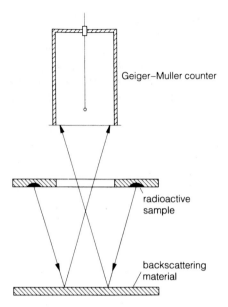

Geiger–Muller counter

radioactive
sample

backscattering
material

Figure 6.10. Set-up for the measurement of back-scattering.

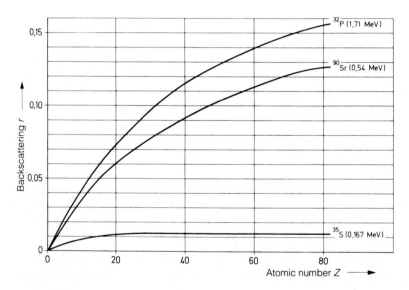

Figure 6.11. Backscattering of β radiation of various energies as a function of the atomic number Z of the absorber.

The maximum energy of β^+ radiation can be determined by plotting an absorption curve as shown in Fig. 6.7, or, more accurately, by use of a magnetic spectrometer, as in the case of α particles, but at much lower flux density B, because of the higher e/m values. At the same energy, the velocity v of electrons is much higher than that of α particles which makes relativistic correction necessary:

$$v = Br\frac{e}{m_0}\sqrt{1 - \left(\frac{v}{c}\right)^2} \tag{6.17}$$

where r is the radius of the β particles in the spectrometer, e their charge, m_0 their rest mass, and c the velocity of light. An example of a β spectrum is given in Fig. 5.7.

Charged particles moving in a substance with a velocity which is higher than the velocity of light in that substance emit Čerenkov radiation. Thus, deep-blue Čerenkov radiation is observed if β radiation passes through a transparent substance, such as water or perspex, because the condition $v \geq c/n$ (n = refactive index of the substance) is already fulfilled for β particles at relatively low energies, in contrast with α particles of the same energy. The electrons can be compared with a source of sound moving faster than the velocity of sound and carrying with it a Mach cone.

With respect to radiation protection, absorbers of low atomic numbers Z are most useful for absorption of β radiation, for instance perspex or aluminium of about 1 cm thickness.

6.4 Gamma Radiation

Gamma rays and X rays have similar properties and are distinguished by their origins: X rays are emitted from the electron shell of atoms, if electrons are passing from states of higher energy to those of lower energy (characteristic X rays) or if electrons are slowed down in the field of nuclei (bremsstrahlung); γ rays are emitted from nuclei, if these pass from excited states to states of lower energy. The energy range of X rays varies from about 100 eV to 100 keV (range of wavelenghts about 10 nm to 10 pm), and that of γ rays from about 10 keV to 10^4 MeV (range of wavelengths about 0.1 nm to 10^{-7} nm). That means there is an overlap in the energy ranges of X rays and γ rays. Electrons with energies >10 MeV striking a substance of high atomic number induce the emission of very energetic ("hard") bremsstrahlung which is also called γ radiation, because of its high energy. In contrast to the γ rays emitted from nuclei, this bremsstrahlung shows a continuous energy distribution.

The absorption of γ rays and X rays is, in principle, different from that of α or β rays. While the latter lose their energy by a succession of collisions, γ-ray photons give off their energy mostly in one process. Because they do not carry a charge, their interaction with matter is small. For the absorption of γ rays an exponential law is valid:

$$I = I_0 e^{-\mu d} \tag{6.18}$$

(μ = absorption coefficient and d = thickness of the absorber). The exact validity of this exponential law is restricted to monoenergetic γ radiation, a narrow pencil of γ rays and a thin absorber.

The relation between the energy of γ radiation and its absorption is characterized by the half-thickness $d_{1/2}$, i.e. the thickness of the absorber by which the intensity of the γ radiation is reduced to half of it. Introducing $I = I_0/2$ in eq. (6.18) gives:

$$d_{1/2} = \frac{\ln 2}{\mu} \tag{6.19}$$

Instead of the linear absorption coefficient, the mass absorption coefficient μ/ρ (cm^2/g) is often used:

$$I = I_0 e^{-\mu d/\rho} \tag{6.20}$$

In this equation, d is in units of g/cm^2. The absorption of the γ radiation of ^{137}Cs as a function of d in g/cm^2 is plotted in Fig. 6.12.

Due to the exponential absorption law, the intensity $I = 0$ is not attained. By $7d_{1/2}$ the initial intensity is reduced to about 1%, and by $10d_{1/2}$ to about 1‰. For the pur-

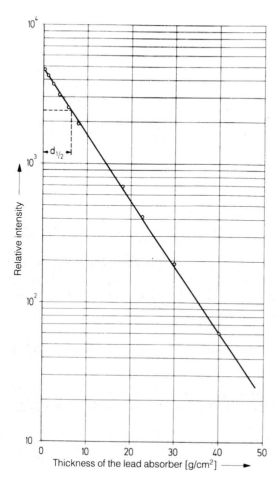

Figure 6.12. Absorption of the γ radiation of ^{137}Cs.

pose of radiation protection, it is useful to know that the intensity of 1 MeV γ radiation is reduced to about 1% by 5 cm of lead or 25 cm concrete.

In the first place, the absorption of γ radiation depends on the density of the absorber, similarly to the absorption of α or β radiation. In Table 6.3 the mass absorption coefficients for various absorbers and various energies of γ radiation are listed.

Table 6.3. Mass absorption coefficient μ/ρ in [g/cm^2] for γ rays of different energy.

E_γ [MeV]	Nitrogen	Water	Carbon	Sodium	Aluminium	Iron	Copper	Lead
0.1022	0.1498	0.165	0.1487	0.1532	0.1643	0.3589	0.4427	5.30
0.2554	0.1128	0.1255	0.1127	0.1086	0.1099	0.1186	0.1226	0.558
0.5108	0.0862	0.096	0.0862	0.0827	0.0833	0.0824	0.0814	0.149
1.022	0.0629	0.0697	0.0629	0.0603	0.0607	0.0590	0.0580	0.0682
2.043	0.0439	0.0488	0.0438	0.0422	0.0427	0.0420	0.0414	0.0442
5.108	0.0270	0.0298	0.0266	0.0271	0.0286	0.0312	0.0315	0.0434
10.22	–	0.0216	–	–	0.0226	–	–	0.0537

According to: Ch. M. Davisson. R. D. Evans; Rev. mod. Physics, **24** (1952) 79.

The half-thickness of γ radiation in lead can be used as a measure of the energy (Fig. 6.13). The fact that at higher energies two values are found for the same half-thickness is due to the overlap of three different main absorption mechanisms:

(a) By the photoeffect, a γ-ray photon transfers its energy to an electron, which is emitted as a photoelectron. The energy of the photoelectron is

$$E_e = E_\gamma - E_B \qquad (6.21)$$

E_γ is the energy of the γ-ray photon and E_B the binding energy of the electron. The recoil energy of the resulting ion can be neglected, because of the small mass of the electron compared with the mass of the ion. E_B is, in general, also relatively small compared with E_γ.

(b) By the Compton effect, a γ-ray photon gives off only a part of its energy to an electron, which is emitted. According to the law of conservation of momentum, the γ-ray photon changes its frequency v and its direction, as shown in Fig. 6.14. The energy of the scattered γ-ray photon is

$$E = \frac{E_0}{1 + q \cdot E_0} \qquad (6.22)$$

where E_0 = initial energy of the photon, $q = (1 - \cos \varphi)/m_0 \cdot c^2$ (φ = angle of scattering, m_0 = rest mass of the electron and c = velocity of light). The energy of the emitted electron is

$$E_e = \frac{q \cdot E_0}{1 + q \cdot E_0} \qquad (6.23)$$

(c) Pair formation is observed at energies $E_\gamma \geq 2m_0c^2 = 1.02\,\mathrm{MeV}$. In the electric field of nuclei the γ-ray photon is transformed into an electron and a positron, provided that the energy is at least equivalent to the sum of the masses of the electron and the positron. Thus pair formation is the reversal of annihilation. Its probability increases sharply with increasing energy of the γ-ray photons and represents the majority of the absorption processes at energies $E_\gamma > 10\,\mathrm{MeV}$. Furthermore, pair formation increases with the square of the atomic number Z of the absorber.

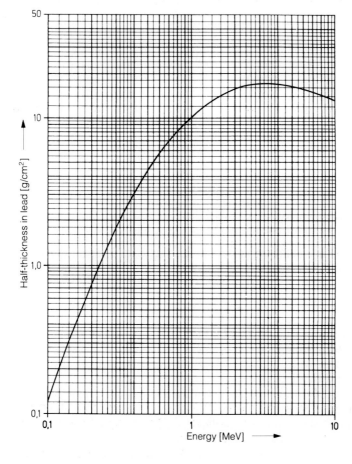

Figure 6.13. Half-thickness of γ radiation in lead as a function of the energy (according to C. M. Davisson, R. D. Evans, Rev. mod. Physics *24*, 79 (1952)).

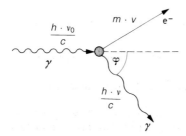

Figure 6.14. Compton effect.

The total absorption coefficient μ is given approximately by the sum of the partial absorption coefficients due to the photoeffect (μ_{Ph}), the Compton effect (μ_C) and pair formation (μ_P):

$$\mu = \mu_{Ph} + \mu_C + \mu_P \qquad (6.24)$$

The contributions of these partial absorption coefficients to the absorption of γ radiation in lead are plotted in Fig. 6.15 as a function of the energy of the γ radiation. The strong increase of the contribution of pair formation at higher energies leads to the bending of the curve for the total absorption coefficient μ in Fig. 6.15 and of the curve in Fig. 6.13.

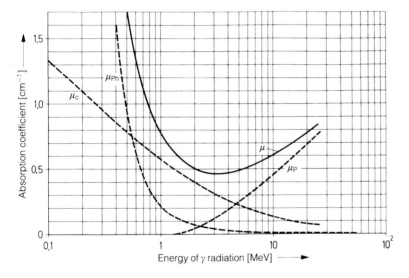

Figure 6.15. Total absorption coefficient μ and partial absorption coefficients of γ radiation in lead as a function of the energy (according to C. M. Davisson, R. D. Evans, Rev. mod. Physics *24*, 79 (1952)).

In Fig. 6.16 the total absorption coefficient of γ radiation in various materials is plotted as a function of the energy of the γ radiation. The strong influence of the atomic number of the absorber is evident from this figure.

Besides the main absorption processes (a) to (c), several other processes are of minor influence:

(d) At low energies of the γ radiation, coherent scattering by the atoms of the absorber is observed, as in the case of X rays.
(e) At high energies of the γ radiation, nuclear reactions are induced (photonuclear reactions).
(f) Thomson and Compton scattering take place on a small scale at the nuclei.

The energy of γ radiation can be found roughly from the half-thickness $d_{1/2}$ by use of the curve in Fig. 6.13. The most accurate method of determination of γ-ray ener-

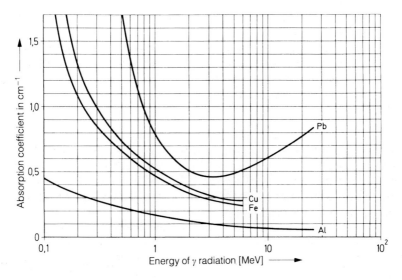

Figure 6.16. Absorption coefficient of γ radiation in various materials as a function of the energy (according to C. M. Davisson, R. D. Evans, Rev. mod. Physics *24*, 79 (1952)).

gies and of identification of radionuclides is γ spectrometry by means of solid-state detectors. In the γ-ray spectra obtained by this method, sharp photopeaks are observed which allow determination of the energy according to eq. (6.21).

Absorption of γ radiation for the purpose of radiation protection requires lead walls or thick walls of concrete, as already mentioned in section 6.1.

6.5 Neutrons

Neutrons are emitted by spontaneously fissioning heavy nuclei. They play an important role in nuclear reactions, in particular in nuclear fission (chapter 8). High fluxes of neutrons are available in nuclear reactors (chapter 11).

The properties of neutrons have been discussed in sections 3.1 and 3.2. Neutrons are not stable in the free state, but undergo β^- decay:

$$\begin{array}{c} {}^1_0\text{n} \xrightarrow[10.25\ \text{min}]{\beta^-\,(0.782)} {}^1_1\text{p} + {}^0_0\text{e}^- + {}^0_0\bar{\nu} \end{array} \qquad (5.33)$$

Because neutrons are electrically neutral, their interaction with electrons is very small and primary ionization by neutrons is negligible. The interaction of neutrons with matter is practically confined to the nuclei and comprises elastic and inelastic scattering and nuclear reactions. In elastic collisions the total kinetic energy remains constant, whereas in inelastic collisions part of the kinetic energy is given off as excitation energy.

The contributions of the different kinds of interaction depend on the energy of the neutrons. The following energy ranges are distinguished:

0–0.1 eV: Thermal neutrons (the energy distribution is comparable to that of gas molecules at ordinary temperature)

0.1–100 eV: Slow neutrons (neutrons with energies of the order of 1 eV to 1 keV are also called resonance neutrons, because maxima of absorption are observed in this energy range)

0.1–100 keV: Neutrons of intermediate energies

0.1–10 MeV: Fast neutrons

In contrast to protons, deuterons, α particles and other particles carrying positive charges, neutrons do not experience Coulomb repulsion by nuclei. Low-energy (thermal and slow) neutrons are very effectively absorbed by a great number of nuclei, giving rise to nuclear reactions. Elements such as B, Cd, Sm, Eu, Gd and Dy are used as excellent neutron absorbers.

Fast neutrons lose their energy mainly in elastic and inelastic collisions. The energy given off in one elastic collision depends on the angle of collision and is at maximum $4AE_0/(A+1)^2$ in the case of central collisions, where E_0 is the energy of the neutron before the collision and A is the mass number of the target nucleus. The lighter the nucleus, the higher the energy loss of the neutron. Hydrogen-containing substances such as water or paraffin are the most effective media to reduce the energy of neutrons. About 20 collisions with protons are necessary to slow a neutron of several MeV down to thermal energies. For this purpose, paraffin about 20 cm thick is sufficient. Graphite is also used as slowing-down (moderating) material for neutrons. About 120 collisions with the nuclei of carbon atoms are necessary to have the same effect as 20 collisions with protons. After having lost the main part of their energy, the neutrons are captured by nuclei, giving rise to nuclear reactions. High-energy (fast) neutrons may also induce nuclear reactions, but the contribution of this kind of interaction is relatively small.

Detection of neutrons is based on ionization processes caused by the products of their interactions (nuclear reactions or collisions) with nuclei.

6.6 Short-lived Elementary Particles in Atoms and Molecules

As already mentioned in section 3.2, absorption of short-lived elementary particles, such as positrons, muons, pions, kaons or sigma particles, may lead to formation of unusual (exotic) kinds of atoms or molecules. A proton in a hydrogen atom may be substituted by a positively charged short-lived elementary particle, such as a positron or a positive muon, or an electron in the electron shell of an atom or molecule may be substituted by a negatively charged short-lived elementary particle, such as a negative muon, a negative pion or an antiproton. Some kinds of hydrogen-like atoms containing short-lived elementary particles are listed in Table 6.4. The lifetime of these species varies between about 10^{-9} and 10^{-6} s. However, during this time their properties can be studied by application of fast and sensitive electronic methods.

Table 6.4. Properties of hydrogen and hydrogen-like atoms containing short-lived elementary particles.

Atom	Binding energy [eV]	Rest mass [u]
(p^+e^-) Hydrogen (H)	13.6	1.0078
(e^+e^-) Positronium (Ps)	6.8	0.0010972
(μ^+e^-) Muonium	13.5	0.11495
$(p^+\mu^-)$ Muonic atom	2531	1.1217
$(p^+\pi^-)$ Pionic atom	3236	1.2717
(p^+K^-)	8618	1.5373
(p^+p^-)	12498	2.0146
$(p^+\Sigma^-)$	14014	2.2843

A positronium atom $(e^+e^- = Ps)$ is comparable with a hydrogen atom and can be considered to represent the lightest form of an atom, in which e^+ and e^- rotate around a common centre of gravity. Positronium atoms are formed by absorption of positrons in matter. For instance, about 30% of the positrons absorbed in argon yield Ps:

$$e^+ + Ar \rightarrow Ps + Ar^+ \tag{5.34}$$

Two ground states of Ps are known, a triplet (ortho) state (o-Ps, $^3S_1$, parallel spins of e^+ and e^-, lifetime $1.4 \cdot 10^{-7}$ s and a singlet (para) state (p-Ps, $^1S_0$, antiparallel spins of e^+ and e^-, lifetime $1.25 \cdot 10^{-10}$ s). o-Ps and p-Ps annihilate by emission of 2 and 3 γ-ray photons, respectively.

By interaction with molecules, the lifetime of Ps is strongly affected. Radicals containing an unpaired electron, such as NO·, cause a conversion of o-Ps to the shorter-lived p-Ps. In many chemical reaction, Ps behaves like a H· radical:

$$Ps + CH\equiv C-CH_2OH \rightarrow CH(Ps)=\dot{C}-CH_2OH \quad \text{(addition)} \tag{5.35}$$

$$Ps + Cl_2 \rightarrow PsCl + Cl\cdot \quad \text{(substitution)} \tag{5.36}$$

$$Ps + Fe^{3+} \rightarrow Fe^{2+} + e^+ \quad \text{(reduction of Fe}^{3+}, \text{oxidation of Ps)} \tag{5.37}$$

$$Ps + e^- \rightarrow Ps^- \quad \text{(reduction of Ps)} \tag{5.38}$$

Reduction of Ps takes place in metals and the resulting Ps^- annihilates very quickly. HPs is an analogue of H_2, in which the centre of gravity is given by the proton surrounded by two electrons and one positron.

Muonium atoms (μ^+e^-) are formed by absorption of positively charged muons in matter. They show similarities to positronium atoms.

Muonic atoms (e.g. $p^+\mu^-$), on the other hand, are obtained by absorption of negatively charged muons. In these atoms, μ^- replaces an electron in the electron shell. Due to the relatively high mass of μ^- compared with that of e^-, their interaction with the nucleus is rather strong, and muons serve as probes to study the properties of nuclei. As the ratio of the atomic orbit is inversely proportional to the mass of the orbiting particle, the μ^- orbitals begin to enter the nuclei of muonic atoms with increasing atomic number and consequently the residence time of the muon in the nucleus increases, too.

For instance, the "Bohr radius" of μ^- in muonic Pb is only about 4 fm, whereas the radius of the nucleus is about 7 fm. Finally, the muon may be captured by the nucleus or it may decay as a free particle. The influence of the charge distribution in nuclei on muons is also greater than that on electrons, and X rays emitted by muonic atoms, in particular from inner orbitals, give information about the charge distribution and surface structure of nuclei. The influence of electron densities and chemical bonds has been studied by use of pionic atoms, such as $p^+\pi^-$.

Finally, another interesting aspect should be mentioned: it is expected that in muonic molecular ions reactions between nuclei are favoured, because of their smaller distance apart, for instance $d^+d^+\mu^- \rightarrow {}^4He^+ + \gamma$ or $d^+t^+\mu^- \rightarrow {}^4He^+ + n + \gamma$. This kind of reactions would offer the possibilty of fusion at relatively low temperatures ("cold fusion") of about 10^3 K in contrast to "hot fusion" at about 10^8 K (section 8.12).

Literature

General

E. Rutherford, J. Chadwick, C. D. Ellis, Radiation from Radioactive Substances, Cambridge University Press, Cambridge, **1930** (reprinted **1951**)

E. Segrè (Ed.), Experimental Nuclear Physics, Vol. 1, Wiley, New York, **1953**

A. H. Compton, S. K. Allison, X-rays in Theory and Experiment, Van Nostrand, London, **1935**

I. G. Draganic, Z. D. Draganic, J. P. Adloff, Radiation and Radioactivity on Earth and Beyond, CRC Press, Boca Raton, FL, **1990**

Interaction with Matter

W. Whaling, The Energy Loss of Charged Particles in Matter, in: Encyclopedia of Physics, Vol. 34 (Ed. E. Flügge), Springer, Berlin, **1958**

S. C. Lind, Radiation Chemistry of Gases, Reinhold, New York, **1961**

A. O. Allen, The Radiation Chemistry of Water and Aqueous Solutions, Van Nostrand, London, **1961**

W. T. Spinks, R. J. Woods, An Introduction to Radiation Chemistry, Wiley, New York, **1964**

I. G. Draganic, Z. D. Draganic, The Radiation Chemistry of Water, Academic Press, New York, **1971**

A. R. Denaro, G. G. Jayson, Fundamentals of Radiation Chemistry, Butterworths, London, **1972**

G. J. Dienes, G. H. Vineyard, Radiation Effects in Solids, Interscience, London, **1957**

A. J. Swallow, Radiation Chemistry of Organic Compounds, Pergamon, Oxford, **1960**

S. C. Lind, Radiation Chemistry of Gases, Reinhold, New York, **1961**

R. O. Bolt, J. G. Carroll, Radiation Effects on Organic Materials, Academic Press, New York, **1963**

A. O. Allen, The Radiation Chemistry of Water and Aqueous Solutions, Van Nostrand, London, **1961**

J. W. T. Spinks, R. J. Woods, An Introduction to Radiation Chemistry, 2nd ed., Wiley, New York, **1976**

J. Kroh (Ed.), Early Developments in Radiation Chemistry, London, **1989**

Y. Tabata, Y. Ito, S. Tagawa (Eds.), Handbook of Radiation Chemistry, CRC Press, Boca Raton, FL, **1991**

I. G. Draganic, Radioactivity and Radiation Chemistry of Water, Radiochim. Acta *70/71*, 317 **(1995)**

Positronium and Muonium Chemistry

J. H. Green, J. Lee, Positronium Chemistry, Academic Press, New York, **1964**

V. H. Hughes, C. S. Wu (Eds.), Muon Physics, Vol. III Chemistry and Solids, Academic Press, New York, **1975**

H. J. Ache (Ed.), Positronium and Muonium Chemistry, Adv. Chem. Ser. 175, American Chemical Society, Washington, DC, **1979**

P. Hautojärvi (Ed.), Positrons in Solids, Topics in Current Physics, Vol. 12, Springer, Berlin, **1979**

P. W. Percival, Muonium Chemistry, Radiochim. Acta *26*, 1 **(1979)**

D. M. Schrader, Y. C. Yeans (Eds.), Positron and Positronium Chemistry, Stud. Phys. Theor. Chem. 57, Elsevier, Amsterdam, **1988**

7 Measurement of Nuclear Radiation

7.1 Activity and Counting Rate

The activity (disintegration rate) A as defined in section 4.2 is a property of radioactive matter and can be measured by various devices which give a certain counting rate I', which depends on the activity A, the overall counting efficiency η of the device and the background counting rate I_0:

$$I' = I + I_0 = \eta A + I_0 \tag{7.1}$$

Usually, the counting rate is measured in counts per minute (cpm). $I = I' - I_0$ is the (net) counting rate caused by the radioactive sample to be measured. I_0 is the reading in the absence of a radioactive sample. It is due to the radiation emitted by the surrounding material and cosmic radiation. At low radioactivity of the sample, the background counting rate I_0 may contribute appreciably to I', and special measures are taken to minimize I_0.

The overall counting efficiency η depends on the properties of the radionuclides and the measuring device. The various factors contributing to η will be discussed in detail later in this section.

At constant η, the net counting rate $I = I' - I_0$ is proportional to the activity A, and for many purposes, such as determination of half-lives or application of radionuclides as tracers, measurement of the relative activity, given by I at constant η, is sufficient.

In Fig. 7.1 the logarithm of the net counting rate I is plotted as a function of time t. The curve obtained is called the decay curve and is used for determination of half-lives.

The counting rate I of a radioactive sample which contains several radionuclides is given by the sum of the counting rates of the individual radionuclides:

$$I = I_1 + I_2 + \cdots = \eta_1 A_1 + \eta_2 A_2 + \cdots \tag{7.2}$$

If the number of radionuclides present in the sample is low, the decay curve can be separated by subtraction into the individual decay curves of the radionuclides, either graphically or arithmetically, as shown in Fig. 7.2. The analysis of decay curves is of practical importance for the investigation of radionuclide purity. As examples, contamination of a sample by a short-lived impurity is shown in Fig. 7.3, and contamination by a long-lived impurity in Fig. 7.4.

The overall counting efficiency η in eq. (7.1) depends on the frequency H of the decay mode measured in relation to the activity, the self-absorption S of the radiation in the radioactive sample, the contribution B of backscattered radiation, the geometrical arrangement G of the sample with respect to the counter, the absorption

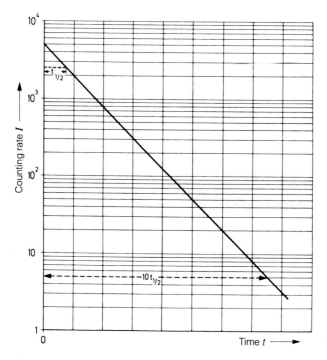

Figure 7.1. Counting rate as a function of time (determination of half-lives).

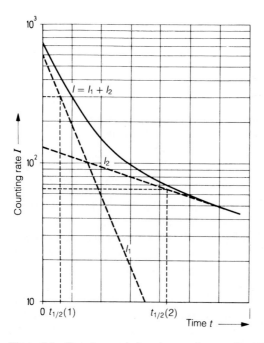

Figure 7.2. Counting rate of a mixture of two radionuclides.

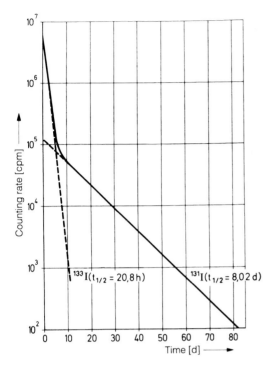

Figure 7.3. Short-lived impurity (^{133}I in ^{131}I).

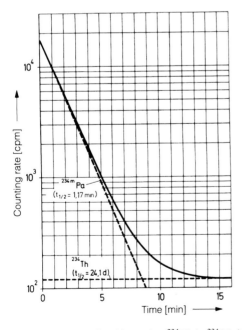

Figure 7.4. Long-lived impurity (^{234}Th in ^{234m}Pa).

W of the radiation in the air and in the window of the counter, the internal counting efficiency η_i of the counter, and the correction *D* for the dead time of the counter:

$$\eta = \Pi \cdot (1 - S) \cdot (1 + B) \cdot G \cdot (1 - W) \cdot \eta_i \cdot (1 - D) \tag{7.3}$$

By self-absorption, absorption in the air and the window, and dead time of the counter, η is reduced, whereas it increases by the influence of backscattering.

The overall counting efficiency η may vary between about 0.01 and 1, depending on the kind of radiation and its energy and the type of counter used. Self-absorption *S* in the sample is high for α and low-energy β radiation, but negligible for γ radiation. As an example, the self-absorption S of the β^- radiation of ^{45}Ca in samples of CaCO$_3$ is plotted in Fig. 7.5 as a function of the thickness of the sample. Backscattering may contribute appreciably to the counting rate in the case of β radiation, if the sample is placed on materials of high atomic number *Z*, as is evident from Fig. 6.11.

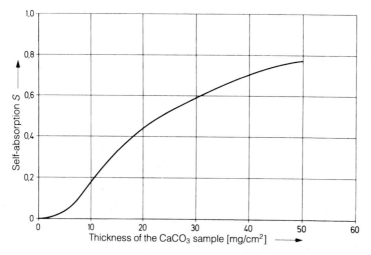

Figure 7.5. Self-absorption *S* of the β^- radiation of ^{45}Ca in CaCO$_3$ as a function of the thickness of the sample.

The influence of the geometrical arrangement is shown in Fig. 7.6. *G* is given by the solid angle $\Omega/4\pi$. Absorption *W* in the air and the window of the counter is very high for α and low-energy β radiation, like self-absorption S in the samples, and thin windows or windowless counters are needed for the measurement of these kinds of radiation. The internal counting efficiency η_i varies appreciably between about 0.01 and 1.0 with the kind of radiation and the type of counter. The dead time of the counters is due to the fact that they need some time for recovery with the result that during a certain period, the dead time, registration of a following event is not possible. Relatively long dead times of the order of 100 to 500 μs are found for Geiger–Muller counters. The number of non-counted events is given by

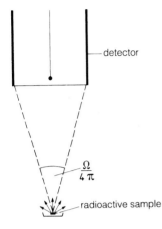

detector

$\frac{\Omega}{4\pi}$

radioactive sample

Figure 7.6. Influence of the geometrical arrangement (factor *G* given by the solid angle $\Omega/4\pi$).

$$I - I^* = \frac{I^{*2} \cdot t}{1 - I^* t} \tag{7.4}$$

I is the "true" counting rate for the dead time $t = 0$, and I^* is the counting rate measured. $I - I^*$ increases appreciably with the number of counted events and with the dead time, as shown in Fig. 7.7.

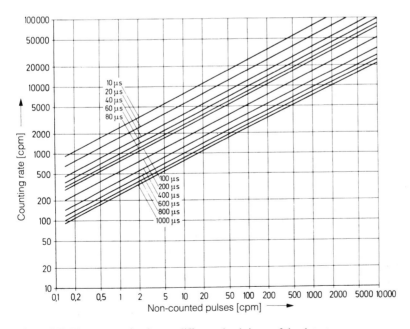

Figure 7.7. Non-counted pulses at different dead times of the detectors.

7.2 Gas-filled Detectors

Gas-filled detectors have been in use since the beginning of radiochemistry. Ionizing radiation passing through a gas creates a trail of ion pairs (positive ions and free electrons), as described in section 6.1. If an electric field is applied, the ions and the electrons move in opposite directions. The motion of the charged particles gives rise to a current that can be measured in an external circuit. A simple arrangement of a gas-filled ionization detector is shown in Fig. 7.8. Commonly the detector has the form of a cylinder, the outer wall serves as the cathode and a wire along the axis of the cylinder acts as the anode. At typical distances of several cm, it takes a few μs for the electrons to reach the anode and a few ms for the slower-moving positive ions to reach the cathode.

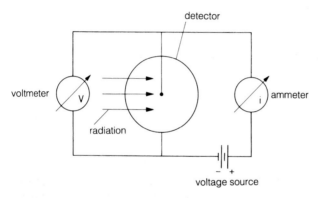

Figure 7.8. Arrangement of a gas-filled ionization detector (schematically).

The total number of ion pairs arriving at the electrodes and hence the height of the pulse observed in the circuit depend on the electric field strength or the voltage, respectively, applied to the detector. In Fig. 7.9 the pulse height is plotted as a function of the voltage. At low voltages, the electrons recombine with the ions (region of partial recombination). With increasing voltage nearly all the electrons are collected at the cathode and a saturation value is observed. This is the range of operation of ionization chambers. Because the specific ionization is appreciably higher in the case of α particles than in the case of β particles, an α particle produces a much higher pulse in a ionization chamber, as indicated in Fig. 7.9.

In general, an α particle gives off its energy while passing through an ionization chamber and produces $N = E_\alpha/E_1$ ions, where E_α is the energy of the α particle and E_1 the energy used for production of one ion pair. As in air $E_1 \approx 35$ eV (section 6.1), it follows that an α particle with $E_\alpha = 3.5$ MeV produces about 10^5 ion pairs, corresponding to a pulse of $\approx 10^{-14}$ A s, which can be measured by means of an efficient amplifier. In contrast to α particles, the pulses produced by β particles in a ionization chamber are smaller by a factor of the order of 10^3 and can hardly be measured.

Further increase of the voltage in the device shown in Fig. 7.8 leads also to an increase of the pulse height (Fig. 7.9). Under these conditions, the electrons are

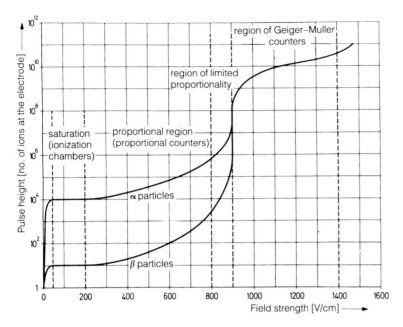

Figure 7.9. Pulse height as a function of the field strength.

accelerated on their way to the cathode and gain so much energy that they are able to produce secondary free electrons by collisions with the gas molecules (gas amplification). Over a wide voltage range the output pulse increases with the voltage applied. This is the range of operation of proportional counters. Thin wires of 20 to 50 μm diameter are used to obtain high field strengths. The multiplication factor, given by the ratio of secondary and primary electrons, depends on the voltage and field strength, respectively, and varies between about 10^3 and 10^5. Because the specific ionization is much higher in the case of α particles, they give higher pulse heights than β particles at the same multiplication factor, and α radiation can be measured in the presence of β radiation, like in ionization chambers. Alpha radiation is measured at low voltages and β radiation at high voltages. The pulse heights measured with proportional counters are usually of the order of several mV. As the number of primary electrons is proportional to the energy of the particles, this energy can be determined.

If the voltage applied in the device shown in Fig. 7.8 is further increased, the number of electrons collected at the cathode becomes independent of the number of initial ion pairs produced by the incident radiation. This is the operation range of Geiger–Muller counters (Fig. 7.9). Under the operation conditions of these counters a single ionization produced in the gas by any kind of radiation leads to a discharge spreading out over the whole counter by a sequence of secondary ionization processes and giving a relatively high pulse of several volts that can be measured directly without amplification. At still higher voltages, continuous discharge takes place without the influence of radiation.

Ionization Chambers
Ionization chambers are constructed in various ways, for instance as grid ion chambers, guard ring chambers, current ion chambers or integrating ion chambers. Usually, the electrodes are parallel plates and enclosed in a gas-tight chamber, filled with air or a noble gas. Because of the low pulse heights to be measured, good electrical isolation of the electrodes is of great importance. Ionization chambers are used to measure α emitters, in particular radioactive gases such as radon, and fission fragments. They are applied for calibration of radioactive sources, radiation monitoring and dosimetry in radiation protection.

Proportional Counters
Proportional counters may consist of a sealed cylinder serving as cathode, a thin wire as anode and a thin window, but they are often constructed as flow counters, as shown in Fig. 7.10. In this type of counter a gas, preferably methane or a mixture of argon and methane, flows through the counter during operation and the sample is brought into the counter. The operational voltage depends on the nature and the pressure of the gas and varies between about 2 and 4 kV.

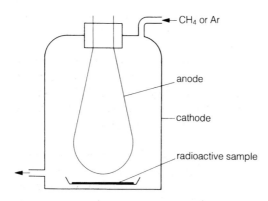

Figure 7.10. Flow counter.

Windowless proportional counters are well suited to measure α and low-energy β radiation. High and well-defined values of G in eq. (7.3) are obtained with 2π and 4π counters, which are shown in cross-section in Fig. 7.11.

In 2π counters half of the total radiation emitted by the sample is recorded, whereas 4π counters are equipped with two anode wires and the radiation emitted by the sample can be counted quantitatively. These types of counters are used for the measurement of the absolute activity A of radioactive samples, because an overall counting efficiency $\eta = 1.0$ in eq. (7.3) can be obtained.

As proportional counters have dead times of only several µs, high counting rates can be measured without losses. Because the internal counting efficiency η_i of proportional counters for γ radiation is low (about 1%), they are not suited to measure γ radiation. However, proportional counters of special design and operating at high gas pressure are applied to X-ray and low energy γ-ray spectrometry.

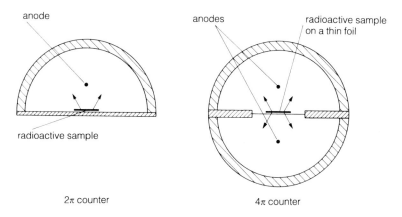

Figure 7.11. Cross sections of 2π and 4π counters.

Geiger–Muller Counters

Geiger–Muller counters are operated at relatively high voltages of several kV. The avalanche-like spreading of the ionization processes leads to the production of a great number of positive ions which are moving more slowly than the electrons. The neutralization of the electrons at the anode wire gives rise to the emission of photons, which react with the gas by emission of photoelectrons. These trigger further avalanches and the processes continue until the build-up of the positive ion sheath in the vicinity of the anode wire reduces the electrical field strength so far that no more events can be counted. It takes about 100 to 500 µs for the ions to reach the cathode, where they can cause secondary electron emission from the surface, thus triggering a new discharge. Therefore, measures must be taken to eliminate the negative influence of the positive ions.

Usually, the processes are stopped by addition of a quench gas to the main filling gas. Vapours of polyatomic molecules such as ethanol, ether, ethyl formate, methane, bromine or chlorine may be applied. Because of the lower ionization energy of these molecules, the positive charge of the ions is transferred to the molecules and these dissipate their energy by dissociation or predissociation. Chlorine and bromine exhibit strong absorption of the photons emitted; they dissociate, recombine and return to the ground state via a series of low-energy excited states.

As already mentioned, the dead time of Geiger–Muller counters varies between about 100 and 500 µs, and therefore the number of non-counted events is relatively high at high counting rates (Fig. 7.7). Organic quench gases are gradually consumed after about 10^8 to 10^9 counts. The advantages of Geiger–Muller counters are their simplicity and the fact that further amplification is not needed. Halogen-quenched counters exhibit longer lifetimes, lower dead times and lower operational voltages. Most Geiger–Muller counters are equipped with windows and are therefore inexpedient for measuring α and low-energy β radiation. Energy discrimination is not possible. Gamma radiation can be counted with a low internal counting efficiency η_i of about 1%.

Various types of Geiger–Muller counters are shown in Fig. 7.12. The end-window counter equipped with mica windows of about 1.5 to 3 mg/cm^2 is a very simple

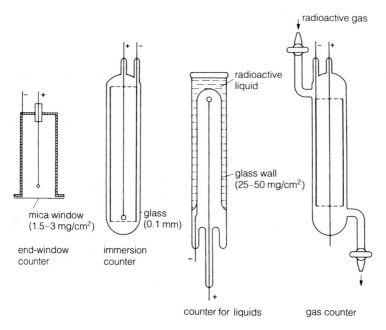

Figure 7.12. Various types of Geiger–Muller counters.

instrument and most frequently used. The immersion counter and the counter with the ring-like glass beaker are used to measure liquids and the gas counter is used for the determination of the activity of gases.

7.3 Scintillation Detectors

The main parts of a scintillation counter are sketched schematically in Fig. 7.13. In the transparent crystal or liquid the radiation is absorbed and photons are emitted. At the photocathode of the photomultiplier tube the photons release electrons which are multiplied by the dynodes of the multiplier to give pulses of several mV. Some examples of solid and liquid scintillators are listed in Table 7.1.

Scintillation counters are applied primarily for measuring γ radiation and low energy β radiation. If γ radiation is to be measured, thick scintillating crystals of high density are used in order to absorb as much γ radiation as possible. NaI or CsI crys-

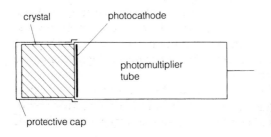

Figure 7.13. Scintillation detector (schematically).

Table 7.1. Some properties of solid and liquid scintillators.

Scintillator	Density [g/cm³]	λ_{max} [nm]	Relative pulse height	Half-life of the excited state [μs]	Suitability
Inorganic crystals					
NaI(Tl) crystals	3.67	410	210	0.175	For γ radiation
CsI(Tl) crystals	4.51	500	55	0.770	For γ radiation
ZnS(Ag)	4.09	450	100	7	For α radiation
Organic crystals					
Anthracene	1.25	440	100	0.022	⎫
trans-Stilbene	1.16	410	60	0.004	⎬ For β radiation
p-Terphenyl	1.23	400	40	0.004	⎭
Liquids					
p-Terphenyl (5g) in 1 l toluene	–	355	35	0.0015	⎫ For low-energy
2,5-Diphenyloxazole (3g) in 1 l toluene	–	382	40	0.0021	⎬ β radiation ⎭

tals doped (activated) by addition of small quantities of Tl are well suited and at a size of 1 to 2 inches (2.5 to 5 cm) they give an internal counting efficiency η_i of 15 to 30%. This counting efficiency is appreciably higher than that obtained with gas counters for γ radiation. Well-type crystals offer a nearly 4π geometry.

For counting low-energy β radiation, the crystal is substituted by a scintillating liquid, and the sample is dissolved in the liquid (internal-source liquid scintillation counting). The method is also used to measure weak X-ray and γ-ray emitters. Under these conditions, self-absorption of the radiation in the sample, absorption of the radiation in the air and the window of the detector, and backscattering of β particles are avoided.

The scintillating liquid is prepared by dissolving a primary and, if necessary, a secondary scintillator in a suitable solvent and adding a solution containing the radioactive sample. A secondary scintillator is needed, if the primary scintillator (e.g. 2,5-diphenyloxazole) emits photons with a wavelength that is too short for the photomultiplier. In this case the secondary scintillator (e.g. p-bis [2-(5-phenyloxazolyl)] benzene) shifts the wavelength to lower values so that the photomultiplier responds. As solvents, mainly organic compounds are used, e.g. toluene, benzene, p-xylene or dioxane. If the sample is added in the form of an aqueous solution, solvents that are miscible with water are applied. However, the sample may also be introduced as an emulsion or a suspension, or even pieces of paper carrying the sample may be added.

Foreign substances, such as water, that are introduced with the sample into the scintillator often reduce the light output and the counting efficiency. The shape of the spectrum may also be changed. This effect is known as quenching. It limits the amount of sample that can be added.

The sample container or vial must be transparent at the wavelength of the scintillator used and resistant to the solvent. Preferably it is put between two photomultipliers to obtain higher counting efficiency. Using this arrangement the thermal noise of the photomultipliers can be reduced to a minimum by means of an anti-coincidence circuit: thermal electrons which are the major source of thermal noise are emitted at random and are registered only by one photomultiplier, whereas the events counted in both photomultipliers are due to the radioactive decay in the sample.

The main advantage of liquid scintillation counting is the relatively high counting efficiency, which can amount to about 90 to 100%.

Transparent organic crystals, such as anthracene, may also be used as scintillators (Table 7.1). They can be used to measure β radiation of medium or high energy, but they exhibit no special advantages.

7.4 Semiconductor Detectors

Semiconductor detectors are sometimes referred to as solid-state detectors to distinguish them from gas-filled detectors, but they differ from other solid-state detectors, such as solid scintillation or solid track detectors, by their semiconductor properties. Semiconductor detectors have found broad application during recent decades, in particular for γ and X-ray spectrometry and as particle detectors. The way of working is similar to that of ion chambers or proportional counters: the radiation is measured by means of the number of charge carriers set free in the detector, which is arranged between two electrodes. Ionizing radiation produces free electrons and holes. If the energy of the electrons is high enough, they are able to produce new electron–hole pairs. This process takes about 1 to 10 ps. The number of electron–hole pairs depends on the energy transmitted by the radiation to the semiconductor. As a result, a certain number of electrons are transferred from the valence band to the conduction band and an equivalent number of holes are left behind in the valence band. Under the influence of an electric field, electrons as well as holes travel to the electrodes, where they give rise to a pulse that can be measured in an outer circuit.

Part of the energy of the incident radiation is used up by the production of electron–hole pairs, and another part by excitation of lattice vibrations, similarly to the situation in gases, where some energy is used for excitation and dissociation processes (section 7.2).

The materials most frequently used as detectors for nuclear radiation are Si and Ge. At room temperature the energy gap between valence and conduction band is 1.09 eV for Si and 0.79 eV for Ge, whereas the total energy required for production of electron–hole pairs and excitation of lattice vibrations is $E \approx 3.6$ eV for Si and $E \approx 2.8$ eV for Ge at room temperature, and about 0.2 eV more at 77 K. Compared with the energy used for production of an ion pair in gas detectors (≈ 35 eV), this energy is very low, and the number of charge carriers produced in a semiconductor is appreciably higher. Consequently, in semiconductor detectors the statistical variation of the pulse height is smaller and the energy resolution is higher. Therefore these detectors are particularly suitable for energy determination of all kinds of nuclear radiation and charged particles. The time resolution is also very good. It depends on the size and the properties of the individual semiconductor detector and varies between about 0.1 ns and 1 μs.

Compared with gas ionization detectors, the density of semiconductor detectors is higher by a factor of the order of 10^3, and charged particles of high energy can give off their energy in a semiconductor detector of relatively small dimensions. The specific energy loss $-\mathrm{d}E/\mathrm{d}x$ can also be determined by use of semiconductor detectors and, by arranging a very thin and a thick semiconductor detector one behind the other, charged particles can easily be identified. For measurement of γ radiation relatively large crystals are needed, because of the low specific ionization of this kind of radiation.

The requirements of high-energy resolution are well met by Si and Ge. The photoelectric effect increases with Z^4 to Z^5, where Z is the atomic number of the substance, and the linear absorption coefficient for 100 keV γ rays in Ge is higher than in Si by a factor of about 40. Therefore, Ge crystals are better suited for measuring γ radiation. Semiconductors with still higher atomic numbers, such as CdTe and HgI_2, have been investigated with respect to their suitability as detector materials, but they are not commonly used.

In the application of semiconductor detectors two influences have to be considered, one caused by temperature and the other by impurities or lattice defects, respectively. The energy gap between the valence and the conduction bands is of the order of 1 eV and small enough to be surmounted by thermal excitation, resulting in thermal noise which increases strongly with temperature. At room temperature the thermal conductivity of Si is about $4 \cdot 10^{-4}$ S/m and that of Ge about $1 \cdot 10^{-2}$ S/m. To avoid high thermal noise, the detectors are operated at low temperatures (liquid nitrogen). This is of special importance in the case of Ge detectors, because for these the energy gap is lower than for Si detectors (0.79 eV compared with 1.09 eV at room temperature). At the temperature of liquid nitrogen, the thermal conductivity of pure Si and Ge is negligible compared with that due to impurities.

Impurities lead to the presence of charge carriers (electrons in the conduction band or holes in the valence band) in the absence of an incident radiation, and cause a leakage current. For instance, an element of group V of the Periodic Table, such as P or As, introduces additional electrons into the lattice of Si or Ge and has the effect of an electron donor. Because of the negative charges, this kind is called a n-type semiconductor. On the other hand, the presence of an element of group III of the Periodic Table, such as B or Ga, gives rise to electron holes and has the effect of an electron acceptor. Due to the positive charges, this kind is called a p-type semiconductor. Additional acceptor or donor levels may be present because of lattice defects. Crystals with defined contents of foreign atoms may be obtained by doping, i.e. by introducing measured amounts of foreign atoms.

The influence of impurities can be ruled out in tow ways: by preparation of high-purity crystals or by elimination of the influence of the charge carriers introduced by the impurities.

High-purity Ge crystals containing only one foreign atom per 10^{10} Ge atoms or less can now be prepared and offer optimal conditions for measuring of γ rays. They are referred to as intrinsic Ge (i-Ge) or high-purity Ge (Hp-Ge) and are operated at liquid-nitrogen temperature, in order to avoid thermal noise. For the purpose of cooling, the crystals are enclosed in a vacuum cryostat. The nuclear radiation entering the crystals produces electron–hole pairs that are collected at the electrodes within microseconds. The charge transported through the crystal is proportional to the energy absorbed. High-purity Ge detectors are highly sensitive and exhibit an excellent energy resolution (up to about 0.2 to 0.5 keV at energies of about 10 to 100 keV, respectively). This makes these detectors very attractive for γ spectrometry.

Various geometrical configurations of Ge detectors are available: planar, coaxial, and well-type detectors. The latter are widely used, because they offer very good geometrical conditions.

The influence of impurities can be eliminated by compensation of the surplus charge carriers or by introducing a p–n barrier, respectively. A p–n barrier is the combination of two zones of semiconductor material, one of the p-type and the other of the n-type. This gives a diode with a p–n junction. At the junction the mobile charge carriers diffuse from regions of higher concentrations to those of lower concentrations. Thus, electrons move from the n-type to the p-type region and combine there with the positive holes. As the result, a depletion layer of high resistance is formed in which there are no charge carriers. This depletion layer is also called an intrinsic (i) layer, and it represents the sensitive (active) volume of the detector which comprises three zones, p–i–n. The depletion zone can be increased by application of a reverse bias (positive electrode on n-type, negative electrode on p-type). For use as nuclear radiation detectors, it is important that the depleted layer of the semiconductor has high sensitivity.

p–n junction detectors containing a depleted layer can be prepared by controlled diffusion, ion implantation or formation of a surface barrier. By thermal diffusion small concentrations of P or As may be introduced into Si or Ge to produce n-type regions, or small concentrations of B or Ga may be added to obtain p-type regions. As an example, a slab of p-type Si is taken as the base material and at one surface of the slab a thin layer (0.1 to 1 µm) of n-type Si is produced by introducing a small concentration of P. By application of a reverse bias and a field of the order of 10^3 V/cm a depleted layer is produced, the thickness of which depends on the magnitude of the applied field.

Ion implantation is carried out by bombarding one surface of the semiconductor with ions accelerated to energies of several hundred keV. Monoenergetic ions have a well-defined range in the semiconductor material and controlled depth profiles of implanted ions can be obtained. The advantage of this kind of detectors is their great stability.

Surface-barrier detectors are prepared from n-type Si. On one side the surface is etched and exposed to air to produce an oxide layer, and a thin gold layer is deposited for electrical contact. By this procedure a p-layer is obtained with a thickness of <1 µm. Another method of preparation of surface-barrier detectors is the vapour deposition of Al on Si. The maximum thickness of the depleted layer is of the order of 1 mm. Surface-barrier detectors are mainly used for charged-particle spectrometry and for α and β spectrometry. Optimal energy resolution for α and β radiation is obtained with thin detectors of small surface area. For instance, at surface areas of about 1 to 4 cm^2, the energy resolution for α and β radiation is about 10 to 30 keV.

Fully depleted Si detectors in which the zone of depletion extends over the whole crystal are also available. The energy loss $-\mathrm{d}E/\mathrm{d}x$ in this kind of detector can be determined by means of another detector which is placed behind the first one and in which the remaining energy is measured.

To obtain high counting efficiencies for γ or X rays, thick depleted layers (thick intrinsic regions) are needed. These are obtained by drifting Li into crystals of Ge or Si. Ge(Li) crystals are used as detectors for γ rays, because of the high density of Ge, and Si(Li) crystals as detectors for X rays. Li atoms act as donor atoms and Li^+ ions are very mobile in the lattice of Ge or Si, moving from one interstitial site to another.

In the process of drifting, an excess of Li is introduced by diffusion into a p-type crystal of Ge or Si producing an n-type region of about 0.01 to 1 mm. By application of a reverse bias and raising the temperature the Li^+ ions are pulled into the p-type region of the crystal where they compensate for the acceptor atoms. In this way, three zones are created, one of n-type, an intrinsic one (i) and a p-type region, as indicated in Fig. 7.14. The intrinsic region extends up to about 15 to 20 mm and exhibits high sensitivity. It defines the active volume of the detector, and the voltage applied is effective across this region.

Because of the high mobility of Li in the lattice of Ge at room temperature, Ge(Li) detectors must be cooled permanently by liquid nitrogen.

A typical γ-ray spectrum taken with a Ge(Li) detector is shown in Fig. 7.15. Because of the different mechanisms of γ-ray absorption (section 6.4) γ-ray spectra

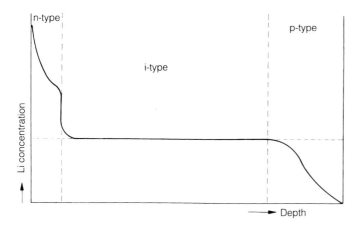

Figure 7.14. Lithium concentration in a Ge (Li) crystal.

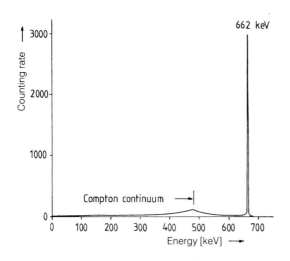

Figure 7.15. γ-ray spectrum of ^{137}Cs taken with a Ge (Li) detector (the γ rays are emitted by the metastable ^{137m}Ba).

show several contributions. A photopeak and a broad Compton continuum are always observed. The photopeak is used for identification, because it gives the full energy of the γ rays. The Compton continuum shows an edge at an energy below that of the photopeak. At energies >1.02 MeV additional contributions are found: the positrons originating from pair production are generally annihilated within the detector, but one or both of the annihilation photons may escape from the detector without interaction, depending on the size of the detector. These photons give rise to an "escape peak" at $E - 0.511$ MeV and a "double escape peak" at $E - 1.022$ MeV, where E is the energy of the photopeak. Additional peaks may be observed at energies above the photopeaks, if two γ rays or a γ ray and an X ray are emitted in cascade, resulting in a small "sum peak". Sum peaks can be distinguished from photopeaks by varying the distance between sample and detector: whereas the intensity of photopeaks varies linearly with the solid angle between sample and detector, the intensity of sum peaks varies with the square of that angle.

7.5 Choice of Detectors

The suitability of the detectors described in sections 7.2 to 7.4 for the measurement of α, β and γ radiation is compiled in Table 7.2.

Gaseous samples are preferably measured in ionization chambers (α radiation), proportional counters or Geiger–Muller gas counters. The samples are introduced into the counters or passed through with a gas stream (flow counters).

For the measurement of liquid samples containing α emitters or low-energy β emitters, the liquid scintillation technique is most favourable. High-energy β emitters or γ emitters in liquid samples can be counted by means of Geiger–Muller counters, but in the case of γ emitters the use of NaI(Tl) scintillation counters is much more favourable, because of the higher counting efficiency. Well-type NaI(Tl) detectors offer optimal geometrical conditions. If the same counting vials and the same volumes of the liquid samples are used, the same counting efficiency is obtained.

Solid samples containing high-energy β emitters can be measured with end-window Geiger–Muller counters, whereas proportional counters are suitable for the measurement of α and β emitters. Self-absorption of α and β rays in solid samples may play an important role and requires special attention in the case of α and low-energy β radiation. Very thin and homogeneous layers can be obtained by electrolytic or vapour deposition of the radionuclides to be measured or by solvent extraction and subsequent evaporation of the solvent.

Surface-barrier detectors are very useful for detection and measurement of α particles. The internal counting efficiency for α particles is $\eta_i = 1.0$. If the geometry G of the arrangement of α source and detector is well-defined and self-absorption S in the sample can be neglected, absolute activities of α emitters can be determined. The performance of surface-barrier detectors is conventionally tested by recording the spectrum of a calibration source, e.g. a ^{241}Am source.

For the measurement of γ emitters in solids NaI(Tl) scintillation detectors or Ge detectors are most suitable, depending upon whether high counting efficiency or high energy resolution is required. For comparison, the spectra of ^{60}Co taken with a NaI(Tl) scintillation detector and with a Ge(Li) detector are plotted in Fig. 7.16.

Table 7.2. Suitability of detectors for the measurement of various kinds of radiation.

Kind of radiation	Ionization chambers	Proportional counters	Geiger–Muller counters	Scintillation detectors	Semiconductor detectors
α radiation	Favourable	Flow counters very favourable	Unfavourable	Liquid scint. favourable	Si barrier detectors favourable
High-energy β radiation ($>1\,\mathrm{MeV}$)	Unsuitable	Suitable	Favourable	Organic cryst. favourable	Si barrier detectors suitable
Low-energy β radiation ($<0.5\,\mathrm{MeV}$)	Unsuitable	Favourable	Unfavourable	Liquid scint. very favourable	Si barrier detectors favourable
High-energy γ radiation ($>0.1\,\mathrm{MeV}$)	Unsuitable	Unsuitable	Unfavourable	NaI(Tl), CsI(Tl) crystals very favourable	i-Ge, Ge(Li) detectors very favourable
Low-energy γ radiation ($<0.1\,\mathrm{MeV}$) and X-rays	Unsuitable	Suitable	X-ray counters favourable	NaI(Tl), CsI(Tl) crystals favourable	Si(Li) detectors very favourable

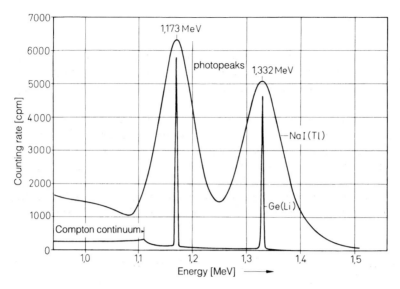

Figure 7.16. γ-ray spectra of ^{60}Co taken with a NaI (Tl) scintillation detector and a Ge (Li) semi-conductor detector.

X rays and γ rays with energies < 50 keV are usually measured with Si(Li) detectors encapsulated in a cryostat with a beryllium window about 20 μm thick. The samples may also be introduced into a windowless vacuum chamber. The internal counting efficiency attains values of ≈ 1.0, if the active layer of the detector is thick enough, and the energy resolution is in the range of about 100 to 200 eV at energies of about 4 to 10 keV. This energy resolution is sufficient to distinguish the characteristic X rays of light elements with neighbouring atomic numbers Z, such as the K_α rays of Al and Si.

Proportional counters may also be applied for detection and measurement of X rays, provided that the energy is low enough to interact effectively with the gas filling. For energies of a few keV, a mixture of He and methane is often used, whereas at higher energies Ar, Kr or Xe are preferable.

Combinations of chemical separation and counting techniques are frequently used in radiochemistry. For instance, activity measurements in an ionization chamber, a proportional counter or a scintillation detector can be made on-line after a gas-chromatographic separation. This combination is known as radio–gas chromatography. A carrier gas flows continuously through the system, and a suitable counting gas is added before the gas stream enters the counter. The simplest method of operation is to use methane as carrier and as counting gas in the proportional counter. After chemical separation of radioactive substances by paper or thin-layer chromatography, the activity distribution on the paper or thin-layer chromatogram can be directly measured, for example by scanning the chromatogram with a proportional counter and recording the activity.

7.6 Spectrometry

Gamma-ray and α-ray spectrometry are important tools of nuclear and radio-chemistry. They are mainly used for identification of radionuclides. Because of the continuous energy distribution of β radiation, β-ray spectrometry is less frequently applied.

For γ-ray spectrometry i-Ge or Ge(Li) detectors are most suitable, because they offer the highest energy resolution (about 0.2 to 0.4 keV at energies of the order of 100 keV). The disadvantages of Ge detectors are their low internal counting efficiency η_i and the fact that they have to be operated at low temperatures (liquid-nitrogen cooling).

To identify unknown γ emitters, the pulse height scale must be calibrated by means of γ-ray sources of known energy. Some γ emitters suitable for calibration are listed in Table 7.3.

Table 7.3. Some γ-ray standards.

Nuclide	γ-ray energy [keV]	Frequency H [number of γ rays per disintegration]	Half-life
^{241}Am	26.345	0.024	432.2 y
	59.536	0.357	
^{57}Co	122.061	0.855	271.8 d
	136.473	0.107	
^{203}Hg	279.188	0.815	46.6 d
^{51}Cr	320.084	0.0983	27.7 d
^{137}Cs	661.66	0.8521	30.17 y
^{54}Mn	834.83	0.99975	312.2 d
^{60}Co	1173.24	0.9990	5.272 y
	1332.50	0.9998	
^{22}Na	1274.53	0.9994	2.603 y
	511.0	1.81	
^{88}Y	898.07	0.927	106.6 d
	1836.08	0.9935	

In general, the detectors are combined with a preamplifier, an amplifier and a multichannel analyser, in which the pulses are sorted according to their pulse heights. Frequently, the multichannel analyser is operated by a computer and a program for peak search, peak net area calculation, energy calibration and radionuclide identification.

Scintillation detectors with NaI(Tl) crystals may also be used for γ spectrometry. Because NaI(Tl) crystals can be made in larger size than Ge crystals and because the atomic number of I is larger than that of Ge, the internal counting efficiency of NaI(Tl) detectors for γ rays is higher than that of Ge crystals, as already discussed in section 7.5 (Fig. 7.16). On the other hand, the energy resolution is appreciably lower (5 to 7% for γ energies of the order of 100 keV). Scintillation detectors are operated in a way similar to that used with Ge detectors, but without cooling.

For α spectrometry silicon surface-barrier detectors are most suitable. They are operated at room temperature in a vacuum chamber to avoid energy losses. The α particles are stopped within a thin depleted region of the detector and the number of electron–hole pairs is directly proportional to the energy of the α particles. The charge pulses are integrated in a charge-sensitive amplifier. Some α emitters used as α sources for the purpose of calibration are listed in Table 7.4.

Table 7.4. Some α standards.

Nuclide	Energy of the α particles [MeV]	Frequency H [number of α particles per disintegration]	Half-life
^{148}Gd	3.1828	1.00	74.6 y
^{232}Th	3.950	0.23	$1.405 \cdot 10^{10}$ y
	4.013	0.77	
^{230}Th	4.621	0.234	$7.54 \cdot 10^4$ y
	4.687	0.763	
^{238}U	4.147	0.23	$4.468 \cdot 10^9$ y
	4.198	0.77	
^{235}U	4.398	0.55	$7.038 \cdot 10^8$ y
^{234}U	4.723	0.275	$2.454 \cdot 10^5$ y
	4.775	0.725	
^{236}U	4.445	0.26	$2.342 \cdot 10^7$ y
	4.494	0.74	
^{231}Pa	4.952	0.228	$3.276 \cdot 10^4$ y
	5.014	0.254	
	5.028	0.20	
	5.058	0.11	
^{239}Pu	5.105	0.106	$2.411 \cdot 10^4$ y
	5.144	0.151	
	5.157	0.732	
^{240}Pu	5.1236	0.2639	$6.563 \cdot 10^3$ y
	5.1681	0.735	
^{243}Am	5.233	0.11	$7.370 \cdot 10^3$ y
	5.275	0.88	
^{210}Po	5.30438	1.00	138.38 d
^{241}Am	5.443	0.128	432.2 y
	5.486	0.852	
^{238}Pu	5.456	0.283	87.74 y
	5.499	0.716	
^{244}Cm	5.762	0.236	18.10 y
	5.805	0.764	

For β spectrometry, semiconductor detectors with a thickness of the intrinsic region exceeding the maximum penetration distance of the electrons to be counted are suitable. Because of the continuous energy distribution of β particles, the spectrum of a pure β emitter is a curve with a maximum at medium energies and extending to a maximum energy E_{max} (Fig. 5.7), whereas conversion electrons give relatively sharp peaks. Some pure β emitters are listed in Table 7.5.

Table 7.5. Some pure β^- emitters

Nuclide	E_{max} [Mev]	Half-life
^{3}H	0.0186	12.323 y
^{14}C	0.156	5730 y
^{32}P	1.710	14.26 d
^{33}P	0.248	25.34 d
^{35}S	0.167	87.51 d
^{36}Cl	0.714	$3.01 \cdot 10^5$ y
^{45}Ca	0.252	163.8 d
^{63}Ni	0.067	100.1 y
^{90}Sr/^{90}Y	0.546/2.27	28.64 y/64.1 h
^{99}Tc	0.292	$2.13 \cdot 10^5$ y
^{147}Pm	0.224	2.623 y
^{204}Tl	0.766	3.78 y

X-ray spectrometry is generally carried out with Si(Li) detectors. The set-up is similar to that applied to γ-ray spectrometry with i-Ge or Ge(Li) detectors: cooling of the detector in a cryostat, operation in combination with a preamplifier, an amplifier and a multichannel analyser. The energy resolution is very good, as already mentioned in section 7.6, and makes it possible to distinguish the characteristic X rays of neighbouring elements. Some X-ray emitters that may be used for calibration purposes are listed in Table 7.6.

The same equipment as for X-ray spectrometry is used for X-ray fluorescence analysis (XFA). In this method, emission of characteristic X rays is induced by excitation with X-ray sources (X-ray tubes or X-ray emitting radionuclides) or with charged particles (PIXE, i.e. particle-induced X-ray emission).

Neutron spectrometry will be considered in section 7.10.

Table 7.6. Some X-ray emitters.

Nuclide	Characteristic X-rays	Energies [keV]	Half-life
^{57}Co	Fe K	6.391; 6.404; 7.058	271.79 d
^{109}Cd	Ag K	21.99; 22.16; 24.93	462.6 d
^{153}Gd	Eu K	40.40; 41.54; 47.00; 48.50	239.5 d
^{241}Am	Np L	11.87; 13.93; 15.86; 17.61; 21.00	432.2 y

7.7 Determination of Absolute Disintegration Rates

As indicated in section 7.1, measurements of relative and absolute activities are to be distinguished. For determination of relative activities, the overall counting efficiency η in eq. (7.3) must be constant, but it need not be known, whereas η must be known exactly for determination of absolute activities. The overall counting efficiency can either be calculated or be determined by calibration. In both cases all factors in eq. (7.3) must be considered.

If the radionuclides in the sample to be measured and in the calibration source are different, differences in the frequency H of the decay mode have to be taken into account. Self-absorption S of the radiation in the sample is, in general, negligible for γ rays, but it may play an important role in the case of α and low-energy β rays, and it cannot be neglected for thick samples in the case of high-energy β rays and X rays. The contribution of backscattering can be avoided by elimination of material of high atomic number Z in the vicinity of the sample. The geometrical factor G may be calculated from the distance between the sample and the window of the counter and the dimensions of both. In 2π or 4π counters the geometrical conditions are well-defined. For the absorption of the radiation in the air and in the window of the counter the same is valid as for the self-absorption S. However, the influence of W can be avoided by use of windowless counters (e.g. 2π or 4π counters).

In many detectors the internal counting efficiency η_i depends on the energy of the radiation. This must also be taken into account, if sources containing other radionuclides are used for calibration. Generally, calibration curves $\eta_i = f(E)$ are determined for the detectors used under defined conditions by application of different radionuclides of known activity as radiation sources.

Finally, the influence of the dead time (D in eq. (7.3)) has to be taken into account, particularly if the dead time of the detector is high (as in the case of Geiger–Muller counters) and if the counting rates of the sample and the calibration source are markedly different.

With respect to the influence of the factors S, B and W in eq. (7.3), determination of the absolute activity A of α and β emitters may cause problems, and thin samples and windowless counters are preferable for these radiations. Thin samples are obtained by electrical or vapour deposition on thin metal sheets or thin polymer foils. By use of windowless 2π- or 4π-proportional counters the influence of W can be neglected. Determination of absolute activities of γ-ray emitters involves fewer problems, because the influence of the factors S, B and W is, in general, negligible.

Absolute activities of radionuclides may also be determined by coincidence measurements, provided that the decay scheme is relatively simple, e.g. only one β transition followed immediately by emission of one or more γ-ray photons. The principles of the use of coincidence circuits are discussed in the following section.

7.8 Use of Coincidence and Anticoincidence Circuits

Many nuclear processes occur one after the other within a very short time of the order of picoseconds or less – for instance α or β decay followed by γ-ray emission or emission of a cascade of γ rays. The events are practically coincident, and for many purposes it is of interest to know whether two particles or photons are emitted practically at the same time or not. For detection and measurement of coincident events two detectors and a coincidence circuit are used. The detectors are chosen according to the coincidences to be measured, e.g. α–γ, β–γ, γ–γ, X–γ, β–e$^-$ or other types of coincidences, and the coincidence circuit records only events occurring within a given short time interval. Scintillation counters and semiconductor detectors are commonly used for these measurements.

On the other hand, by application of an anticoincidence circuit only those events are recorded that are not in coincidence with others.

Coincidence studies are very useful for detailed investigation of decay schemes. For that purpose, in both detectors the pulse heights are determined simultaneously, giving the energies of the coincident particles or photons, respectively.

An application of anticoincidence circuits is the anti-Compton spectrometer. The Compton continuum in γ spectra can be reduced relative to the photopeaks by placing the Ge detector inside a second detector, usually a scintillation detector, connected in anticoincidence, so that only pulses in the central detector that are not coincident with those in the outer detector are recorded. Anti-Compton spectrometers are very useful for measurement of γ rays of very high energy.

7.9 Low-level Counting

If samples of very low activity are to be measured, the contribution of the background to the counting rate and hence the error of the measurement are relatively high. The influence of the background can be reduced by intensification of the detector shielding and by coincidence or anticoincidence circuits.

Usually, detectors are shielded by housings of lead or of lead outside and steel inside. Steel is used to absorb the radiation from radioactive impurities in lead. In order to absorb more background radiation, thicker walls may be used for these housings. In addition, low counting rates may be measured in underground locations to reduce the contribution of high-energy cosmic radiation. With regard to the use of lead and steel as shieldings, it has to be taken into account that freshly prepared Pb contains ^{210}Pb, and steel produced after 1940 may contain ^{60}Co. Therefore, old lead and steel produced before 1940 are preferable as shielding materials.

To reduce airborne radioactivity (e.g. ^{222}Rn), the space between shielding and detector may be filled with paraffin. Cadmium and copper may be added to the shielding as neutron absorbers. Because (n, γ) reactions in these materials give rise to a secondary γ-ray background, the detector may be surrounded by mercury, which absorbs γ rays very effectively and may be refined by distillation.

By coincidence or anticoincidence methods the background of a detector may by decreased by a factor of about 100 or more. The application of an anticoincidence circuit is the same as in an anti-Compton spectrometer.

7.10 Neutron Detection and Measurement

Neutron detection and measurement are based largely on the production of secondary ionizing radiation by the neutrons. Low-energy (slow) neutrons, in particular thermal neutrons, are measured with high efficiency by means of the charged particles emitted in neutron-induced reactions, such as (n, p) or (n, α) reactions or nuclear fission (chapter 8). High-energy (fast) neutrons are detected via recoiling ions, preferably protons, produced by collision of the high-energy neutrons with other substances, preferably hydrogen or hydrogen-containing compounds. The charged particles (e.g. protons or α particles) can be measured in gas-filled detectors, scintillation detectors or semiconductor detectors.

Thermal neutrons are most frequently detected by means of the nuclear reactions $^{10}B(n, \alpha)^7Li$, $^3He(n, p)^3H$ or neutron fission of ^{235}U. For this purpose, ionization chambers are covered on the inner surface with a thin layer of material containing ^{10}B or ^{235}U, or they may be filled with $^{10}BF_3$ or ^3He. Fast neutrons can be detected by use of a filling gas containing hydrogen; the recoiling protons produced by collisions of the neutrons with the hydrogen nuclei are measured. With these kinds of ionization chambers integral fluxes can be determined at high neutron fluxes over large ranges of intensity. The ionization chambers are operating at high γ-ray intensities, because the internal counting efficiency for γ rays is very low. Gridded ionization chambers filled with ^3He are applied as neutron spectrometers in the energy range of about 10 keV to 2 MeV.

Proportional counters filled with $^{10}BF_3$ or ^3He are used for integral measurement of thermal and epithermal neutrons. Fluxes and spectra of neutrons of intermediate or high energy can be measured with proportional counters filled with ^3He or H_2. Instead of H_2, solid hydrogen-containing compounds such as polyethylene can be used.

In scintillators the counting efficiency for neutrons is relatively high, but the discrimination against γ rays is low. Slow neutrons may be detected via the $^6Li(n, \alpha)^3H$ or the $^{10}B(n, \alpha)^7Li$ reactions. ^6Li or ^{10}B is incorporated in a ZnS scintillator or in liquid or glass scintillators. Crystals of ^6LiI are also used. Fast neutron spectra can be measured via proton recoil in large organic solid or liquid scintillators. Very low neutron fluxes can be determined with scintillators containing Gd by measuring the total energy of the γ rays from the interaction of the neutrons with Gd.

Semiconductor detectors are used as neutron counters by deposition of a suitable "converter" material on the surface of a semiconductor. For instance, a compound containing ^6Li or ^{235}U may be deposited by evaporation, or ^3He may be sealed into the vacuum-tight detector housing. Advantages of these detectors are their small size and their high counting efficiency for thermal neutrons. However, they are not applicable at high neutron fluxes, because of high leakage currents and deterioration.

Instead of direct counting, neutron fluxes can also be determined by activation methods. Activation by (n, γ) reactions (chapter 8) and subsequent measurement of the induced activity is a widely used technique. Au, In and Co are frequently applied as flux monitors. The presence of epithermal neutrons makes corrections necessary. Epithermal neutrons may be measured independently by wrapping the flux monitors in Cd or Gd which absorb the thermal neutrons. Fluxes of high-energy (fast) neutrons can be determined by means of reactions with energy thresholds, such as (n, p) or (n, α) reactions. By application of several detectors with different thresholds information about neutron spectra may be obtained.

7.11 Statistics and Errors of Counting

Radioactive decay is governed by the laws of statistics. Thus, the decay of a single atom cannot be predicted and every counting rate has a statistical error which decreases with the number of counts measured. If the counting rate of a long-lived radionuclide is measured several times under identical conditions, a distribution of the counting rates x_i about a mean value $\bar{x}$ is observed:

$$\bar{x} = \frac{x_1 + x_2 + x_3 + \ldots x_n}{n} = \frac{1}{n} \sum x_i \tag{7.5}$$

The width of the distribution is characterized by the standard deviation σ which is defined by

$$\sigma^2 = \frac{\sum (x - \bar{x})^2}{n - 1} \tag{7.6}$$

σ^2 is also referred to as variance. For a great number of countings $(n \gg 1)$ it follows

$$\sigma^2 = \frac{\sum (x - \bar{x})^2}{n} = \overline{x^2} - \bar{x}^2 \tag{7.7}$$

and at $n \to \infty$ the Poisson distribution is valid:

$$W(x) = \frac{(x)^x}{x!} \exp(-\bar{x}) \tag{7.8}$$

The Poisson distribution is plotted in Fig. 7.17 for $\bar{x} = 5$.

At high values of x $(x \gg 1)$ the asymmetric Poisson distribution becomes identical with the symmetric Gaussian distribution

$$W(x) = \frac{1}{\sigma \sqrt{2\pi}} \exp \left(\frac{(x - \bar{x})^2}{2\sigma^2} \right) \tag{7.9}$$

According to this distribution, there is a probability of 68.3% that the counting is within the standard deviation $\sigma (|x - \bar{x}| \leq \sigma)$, a probability of 95.5% that it is within 2σ and a probability of 99.9% that it is within 3σ.

From eqs. (7.7) and (7.8) it follows

$$\sigma = \pm \sqrt{\bar{x}} \tag{7.10}$$

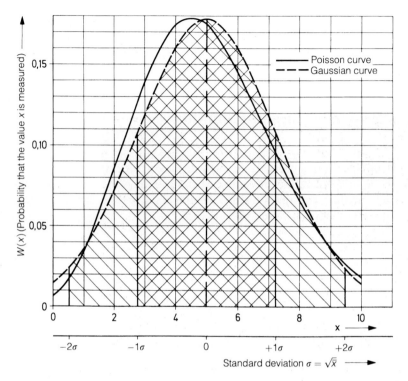

Figure 7.17. Poisson and Gaussian distribution for $\bar{x} = 5$.

If x counts are recorded in a single measurement, the approximate relation

$$\sigma \approx \pm \sqrt{x}$$

is valid (with $x \approx \bar{x}$). For example, σ is about 1% of the result if 10^4 counts are measured.

7.12 Track Detectors

Photographic Emulsions and Autoradiography
The oldest track detectors are photographic plates. They led to the detection of radioactivity by Becquerel in 1896. Photographic emulsions on plates or films indicate the position of radionuclides (autoradiography). The main advantage of autoradiography is the possibility of exact localization of radionuclides emitting α or β rays.

Smooth surfaces are required for autoradiography. The samples can be metals, polished surfaces of minerals, paper chromatograms or thin sections of tissues of biological or medical origin. Autoradiography is often used in mineralogy and biology. It may also be applied for studying chemical processes. As an example, the autoradiograph of an iron surface is shown in Fig. 7.18, indicating the deposition, by corrosion, of extremely small amounts of iron hydroxide labelled with ^{59}Fe in a drop of water.

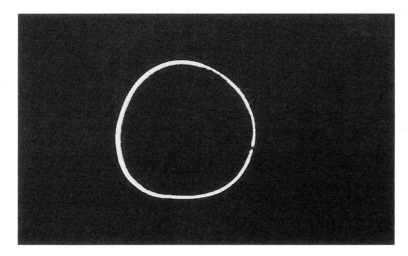

Figure 7.18. Autoradiograph of a sheet of iron showing the very early stage of corrosion at the edge of a drop of water by labelling with ^{59}Fe.

The local resolution of an autoradiograph depends on the thickness of the layer containing the radionuclide, the distance between that layer and the photographic emulsion, the thickness of the photographic emulsion and the radiation emitted by the radionuclide. The influence of the distance between the radioactive layer and the photographic emulsion is illustrated in Fig. 7.19. The thickness of the photographic emulsion should not be greater than 10 µm. Fine-grain emulsions are preferable, but they need longer exposure. Special nuclear emulsions are available. Short-range radiation such as α or low-energy β radiation gives good contrast. The influence of the energy of β radiation is shown in Fig. 7.20. At lower energy of the β particles the autoradiograph is appreciably sharper.

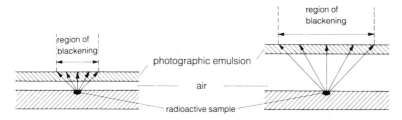

Figure 7.19. Autoradiography: influence of the distance between a radioactive sample and the photographic emulsion.

Figure 7.20. Autoradiograph of two β emitters of different energies: a) ^{35}S ($E_{max} = 0.167\,MeV$); b) ^{32}P ($E_{max} = 1.71\,MeV$).

Autoradiographs may be obtained by pressing a photographic plate or film on the surface of the sample or by the stripping-film or liquid emulsion techniques.

In these techniques, a strip of a thin photographic layer or a liquid emulsion are directly placed on the surface of the sample. After exposure and developing, sample and photographic emulsion can be investigated by means of a microscope (microautoradiography).

The time of exposure depends on the activity and the energy of the radionuclide, and on the sensitivity of the photographic emulsion. At activities of the order of $10^5\,Bq/cm^2$ several hours may be needed. However, the most favourable exposure times have to be found in separate experiments.

Photographic films with nuclear emulsions are used successfully in order to record very rare events induced by cosmic radiation. For this purpose, light-shielded packages of films have been carried by balloons to high altitudes. Some elementary particles, e.g. the positron and the π meson, have been detected by use of photographic emulsions.

Dielectric Track Detectors

Heavy ionizing particles produce tracks of radiation damage in insulating or semi-conducting solids. The tracks have the shape of a cylindrical channel with dimensions of about 1 to 10 nm. Without further treatment, they can only be observed in an electron microscope, but after etching (e.g. by HF) they are visible under a normal microscope. This is due to the fact that chemical attack is much faster in the region where the material is damaged by the ions. Various etching procedures have been developed and tracks of heavy particles have been studied in many materials, such as minerals, glasses, inorganic crystals and plastics. Mica is frequently applied. Each material can be characterized by a certain value of linear energy transfer (LET, given by dE/dx), below which tracks are not observed. For instance, this value related to the density is $(1/\rho)(dE/dx) \approx 13\,(MeV/mg)\,cm^2$ for mica, which means that mica does not register ions with an atomic mass $A < \approx 30$, and fission fragments can be detected by mica in the presence of much higher fluxes of lighter ions. It has been found that the density of ions and excited atoms along the track is directly related to the specific ionization of the radiation, which increases with the atomic number Z of the ions that have produced the tracks. As the rate of etching depends on the density of ionization, the atomic number Z of the particles can be identified. Information about the mass number A may also be obtained.

Solid-track detectors have found application in the investigation of spontaneous fission of transuranium nuclides, of cosmic radiation at high altitudes of the order of 20 km and in dating of minerals by counting the number of tracks.

Another application of track detectors is dosimetry of α particles and neutrons. For neutron dosimetry the track detectors may be covered with uranium foils in which the neutrons induce fission. Alternatively, the detectors may be covered with a foil containing B or Li, and the α particles produced by (n, α) reactions are recorded.

Since tracks caused by radiation damage are very stable, they can be investigated after very long periods of time. Many minerals contain a record of damage by fission products or cosmic rays that has been conserved over millions of years. This makes track detectors very valuable for geochemistry and space science.

Cloud Chambers

In cloud chambers (Wilson chambers) the tracks of ionizing particles are visible by condensation of droplets on the ions produced. The gas in the chamber is saturated with the vapour of water, alcohol or other volatile liquids. By sudden expansion supersaturation is obtained and condensation occurs along the ion tracks. Dust or other condensation centres must be eliminated, to avoid interferences. Cloud chambers can be operated in cycles by a piston or diaphragm (expansion chamber) or by diffusion of a saturated vapour into a colder region (diffusion cloud chamber).

Bubble Chambers

The bubble chamber makes use of the fact that liquids can be heated for short times above their boiling points without actually boiling. Charged particles passing through such a superheated liquid induce the formation of vapour bubbles along their tracks. Because superheated liquids are not stable for long periods of time, bubble chambers are operated intermittently by variation of the pressure. At normal pressure the liquid is just below the boiling point, and by reducing the pressure supersaturation is obtained.

The advantage of bubble chambers is the higher density compared with that in cloud chambers. This makes them particularly useful for the detection of high-energy particles in high-energy accelerators. Bubble chambers with volumes of several cubic metres have been built. They are preferably operated with liquid hydrogen.

Spark Chambers

Spark chambers contain a system of parallel plates or wires that are alternately at a positive high voltage and at earth potential. The chamber is filled with a noble gas. An ionizing particle crossing the gap between two plates or wires causes the production of a spark near its trajectory. The high voltage is applied in short pulses which are triggered by counters surrounding the chambers. In this way, the chambers are sensitive only for selected types of events. The detection efficiency of spark chambers is very high and the resolution is very good. The selective operation is a great advantage over bubble chambers. Wire spark chambers with dimensions of many metres containing of the order of 10^5 wires have been constructed.

7.13 Detectors Used in Health Physics

For the monitoring of personnel radiation exposures, measurement of radioactive contamination and surveying of laboratories and equipment, and for the detection of radionuclides incorporated in the human body, various detectors and instruments are used. The principles of operation of these detectors have been discussed in the previous sections of this chapter.

Portable Counters and Survey Meters
The dose rate in a given radiation field can be determined by means of sensitive portable detectors such as Geiger–Muller counters or ionization chambers in combination with batteries and compact DC amplifiers. A movable shield may be used to distinguish between hard and soft radiation. The same types of instruments are suitable for surveying laboratories and instruments. Geiger–Muller counters have higher sensitivity and may trigger audible signals. They are preferably used for rapid surveys and detection of small amounts of high-energy β radiation. In the usual form as end-window counters they are not suitable for the detection of α or low-energy β radiation. Ionization chambers can be equipped with thin windows ($<3\,\mathrm{mg/cm^2}$) or open screens for detection of α and low-energy β radiation. As monitors for γ rays, portable scintillation counters are the most sensitive.

Film Badges
Persons who are handling radioactive material or are exposed to nuclear radiation should wear badge-type holders containing a photographic film which records the general exposure to radiation over a certain period of time (usually one week). X-ray film is generally used for β and γ dosimetry. Information about the type and energy of the radiation is obtained by placing various filters (plastic, aluminium and cadmium foils) over certain areas of the film. Exposure to thermal neutrons is measured with boron-loaded films. Track counters are used as monitors of fast neutrons and radiation of very high energy.

Pocket Ion Chambers
Pocket ionization chambers are small enough to be worn like a ball-pen. They are charged by a temporary connection to a voltage source, and the residual charge after exposure is read on the scale of an electrometer fitted in the ionization chamber. Pocket ion chambers are not sensitive to α and low-energy β radiation.

Thermoluminescence Dosimeters
In certain crystals such as LiF or CaF_2 containing Mn as impurity, the electrons and holes produced by nuclear radiation are trapped on the impurities. In this way energy is stored in the crystals. It can be released in the form of light by heating the crystals. This property of thermoluminescence can be used for radiation dosimetry.

The crystals are heated electrically and the light output is measured by means of an photomultiplier. In the case of LiF(Mn) crystals the response is nearly independent of the energy in the range between about 30 keV and 3 MeV, and the effective atomic number of LiF is similar to that of soft tissue. On the other hand, Li(Mn) is only suitable for doses ≥ 10 mrad, whereas $CaF_2(Mn)$ is sensitive down to ≈ 0.1 mrad.

Contamination Monitors

Contamination on the hands can be detected with Geiger–Muller or proportional counters. For detection and measurement of air-borne contamination, filters may be used. The activity on the filters can be measured continuously by passing the filters through a proportional counter. Radioactive contamination on laboratory benches, instruments or floors can be detected by the simple method of wiping: the surface to be checked for contamination is wiped with a piece of filter paper, and the filter paper is measured with a suitable detector for α, β or γ activity.

Whole-body Counters

Whole-body counters consist of a heavily shielded space. The person to be examined is placed inside and surrounded by a large number of scintillation detectors. In this way, γ-emitting radionuclides in the body can be detected with high sensitivity and identified. In the absence of contamination by artificial radionuclides, the γ radiation from ^{40}K is observed. The uptake of small amounts of artificial γ-ray emitters such as ^{137}Cs can be determined effectively, whereas pure α or β emitters cannot be detected in the body.

Literature

General

G. B. Cook, J. F. Duncan, Modern Radiochemical Practice, Oxford University Press, Oxford, **1952**
S. Flügge, E. Creutz (Eds.), Instrumentelle Hilfsmittel der Kernphysik II, Handbuch der Physik XLV, Springer, Berlin, **1958**
R. T. Overman, H. M. Clark, Radioisotope Techniques, McGraw-Hill, New York, **1960**
L. C. L. Yuan, C. S. Wu (Eds.), Methods of Experimental Physics, Vol. 5A Nuclear Physics, Academic Press, New York, **1961**
A. H. Snell (Ed.), Nuclear Instruments and their Uses, Wiley, New York, **1962**
G. D. O'Kelley, Detection and Measurement of Nuclear Radiation, NAS-NS 3105, Washington, DC, **1962**
W. J. Price, Nuclear Radiation Detection, McGraw-Hill, New York, **1964**
K. Siegbahn (Ed.), Alpha-, Beta- and Gamma-Ray Spectroscopy, 2 Vols., North-Holland, Amsterdam, **1966**
K. Bächmann, Messung radioaktiver Nuklide (Ed. K. H. Lieser), Verlag Chemie, Weinheim, **1970**
J. H. Hamilton (Ed.), Radioactivity in Nuclear Spectroscopy, Modern Techniques and Applications, Vols. I and II, Gordon and Breach, New York, **1972**
J. Krugers (Ed.), Instrumentation in Applied Nuclear Chemistry, Plenum Press, New York, **1973**
J. Cerny (Ed.), Nuclear Spectroscopy and Reactions, Vols. A, B and C, Academic Press, New York, **1974**
P. J. Ouseph, Introduction to Nuclear Radiation Detection, Plenum Press, New York, **1975**
W. B. Mann, R. L. Ayres, S. B. Garfinkel, Radioactivity and its Measurement, 2nd ed., Pergamon, Oxford, **1980**
W. H. Tait, Radiation Detection, Butterworths, London, **1980**

R. A. Faires, G. G. J. Boswell, Radioisotope Laboratory Techniques, 4th ed., Butterworths, London, **1981**

G. F. Knoll, Radiation Detection and Measurement, 2nd ed., Wiley, New York, **1989**

More Special

H. Yagoda, Radioactive Measurements with Nuclear Emulsions, Wiley, New York, **1949**

B. B. Rossi, H. H. Staub, Ionization Chambers and Counters, Nat. Nuclear Energy Series V, Vol. 2, McGraw-Hill, New York, **1949**

D. H. Wilkinson, Ionization Chambers and Counters, Cambridge University Press, Cambridge, **1950**

M. Blau, Photographic Emulsions, in: Methods of Experimental Physics, Vol. 5A Nuclear Physics (Eds. L. C. L. Yuan, C. S. Wu), Academic Press, New York, **1961**

W. H. Barkas, Nuclear Research Emulsions, Academic Press, New York, **1963**

E. Schram, R. Lombaert, Organic Scintillation Detectors, Elsevier, Amsterdam, **1963**

J. B. Birks, The Theory and Practice of Scintillation Counting, Pergamon, Oxford, **1964**

H. H. Chiang, Basic Nuclear Electronics, Wiley, New York, **1969**

G. Bertolini, A. Coche (Eds.), Semiconductor Detectors, North-Holland, Amsterdam, **1968**

G. T. Ewan, Semiconductor Spectrometers, in: Progress in Nuclear Techniques and Instrumentation, Vol. III (Ed. F. J. M. Farley), North Holland, Amsterdam, **1968**

R. J. Brouns, Absolute Measurement of Alpha Emissions and Spontaneous Fission, NAS-NS 3112, Washington, D.C., **1968**

E. D. Bransome (Ed.), Liquid Scintillation Counting, Grune and Stratton, New York, **1970**

E. Kowalski, Nuclear Electronics, Springer, Berlin, **1970**

F. Adams, R. Dams, Applied Gamma-Ray Spectrometry, 2nd ed. Pergamon, Oxford, **1970**

C. A. Crouthamel, F. Adams, R. Dams, Applied Gamma-Ray Spectrometry, Pergamon, Oxford, **1970**

H. A. Fischer, G. Werner, Autoradiography, Walter de Gruyter, Berlin, **1971**

A. Dyer (Ed.), Liquid Scintillation Counting, Vol. 1, Heyden, London, **1971**

S. Deme, Semiconductor Detectors for Nuclear Radiation Measurement, Hilger, London, **1971**

P. Quittner, Gamma-Ray Spectroscopy with Particular Reference to Detector and Computer Evaluation Techniques, Hilger, London, **1972**

D. L. Horrocks, Applications of Liquid Scintillation Counting, Academic Press, New York, **1974**

K. D. Neame, C. A. Homewood, Liquid Scintillation Counting, Wiley, New York, **1974**

Semiconductor Radiation Detectors, in: Nuclear Spectroscopy and Reactions, Vol. A (Ed. J. Cerny), Academic Press, New York, **1974**

Users Guide for Radioactivity Standards, NAS-NS 3115, Washington, D. C., **1974**

R. L. Fleischer. P. B. Price, R. M. Walker, Nuclear Tracks in Solids, University of California Press, Berkeley, CA, **1975**

H. Morinaga, T. Yamazaki, In-Beam Gamma-Ray Spectroscopy, North-Holland, Amsterdam, **1976**

G. Erdtmann, W. Soyka, The Gamma Rays of the Radionuclides (Ed.: K. H. Lieser), Verlag Chemie, Weinheim **1979**

A. Dyer, Liquid Scintillation Counting Practice, Heyden, London, **1980**

K. Debertin, R. G. Helmer, Gamma- and X-Ray Spectrometry with Semiconductor Detectors, North-Holland, Amsterdam, **1988**

S. deFilippis, Activity Analysis in Liquid Scintillation Counting, in: Radioactivity and Radiochemistry *1*, 4, 22 **(1990)**

Statistics and Data Analysis

W. Feller, Probability Theory and its Applications, Wiley, New York, **1950**

C. A. Bennett, N. L. Franklin, Statistical Analysis in Chemistry and Chemical Industry, Wiley, New York, **1954**

K. A. Brownlee, Statistical Theory and Methodology in Science and Engineering, Wiley, New York, **1960**

G. D. O'Kelley (Ed.), Applications of Computers to Nuclear and Radiochemistry, NAS-NRC, Washington, D. C., **1963**

S. L. Meyer, Data Analysis for Scientists and Engineers, Wiley, New York, **1975**

8 Nuclear Reactions

8.1 Mono- and Binuclear Reactions

The simplest form of a nuclear reaction is radioactive decay according to eq. (4.2):

$$A \rightarrow B + x + \Delta E$$

This is a mononuclear or first-order reaction which is often referred to in chemical kinetics as an example of a real first-order reaction. In nuclear science, however, binuclear reactions are generally understood by the term "nuclear reactions". They are described by the equation:

$$A + x \rightarrow B + y + \Delta E \tag{8.1}$$

where A is the target nuclide, x the projectile, B the product nuclide, and y the particle or photon emitted. The energy ΔE is also called the Q value of the reaction.

The first binuclear reaction was observed in a cloud chamber in 1919 by Rutherford:

$$^{14}_{7}N + ^{4}_{2}He \rightarrow ^{17}_{8}O + ^{1}_{1}H \tag{8.2}$$

The sum of the mass numbers and the sum of the atomic numbers must each be the same on the left- and right-hand sides of the equation. The short form of eq. (8.2) is

$$^{14}N(\alpha, p)^{17}O \tag{8.3}$$

or, generally, instead of eq. (8.1)

$$A(x, y)B \tag{8.4}$$

For all nuclear reactions the conservation laws (number of nucleons, charge, sum of mass and energy equivalent of mass, momentum, angular momentum, parity) are valid. The main differences between chemical reactions and nuclear reactions are the following:

- In chemical reactions the conversion of weighable amounts of matter is considered (e.g. grams or moles), but in nuclear reactions the conversion of atoms is involved.
- In chemical reactions the nuclides remain the same, whereas they are changed in nuclear reactions.
- For chemical reactions the law of conservation of mass is valid (the extremely small variations caused by the changes in bonding are negligible), whereas in the case of nuclear reactions the sum of the energy equivalents of the masses and the energies $(\Sigma(mc^2 + E))$ remains constant.

- The energies of chemical reactions are comparable with the energies of chemical bonds and relatively small (of the order of eV), whereas the energies involved in nuclear reactions are about 6 orders of magnitude higher (of the order of MeV).

Many nuclear reactions pass over a transition state, similarly to chemical reactions:

$$A + x \rightarrow (C) \rightarrow B + y + \Delta E \tag{8.5}$$

The transition state (C) is also called a compound nucleus. Its lifetime is very short ($<10^{-13}$ s).

Whereas the probability of a mononuclear reaction (i.e. for radioactive transmutation) is given by the decay constant λ, two probabilities are decisive in the compound nucleus model: the probability that the projectile x will react with the nuclide A (the first step of reaction (8.5)) and the probability that the nuclide B is produced (the second step of reaction (8.5)).

The time taken by a nuclear reaction varies between about 10^{-23} and 10^{-13} s. The lower limit is given by the time needed by a particle with the velocity of light to cross the nucleus, and the higher limit holds for slow nuclear reactions, e.g. reactions with thermal neutrons.

Scattering processes, elastic scattering

$$A + x \rightarrow A + x \tag{8.6}$$

and inelastic scattering

$$A + x \rightarrow A^* + x \tag{8.7}$$

will not be considered in detail, because the nuclides remain the same. In elastic scattering no energy is transferred, whereas in inelastic scattering nuclide A goes into an excited state. Investigation of inelastic scattering makes it possible to set up energy diagrams of the nuclides.

8.2 Energetics of Nuclear Reactions

The energy ΔE (Q value) of a nuclear reaction (eq. (8.1)) can be calculated by comparison of the nuclide masses, as in the case of radioactive decay:

$$\Delta E = (m_A + m_x - m_B - m_y) \cdot c^2 \tag{8.8}$$

Introducing the nuclide masses $M = m + Z \cdot m_e$, where Z is the atomic number and m_e the mass of an electron, gives

$$\Delta E = (M_A + M_x - M_B - M_y) \cdot c^2 \tag{8.9}$$

Thus, the electron masses cancel out in all nuclear reactions with the exception of β^+ decay (section 5.3). If atomic mass units u are used, ΔE is obtained in MeV by setting 931.5 MeV for c^2 (section 2.5).

If $\Delta E < 0$ (endoergic or endoenergetic reactions), the missing energy must be introduced by the projectile x. By application of the law of conservation of momentum, the minimum energy of the projectiles for endoergic reactions is calculated to be

$$E_x(\text{th}) = -\Delta E \left(1 + \frac{M_x}{M_A} \right) \tag{8.10}$$

This energy is called the threshold energy of the reaction.

The initial kinetic energies of the reactants (nucleus A and projectile x) and the energy ΔE appear in the form of the kinetic energies of the products B and y and of the excitation energy of B:

$$E_A + E_x + \Delta E = E_B + E_y + E_B^* \tag{8.11}$$

If nuclide A is at rest, $E_A = 0$ and

$$E_x + \Delta E = E_B + E_y + E_B^* \tag{8.11a}$$

The energy ΔE can be calculated from E_x, E_B and E_y, if the angles are known under which the products are leaving (Fig. 8.1). From the law of conservation of momentum it follows that

$$\Delta E = E_y \left(1 + \frac{M_y}{M_B} \right) - E_x \left(1 - \frac{M_x}{M_B} \right) \frac{2}{M_B} \sqrt{E_x E_y M_x M_y} \cos \vartheta + E_B^* \tag{8.12}$$

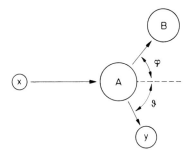

Figure 8.1. Angular distribution of the products of a nuclear reaction.

An example of an exoergic (exoenergetic) reaction is the production of neutrons by the reaction of α particles with Be:

$$^9_4\text{Be} + ^4_2\text{He} \rightarrow ^{12}_6\text{C} + ^1_0\text{n} \tag{8.13}$$

On the basis of this reaction neutrons were discovered in 1932 by Chadwick. By application of eq. (8.9) $\Delta E = 5.70\,\text{MeV}$ is obtained. In contrast to reaction (8.13), the reaction

$$^{12}_{6}C + ^{4}_{2}He \rightarrow ^{15}_{7}N + ^{1}_{1}H \qquad (8.14)$$

is endoergic. Eq. (8.9) gives $\Delta E = -4.97\,\mathrm{MeV}$, and the threshold energy according to eq. (8.10) is $E_x(\mathrm{th}) = 6.63\,\mathrm{MeV}$.

8.3 Projectiles for Nuclear Reactions

Neutrons

Neutrons are the most frequently used projectiles for nuclear reactions. As they do not carry a positive charge, they do not experience Coulomb repulsion, and even low-energy (thermal and slow) neutrons can easily enter the nuclei. Neutrons with energies of the order of 1 to 10 eV (resonance neutrons) exhibit relatively high absorption maxima. Furthermore, neutrons are available in large quantities in nuclear reactors with fluxes of the order of about 10^{10} to $10^{16}\,\mathrm{cm}^{-2}\,\mathrm{s}^{-1}$.

Neutrons may also be produced by nuclear reactions, such as $^9\mathrm{Be}(\alpha, \mathrm{n})^{12}\mathrm{C}$ (reaction (8.13)), $^9\mathrm{Be}(\gamma, \mathrm{n})2\alpha$, $\mathrm{d}(\gamma, \mathrm{n})\mathrm{p}$, $^9\mathrm{Be}(\mathrm{d}, \mathrm{n})^{10}\mathrm{B}$, $\mathrm{d}(\mathrm{d}, \mathrm{n})^3\mathrm{He}$ or $\mathrm{t}(\mathrm{d}, \mathrm{n})\alpha$. Alpha particles are available from α emitters such as $^{226}\mathrm{Ra}$ or $^{210}\mathrm{Po}$, and Ra–Be neutron sources have very often been applied in the early stages of experiments with neutrons. Gamma-ray photons of sufficiently high energy may also be supplied by radionuclides, such as $^{124}\mathrm{Sb}$. Neutrons are also available from spontaneously fissioning nuclides, such as $^{252}\mathrm{Cf}$ (1 µg of $^{252}\mathrm{Cf}$ emits $2.3 \cdot 10^6$ neutrons per second). In neutron generators deuterons with energies between 0.1 and 10 MeV are produced in small accelerators and directed on a suitable target, e.g. a tritium target. The neutron fluxes available from neutron sources and neutron generators vary between about 10^5–10^8 and 10^8–$10^{11}\,\mathrm{cm}^{-2}\,\mathrm{s}^{-1}$, respectively. Neutrons produced by the reaction t (d, n) α have energies of about 14 MeV.

Reactions with neutrons are used to produce nuclides on the right-hand side of the line of β stability, i.e. nuclides exhibiting β^- decay.

Charged Particles

Charged particles, such as protons, deuterons or ions with higher atomic numbers Z, must have a minimum energy to pass the Coulomb barrier of the nuclei. Approximately, the Coulomb barrier can be calculated from the equation

$$U \approx \frac{Z_A Z_x e^2}{4\pi\varepsilon_0 r} \approx \frac{Z_A Z_x}{A_A^{1/3} + A_x^{1/3}} \qquad (8.15)$$

where Z_A and Z_x are the numbers of the positive charges of the nuclide A and the projectile x, respectively, e is the unit charge, ε_0 the electric field constant, and r the distance, which in this approximation is set to be $r \approx r_0(A_A^{1/3} + A_x^{1/3})$. A_A and A_x are the mass numbers of the nuclide A and the projectile x. By application of eq. (8.15) the following values are calculated: $U \approx 1.8\,\mathrm{MeV}$ for the reaction of a proton with $^{12}\mathrm{C}$, $U \approx 13\,\mathrm{MeV}$ for the reaction of a proton with $^{238}\mathrm{U}$, $U \approx 24\,\mathrm{MeV}$ for the reaction of an α particle with $^{238}\mathrm{U}$, $U \approx 130\,\mathrm{MeV}$ for the reaction of $^{12}\mathrm{C}$ with $^{238}\mathrm{U}$ and $U \approx 700\,\mathrm{MeV}$ for the reaction of $^{238}\mathrm{U}$ with $^{238}\mathrm{U}$. At higher atomic numbers the approximation formula (8.15) gives inexact values of U. A more exact value for the Coulomb barrier in the reaction of $^{238}\mathrm{U}$ with $^{238}\mathrm{U}$ is $U \approx 1500\,\mathrm{MeV}$.

If the collision between the projectile and the nucleus is not central, angular momentum is also transmitted to the nucleus, and in addition to the Coulomb barrier a centrifugal barrier has to be taken into account which is given by

$$V = \frac{\hbar^2 l(l+1)}{8\pi^2 \mu r_0^2 (A_A^{1/3} + A_x^{1/3})^2} \tag{8.16}$$

Angular momentum can only be transmitted in multiples l of $h/2\pi$ ($l = 0$ for central collisions). $\mu = m_A \cdot m_x / (m_A + m_x)$ is the reduced mass of the system.

If the energy E of the projectile is at least equal to the sum of U and V, a nuclear reaction can be expected. However, there is also the possibility of a tunnel effect as in the case of α decay (section 5.2), and particles with energies $E < U + V$ may also lead to nuclear reactions.

Charged particles are produced in ion sources by bombarding a gas with energetic electrons. The positive ions are extracted by means of an electrode to which a negative voltage of about 1–10 keV is applied.

For acceleration of the ions, various types of installations are used. The two main groups are linear and circular accelerators, and in the case of linear accelerators (linacs) single- and multiple-stage machines are distinguished. The final energy of the charged particles obtained in linear accelerators depends on the voltage applied and the number of subsequent sections of acceleration. In circular accelerators (cyclotrons and synchrotrons) it depends on the strength of the magnetic field. In practice, voltages of up to several million volts per metre and magnetic field strengths of up to several tesla can be applied. Highest energies can be obtained in linear accelerators by increasing the length and in circular accelerators by increasing the radius.

Accelerators of the Cockroft–Walton type are single-stage linear accelerators by which proton or deuteron beams of up to about 10 mA with energies of up to about 4 MeV are obtained. These accelerators are often applied as injectors in machines designed for the production of high-energy particles. Van de Graaff accelerators are built with one accelerator tube or as tandem accelerators with two or three acceleration stages. Their main advantage is the precisely defined energy of the particles. In two-stage tandem-type Van de Graaff accelerators proton beams of about 20 MeV and α particles of about 30 MeV are obtained. The intensity of the beam current varies between about 10 and 100 μA. By use of three stages the energy of the protons may be increased to ≈ 45 MeV.

In multistage linear accelerators the ions pass a series of sections of increasing length (Fig. 8.2). In cyclotrons a constant magnetic field is applied and the ions move

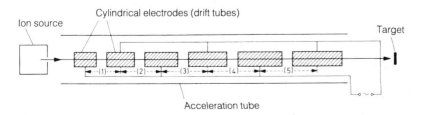

Figure 8.2. Schematic diagram of a linear accelerator.

on spiral paths of increasing radius (Fig. 8.3). The maximum energy to which ions can be accelerated in standard cyclotrons is limited by the relativistic mass increase (for protons to about 25 MeV). Higher energies are obtained in section-focused (iso-chronous) cyclotrons, in which the magnetic field strength is varied over the ion path but stays constant with time. Another way of overcoming the limitations of standard cyclotrons is by modulation of the oscillator frequency in machines called synchro-cyclotrons, in which protons may be accelerated to energies of up to about 700 MeV.

In synchrotrons (Fig. 8.4) the radius of the orbit of the charged particles is kept

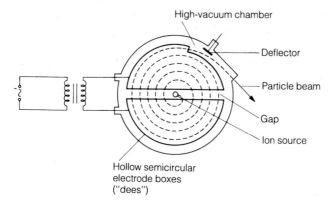

Figure 8.3. Schematic diagram of a cyclotron.

constant by a magnetic field that increases with the momentum of the particles. Typical beam intensities in proton synchrotrons are 10^{11} to 10^{12} per pulse. By application of alternating-gradient focusing of the beam an appreciable reduction of the cost of the installations is achieved.

In modern machines, protons, deuterons and α particles with energies of several 100 MeV up to about 1 GeV are available. Proton linacs serve frequently as injectors of 50 to 200 MeV protons into proton synchrotrons. For the production of radio-nuclides, relatively small cyclotrons are used by which particle energies of the order of 10 to 30 MeV and ion currents of the order of 100 µA are available. Radionuclides obtained by reactions with protons exhibit β^+ decay or electron capture (ε).

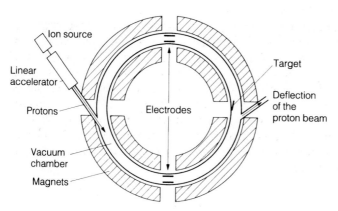

Figure 8.4. Schematic diagram of a proton synchrotron.

Heavy Ions

Heavy ions have found broad application as projectiles in recent decades. The term "heavy ions" is used in this context for all ions heavier than α particles and includes "light ions" such as the ions of carbon or oxygen as well as "heavy ions" of elements up to uranium. The main features of heavy ions are as follows:

- They contain a great number of nucleons which they can transmit to the target nuclide in one reaction. Thus, nuclides may be obtained that are far away from the target nuclide in the chart of the nuclides.
- They are able to transfer high linear momentum and high angular momentum to the target nucleus. This may lead to a state of high compression of nuclear matter, to high excitation of the system and to a high rotation frequency.
- The nuclei of heavy ions have high charges, and if a great number of the electrons are stripped off, highly charged ions are obtained which exhibit interesting properties: the spectra of electronic transitions are in the energy range of UV and X rays. As the linear energy transfer (LET) increases with the square of the ion charge, the high charges of the ions offer new applications in solid-state physics, biology and medicine.

Heavy ions are produced in special types of linear or circular accelerators. One or several electrons are removed from the atoms in the ion source. Because the acceleration by a given electric field depends on the value of n/A, where n is the number of positive charges of the ion and A is the mass number, special efforts are made to obtain ions with high values of n. After the first step of acceleration, n may be increased by stripping off further electrons. This is achieved by passing the ions through foils or gas beams.

Powerful machines for the investigation of heavy-ion reactions are available at the GSI, Darmstadt. They comprise the linear heavy-ion accelerator UNILAC, the heavy-ion synchrotron SIS and the heavy-ion storage ring ESR. Energies of about 25 MeV/u can be obtained in the UNILAC, and about 2 GeV/u in the SIS. A segment separator allows application of isotopically pure heavy-ion beams.

The main fields of application of heavy ions are: synthesis of new elements (superheavy elements), production of nuclides far away from the line of β stability (exotic nuclides), investigation of nuclear matter at high densities, production of small holes of certain diameters in thin foils and irradiation of tumours in medicine.

Storage Rings and Colliding Beams

Storage rings are used in nuclear physics to study the collisions of high-energy particles. The reason is that in collisions of high-energy particles with particles at rest only a part of the kinetic energy of the projectiles is available for the production of new particles. Under these conditions, the centre of mass motion has to be considered, and due to the conservation of momentum a certain fraction of the momentum of the projectile is transformed into momenta of the products. This fraction increases strongly with the energy of the projectile, in particular if its velocity approaches the velocity of light. This drawback is avoided by use of storage rings and colliding beams. For instance, if two protons of the same energy collide head-on, no energy is lost due to centre-of-mass motion.

Because the particles in these storage rings lose energy by emission of synchrotron radiation, further acceleration is necessary. The energy loss by synchrotron radiation is relatively high in the case of electrons and low in the case of heavy ions.

Electrons and γ Rays

For the acceleration of electrons to high energies and velocities approaching the velocity of light, much less energy is needed than for the acceleration of ions. Electrons with an energy of 2 MeV already have 98% of the velocity of light. They are accelerated in linear accelerators, in betatrons or in synchrotrons. The betatron operates like a transformer: the electrons moving in a ring-like vacuum chamber represent the secondary winding. The magnetic flux density is regulated according to the increasing energy. Depending on the size of the installations, energies of the order of 10 to 300 MeV are obtained in linear accelerators and betatrons, whereas in electron synchrotrons electrons with energies up to the order of 10 GeV are available. The upper energy limit of the electrons is about 50 GeV. It is caused by the energy loss due to the emission of synchrotron radiation, which increases with the fourth power of the ratio E/m_0c^2, where m_0 is the rest mass of the electron.

High-energy electrons are used in nuclear physics for the investigation of elementary and fundamental particles. Energies of the order of several GeV are sufficient to produce other elementary particles and antiparticles and to study their interactions. Collisions of electrons and positrons are investigated by use of storage rings.

Electrons are not used for the production of radionuclides. However, the high-energy bremsstrahlung emitted when the electrons hit a target of high atomic number, such as tungsten, is applied for nuclear reactions induced by photons (photonuclear reactions). The maximum energy of this bremsstrahlung corresponds to the energy of the incident electrons and is in the range of high-energy γ-ray photons. Gamma rays are also available from γ emitters, such as ^{60}Co or ^{124}Sb, but with relatively low energies and low flux densities.

The monoenergetic synchrotron radiation emitted by electrons moving in the magnetic field of a synchrotron has found many applications, because of the high intensities available. The energy of the synchrotron radiation is in the range of the energy of X rays.

8.4 Cross Sections of Nuclear Reactions

The probability that a nuclear reaction may occur is given by the cross section of the reaction, which is comparable with the rate constant of a bimolecular chemical reaction. Considering the general equation for a binuclear reaction

$$A + x \rightarrow B + y \tag{8.17}$$

the production rate of nuclide B is given by

$$\frac{dN_B}{dt} = \sigma \Phi N_A \tag{8.18}$$

where σ is the cross section of reaction (8.17), Φ is the flux density of the projectiles, and N_A is the number of the atoms of the target nuclide A.

If the flux density is measured in $cm^{-2}\,s^{-1}$, σ is in units of cm^2. Thus the cross section can be imagined to represent the geometric area of one nucleus of A that may be hit by the incident projectiles x, as indicated in Fig. 8.5. If, according to this rough picture, nuclear forces and Coulomb interaction are neglected, the cross section should be comparable with the real geometric cross section of a nucleus, which is of the order of $10^{-24}\,cm^2$. This value is chosen as the unit of cross section and called 1 barn: 1 barn (b) $= 10^{-24}\,cm^2 = 10^{-28}\,m^2$. The cross sections for the scattering of fast neutrons ($E > 10\,Mev$) agree rather well with the geometric cross sections of the nuclei. For many (n, γ) reactions, the order of magnitude of $\sigma_{n,\gamma}$ is 1 b. However, for some nuclides $\sigma_{n,\gamma}$ is of the order of only 10^{-4} b, and for others it may be in the order of 10^4 b. This indicates the influence of repulsive and attractive interactions between nuclei and projectiles and the limitations of the simple geometric model.

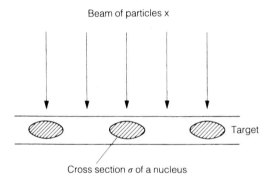

Beam of particles x

Target

Cross section σ of a nucleus

Figure 8.5. Schematic explanation of the cross section of a nuclear reaction.

Cross sections are defined only for certain nuclear reactions. For example, $\sigma_{n,\gamma}^A$ is the cross section for the (n, γ) reaction of the nuclide A, and σ_a^A is the absorption cross section of the nuclide A in which all absorption processes of certain projectiles, in particular neutrons, in A are summarized, independently of the kind of reaction they are inducing. Another quantity that is often used is the total cross section σ_t^A,

$$\sigma_t^A = \sum_i \sigma_i^A \tag{8.19}$$

which is the sum of all the individual cross sections of certain projectiles, in particular neutrons, with a target nuclide A, comprising absorption and scattering processes. The macroscopic cross section Σ_i is defined by the equation

$$\Sigma_i = \sum_j \sigma_i^j N_j \tag{8.20}$$

where σ_i are the cross sections of a certain reaction i in all nuclides j present in the target material, and N_j are the numbers of the atoms of the nuclides j. For absorption processes the macroscopic cross section is

$$\Sigma_a = \sum_j \sigma_a^j N_j \tag{8.21}$$

and if only one nuclide A is present in the absorber material,

$$\Sigma_a = \sigma_a^A N_A \tag{8.22}$$

If scattering processes are left out of consideration, Σ_a is equal to the absorption coefficient μ for certain particles or photons. $1/\Sigma_a$ is then equal to the mean free path of the particles or photons in the absorber. If absorption and scattering are taken into account, the absorption coefficient μ is given by the total macroscopic cross section Σ_t.

All cross sections depend on the energy of the particles or photons. This energy dependence $\sigma = f(E)$ is called the excitation function, because the energy of the incident particle or photon is transferred to the nucleus as excitation energy. The dependence of the neutron absorption cross section σ_a on the neutron energy is shown schematically in Fig. 8.6. Even neutrons with kinetic energies corresponding to liquid-air temperature $(E_n \approx 0.01\,\mathrm{eV})$ are able to enter a nucleus without any

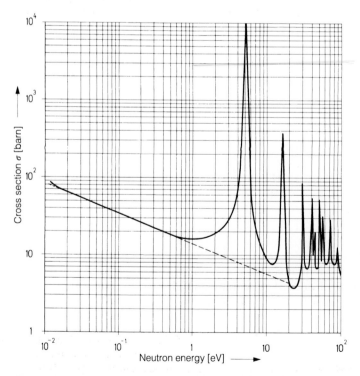

Figure 8.6. Cross section σ_a of Ag for neutrons as a function of the energy of the neutrons (according to AECU-2040 and its supplement).

energy barrier, and the cross sections are relatively high at low energies. With increasing velocity v_n of the neutrons, the residence time near the nuclei and therefore the probability of entering the nuclei become smaller:

$$\sigma_a \sim \frac{1}{v_n} \tag{8.23}$$

This relation is valid in the range of low neutron energies up to about 1 eV. At higher energies, in the range between about 1 eV and about 10 keV (epithermal neutrons), absorption maxima or resonances are observed. The absorption maxima correspond to the excitation energy levels of the nucleus, and neutrons carrying just the excitation energies are absorbed with extremely high probabilities (resonance).

Excitation functions of proton-induced reactions are plotted schematically in Fig. 8.7. Up to the threshold energy, which is due to Coulomb repulsion, no reaction takes place. Above the threshold energy, the cross sections increase sharply and reach a maximum value at about twice the threshold energy. At higher energies the cross sections drop again to smaller values. With increasing energy, more and more reactions become possible. Excitation functions similar to those shown in Fig. 8.7 are found for other charged particles, such as deuterons or α particles.

Excitation functions for γ-ray induced reactions (photonuclear reactions) start also at a certain threshold energy which is given by the excitation energy necessary to emit a neutron or proton and reach a flat maximum at higher energies. Again, in the range of higher energies, the excitation functions for (n, γ) reactions are superimposed by those for $(\gamma, 2n)$ and other γ-ray induced reactions.

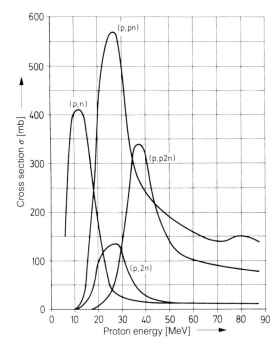

Figure 8.7. Cross sections of several nuclear reactions of protons with ^{63}Cu as a function of their energy (excitation functions; according to J. W. Meadows; Physic. Rev. **91**, 885 (1953)).

8.5 Yield of Nuclear Reactions

The yield of a nuclear reaction can be calculated if the cross secion σ, the flux density Φ, and the number of atoms N_A of a certain nuclide A in eq. (8.18) are known. If the product nuclide B is radioactive, its decay rate

$$-\frac{dN_B}{dt} = \lambda N_B \tag{8.24}$$

must be taken into account, and the net production rate is

$$-\frac{dN_B}{dt} = \sigma \Phi N_A - \lambda N_B \tag{8.25}$$

In order to solve this differential equation, it is assumed that Φ is constant within the target and with time and that N_A does not change appreciably with time. Then integration of eq. (8.25) gives

$$N_B = \frac{\sigma \Phi N_A}{\lambda}(1 - e^{-\lambda t}) \tag{8.26}$$

N_B is the number of the atoms of nuclide B produced after irradiation time t. The activity of B is

$$A = -\frac{dN_B}{dt} = \lambda N_B = \sigma \Phi N_A (1 - e^{-\lambda t}) \tag{8.27}$$

By this equation the activity is obtained in s^{-1} or Bq. The relation between the number of atoms N_A of the nuclide A and the mass m of the element containing the nuclide A is

$$N_A = \frac{N_{Av}}{M} Hm \tag{8.28}$$

where N_{Av} is Avogadro's number ($N_{Av} = 6.022 \cdot 10^{23}$), M is the atomic mass of the element, and H is the isotopic abundance of the nuclide A in the element. Substitution of eq. (8.28) in eq. (8.27) gives

$$A = \sigma \Phi Hm \frac{N_{Av}}{M}(1 - e^{-\lambda t}) \tag{8.29}$$

Introducing $N_{Av} = 6.022 \cdot 10^{23}$, and expressing σ in barns ($1\,b = 10^{-24}\,cm^2$) and Φ in $cm^{-2}\,s^{-1}$ leads to

$$A = 0.6022\sigma(\text{in b})\Phi\frac{Hm}{M}(1 - e^{-\lambda t}) \tag{8.30}$$

or

$$A = 0.6022\sigma(\text{in b})\Phi\frac{Hm}{M}\left[1 - \left(\frac{1}{2}\right)^{t/t_{1/2}}\right]$$ (8.31)

Equations (8.27) and (8.29) to (8.31) are different forms of the equation of activation and may be used to calculate the yield of a nuclear reaction.

The turnover of a nuclear reaction is given by the ratio

$$N_B/N_A = \frac{\sigma\Phi}{\lambda}(1 - e^{-\lambda t})$$

The net rate of production is positive if $\sigma\Phi > \lambda(N_B/N_A)$. If a high turnover is required ($N_B \approx N_A$), it is necessary that $\sigma\Phi > \lambda$.

The equation of activation has broad application in nuclear science. It is used to determine the cross section σ if the flux density Φ is known, or the flux density Φ if the cross section σ is known. Alternatively, if σ and Φ are known, m can be determined; this is the field of activation analysis.

The influence of irradiation time on activation is illustrated in Fig. 8.8. For production of radionuclides, irradiation times of up to several half-lives are sufficient. About 99% of the saturation activity is obtained after irradiating for seven half-lives.

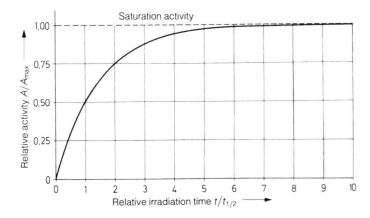

Figure 8.8. Activity as a function of irradiation time.

On the other hand, if the half-life of the radionuclide is very long compared with the irradiation time that can be applied for practical reasons ($t \ll t_{1/2}$), the activity increases linearly with the irradiation time. Advantage may be taken of this time dependence in the following ways: If activation of short-lived radionuclides is required, whereas activation of long-lived radionuclides is not desired, short irradiation times are chosen and the samples are measured or used immediately after the end of irradiation. In the opposite case, long irradiation times are chosen and the sample is allowed to decay until the short-lived activity has become negligible. With increasing decay time, the activity ratios of the radionuclides that have been produced in the sample change according to the half-lives. Optimal conditions with respect to the time of irradiation and the time of decay after the end of irradiation can be calculated by application of the formula

$$A = \sigma \Phi N_A (1 - e^{-\lambda t_i}) e^{-\lambda t_d} \tag{8.32}$$

where t_i is the time of irradiation and t_d the time of decay after the end of irradiation.

Above, it was assumed that the flux density Φ is constant within the target and during the irradiation time t. These conditions are not always fulfilled. Flux densities have gradients in all irradiation facilities. For appropriate application of the equation of activation the samples should therefore be small and they should be irradiated in positions in which the flux gradients are small.

If the absorption cross section σ_a of the projectiles in the sample is high, the flux density in the sample decreases appreciably with the thickness of the sample, as shown in Fig. 8.9. Furthermore, the energy distribution may change with the thickness of the sample. In order to avoid these effects the samples should be thin. If the samples cannot be made thin enough, a mean flux density Φ_m obtained by correction for absorption may be used instead of the flux density Φ_0 which is measured in absence of the sample. Φ_m may be calculated by the formula

$$\Phi_m = \frac{1}{x} \int_0^x \Phi \, dx = \frac{\Phi_0}{x} \int_0^x e^{-\Sigma_t x} dx = \frac{\Phi_0}{\Sigma_t x} (1 - e^{-\Sigma_t x}) \tag{8.33}$$

where Σ_t is the total macroscopic cross section of the sample, which is identical to the absorption coefficient μ, and x is the thickness of the sample.

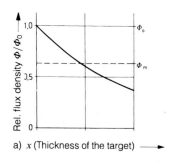
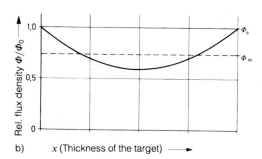

a) x (Thickness of the target) ⟶   b) x (Thickness of the target) ⟶

Figure 8.9. Relative flux density in a "thick" target – (a) beam from one direction; (b) radiation field.

Variation of the flux density with time is taken into account by introducing the integral flux $\int_0^t \Phi \, dt$ (unit cm^{-2}) and using the mean flux density

$$\Phi_t = \frac{1}{t} \int_0^t \Phi \, dt$$

instead of Φ in the equations of activation.

Yields of nuclear reactions can also be calculated, if binuclear reactions and mononuclear reactions (decay) occur simultaneously. A simple case is the decay of the nuclide produced by a binuclear reaction:

$$(A) \xrightarrow{\text{binuclear reaction}} (1) \xrightarrow{\text{decay}} (2) \tag{8.34}$$

The net production rate of nuclide 2 is

$$\frac{dN_2}{dt} = \sigma \Phi N_A (1 - e^{-\lambda_1 t}) - \lambda_2 N_2 \tag{8.35}$$

If N_A can be assumed to remain constant and $N_1 = N_2 = 0$ at $t = 0$, the following equation is obtained:

$$N_2 = \sigma \Phi N_A \left[\left(\frac{1 - e^{-\lambda_2 t_i}}{\lambda_2} + \frac{e^{-\lambda_1 t_i} - e^{-\lambda_2 t_i}}{\lambda_1 - \lambda_2} \right) e^{-\lambda_2 t_d} + \frac{(1 - e^{-\lambda_1 t_i})(e^{-\lambda_1 t_d} - e^{-\lambda_2 t_d})}{\lambda_2 - \lambda_1} \right] \tag{8.36}$$

where t_i is the time of irradiation and t_d the decay time after the end of irradiation.

If the nuclide 2 is produced by decay of nuclide 1 and additionally by binuclear reaction from another nuclide B,

$$\begin{array}{ccc} (A) & & (B) \\ \text{binuclear} & \Big\downarrow & \Big\downarrow \quad \text{binuclear} \\ \text{reaction} & & \quad \text{reaction 2} \\ & (1) \xrightarrow{\text{decay}} (2) & \end{array} \tag{8.37}$$

the following equation is obtained for the number of atoms of nuclide 2 at the end of irradiation:

$$N_2 = \sigma_2 \Phi_2 N_B \frac{1 - e^{-\lambda_2 t}}{\lambda_2} + \sigma_1 \Phi_1 N_A \left[\frac{1 - e^{-\lambda_2 t}}{\lambda_2} + \frac{e^{-\lambda_1 t} - e^{-\lambda_2 t}}{\lambda_1 - \lambda_2} \right] \tag{8.38}$$

provided that N_A and N_B are constant and $N_1 = N_2 = 0$ at $t = 0$.

If the assumption $N_A = $ constant is not fulfilled, because of noticeable transmutation of A by mononuclear or binuclear reactions, the following equation is applied:

$$-\frac{dN_A}{dt} = N_A + \sigma \Phi N_A = (\lambda + \sigma \Phi) N_A = \Lambda N_A \tag{8.39}$$

in which the transmutation of A by mononuclear and binuclear reactions is taken into account by the constant Λ. Integration of eq. (8.39) gives

$$N_A = N_A^0 e^{-\Lambda t} \tag{8.40}$$

which is analogous to the equation of radioactive decay (eq. (4.4)).

Considering a group of nuclides transmutating into each other,

$$(1) \xrightarrow{\text{decay}} (2) \xrightarrow{\text{decay}} (3)$$

$$\downarrow \qquad\qquad \downarrow$$

binuclear   binuclear
reaction 1   reaction 2

the following equation is obtained for the number of atoms of nuclide 2 after irradiation time t for the initial conditions $N_1 = N_1^0$ and $N_2 = 0$ at $t = 0$:

$$N_2 = \frac{\lambda_1}{\Lambda_2 - \Lambda_1} N_1^0 (e^{-\Lambda_1 t} - e^{-\Lambda_2 t}) \tag{8.42}$$

The situation becomes more complicated if the nuclides 1 and 2 in the reaction scheme (8.41) are not only produced by radioactive decay, but also by binuclear reactions:

$$(1) \underset{\text{decay}}{\overset{\text{decay}}{\rightleftharpoons}} (2) \underset{}{\overset{\text{decay}}{\rightleftharpoons}} (3)$$

binuclear   binuclear
reaction 12   reaction 2

$$\tag{8.43}$$

binuclear   binuclear
reaction 1   reaction 2

For the same initial conditions $N_1 = N_1^0$ and $N_2 = 0$ at $t = 0$, the number of atoms of nuclide 2 after irradiation time t is

$$N_2 = \frac{\Lambda_1^*}{\Lambda_2 - \Lambda_1} N_1^0 (e^{-\Lambda_1 t} - e^{-\Lambda_2 t}) \tag{8.44}$$

Λ_1 and Λ_2 include all reactions leading to a decrease of N_1 and N_2, respectively, whereas in Λ_1^* only reactions leading to the formation of nuclide 2 from nuclide 1 are considered:

$$\Lambda_1 = \lambda_1 + \sigma_1 \Phi_1 + \sigma_{12} \Phi_{12}$$
$$\tag{8.45}$$
$$\Lambda_2 = \lambda_2 + \sigma_2 \Phi_2 + \sigma_{23} \Phi_{23}$$

$$\Lambda_1^* = \lambda_1 + \sigma_{12} \Phi_{12} \tag{8.46}$$

Generally, for any nuclide i in a chain of mononuclear and binuclear transmutations the following relation is valid:

$$\frac{dN_i}{dt} = \Lambda_{i-1}^* N_{i-1} - \Lambda_i N_i \tag{8.47}$$

This equation is similar to eq. (4.47). The solution for the initial conditions $N_1 = N_1^0$, $N_2 = N_3 = \ldots N_n = 0$ at $t = 0$ is

$$N_n = C_1 e^{-\Lambda_1 t} + C_2 e^{-\Lambda_2 t} + \ldots C_n e^{-\Lambda_n t} \tag{8.48}$$

with the coefficients

$$C_1 = \frac{\Lambda_1^* \Lambda_2^* \ldots \Lambda_{n-1}^*}{(\Lambda_2 - \Lambda_1)(\Lambda_3 - \Lambda_1) \ldots (\Lambda_n - \Lambda_1)} N_1^0 \tag{8.49}$$

and so forth, in analogy to eqs. (4.49).

8.6 Investigation of Nuclear Reactions

Investigation of nuclear reactions comprises identification of the products and deter-mination of the cross sections σ, the energy of the products and their angular distri-bution. As far as possible, all values are to be measured as a function of the energy E of the projectiles with the aim to determine the excitation functions $\sigma = f(E)$, as plotted in Figs. 8.6 and 8.7. This can be done by measuring the yields of the products B or y (eq. (8.1)) or by measuring the change of the flux density due to the nuclear reaction in the target. If B is radioactive, the activity produced in the target may be determined by application of the equations in section 8.5 (activation method).

The activation method requires the use of high-purity targets, in order to exclude the influence of other nuclear reactions. Chemical procedures may be applied to separate the reaction products and to identify their atomic number Z. If short-lived radionuclides are to be measured, fast separation methods are required, for instance on-line separation in a gas stream that passes a temperature gradient (thermochro-matography). In the case of half-lives of the order of milliseconds or less, however, only physical methods are applicable, in particular separation by a sequence of elec-tric and magnetic fields. Stable or long-lived products may be determined by use of mass spectrometry, provided sufficient masses are available.

Flux densities are determined by means of monitor foils for which the cross sec-tions are well known. Sample and monitor have to be irradiated under exactly the same conditions. Sandwich arrangements or stacks of sample and monitor foils are preferred.

In all nuclear reactions the product nucleus suffers a recoil, which may be marked if nucleons or α particles are emitted (chapter 9). Due to this recoil, a certain amount of the product nuclei are thrown out from the target and may be collected in catcher foils. Stacks of samples and thin monitor and catcher foils are used to obtain as many data as possible.

Measuring the change of the flux density within the target provides only informa-tion about the total cross section of all interactions in the target. This method is mainly used for determination of neutron absorption. In the case of charged parti-cles it has to be taken into account that these are also losing energy by ionization processes.

8.7 Mechanisms of Nuclear Reactions

Most low-energy nuclear reactions proceed via formation of a compound nucleus (eq. (8.5)). In the compound nucleus model that was proposed in 1936 by Bohr it is assumed that the energy of the incident particle and its binding energy are distributed evenly or nearly evenly to all nucleons of the target nucleus. The excitation energy of the compound nucleus is

$$E^* = E_x + E_B - E_C = E_x \left(\frac{M_A}{M_A + M_x} \right) + E_B \qquad (8.50)$$

where E_x is the kinetic energy of the particle x, E_B its binding energy in the nucleus and E_C the kinetic energy of the compound nucleus C. The compound nucleus may accept various excited states, and if sufficient energy is transformed to one nucleon or to a group of nucleons these may pass the energy barrier and leave the nucleus. The mean lifetime τ of the excited states can be calculated by application of Heisenberg's uncertainty principle,

$$\tau \approx \frac{h}{2\pi\Gamma} \qquad (8.51)$$

where h is the Planck constant and Γ is the half-width of the resonance lines in the excitation functions. As Γ varies between about 0.01 and 10 eV, τ is of the order of 10^{-16} to 10^{-13} s.

The fate of a compound nucleus depends on its composition and its excitation energy, not on the way it is formed. It may be formed and decay in different ways:

$$
\begin{matrix}
A_1 + x_1 \searrow & & \nearrow B_1 + y_1 \\
A_2 + x_2 \text{---} & (C) \text{---} & B_2 + y_2 \\
A_3 + x_3 \nearrow & & \searrow B_3 + y_3 \\
\cdots & & \cdots
\end{matrix}
\qquad (8.52)
$$

In the resonance region, where the excitation functions exhibit sharp resonances, the cross sections of the individual reactions can be calculated from the line widths of the resonance lines by application of the Breit–Wigner formulas derived in 1936. Emission of neutrons from compound nuclei is preferred over emission of protons and relatively high cross sections are expected for (p, n) and (α, n) reactions if the energy of the incident particles is >1 MeV.

If the excited states of compound nuclei overlap, statistical methods are applied which also allow prediction of the emission of particles by a compound nucleus and of the cross section of a certain nuclear reaction.

With increasing energy of the projectiles direct interactions between the incident particles and the nucleons have to be taken into account. The energy is not transferred to the entire nucleus and the compound nucleus model is not applicable any more. The angular distribution of the products becomes asymmetric and, in general, more reaction products are observed in the forward direction of the projectile. The fraction of reaction products with higher energies increases, and single resonances are not observed in the excitation functions.

Direct interactions proceed faster than compound–nucleus reactions. The time of direct interactions is given by the time needed by the projectile to enter the nucleus and the time needed to leave the nucleus after collision. Reactions at the surface of the nucleus proceed within 10^{-23} s and reactions within the nucleus in about 10^{-21} s.

Two types of direct interactions are distinguished, knock-on reactions and transfer reactions. Knock-on reactions may proceed in various ways: the incident particle may transfer a part of its energy to a nucleon and continue on its way (inelastic scattering), it may induce collective motion of the nucleus (vibration or rotation) or it may be captured by the nucleus and transfer its energy to one or several nucleons which leave the nucleus. The number of nucleon–nucleon collisions increases with increasing energy of the incident particles.

In transfer reactions one or several nucleons are transferred from the projectile to the target nucleus, or vice versa. The first kind is also called a "stripping reaction" and the second a "pick-up reaction". The simplest forms of stripping reactions are the (d, p) and (d, n) reactions with deuterons, which proceed with relatively high cross sections:

$$^{A}Z + d \rightarrow {}^{A+1}Z + p \tag{8.53}$$

$$^{A}Z + d \rightarrow {}^{A+1}(Z+1) + n \tag{8.54}$$

In these reactions only one of the two nucleons of the deuteron is transferred to the target nucleus. The fact that (d, p) reactions are observed if the energy of the deuterons is lower than the Coulomb repulsion by the nucleus leads to the conclusion that the deuteron is split off under the influence of the positive charge of the nucleus, as indicated in Fig. 8.10.

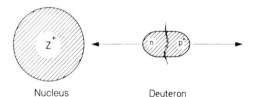

Nucleus   Deuteron

Figure 8.10. Stripping reaction with deuterons (Oppenheimer–Phillips reaction).

Furthermore, the binding energy of a nucleon in the target nucleus (about 8 MeV) is appreciably higher than the binding energy of the two nucleons in the deuteron (2.23 MeV), which favours splitting of the deuteron. The stripping reactions (8.53) and (8.54) are also referred to as Oppenheimer–Phillips reactions.

An example of a pick-up reaction is

$$^{A}Z + {}^{3}He \rightarrow {}^{A-1}Z + {}^{4}He \tag{8.55}$$

In transfer reactions one nucleon may be transferred, as in (d, p), (d, n), (d, t), (t, α), $(^{3}He, \alpha)$ reactions, two nucleons, as in (d, α) reactions, or even clusters of nucleons.

The probability of direct interactions increases with the energy of the projectiles and it is higher in peripheral collisions, because in this case single nucleons at the surface of the nucleus may be hit. Direct interactions are also more probable if heavy nuclides ($A > 200$) are involved.

Furthermore, emission of charged particles is strongly favoured in direct interactions, because single nucleons or groups of nucleons receive much higher energies than in compound nucleus reactions with the consequence that charged particles may surmount the energy barrier much more easily to leave the nucleus.

8.8 Low-energy Reactions

A great variety of nuclear reactions is available by application of different projectiles and variation of their energy. In this section nuclear reactions induced by projectiles with energies of up to about 50 MeV are considered. This energy range is preferred for the production of radionucldies. If the energies are not too high, the concept of formation of a compound nucleus (C) as an intermediate according to eq. (8.5) is applicable. In this concept it is assumed that the energy of the incident projectile is distributed equally to all nucleons. At higher projectile energies, however, the concept of formation of a compound nucleus is not valid any more, and direct interactions between the projectile and the individual nucleons prevail.

Fig. 8.11 gives a survey of the location in the chart of the nuclides of the products obtained by various low-energy nuclear reactions. By reactions with neutrons an isotopic compound nucleus is formed which may emit particles or photons, depending on its structure and excitation. In (n, γ) reactions the excitation energy of the compound nucleus is given off in the form of γ-ray photons. In (n, p) reactions the compound nucleus emits a proton and the product is an isobar of the target nuclide.

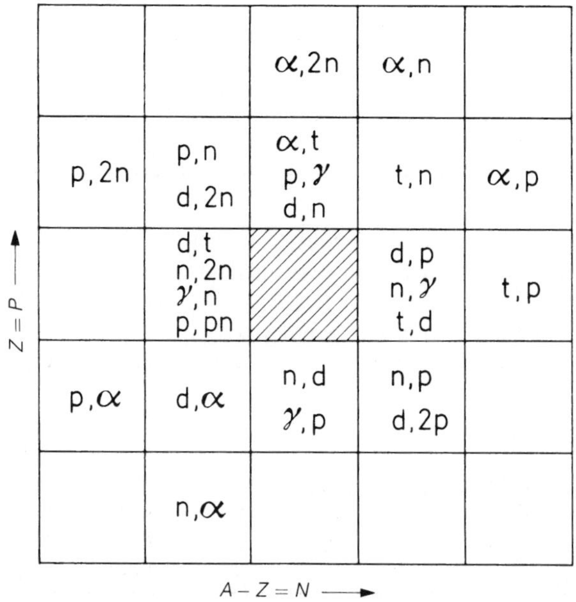

Figure 8.11. Survey of transmutations of nuclides by nuclear reactions.

Other reactions with charged particles may be discussed in a similar way. In reactions with γ-ray photons the compound nuclei are excited states of the target nuclei and may give off their excitation energy by emission of particles. It is evident from Fig. 8.11 that the same nuclides can be produced by application of various nuclear reactions, and by the method of "cross bombardment" radionuclides can be identified.

Nuclear reactions may lead to stable or unstable (radioactive) products. In general, (n, γ), (n, p), and (d, p) reactions give radionuclides on the right-hand side of the line of β stability that exhibit β^- decay, whereas (p, n), $(d, 2n)$, $(n, 2n)$, (γ, n), (d, n) and (p, γ) reactions lead to radionuclides on the left-hand side of the line of β stability that exhibit β^+ decay or electron capture (ε). (n, γ), (d, p), $(n, 2n)$ and (γ, n) reactions give isotopic nuclides, and these cannot be separated from the target nuclides by chemical methods, except for the application of the chemical effects of nuclear transformations which will be discussed in chapter 9.

Rules for nuclear reactions between neutrons and charged particles with energies up to 50 MeV and nuclides of medium and high mass $(25 < A < 80$ and $80 < A < 250$, respectively) are summarized in Table 8.1. In the case of endoergic reactions, the projectiles must introduce at least the threshold energy (section 8.2), and charged particles must have a minimum energy to surmount the Coulomb repulsion of the nuclei, which increases with the atomic number Z (section 8.3). If higher excitation energies are transmitted by the projectiles, emission of more than one nucleon is observed. The same holds for photonuclear reactions which are not listed in Table 8.1: with increasing energy of γ-ray photons, besides (γ, n) reactions $(\gamma, 2n)$, $(\gamma, 3n)$ and other reactions take place.

By reactions with light nuclei $(A < 25)$ a variety of products may be obtained. The first nuclear reaction observed in 1919 by Rutherford and the discovery of the neu-

Table 8.1. Predominant nuclear reactions for various energy ranges.

Energy of incident radiation	Nuclides of medium mass $(25 < A < 80)$				Heavy nuclides $(80 < A < 250)$			
	n	p	d	α	n	p	d	α
0–1 keV	n, γ	–	–	–	n, γ	–	–	–
1–500 keV	n, γ	p, n p, γ p, α	d, p d, n	α, n α, γ α, p	n, γ	–	–	–
0.5–10 MeV	n, α n, p	p, n p, α	d, p d, n d, pn d, 2n	α, n α, p	n, p n, γ	p, n p, γ	d, p d, n d, pn d, 2n	α, n α, p α, γ
10–50 MeV	n, 2n n, p n, np n, 2p n, α	p, 2n p, n p, np p, 2p p, α	d, p d, 2n d, pn d, 3n d, t	α, 2n α, n α, p α, np α, 2p	n, 2n n, p n, pn n, 2p n, α	p, 2n p, n p, np p, 2p p, α	d, p d, 2n d, np d, 3n d, t	α, 2n α, n α, p α, np α, 2p

tron in 1932 by Chadwick have already been mentioned (eqs. (8.2) and (8.13), respectively).

$^{7}_{3}\text{Li}$ may react with protons in various ways:

$$\begin{array}{l} ^{7}_{3}\text{Li} + ^{1}_{1}\text{H} \searrow \qquad\qquad \nearrow 2\,^{4}_{2}\text{He}\,(2\,\alpha) \\ ^{7}_{3}\text{Li} + ^{1}_{1}\text{H} \longrightarrow (^{8}\text{Be}) \longrightarrow ^{8}_{4}\text{Be} + \gamma \to 2\alpha + \gamma \\ ^{7}_{3}\text{Li} + ^{1}_{1}\text{H} \nearrow \qquad\qquad \searrow ^{6}_{3}\text{Li} + ^{2}\text{H} \end{array} \qquad (8.56)$$

In the first reaction the compound nucleus decays immediately into two α particles and the energy $\Delta E = 17.35\,\text{MeV}$ which is set free in the reaction is distributed equally to both α particles. In the second reaction excitation energy of the compound nucleus is given off in the form of a γ-ray photon and the unstable ^{8}Be decays with the very short half-life of $2 \cdot 10^{-16}\,\text{s}$ into two α particles.

The compound nucleus formed by the reaction of protons with ^{11}B may also decay in different ways:

$$^{11}_{5}\text{B} + ^{1}_{1}\text{H} \longrightarrow (^{12}_{6}\text{C}) \begin{array}{l} \nearrow ^{8}_{4}\text{Be} + ^{4}_{2}\text{He} \to 3\,\alpha \\ \searrow ^{11}_{6}\text{C} + ^{1}_{0}\text{n} \end{array} \qquad (8.57)$$

Reactions with deuterons, in particular (d, p) and (d, n) reactions, have relatively high cross sections, as already mentioned in section 8.7. Examples are

$$^{12}_{6}\text{C} + ^{2}_{1}\text{H} \to ^{13}_{6}\text{C} + ^{1}_{1}\text{H} \qquad (8.58)$$

$$^{12}_{6}\text{C} + ^{2}_{1}\text{H} \to ^{13}_{7}\text{N} + ^{1}_{0}\text{n} \qquad (8.59)$$

Tritium and ^{3}He are obtained by the reactions d(d, p)t and d(d, n)^{3}He, respectively, if solid D_2O is irradiated with deuterons. However, tritium is produced with much higher yield by irradiation of ^{6}Li with neutrons ($\sigma_{n,\gamma} = 940\,\text{b}$):

$$^{6}_{3}\text{Li} + ^{1}_{0}\text{n} \to (^{7}_{3}\text{Li}) \to ^{3}_{1}\text{H} + ^{4}_{2}\text{He} \qquad (8.60)$$

This reaction is exoergic, in contrast to most other (n, α) reactions. The same holds for the (n, p) reaction with ^{14}N,

$$^{14}_{7}\text{N} + ^{1}_{0}\text{n} \to (^{15}_{7}\text{N}) \to ^{14}_{6}\text{C} + ^{1}_{1}\text{H} \qquad (8.61)$$

by which ^{14}C is produced in the atmosphere and artificially in nuclear reactors. Most other (n, p) reactions are endoergic. All (n, 2n) reactions are endoergic and occur only with neutrons of higher energy ($>10\,\text{MeV}$).

For the production of radionuclides, (n, γ) reactions are of greatest practical importance, because in many cases the cross sections are relatively high and high fluxes of neutrons are available in nuclear reactors.

If a certain nuclide is required, it can often be produced by different nuclear reactions, for example

$$^{23}\text{Na}(n, \gamma)^{24}\text{Na}$$

$$^{24}\text{Mg}(n, p)^{24}\text{Na}$$

$$^{27}\text{Al}(n, \alpha)^{24}\text{Na}$$

$$^{23}\text{Na}(d, p)^{24}\text{Na}$$

$$^{26}\text{Mg}(d, \alpha)^{24}\text{Na} \qquad (8.62)$$

$$^{27}\text{Al}(d, p\alpha)^{24}\text{Na}$$

$$^{25}\text{Mg}(\gamma, p)^{24}\text{Na}$$

$$^{27}\text{Al}(\gamma, 2pn)^{24}\text{Na}$$

$$^{27}\text{Al}(p, 3pn)^{24}\text{Na}$$

On the other hand, different products may be obtained from the same target and the same projectile, for example

$$^{27}_{13}\text{Al} + {}^{2}_{1}\text{H} \longrightarrow \left({}^{29}_{14}\text{Si} \right) \begin{cases} {}^{25}_{12}\text{Mg} + {}^{4}_{2}\text{He} \\ {}^{28}_{13}\text{Al} + {}^{1}_{1}\text{H} \\ {}^{28}_{14}\text{Si} + {}^{1}_{0}\text{n} \\ {}^{24}_{11}\text{Na} + {}^{1}_{1}\text{H} + {}^{4}_{2}\text{He} \end{cases} \qquad (8.63)$$

The prevailing reaction channel depends on the energy of the projectiles, because the excitation energy transferred to the compound nucleus determines the way it is transmuted. With increasing excitation energy, the number of reaction channels increases markedly.

8.9 Nuclear Fission

In section 5.6 it has been shown that nuclides with mass numbers $A > 100$ are energetically unstable with respect to fission. The fact that fission is not observed is due to the fission barrier. However, if enough excitation energy is transferred to heavy nuclides, the fission barrier can be surmounted and the nuclides undergo fission.

Fission of uranium was discovered by Hahn and Strassmann in their attempts to produce transuranium elements by irradiation of uranium with neutrons followed by β^- transmutation of the products. Instead of the expected transuranium elements they found radioactive products with appreciably lower atomic mass such as ^{140}Ba, indicating the fission of the uranium nuclei.

Low-energy fission induced by projectiles with energies up to about 10 MeV and high-energy fission by particles with energies of the order of 100 MeV and above show different features. Low-energy fission by neutrons may be described by the following general equation:

$$A + n \rightarrow B + D + \nu n + \Delta E \qquad (8.64)$$

or in a short form $A(n, f)B$, where f stands for fission. The fission products B and D have mass numbers in the range between about 70 and 160, and the numbers of neutrons emitted is $\nu \approx 2$–3. The energy ΔE set free by fission is relatively high ($\Delta E \approx 200\,\text{MeV}$), because the binding energy per nucleon is higher for the fission products than for the heavy nuclei (Fig. 2.6). In the case of even–odd heavy nuclei,

Table 8.2. Cross sections $\sigma_{n,f}$ of nuclear fission by thermal neutrons (energy 0.025 eV) and mean number $\bar{\nu}$ of neutrons set free by fission.

Nuclide	$\sigma_{n,f}$ [barn]	$\bar{\nu}$	Nuclide	$\sigma_{n,f}$ [barn]	$\bar{\nu}$
Th-227	≈ 200		Am-241	3.1	
228	<0.3		242	2100	3.22 ± 0.04
229	30		242 m$_1$	7000	
230	≤ 0.0005	2.08 ± 0.02	243	0.074	3.26 ± 0.02
232	0.000003		244	2200	
233	15		244 m	1600	
234	<0.01		Cm-242	≈ 5	2.65 ± 0.09
Pa-230	1500		243	620	
231	<0.020		244	1.1	3.43 ± 0.05
232	≈ 700		245	2100	
233	<0.1		246	0.16	3.83 ± 0.03
234	<5000		247	82	
234 m	<500		248	0.36	
U-230	≈ 25		249	≈ 1.6	
231	250		Bk-250	1000	
232	74		Cf-249	1700	
233	530	3.13 ± 0.06	250	<350	
235	586	2.432 ± 0.066	251	4500	
238	0.000003		252	32	3.86 ± 0.07
239	15		253	1300	
Np-234	≈ 900		Es-254	2000	
236	2600		254 m	1800	
237	0.020		Fm-255	3300	
238	2100		257	2950	
239	<1				
Pu-236	160	2.30 ± 0.19			
237	2300				
238	17	2.33 ± 0.08			
239	752	2.874 ± 0.138			
240	≈ 0.044	2.884 ± 0.007			
241	1010	2.969 ± 0.023			
242	<0.2	2.91 ± 0.02			
243	200				

such as ^{233}U, ^{235}U, ^{239}Pu and ^{241}Pu, the binding energy of an additional neutron is particularly high, and the fission barrier is easily surmounted. Therefore, these nuclides have high fission cross sections $\sigma_{n,f}$ for fission by thermal neutrons. Some data for nuclear fission by thermal neutrons are complied in Table 8.2. The fission cross sections $\sigma_{n,f}$ depend strongly on the energy of the neutrons.

Fission of heavy nuclei always results in a high neutron excess of the fission products, because the neutron-to-proton ratio in heavy nuclides is much larger than in stable nuclides of about half the atomic number, as already explained for spontaneous fission (Fig. 5.15). The primary fission products formed in about 10^{-11} s by fission and emission of prompt neutrons and γ rays decay by a series of successive β^- transmutations into isobars of increasing atomic number Z. The final products of these decay chains are stable nuclides.

Fission by thermal neutrons is induced by the binding energy of the neutron (≈ 6 MeV) which leads to excitation of the nucleus, oscillation and finally splitting into two parts, if a critical deformation is exceeded. The stages and the time scale of fission by thermal neutrons are illustrated in Fig. 8.12. Critical deformation is

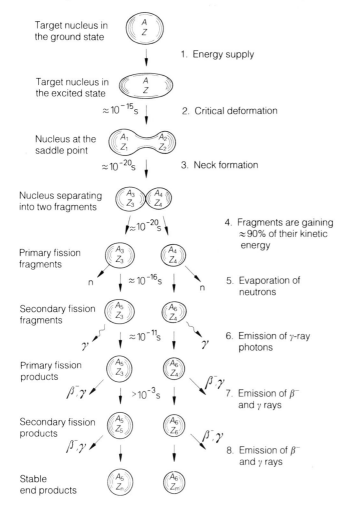

Figure 8.12. The phases of low-energy nuclear fission.

attained about 10^{-15} s after neutron capture. Then fission occurs very fast, followed by emission of neutrons and γ rays by the highly excited fission products and finally by a sequence of β^- transformations. Due to their high positive charges, the fission fragments repel each other very strongly and attain high kinetic energies. Immediately after fission the primary and secondary fission fragments emit prompt neutrons and γ-ray photons, respectively, and the resulting primary and secondary fission products emit β^- radiation, γ radiation and β^--delayed neutrons. These β^--delayed neutrons are emitted about 0.1 to 100 s after fission. Their percentage in relation to the total number of neutrons set free by fission is 0.26% for ^{233}U, 0.65% for ^{235}U and 0.21% for ^{239}Pu.

In Fig. 8.13 the yield of fission products obtained by thermal fission of ^{235}U is plotted as a function of the mass number A (mass distribution). The maxima of the yields are in the ranges of mass numbers 90–100 and 133–143. In these ranges the fission yields are about 6%, whereas symmetrical fission occurs with a yield of only about 0.01%. The peaks in the mass distribution curve at $A = 100$ and at $A = 134$ are explained by the fact that formation of even–even nuclei is preferred in the fission of the even–even compound nucleus ^{236}U. It should be taken into account that the sum of the fission yields is 200%, because each fission gives two fission products.

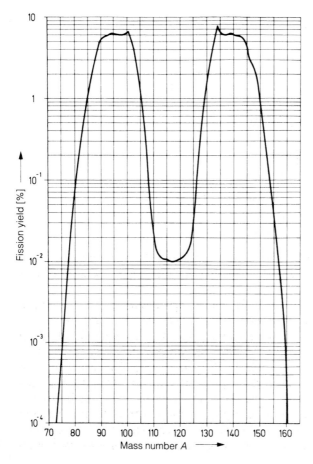

Figure 8.13. Fission yields for the fission of ^{235}U by thermal neutrons (according to AECI-1054).

The mass distribution obtained by fission of ^{233}U and ^{239}Pu with thermal neutrons (Fig. 8.14) is similar to that observed for ^{235}U. Whereas the maximum for heavy fission products is nearly at the same place in the case of ^{233}U and ^{239}Pu, the maximum for light fission products is shifted to the right in the case of ^{239}Pu. This tendency continues with increasing mass of the fissioning nuclei, and in thermal-neutron fission of ^{258}Fm the two maxima merge into one another.

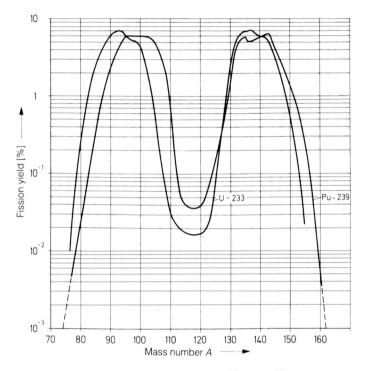

Figure 8.14. Fission yields for the fission of ^{233}U and ^{239}Pu by thermal neutrons (according to S. Katcoff, Nucleonies [New York] **16/4**, 78 (1958)).

The influence of the energy of the neutrons on the mass distribution of the fission products is shown in Fig. 8.15: at higher neutron energies, the probability of symmetric fission increases strongly.

Increase of symmetric fission is also observed at lower atomic numbers Z. It prevails at $Z \leq 85$, and at $Z = 89$ (^{227}Ac) symmetric and asymmetric fission have nearly the same probability, which results in three maxima in the mass distribution. Three maxima are also observed in the fission of ^{226}Ra by 11 MeV protons or by γ rays.

The mass distribution curves in Figs. 8.13 to 8.15 give the total yields of the decay chains of mass numbers A. The independent yields of members of the decay chains, i.e. the yields due to direct formation by the fission process, are more difficult to determine, because the nuclides must be rapidly separated from their precursors. Only a few so-called shielded nuclides (shielded from production via β^- decay by a stable isobar one unit lower in Z) are unambiguously formed directly as primary

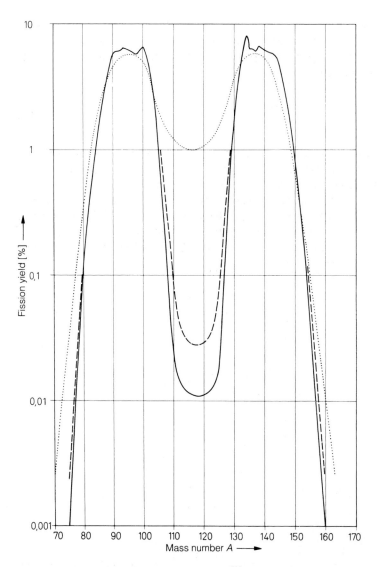

Figure 8.15. Fission yields for the fission of ^{235}U by neutrons of various energies: ——, thermal neutrons; - - - -, neutrons produced by fission; ······, 14 MeV neutrons (according to K. F. Flynn, L. E. Glendenin, Rep. ANL-7749 Argonne Nat. Lab., Argonne, Ill. 1970).

products. In spite of these problems, independent yields have been obtained for many decay chains. They show a narrow Gaussian distribution of the isobaric yields around a most probable charge Z_p. The probability of a primary fission product with atomic number Z is given by

$$p(Z) = \frac{1}{\sqrt{\pi C}} \exp\left(-\frac{(Z - Z_p)^2}{C}\right) \tag{8.65}$$

$P(Z)$ is the relative independent yield and C is a constant with a mean value of 0.80 ± 0.14. This charge distribution is plotted in Fig. 8.16 for the fission of ^{235}U by thermal neutrons and holds for all mass numbers. For even numbers of Z the yields are systematically higher than those for odd numbers of Z. Z_p, the most probable value of Z, is about 3 to 4 units lower than the atomic number of the most stable nuclide in the sequence of isobars. Nuclides with Z_p are obtained with about 50% of the total isobaric yield, nuclides with $Z = Z_p \pm 1$ with about 25% each and nuclides with $Z = Z_p \pm 2$ with about 2% each.

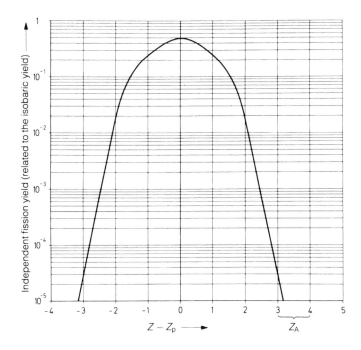

Figure 8.16. Independent fission yields for the fission of ^{235}U by thermal neutrons (charge distribution).

Fission by thermal neutrons proceeds also via a double-humped barrier as in spontaneous fission (Fig. 5.19). The excitation energy acquired by the uptake of an additional neutron enables easily fissionable nuclei like ^{233}U, ^{235}U and ^{239}Pu to surmount the two barriers A and B immediately. At lower binding energies of additional neutrons, as in the case of ^{238}U, the excitation energy acquired by the uptake of thermal neutrons is too small to pass the fission barriers easily.

Going from Th to Cf, the height of the barrier A remains nearly constant with increasing atomic number Z, whereas the barrier B decreases from about 6 MeV to about 4 MeV. Therefore, isomeric states of nuclei with $Z > 92$, being highly deformed and in the valley between barriers A and B, can easily surmount the barrier B.

The total energy ΔE set free by fission appears as kinetic energy and excitation energy of the primary fission fragments:

$$\Delta E = E_{kin} + E_{exc} \tag{8.66}$$

and E_{kin} is the major part of ΔE. In the case of low-energy fission E_{kin} is given by the empirical relation

$$E_{kin} \sim \frac{Z^2}{A^{1/3}} \tag{8.67}$$

where Z and A are the atomic number and the mass number of the fissioning nucleus. This relation is in agreement with the assumption that the energy results from the Coulomb repulsion of two fission fragments with atomic numbers $Z/2$ and charges $Ze/2$ at a distance given by the sum of their radii $(r_1 + r_2)$:

$$E_{kin} = \frac{Z^2 e^2}{4(r_1 + r_2)} = \frac{Z^2 e^2}{8(1/2)^{1/3} r_0 A^{1/3}} \tag{8.68}$$

The excitation energy of the primary fission fragments is given off by emission of prompt neutrons with energies varying between about 0 and 10 MeV (mean value ≈ 2 MeV) and of prompt γ-ray photons. The number of prompt neutrons emitted by the primary fission fragments depends mainly on their excitation energy. It increases with the mass number of the fissioning nuclei (Table 8.2). In Fig. 8.17 this number is plotted as a function of the mass of the fission fragments. It is relatively low for fragments with filled neutron shells ($N = 50$, $N = 82$).

On average, 7.5 γ-ray photons with a mean energy of about 1 MeV are emitted per fission, corresponding to a total energy of prompt γ rays of about 7.5 MeV per fission. Low-energy transitions of the excited fission fragments occur by emission of conversion electrons and X rays (about one X-ray photon per fission).

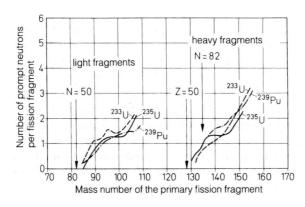

Figure 8.17. Neutron yields as a function of the mass of the primary fission fragments (according to J. Terrell, Proc. IAEA Symp. Phys. Chem. Fission, Salzburg 1965, IAEA, Vienna, Vols. 2, 3).

High-energy α particles are also observed in low-energy fission, but relatively seldom (one α particle in about 300 to 500 fissions). They are emitted at an early stage of the fission process, when the fission fragments are still very near each other. This leads to the concept of ternary (in contrast to binary) fission, i.e. formation of three (instead of two) fragments. Besides α particles, p, d, t, ^{3}He, ^{7}Li, ^{8}Li, ^{9}Li, ^{9}Be, ^{10}Be and isotopes of B, C, N and O may also be emitted in the course of fission by thermal neutrons, but very rarely (in about $10^{-6}-10^{-5}$ of all fission events). However, the probability of ternary fission increases strongly with the excitation energy of the fissioning nuclei. For example, high-energy fission of ^{232}Th with 400 MeV argon ions leads to a ratio of ternary to binary fission of about 1 : 30.

Ternary fission into three fragments of similar mass may proceed in two different ways (Fig. 8.18): (a) is a sequence of two binary fissions, also called cascade fission, whereas (b) illustrates a true ternary fission. Theoretical considerations lead to the conclusion that, in general, cascade fission is more probable, with the exception of ternary fission at low Z^2/A values and at low excitation energies.

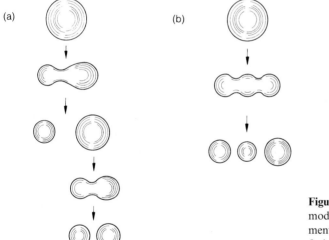

(a)   (b)

Figure 8.18. The two different modes of fission into three fragments (schematic) (a) cascade fission; (b) true ternary fission.

High-energy fission induced by neutrons or other particles of high energy leads to an appreciable change in the spectrum of fission products, as shown in Fig. 8.19. The probability of symmetrical fission increases considerably with increasing excitation energy transferred to the nuclei by high-energy particles: for example, by bombarding heavy nuclei with 100 to 600 MeV protons, the minimum in the fission yield flattens out and the two maxima are replaced by one flat maximum somewhat below half the mass number of the target nuclide (Fig. 8.19). In contrast to low-energy fission, neutron-deficient fission products are also observed. This is explained by the fact that the high excitation energy is preferably given off by emission of neutrons. If protons of still higher energy (about 2–30 GeV) are applied, the yield in the medium mass range decreases, whereas it increases in light and heavy mass ranges.

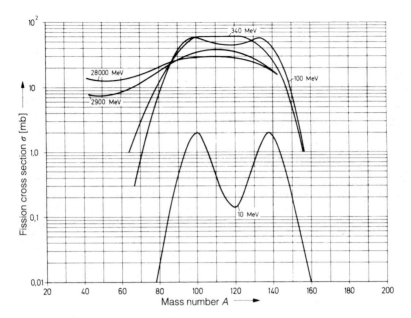

Figure 8.19. Fission cross sections for the fission of ^{238}U by protons of various energies. (The Coulomb barrier for the fission by protons is 12.3 MeV, which explains the low values for 10 MeV protons.) (According to P. C. Stevenson, Physic. Rev. **111** (1958) 886; G. Friedlander: BNL 8858 (1965)).

8.10 High-energy Reactions

From Table 8.1 it is evident that at higher projectile energies more than one particle is emitted. Examples are (x, 2n), (x, np) and (x, 2p) reactions (x = n, p, d, α) at energies >10 MeV. The high excitation energy of the nuclei is given off by emission (evaporation) of nucleons, preferably neutrons. The distinction between high-energy and low-energy reactions is arbitrary. In general, reactions at projectile energies >100 MeV are understood by the term "high-energy reactions".

Compared with low-energy reactions, the number of possible reactions (reaction channels) and of products is higher in high-energy reactions. By spallation and fragmentation, products with appreciably lower atomic numbers are obtained. High-energy reactions are induced by cosmic radiation and play an important role in cosmochemistry, astrophysics and investigation of meteorites. Reactions with heavy nuclei as projectiles are, in general, high-energy reactions, because high energies are needed to surmount the Coulomb repulsion; they are discussed in more detail in section 8.11.

For the investigation of high-energy nuclear reactions, photographic emulsions or other solid track detectors may be applied. The emulsions are selected with respect to their sensitivity to ionizing radiation of high or low LET values (e.g. highly charged ions such as recoiling ions or protons). By means of this technique, charges, masses, kinetic energies and angles with respect to the incident radiation can be determined at relatively low integral fluxes. Time and experience are needed for evaluation of the results.

Solid track detectors of mica, glass or plastic are used to study the tracks of highly charged ions, either directly by means of an electron microscope or after etching under a normal microscope. The main advantage of the use of solid track detectors is the fact that only particles carrying a certain charge and thus producing a minimum of specific ionization give rise to a visible track. For instance, the tracks of low-energy α particles with energies <3 MeV can be detected in plastic foils, whereas the tracks of particles with $Z < 14$ cannot be detected in mica. Therefore, mica is well suited for investigation of high-energy fission. The target material (e.g. U, Bi, Au, Ag) can be put on mica in the form of thin films (5–100 µm) by vapour deposition, and by a sandwich arrangement the tracks of all fission fragments can be investigated.

Mass spectrometry is another valuable method for the investigation of high-energy nuclear reactions, in particular if an on-line arrangement is used in such a way that the reaction products are immediately transported from the target into the mass spectrometer or mass separator. The transport may be combined with a chemical separation in the gas phase.

Some radionuclides produced in high-energy reactions exhibit characteristic properties and can be detected without chemical separation. Examples are ^{9}Li, ^{16}C, ^{17}N, which emit delayed neutrons, or ^{11}Be, ^{12}B, ^{15}C, ^{12}N, ^{16}N, ^{24}Al, ^{28}P, which emit high-energy β^- or β^+ radiation, respectively, and can be measured in the presence of other radionuclides.

Examples of monitor reactions that are applied for determination of cross sections with protons are ^{12}(p, pn)^{11}C and ^{27}Al(p, 3pn)^{24}Na. Cross sections for these reactions in the energy range between 50 MeV and 30 GeV are listed in Table 8.3. ^{12}C and ^{27}Al monitors are used in the form of plastic or Al foils, respectively.

Table 8.3. Cross sections of the monitor reactions ^{12}C(p, pn)^{11}C and ^{27}Al(p, 3pn)^{24}Na for various proton energies.

Proton energy [MeV]	^{12}C(p, pn)^{11}C σ [mb]	^{27}Al(p, 3pn)^{24}Na σ [mb]
50	86.4	6.2
60	81.1	8.7
80	70.5	10.0
100	61.3	10.2
150	45.0	9.4
200	39.0	9.3
300	35.8	10.1
400	33.6	10.5
600	30.8	10.8
1000	28.5	10.5
2000	27.2	9.5
3000	27.1	9.1
6000	27.0	8.7
10 000	26.9	8.6
28 000	26.8	8.6

The relative mass distributions obtained by bombarding nuclides of medium mass with protons of various energies are plotted in Fig. 8.20. At proton energies up to about 50 MeV the nuclear reactions lead to the emission of one or several particles and the mass number is not changed appreciably. At energies >100 MeV many products of the nuclear reactions have much smaller mass numbers, because many nucleons and particles containing several nucleons are split off from the target nucleus. This process is called spallation. At still higher energies (>1 GeV), the number of spallation products increases, and high-energy nuclear reactions can be described by the equation

$$A + x \rightarrow B + D + F + \ldots + \nu_1 n + \nu_2 p + \ldots + \Delta E \qquad (8.69)$$

The short notation of spallation reactions is (x, s).

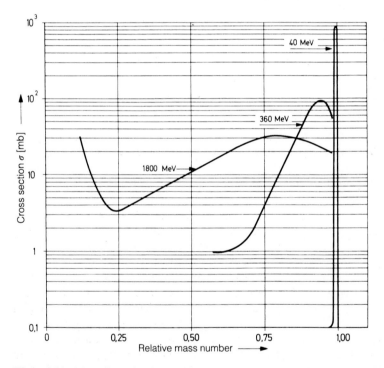

Figure 8.20. Mass dispersion for nuclear reactions of protons with nuclides of medium mass. Example: reaction of protons of various energies with copper (according to J. M. Miller, J. Hudis, Annu. Rev. Sci. **9**, 159 (1959)).

High-energy reactions with heavy nuclides ($Z > 70$) lead to different relative mass distributions, because nuclear fission occurs in addition to other reactions. This is shown in Fig. 8.21 for reactions of high-energy protons with bismuth. As for target nuclei of medium mass, at proton energies of up to about 50 MeV the masses of the products do not differ markedly from the mass of the target nuclide. Only easily fissionable nuclides such as those listed in Table 8.2 undergo fission. But at proton

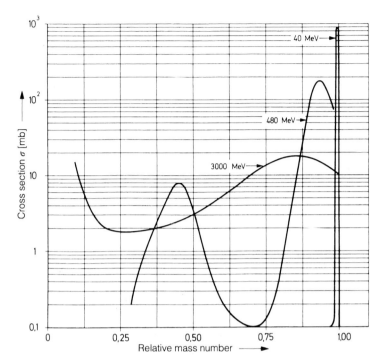

Figure 8.21. Mass dispersion for nuclear reactions of protons with heavy nuclides. Examples: reaction of protons of various energies with bismuth (according to J. M. Miller, J. Hudis, Annu. Rev. Nucl. Sci. **9** 159 (1959)).

energies of ≈ 400 MeV, nuclides with atomic numbers $Z < 90$ undergo fission also, and besides the spallation products with relative mass numbers $A/A_0 > 0.75$, fission products with relative mass numbers $A/A_0 \approx 0.5$ are obtained. Finally, at proton energies >1 GeV, the difference between spallation and fission vanishes.

The broad spectrum of products observed in high-energy reactions (Figs. 8.20 and 8.21) cannot be explained by the compound nucleus mechanism and the simple model of direct interactions is not applicable any more either. At projectile energies of ≥ 3 GeV all mass yields are nearly the same within one order of magnitude. The angular distribution of the products leads to the conclusion that spallation is a two-stage process: in the first stage nucleons, α particles or other nuclear fragments are emitted within about 10^{-22}–10^{-21} s by cascades of collisions, as illustrated in Fig. 8.22, and in a second stage the remaining highly excited nucleus gives off its excitation energy by evaporation processes and/or by fission. The formation of neutron-deficient products, such as ^{18}F, ^{22}Na, at projectile energies >1 GeV is explained by fast fragmentation processes.

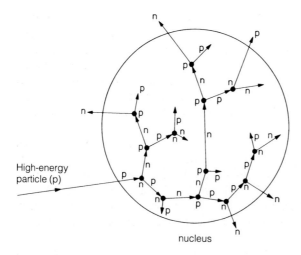

High-energy
particle (p)

nucleus

Figure 8.22. Reaction cascades in nuclei set off by high-energy particles (schematically; p = proton, n = neutron).

8.11 Heavy-ion Reactions

The use of heavy ions as projectiles has opened up new fields of nuclear reactions, as already mentioned in section 8.3. The general formula of a heavy-ion reaction is

$$A + B \rightarrow D + E + \ldots + \nu_1 n + \nu_2 p + \ldots + \Delta E \tag{8.70}$$

where B is the heavy ion used as the projectile. B must have a minimum energy to surmount the Coulomb repulsion of the nucleus A which brings heavy-ion reactions into the range of high-energy reactions (section 8.3). Because heavy ions consist of a bundle of nucleons and have, in general, an angular momentum, heavy-ion reactions are more complicated and transfer of angular momentum plays an important role. Heavy ions with energies of 1 MeV/u up to several hundred GeV/u are used as projectiles (u is the atomic mass unit).

Different types of interaction are distinguished, as illustrated in Fig. 8.23. (The spherical form is a simplification which is only applicable for nuclei with nuclear spin $I = 0$.) On path 1 the nuclei are not touching each other; elastic scattering and Coulomb excitation are expected. On path 2 the nuclei are coming into contact with each other and nuclear forces become effective; inelastic scattering and transfer reactions

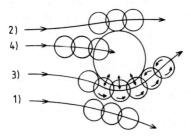

2)

4)

3)

1)

Figure 8.23. Heavy ion reactions: different types of interaction (schematically) path 1): elastic scattering; path 2): quasielastic collision; path 3): deeply inelastic collision; path 4): frontal collision.

will take place. These processes are also called "quasielastic". On path 3 strong interaction between the two nuclei and transfer of large amounts of the kinetic energy of the incident heavy ion as well as transfer of many nucleons (multinucleon transfer, diffusion of nucleons) are expected. This process is called "deeply inelastic collision". On path 4 more or less frontal collision and "fusion" of the two nuclei to a highly excited compound nucleus will occur. Depending on the excitation, nucleons and γ-ray photons will be emitted, and fusion may be followed by fission or fragmentation.

The experimental results are in agreement with this concept. At energies below the Coulomb barrier elastic scattering, inelastic scattering and transfer of nucleons are observed and excited states of nuclei are produced. Fusion to a compound nucleus may occur by tunnelling. With increasing energies of the order of several MeV/u, deeply inelastic processes and fusion become prevalent. The excitation energy increases with the time of contact and it is spread over the whole system in a kind of equilibration within about 10^{-21} s.

At energies of 30–50 MeV/u interactions between the individual nucleons begin to play a role and fragmentation is observed. In the so-called relativistic range above about 200 MeV/u, where the velocity of the projectiles approaches the velocity of light, collisions lead to a marked compression of nuclear matter to states of high density and disintegration to nucleons, light fragments and a variety of new particles.

Depending on the geometrical conditions of the collision, smaller or greater amounts of angular spin are transferred, varying between $l = 0$ for central collision and l_{max} for peripheral collisions; l_{max} increases with the mass of the nuclei. In the range of lower energies, the angular momentum transmitted in heavy-ion reactions has a dominating influence. In peripheral collisions (path 3 in Fig. 8.23) fusion is prevented by the high angular momentum and the rotation of the nuclei.

The methods used in the investigation of heavy-ion reactions are similar to those described in section 8.6. The high linear energy transfer (LET) and the relatively short range of heavy ions have to be taken into account. On-line separation of short-lived products is of special importance.

As an example, the mass distribution of the products obtained by the bombardment of ^{238}U with ^{40}Ar is plotted in Fig. 8.24. The curve is explained by superposition of the processes described above: only few nucleons are transferred by quasielastic reactions (a), and many nucleons by deeply inelastic processes (b). Fusion followed by fission of highly excited products leads to a broad distribution of fission products around $1/2(A_1 + A_2)$, where A_1 and A_2 are the mass numbers of ^{238}U and ^{40}Ar, respectively (c), and asymmetric fission of heavy products of low excitation energy gives two small maxima (d).

Within the very short time in which the distance between incident heavy ions and target atoms decreases and increases again without collision of the nuclei (path 1 in Fig. 8.23), systems are formed consisting of the two nuclei and their electron shells. These systems are described as "quasimolecules" if the nuclei have only a few electrons in common, and as "quasiatoms" if they share all the electrons in the intermediate stage of short distances. The quasimolecules may exchange electrons between the electron shells of the nuclei, and electrons as well as X-ray photons may be emitted. From the energy of these electrons and X rays the energy levels of the quasimolecules can be determined, provided that the time of approach is long enough to allow population of the new electron orbitals.

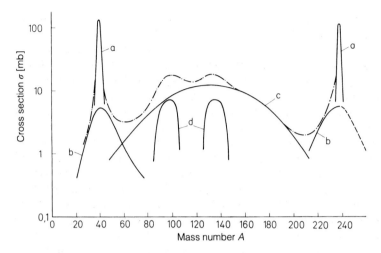

Figure 8.24. Cross sections (mass yields) for the heavy ion reaction ^{40}Ar + ^{238}U. –·–·– experimental values (thick target, chemical separation); a: quasielastic processes; b: multinucleon transfer; c: fusion followed by fission; d: fission of heavy nuclei produced from ^{238}U by transfer reactions. (According to J. V. Kratz, J. O. Liljenzin, A. E. Norris, G. T. Seaborg, Phys. Rev. **C 13** 2347 (1976).)

Quasiatoms containing two nuclei of uranium and a common electron shell may be produced by bombarding uranium with uranium ions. In such a quasiatom an intermediate apparent atomic number $Z = 92 + 92 = 184$ may be obtained. At these high atomic numbers, the K electrons are near the nuclei and at $Z = 184$ their mean distance from the nucleus is of the order of its diameter. Therefore, both nuclei must approach each other to distances of the order of the diameter of a nucleus in order that K electrons "observe" both nuclei as only one.

As the binding energy of K electrons increases with Z^2 (and even more if relativistic effects are taken into account), the binding energy of electrons in these quasiatoms appreciably exceeds the value of 0.511 MeV, the energy equivalent of the mass of an electron. Thus, quasiatoms allow the investigation of the behaviour of relativistic electrons, i.e. electrons approaching the velocity of light.

The electrical field around the nuclei changes very rapidly if two nuclei approach each other within about 10^{-21} s. This leads to the production of electron–positron pairs, and spectra of electrons and positrons with a broad energy distribution superimposed by peaks of well-defined energy are observed.

"Soft" heavy-ion reactions observed somewhat above the Coulomb barrier (about 6 MeV/u; UNILAC, GSI), and "hard" heavy-ion reactions occurring at relativistic energies of about 1 GeV/u (SIS, GSI) are distinguished. In the case of central collision, the latter proceed in three stages:

– approach of the nuclei within about 10^{-23} s,
– compression of nuclear matter to about three- to fivefold density,
– explosion ("fireball").

About one-third of the nucleons in such a "fireball" are highly excited ("resonance matter").

For example, at the European research centre CERN, high-energy heavy ions are used as projectiles to study the behaviour of nuclear matter under extreme conditions of temperature and pressure. Quasi-liquid and quasi-vapourized nuclear matter has been obtained at GSI by interaction of high-energy ^{197}Au ions with gold. The background of these investigations is the search for a quark–gluon plasma, the formation of which is assumed at the first stage of the "big bang" (section 15.3).

In the liquid drop model, nuclei are compared with a drop of a liquid. With increasing temperature, the molecules in such a drop of a liquid pass to the gaseous state, then the bonds in the molecules are broken and finally a plasma of ions and free molecules is obtained. Similarly, nuclei are expected to form free nucleons and finally a quark–gluon plasma under extreme conditions of temperature and pressure. Extreme conditions ($T > 10^{13}$ K, density $\approx 10^{15}$ g/cm^3) have recently been generated at CERN by shooting lead ions of $33 \cdot 10^{12}$ eV (160 GeV per nucleon) on a lead foil, and a quark-gluon plasma containing free quarks has been produced. This new state of matter existed only for a very short time, the plasma expanded quickly, and with decreasing temperature recombination to hadrons and atomic nuclei took place.

The concept of the application of heavy ions for cancer therapy is based on the fact that specific ionization due to monoenergetic heavy ions increases sharply at the end of their path, in contrast to γ rays. This offers the possibility of delivering high radiation doses to selected places inside the body by application of heavy ions of definite energy without transmitting high doses to other parts. By appropriate moving of the heavy-ion beam or the patient, the radiation dose delivered to other parts of the body can be minimized. For treatment of inoperable tumors, heavy ions have been applied successfully.

8.12 Nuclear Fusion – Thermonuclear Reactions

Nuclear fusion of heavy atoms has been discussed in the previous section. Exoergic nuclear fusion of light atoms resulting in a gain of energy according to Fig. 2.6 is considered in this section.

The general formula of nuclear fusion is

$$A + B \rightarrow D + \Delta E \tag{8.71}$$

With respect to the production of energy, the fusion of hydrogen to helium is of special interest:

$$4\,p \rightarrow {}^4He + 2\,e^+ + 2\,\nu_e + \Delta E \tag{8.72}$$

The energy ΔE can be calculated from the mass differences to be $\Delta E = 24.69$ MeV. After addition of the energy set free by annihilation of the positrons ($2 \cdot 1.02$ MeV) and subtraction of the energy of the escaping neutrinos (about 2% of ΔE) a value of 26.2 MeV for the utilizable energy is obtained, corresponding to ≈ 6.5 MeV/u. The latter value is appreciably higher than the value of ≈ 0.8 MeV/u set free by nuclear fission.

In the sun and in the stars energy is produced mainly by nuclear fusion according to eq. (8.72). Two mechanisms are discussed:

(a) The deuterium cycle (Salpeter, 1952):

$$p + p \rightarrow d + e^+ + \nu_e \quad \text{(slow)} \tag{8.73}$$

$$d + p \rightarrow {}^3\text{He} + \gamma \quad \text{(fast)} \tag{8.74}$$

$${}^3\text{He} + {}^3\text{He} \rightarrow {}^4\text{He} + 2p \quad \text{(fast)} \tag{8.75}$$

(b) The carbon–nitrogen cycle (Bethe, 1938):

$${}^{12}\text{C} + p \rightarrow {}^{13}\text{N} + \gamma$$
$$ \hookrightarrow {}^{13}\text{C} + e^+ + \nu_e \tag{8.76}$$

$${}^{13}\text{C} + p \rightarrow {}^{14}\text{N} + \gamma \tag{8.77}$$

$${}^{14}\text{N} + p \rightarrow {}^{15}\text{O} + \gamma$$
$$ \hookrightarrow {}^{15}\text{N} + e^+ + \nu_e \tag{8.78}$$

$${}^{15}\text{N} + p \rightarrow {}^{12}\text{C} + {}^4\text{He} \tag{8.79}$$

By summing up the respective stages, eq. (8.72) is obtained for both cycles.

An important prerequisite for all these reactions is sufficiently high kinetic energy of the reacting particles. According to section 8.3, the Coulomb barrier for the reaction between two protons is ≈ 0.5 MeV and that for the reaction of a proton with ${}^{12}\text{C}$ is ≈ 1.8 MeV. According to the kinetic theory of gases, the mean kinetic energy of gas molecules $\bar{E}_{\text{kin}}$ is proportional to the temperature T,

$$\bar{E}_{\text{kin}} = \frac{m}{2}\overline{v^2} = \frac{3}{2}k_\text{B}T \tag{8.80}$$

where k_B is the Boltzmann constant, and the most probable velocity v is given by the maximum of the Maxwell velocity distribution

$$v = \sqrt{\frac{2}{3}\overline{v^2}} \tag{8.81}$$

Then the most probable kinetic energy is

$$E_{\text{kin}} = \frac{m}{2}v^2 = k_\text{B}T \tag{8.82}$$

and a mean energy of 1 eV corresponds to a temperature of $1.16 \cdot 10^4$ K. A mean energy of 0.5 MeV is attained at a temperature of $\approx 5.8 \cdot 10^9$ K. For comparison, the temperature at the surface of the sun is ≈ 6000 K, and inside the sun it is $\approx 1.5 \cdot 10^7$ K. Due to the Maxwell energy distribution, some particles have relatively high energies and are able to react according to the reaction cycles (a) or (b).

It is evident that fusion reactions become possible only at very high temperatures, and they are therefore called thermonuclear reactions. It is assumed that the deuterium cycle (a) prevails in the sun and in relatively cold stars, whereas the carbon cycle (b) dominates in hot stars. In the centre of stars densities of the order of $10^5 \, \text{g/cm}^3$ and temperatures of the order of $10^9 \, \text{K}$ may exist, and under these conditions other thermonuclear reactions become possible:

$$^4\text{He} + {}^4\text{He} \rightarrow {}^8\text{Be} \tag{8.83}$$

$$^8\text{Be} + {}^4\text{He} \rightarrow {}^{12}\text{C} + \gamma \tag{8.84}$$

$$^{12}\text{C} + {}^4\text{He} \rightarrow {}^{16}\text{O} + \gamma \tag{8.85}$$

^8Be is unstable, but the low equilibrium concentration is sufficient to allow further reactions. At temperatures of the order of $10^9 \, \text{K}$, thermonuclear reactions between ^{12}C and ^{16}O may also occur:

$$^{12}\text{C} + {}^{12}\text{C} \rightarrow {}^{24}\text{Mg} + \gamma \tag{8.86}$$

$$^{12}\text{C} + {}^{12}\text{C} \rightarrow {}^{23}\text{Na} + \text{p} \tag{8.87}$$

$$^{12}\text{C} + {}^{12}\text{C} \rightarrow {}^{20}\text{Ne} + {}^4\text{He} \tag{8.88}$$

$$^{12}\text{C} + {}^{16}\text{O} \rightarrow {}^{28}\text{Si} + \gamma \tag{8.89}$$

$$^{12}\text{C} + {}^{16}\text{O} \rightarrow {}^{24}\text{Mg} + {}^4\text{He} \tag{8.90}$$

$$^{16}\text{O} + {}^{16}\text{O} \rightarrow {}^{32}\text{S} + \gamma \tag{8.91}$$

$$^{16}\text{O} + {}^{16}\text{O} \rightarrow {}^{28}\text{Si} + {}^4\text{He} \tag{8.92}$$

The high-energy α particles originating from these reactions may induce further reactions, such as

$$^{32}\text{S} + {}^4\text{He} \rightarrow {}^{36}\text{Ar} + \gamma \tag{8.93}$$

By those exoergic thermonuclear reactions elements of increasing atomic number Z are produced. Elements of higher atomic numbers are formed by neutron capture followed by β^- decay.

Because of the high temperatures needed and the problems encountered in confining a plasma of ions and electrons at these temperatures, many efforts in research and development are necessary until production of energy by controlled thermonuclear reactions on a technical scale becomes possible.

Thermonuclear reactions between deuterons (D–D reaction), between deuterons and tritons (D–T reaction), and between tritons (T–T reaction), respectively, have appreciably higher cross sections than those between protons (P–P reaction); they are therefore of special interest:

$$d + d \rightarrow t + p + 4.03\,\text{MeV} \tag{8.94}$$

$$d + t \rightarrow {}^{4}\text{He} + n + 17.6\,\text{MeV} \tag{8.95}$$

or

$$d + d \rightarrow {}^{3}\text{He} + n + 3.27\,\text{MeV} \tag{8.96}$$

$$d + {}^{3}\text{He} \rightarrow {}^{4}\text{He} + p + 18.3\,\text{MeV} \tag{8.97}$$

and

$$t + t \rightarrow {}^{4}\text{He} + 2\,n + 11.3\,\text{MeV} \tag{8.98}$$

The cross sections of the D–D and D–T reactions are plotted in Fig. 8.25 as a function of the energy of the deuterons. From this figure it can be concluded that energies of the order of 10 keV (corresponding to temperatures of the order of 10^{8} K) are needed to get these thermonuclear reactions going and to produce utilizable energy. The starting temperature for the D–T reaction is about $0.5 \cdot 10^{8}$ K and that for the D–D reaction about $5 \cdot 10^{8}$ K.

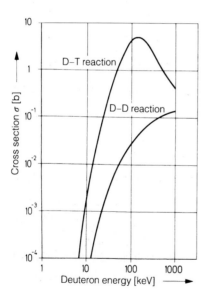

Figure 8.25. Cross sections of the D–D and D–T reactions as a function of the energy of the deuterons. (According to A. S. Bishop: Project Sherwood – The U.S. Program in Controlled Fusion. Addison-Wesley Publ. Comp., Reading, Mass., 1958.)

Literature

General

O. Frisch (Ed.), Progress in Nuclear Physics, Vol. 3, Pergamon, Oxford, **1953**
R. D. Evans, The Atomic Nucleus, McGraw-Hill, New York, **1955**
P. M. Endt, M. Demeur, Nuclear Reactions, North-Holland, Amsterdam, **1959**
I. Kaplan, Nuclear Physics, 2nd ed., Addison-Wesley, Reading, MA, **1964**
M. Lefort, La Chimie Nucléaire, Dunod, Paris, **1966**; English Translation: Nuclear Chemistry, Van Nostrand, London, **1968**
L. Yaffe (Ed.), Nuclear Chemistry, 2 Vols., Academic Press, New York, **1968**
J. Cerny (Ed.), Nuclear Spectroscopy and Reactions, Parts A, B, C, Academic Press, New York, **1974**
Advances in Nuclear Physics, Vol. 9, Plenum Press, New York, **1977**
G. Friedlander, J. W. Kennedy, E. S. Macias, J. M. Miller, Nuclear and Radiochemistry, 3rd ed., Wiley, New York, **1981**
G. T. Seaborg, W. D. Loveland, Nuclear Chemistry, Hutchinson, Stroudsburg, PA, **1982**
H. J. Arnikar, Essentials of Nuclear Chemistry, 2nd ed., Wiley, New York, **1987**
A. Vertes, I. Kiss, Nuclear Chemistry, Elsevier, Amsterdam, **1987**

More Special

R. Rubinson, The Equations of Radioactive Transformation in a Neutron Flux, J. Chem. Phys. *17*, 542 **(1949)**
A. C. Graves, D. K. Froman (Eds.), Miscellaneous Physical and Chemical Techniques of the Los Alamos Project, National Nuclear Energy Series, Div.V., Vol. 3, McGraw-Hill, New York, **1952**
R. Huby, Stripping Reactions, in: Progress in Nuclear Physics, Vol. III (Ed. O. Frisch), Pergamon, London, **1953**
D. J. Hughes, J. A. Harvey, Neutron Cross Sections, United States Atomic Energy Commission, McGraw-Hill, New York, **1955**
F. L. Friedman, V. F. Weisskopf, The Compound Nucleus, in: Niels Bohr and the Development of Physics (Ed. W. Pauli), McGraw-Hill, New York, **1955**
B. B. Kinsey, Nuclear Reactions, Levels and Spectra of Heavy Nuclei, in: Handbuch der Physik (Ed. S. Flügge), Vol. 40, Springer, Berlin, **1957**
J. Rainwater, Resonance Processes by Neutrons, in: Handbuch der Physik (Ed. S. Flügge), Vol. 40, Springer, Berlin, **1957**
E. Segrè (Ed.), Experimental Nuclear Physics, Vol. III, Wiley, New York, **1959**
M. S. Livingstone, J. P. Blewett, Particle Accelerators, McGraw-Hill, New York, **1962**
M. S. Livingstone, The Development of High-Energy Accelerators, Dover, New York, **1966**
E. Henly, E. Johnson, The Chemistry and Physics of High Energy Reactions, University Press, Washington, DC, **1969**
R. Klapisch, Mass Separation for Nuclear Reaction Studies, Annu. Rev. Nucl. Sci. *19*, 33 **(1969)**
G. Herrmann, O. Denschlag, Rapid Chemical Separations, Annu. Rev. Nucl. Sci. *19*, 1 **(1969)**
F. W. K. Firk, Low-Energy Photonuclear Reactions, Annu. Rev. Nucl. Sci. *20*, 39 **(1970)**
K. Bethge, Alpha-Particle Transfer Reactions, Annu. Rev. Nucl. Sci. *20*, 255 **(1970)**
K. A. Keller, J. Lange, H. Münzel, G. Pfennig, Excitation Functions for Charged-Particle Induced Nuclear Reactions, Springer, Berlin, **1973**
E. G. Fuller, E. Hayward (Eds.), Photonuclear Reactions, Dowden, Hutchinson and Ross, New York, **1976**
R. Guillaumont, D. Trubert, On the Discovery of Artificial Radioactivity, Radiochim. Acta *70/71*, 39 **(1995)**
S. M. Qaim, Radiochemical Studies of Complex Particle Emission in Low and Intermediate Energy Reactions, Radiochim. Acta *70/71*, 163 **(1995)**

Nuclear Fission

O. Hahn, F. Strassmann, Über den Nachweis und das Verhalten der bei der Bestrahlung des Urans mittels Neutronen entstehenden Erdalkalimetalle, Naturwiss. *27*, 11 **(1939)**

E. K. Hyde, The Nuclear Properties of the Heavy Elements, III Fission Phenomena, Prentice Hall, Englewood Cliffs, N.J., **1964**

G. Herrmann, 25 Jahre Kernspaltung, Radiochim. Acta *3*, 169 **(1964)** and *4*, 173 **(1965)**

International Atomic Energy Agency, Physics and Chemistry of Fission, IAEA, Vienna, **1965**, **1969**, **1973**

R. Vandenbosch, J. R. Huizenga, Nuclear Fission, Academic Press, New York **1973**

J. R. Nix, Calculation of Fission Barriers for Heavy and Superheavy Nuclei, Annu. Rev. Nucl. Sci. *22*, 65 **(1972)**

H. J. Specht, Nuclear Fission, Rev. Mod. Phys. *46*, 773 **(1974)**

D. C. Hoffman, M. M. Hoffman, Post-Fission Phenomena, Annu. Rev. Nucl. Sci. *24*, 151 **(1974)**

H. Dreisvogt, Spaltprodukt-Tabellen, Bibliogr. Inst. Mannheim, **1974**

G. Herrmann, Fifty Years ago: From the "Transuranics" to "Nuclear Fission", Angew. Chem. Int. Ed. *29*, 481 **(1990)**

G. Herrmann, The Discovery of Nuclear Fission – Good Solid Chemistry Got Things on the Right Track, Radiochim. Acta *70/71*, 51 **(1995)**

Heavy-Ion Reactions

P. E. Hodgson, Nuclear Heavy Ion Reactions, Clarendon Press, Oxford, **1978**

W. U. Schröder, J. R. Huizenga, Damped Heavy Ion Collisions, Annu. Rev. Nucl. Sci. *27*, 465 **(1977)**

A. S. Goldhaber, H. H. Heckman, High-Energy Interactions of Nuclei, Annu. Rev. Nucl. Sci. *28*, 161 **(1978)**

R. Bass, Nuclear Reactions with Heavy Ions, Springer, Berlin, **1979**

R. F. Post, Controlled Fusion Research and High-Energy Plasmas, Annu. Rev. Nucl. Sci. *20*, 509 **(1970)**

R. Bock (Ed.), Heavy Ion Collisions, 3 Vols., North-Holland, Amsterdam, **1979–1981**

D. A. Bromley (Ed.), Treatise on Heavy Ion Science, Vol. 4, Plenum Press, New York, **1985**

R. Bock, G. Herrmann, G. Siegbert, Schwerionenforschung, Wiss. Buchges., Darmstadt, **1993**

J. V. Kratz, Radiochemical Studies of Complex Nuclear Reactions, Radiochim. Acta *70/71*, 147 **(1995)**

G. Kraft, The Radiobiological and Physical Basis for Radiotherapy with Protons and Heavier Ions, Strahlenther. Onkol. *166*, 10 **(1990)**

9 Chemical Effects of Nuclear Reactions

9.1 General Aspects

The binding energies between atoms vary between about 40 and 400 kJ/mol, corresponding to about 0.4 to 4 eV (1 eV $\simeq$ 96.5 kJ/mol). The energies involved in nuclear reactions are of the order of up to several MeV, and parts of this energies are transmitted to the atoms in the form of recoil and of excitation energy. Therefore, chemical bonds are strongly affected by nuclear reactions.

High kinetic energy of single atoms does not mean high temperature, because the temperature of a system is given by the mean kinetic energy of all the atoms or molecules, $\bar{E}_{kin} = \frac{3}{2} k_B T$ (for three degrees of freedom; eq. (8.80)). However, deviating from the usual concept of temperature, the temperature equivalent of a single particle may be related to its kinetic energy by the equation

$$T \simeq E_{kin}/k_B = 1.16 \cdot 10^4 E_{kin} \quad [\text{K}/\text{eV}] \tag{9.1}$$

Because energies of the order of 1 eV to 1 MeV are transmitted to the atoms by nuclear reactions, corresponding to temperature equivalents of the order of 10^4 to 10^{10} K, these atoms are called "hot atoms" and their chemistry is called "hot atom chemistry", or "recoil chemistry" if the recoil effects are considered.

Chemical effects of nuclear reactions were first observed by Szilard and Chalmers in 1934 when irradiating ethyl iodide with neutrons. They found several chemical species containing ^{128}I that are produced by the chemical effects of the nuclear reaction $^{127}I(n, \gamma)^{128}I$. In the following years, chemical effects of radioactive decay were observed in gaseous compounds, liquids and solids.

The chemical effects of mononuclear and binuclear reactions can be divided into primary effects taking place in the atom involved in the nuclear reaction, secondary effects in the molecules or other associations of atoms, and subsequent reactions. Primary and secondary effects are observed within about 10^{-11} s after the nuclear reaction.

Primary effects comprise recoil of the nucleus and excitation of the electron shell of the atom. The excitation may be due to recoil of the nucleus, change of atomic number Z or emission of electrons from the electron shell. Secondary effects and subsequent reactions depend on the chemical bonds and the state of matter. Chemical bonds may be broken by recoil or excitation. In gases and liquids mainly the bonds in the molecules are affected. The range of recoil atoms is relatively large in gases and relatively small in condensed phases (liquids and solids). Fragments of molecules are mobile in gases and liquids, whereas they may be immobilized in solids on interstitial sites or lattice defects and become mobile if the temperature is increased.

The chemical reactions taking place after nuclear reactions may be distinguished as "hot", "epithermal" and "thermal" reactions. Hot reactions proceed at high energies of the atoms in a statistical way, i.e. without preference for certain bonds or reaction partners. In epithermal reactions certain bonds or reaction partners, respectively, are preferred, but the reactions exhibit unusual chemical features. Hot and epithermal reactions are comparable with reactions induced by ionizing radiation; ions and radicals play an important role. Thermal reactions proceed at relatively low energies (<1 eV) and are similar to chemical reactions observed at temperatures up to several hundred degrees Celsius. Recoiling atoms of high energy break their bonds, leave their position as ions and give off their energy in a cascade of hot, epithermal and thermal reactions.

The fraction of the atoms produced in a nuclear reaction and found in the form of the original chemical compound is called retention. Retention can be due to non-breaking of the chemical bonds (primary retention) or to breaking of the bonds followed by recombination or substitution reactions (secondary retention).

9.2 Recoil Effects

Mononuclear and binuclear reactions involve recoil effects. The energy ΔE set free in the reactions is shared out between the reaction products according to the law of conservation of momentum,

$$E_1 m_1 = E_2 m_2 \tag{9.2}$$

E_1 and m_1 are the energy and the mass of the recoiling atom, and E_2 and m_2 the energy and the mass of the particle emitted in the course of the nuclear reaction. If the velocity of the emitted particle approaches the velocity of light, the increase of m_2 according to eq. (2.7) has to be taken into account and the recoil energy E_1 is

$$E_1 = \frac{m_2^0}{m_1} E_2 + \frac{E_2^2}{2\,m_1 c^2} \tag{9.3}$$

By inserting the nuclide masses M and the value of E_2 in MeV the following relation is obtained:

$$E_1 = \frac{E_2}{M_1} (M_2 + 5.37 \cdot 10^{-4} E_2) \tag{9.4}$$

The second term in the parentheses on the right-hand side is negligible if $M_2 \geq 1$ u. In the case of emission of electrons or positrons, eq. (9.4) becomes

$$E_1 = \frac{E_2}{M_1} (5.49 + 5.37 E_2) \cdot 10^{-4} \qquad (M_1 \text{ in } u, \text{ energies in MeV}) \tag{9.5}$$

In Fig. 9.1 the recoil energy for β decay is plotted as a function of the mass number for various energies of the β particles.

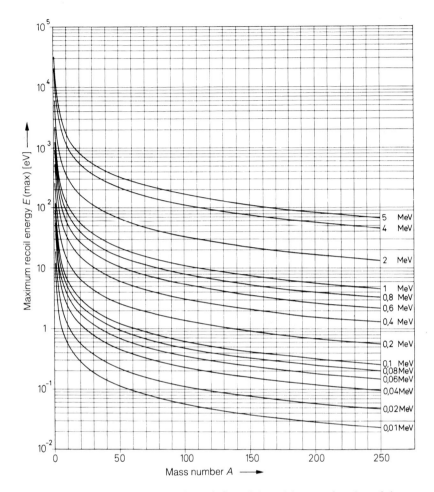

Figure 9.1. Recoil energy due to the emission of β particles as a function of the mass number for various energies of the β particles.

If a γ-ray photon is emitted, the law of conservation of momentum gives

$$m_1 v_1 = \frac{E_\gamma}{c} \tag{9.6}$$

and the energy of the recoiling atom is

$$E_1 = 5.37 \cdot 10^{-4} \frac{E_\gamma^2}{M_1} \quad (\text{M}_1 \text{ in u, energies in MeV}) \tag{9.7}$$

The recoil energy for emission of γ-ray photons of various energies is plotted in Fig. 9.2 as a function of the mass number.

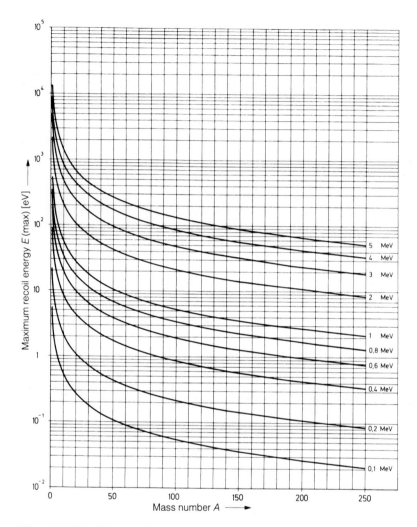

Figure 9.2. Recoil energy due to the emission of γ-ray photons of various energies as a function of the mass number.

Application of eqs. (9.2) to (9.7) is illustrated by some examples:

(a) α decay of ^{212}Po: The energy of the α particles is $E_2 = 8.785\,\mathrm{MeV}$ and the recoil energy is $E_1 = 0.169\,\mathrm{MeV}$. As the recoil energy due to α decay is always many orders of magnitude higher than the energy of chemical bonds, α decay will always cause breaking of chemical bonds.

(b) β^- decay of ^{90}Sr: The energy of the β^- particles (electrons) varies between zero and the maximum value of $0.546\,\mathrm{MeV}$. Together with the electron an anti-neutrino is emitted ($E_e + E_{\bar{\nu}} = 0.546\,\mathrm{MeV}$). The recoil energy of the daughter nuclide ^{90}Y depends on the masses of the electron and the neutrino and on the angle under which they are emitted (Fig. 9.3). The maximum recoil energy calculated by eq. (9.5) is $E_1(\text{max}) = 5.11\,\text{eV}$. This value is greater than the energy of chemical bonds. However, most frequently the recoiling atom will have only a part of E_1 (max) and the chemical bonds may not be broken. Generally, the following statement can be made in the case of β decay: with decreasing energy of β decay, the probability increases that chemical bonds are not broken by recoil effects.

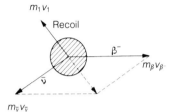

Figure 9.3. Recoil effect due to the emission of a β^- particle and an antineutrino.

(c) Emission of γ-ray photons in isomeric transition: In isomeric transition only γ-ray photons are emitted – in the case of isomeric transition of ^{80m}Br photons of 0.049 and 0.037 MeV. Eq. (9.7) gives for the recoil energy due to emission of 0.049 MeV photons $E_1 = 0.016\,\text{eV}$, which is not sufficient to break chemical bonds. This holds for all isomeric transitions in which low-energy photons are emitted.

(d) Emission of γ-ray photons immediately after α or β decay or emission of several γ-ray photons: In this case the recoil effects overlap as indicated in Fig. 9.4.

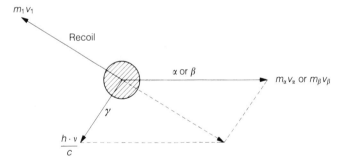

Figure 9.4. Recoil effect due to the simultaneous emission of an α or β particle, respectively, and a γ-ray photon.

(e) Emission of γ-ray photons in binuclear reactions: The energy of the exoergic reaction ^{37}Cl $(n, \gamma)^{38}$Cl is $\Delta E = 6.11\,\text{MeV}$. By absorption of thermal neutrons negligible amounts of kinetic energy are transmitted by the neutrons. The excitation energy ΔE is given off by emission of one or several γ-ray photons, and by application of eq. (9.7) the maximum recoil energy due to the emission of one 6.11 MeV photon is calculated to be $E_1(\text{max}) = 528\,\text{eV}$. This energy is appreciably higher than the energy of chemical bonds, and even fractions of this energy resulting from the emission of several photons are high enough to break chemical bonds. Therefore, in most (n, γ) reactions chemical bonds will be broken. In Szilard–Chalmers reactions use is made of this recoil effect to separate isotopic product nuclides from target nuclides.

(f) Emission of protons, neutrons or α particles in binuclear reactions: The energy of the exoergic reaction $^{14}N(n, p)^{14}C$ is $\Delta E = 0.626\,MeV$. If the kinetic energy of the neutrons is neglected, ΔE is shared out among the reaction products, and from eq. (9.2) it follows that

$$E_1 = \frac{m_2}{m_1}(\Delta E - E_1) = 0.042\,MeV$$

This recoil energy is appreciably higher than the energy of any chemical bond. Generally, emission of protons, neutrons or α particles leads to breaking of chemical bonds.

Recoil energies due to emission of α particles, protons, neutrons, β particles and γ-ray photons are listed in Table 9.1 for various mass numbers and various energies of the emitted particles. It is evident from this table that by emission of particles with $M \geq 1$ chemical bonds will be broken, whereas recoil energies due to emission of

Table 9.1. Recoil energy due to emission of α particles, protons, neutrons, electrons and γ-ray photons.

Energy of the particle emitted [MeV]	Mass number A	Recoil energy			
		α	p or n	β	γ
0.1	10	40 keV	10 keV	6.0 eV	0.54 eV
	50	8 keV	2 keV	1.2 eV	0.11 eV
	100	4 keV	1 keV	0.6 eV	0.05 eV
	200	2 keV	0.5 keV	0.3 eV	0.03 eV
0.3	10	120 keV	30 keV	21.3 eV	4.83 eV
	50	24 keV	6 keV	4.3 eV	0.97 eV
	100	12 keV	3 keV	2.1 eV	0.48 eV
	200	6 keV	1.5 keV	1.1 eV	0.24 eV
1.0	10	400 keV	101 keV	109 eV	53.7 eV
	50	80 keV	20 keV	22 eV	10.7 eV
	100	40 keV	10 keV	11 eV	5.4 eV
	200	20 keV	5 keV	5 eV	2.7 eV
3.0	10	1201 keV	302 keV	648 eV	483 eV
	50	240 keV	60 keV	130 eV	97 eV
	100	120 keV	30 keV	65 eV	48 eV
	200	60 keV	15 keV	32 eV	24 eV
10.0	10	4003 keV	1008 keV	5.92 keV	5.4 keV
	50	800 keV	202 keV	1.18 keV	1.1 keV
	100	400 keV	101 keV	0.59 keV	0.5 keV
	200	200 keV	50 keV	0.30 keV	0.3 keV

electrons, positrons or γ-ray photons may be higher or lower than the energies of chemical bonds.

Effects similar to recoil effects are observed if nuclei are bombarded with high-energy projectiles. The momentum transmitted to the target nucleus results in a kinetic energy given by

$$E_1 = E_x \frac{M_x}{M_A + M_x} = E_x - E^* \tag{9.8}$$

where E_x is the energy of the projectile and E^* the excitation energy of the product nucleus (section 8.2). Neutrons, protons or α particles with energies of about 1 MeV hitting a nucleus transmit energies of the order of several up to several hundred keV with the result that chemical bonds are always broken. On the other hand, the energies transmitted by absorption of low-energy photons will not rupture chemical bonds.

9.3 Excitation Effects

Electron shells are influenced by mono- or binuclear reactions in various ways:

– excitation due to recoil;
– excitation due to change of the atomic number;
– excitation due to electron capture or internal conversion.

These effects overlap and lead to ionization, emission of electrons from the electron shell and fluorescence. They may cause secondary reactions in molecules and subsequent reactions of the ions or excited atoms or molecules produced by these effects.

If a nucleus suffers a recoil, parts of the electron shell, in particular valence electrons, may be stripped off and stay behind, resulting in ionization of the atom. This ionization depends on the recoil energy, the strength of the chemical bonds and the state of matter. Ions carrying from 1 to about 20 positive charges have been observed.

Alpha decay leads to a decrease of the atomic number by two units, $Z' = Z - 2$, and causes an expansion of the electron shell, as illustrated in Fig. 9.5 for the α decay of radioisotopes of Bi. Differences in the binding energies are marked for electrons in the inner shells. Furthermore, there are two surplus electrons after α decay. However, in the case of α decay the excitation effects due to the change of the atomic number are relatively small compared with the recoil effects that have been discussed in the previous section, with the result that the recoil effects dominate.

In the case of β decay, on the other hand, the effects due to the change of the atomic number may be significant, in particular if the energy of the β particles is low. β^- decay leads to an increase in the atomic number by one unit, $Z' = Z + 1$, and to a contraction of the electron shell, as illustrated in Fig. 9.6 for β^- decay of radioisotopes of Sr. Furthermore, one electron is missing after β^- decay. Immediately after β^- decay all the electrons of the atom are at energy levels that are higher than those corresponding to the new atomic number, which means that all the electrons are in an excited state. The total excitation energy of the electron shell amounts to

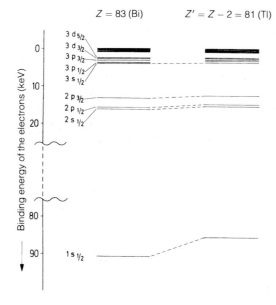

Figure 9.5. Expansion of the electron shell due to α decay.

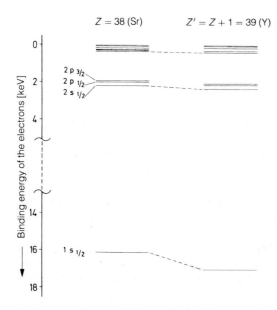

Figure 9.6. Contraction of the electron shell due to β⁻ decay.

many eV (e.g. 107 eV in the case of β^- decay of a radioisotope of tin ($Z = 50$)). It is given off by emission of electrons from the electron shell. These electrons are called "Auger" electrons. The remaining vacancies in the inner orbitals are filled by other electrons, and X-ray photons are emitted that are characteristic for the new atomic number.

In contrast to β^- decay, β^+ decay, just as α decay, leads to an expansion of the electron shell, the electrons take up energy from neighbouring atoms, and the excitation effects discussed for β^- decay are not observed.

Another effect due to the change of the atomic number has to be considered in case of β decay, in particular if the recoil energy is lower than the energy of the chemical bonds. The chemical properties of the isobaric daughter nuclide resulting from β decay are, in general, different from those of the mother nuclide. If the recoil energy is not sufficient to break the chemical bonds, the daughter nuclide stays in the same position, but in an unusual chemical surrounding. Examples are

$$^{14}CH_3-CH_3 \xrightarrow[5730\,y]{\beta^-\,(0.156)} (^{14}NH_3-CH_3)^+ \qquad (9.9)$$
$$^{14}NH_3 + CH_3^+ \quad or \quad ^{14}NH_2^+ + CH_4$$

and

$$^{35}SO_2Cl_2 \xrightarrow[87.2\,d]{\beta^-\,(0.167)} (^{35}ClO_2Cl_2)^+ \qquad (9.10)$$
$$^{35}ClO_2^+ + Cl_2 \quad or \quad ^{35}ClO_2 + Cl_2^+$$

Occasionally, after β^- decay the daughter nuclide is in a chemically stable state:

$$^{212}Pb^{2+} \xrightarrow{\beta^-} {}^{212}Bi^{3+} \qquad (9.11)$$

$$^{14}CO \xrightarrow{\beta^-} {}^{14}NO^+ \qquad (9.12)$$

With respect to the lifetime of the excited states resulting from changes in the atomic number, isothermal and adiabatic decay may be discussed. All the experimental results indicate an adiabatic decay, which means transfer of the excitation energy to all the electrons, resulting in a certain lifetime of the excited state of the daughter nuclide of the order of about 1 µs.

Ionization and excitation of daughter nuclides due to β^- decay have been proved by experimental results. For example, in the case of β^- decay of ^{85}Kr, 79.2% of the resulting ^{85}Rb was found in the form of Rb^+, 10.9% in the form of Rb^{2+} and the rest in the form of Rb ions with higher charges up to 10+. Decay of ^3HH gives a yield of 90% $^3HeH^+$, and recoil effects cannot be responsible for formation of these ions, because the maximum recoil energy is 0.82 eV, whereas the ionization energy is ≈ 2 eV.

By electron capture (ε) or internal conversion, electrons are taken away from inner orbitals and the vacancies are filled with electrons from outer orbitals with resulting emission of characteristic X rays. Electrons may also be emitted by an internal photoeffect. Finally, at least one electron is missing, and this may also cause breaking of the chemical bond. As electron capture leads also to a change of the atomic number $(Z' = Z - 1)$, it is not possible to distinguish the effects due to the capture of an electron from those that are caused by the change of Z. In internal conversion, however, the atomic number is not changed, and the chemical effects observed in this case can be due either to recoil or to disappearance of the electron. With respect to the chemical effects, the isomeric transition of ^{80m}Br has been investigated in detail:

$$^{80m}Br \xrightarrow[4.42\ h]{IT(0.049;0.037)} {}^{80}Br \tag{9.13}$$

The energies of isomeric transition are only 49 and 37 keV, and the conversion coefficients are 1.6 and ≈ 300, respectively. The recoil energy due to emission of a 49 keV γ-ray photon is only 0.016 eV (section 9.2). It is too low to break a C–Br bond (247 kJ $\simeq$ 2.6 eV). The recoil energy due to emission of an electron from the 1s orbital (0.45 eV) is also too small to break the C–Br bond. Actually, breaking of the C–Br bond due to isomeric transition of ^{80m}Br is observed, for instance, if butyl bromide labelled with ^{80m}Br is shaken with water. In this experiment the main fraction of ^{80}Br is found free of ^{80m}Br in the aqueous phase. Results of experiments with various compounds labelled with ^{80m}Br are compiled in Table 9.2. The high retention in the case of solid $(NH_4)_2[PtBr_6]$ is due to the fact that in the solid state the missing electron is quickly substituted.

Table 9.2. Breaking of bonds due to internal conversion of ^{80m}Br.

Compound	Free ^{80}Br [%]	Retention [%]
HBr (g)	75	25
CF$_3$Br (g)	99	1
CH$_3$Br (g)	94	6
CH$_3$Br + Br$_2$ (l)	94	6
CCl$_3$Br (g)	93	7
CCl$_3$Br + Br$_2$ (l)	87	13
C$_6$H$_5$Br + Br$_2$ (l)	87	13
$[Co(NH_3)_5Br]^{2+}$ (aq)	100	0
$[Co(NH_3)_5Br]$ (NO$_3$)$_2$ (s)	86	14
$[PtBr_6]^{2-}$ (aq)	47	53
$(NH_4)_2[PtBr_6]$ (s)	0	100

The effects due to the vacancies in the inner orbitals after β^- decay, electron capture or internal conversion can be summarized as follows: The vacancies are filled with electrons from outer orbitals and characteristic X-ray photons are emitted. These photons may transmit their energy or a part of their energy to electrons in the same electron shell by an internal photo- or Compton effect, respectively, and these electrons leave the atom as Auger electrons. The resulting vacancies are filled again with electrons from outer orbitals, and additional photons are emitted. This process is called internal fluorescence. The photons may again liberate Auger electrons by an internal photoeffect, and so forth. The ratio of the number of Auger electrons to the number of photons emitted depends on the atomic number of the atom. The number of K-X rays emitted per vacancy in the K shell is called the fluorescence yield ω_K, and $1 - \omega_K$ is the Auger yield for the K shell. The fluorescence yields and the Auger

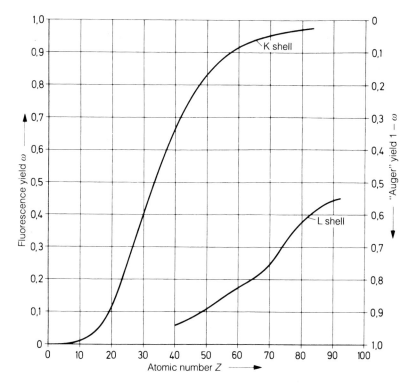

Figure 9.7. Fluorescence yield ω and "Auger" yield $1 - \omega$ for the K and L shells as a function of the atomic number. (According to E. H. S. Burshop: The Auger Effect and other Radiationless Transitions. Cambridge University Press, London 1952.)

yields for the K shell and the L shell are plotted in Fig. 9.7 as a function of the atomic number Z.

The result of an Auger effect is always an ionization of the atom. For example, due to the isomeric transition of ^{131m}Xe, ions of ^{131}Xe are found carrying on the average a charge of 8.5+.

9.4 Gases and Liquids

In gases and liquids, intramolecular bonds are only affected to a certain degree by the recoil due to a mononuclear or binuclear reaction occurring in an atom of a molecule or by the kinetic energy transmitted to an atom by an incident projectile. Molecules in gases and liquids are mobile and the intermolecular binding forces are small, provided that the pressure is not too high.

The situation in a gas molecule is illustrated schematically in Fig. 9.8. The effect of the recoil of the atom (1) on the bond between that atom and the rest of the molecule (R) depends on the direction of the recoil and the inertia of R. If the recoil of 1 is not

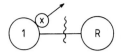

Figure 9.8. Recoil effect in a gas molecule; x: particle or photon emitted; 1: nucleus suffering recoil; R: rest of the molecule.

in the direction to R, the recoil energy E_1 is split up into kinetic energy of the molecule and the energy E_B affecting the chemical bond:

$$E_1 = E_B + \frac{m_1 + m_R}{2} v_s^2 \tag{9.14}$$

where v_s is the velocity of the centre of gravity of the molecule. Application of the law of conservation of momentum gives

$$E_B = E_1 \left(1 - \frac{m_1}{m_1 + m_R} \right) = E_1 \frac{m_R}{m_1 + m_R} \tag{9.15}$$

This equation shows that the influence of the recoil on the chemical bond increases with the mass of the rest R.

If the recoil is in the direction towards R, part of the kinetic energy E_1 may be transformed by inelastic collision into excitation energy of the recoiling atom 1 and the rest R:

$$E_1 = E_1' + E_R + E_1^* - E_R^* \tag{9.16}$$

E_1' and E_R are the kinetic energies, and E_1^* and E_R^* the excitation energies of 1 and R, respectively. In this case, the effect of the recoil on the chemical bond depends on the kinetic energy E_R transmitted to R and on the excitation energies E_1^* and E_R^*.

If liquids are considered, intermolecular forces have to be taken into account. These forces cause a greater inertia of R, apparently greater values of m_R and greater values of E_B (eq. (9.15)). In the limiting case the rest R is so strongly bound to neighbouring molecules that $m_R \gg m_1$ and $E_B \approx E_1$.

Chemical effects of nuclear reactions in gases are preferably investigated by use of mass spectrometry, for example:

- Radioactive decay of tritium in ³H-labelled ethane leads to the formation of ethyl ions ($\approx 80\%$) and fragments of ethane ($\approx 20\%$). Because the recoil energy is too low to break C–C bonds, fragmentation of the ethane molecule must be due to excitation effects.
- The charge distributions of the ions found after β^- decay of ^{133}Xe and after isomeric transition of ^{131m}Xe are plotted in Fig. 9.9. The rather similar curves found for the ions of ^{133}Cs and of ^{131}Cs result mainly from excitation effects.
- The observation that the decay of T in $C_6H_5CH_2T$ and in $C_6H_4TCH_3$ gives very similar spectra of products is explained by the fact that tropylium ions ($C_7H_7^+$) are formed as intermediates.

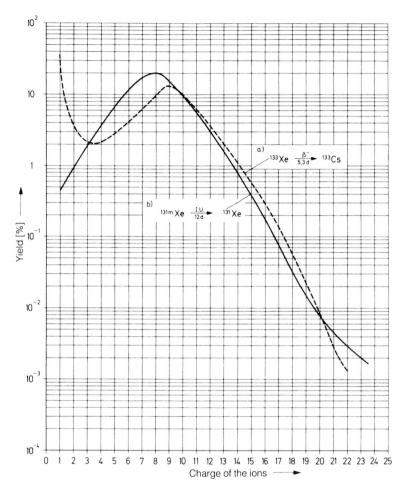

Figure 9.9. Charge distribution of the ions: a) after β^- decay of ^{133}Xe; b) after isomeric transition of ^{131m}Xe. (According to A. H. Snell, F. Pleasonton, T. A. Carlson: Proceedings Series, Chemical Effects of Nuclear Transformations, Vol. I. IAEA Vienna 1961, S. 147.)

The chemical effects of nuclear reactions in liquids have been investigated in great detail with alkyl halides. The first example was studied by Szilard and Chalmers in 1934. They irradiated ethyl iodide with neutrons and were able to extract about half of the ^{128}I produced by the nuclear reaction ^{127}I$(n, \gamma)^{128}$I into an aqueous phase. Similar results are obtained in the case of (d, p), (n, 2n) and (γ, n) reactions and of other alkyl or aryl halides: appreciable amounts of the radioisotopes of iodine or other halogens obtained by these reaction can be extracted into aqueous solutions.

The chemical effects of the nuclear reaction ^{127}I$(n, \gamma)^{128}$I are explained as follows: In the first stage (primary effect) the chemical bond between ^{128}I and C is broken by the recoiling "hot" ^{128}I atom which loses its energy in a sequence of collisions with other molecules ("hot reactions"). By these reactions various fragments of the molecules are produced, which are difficult to detect because their concentration is extremely small and because they disappear quickly by subsequent reactions. After the "hot" atom has given off the main part of its kinetic energy by hot and epithermal reactions, various thermal reactions are possible, in particular recombination or substitution reactions:

$$C_2H_5{}^{127}I + {}^{128}I \cdot \rightarrow C_2H_5 \cdot + {}^{128}I \cdot + {}^{127}I \cdot \qquad (9.17)$$

$$^{127}I \cdot + {}^{128}I \cdot \rightarrow {}^{127}I^{128}I \qquad (9.18)$$

$$C_2H_5 \cdot + {}^{128}I \cdot \rightarrow C_2H_5{}^{128}I \qquad (9.19)$$

$$C_2H_5{}^{127}I + {}^{128}I \cdot \rightarrow C_2H_5{}^{128}I + {}^{127}I \cdot \qquad (9.20)$$

or, less frequently,

$$C_2H_5{}^{127}I + {}^{128}I \cdot \rightarrow CH_2{}^{128}ICH_2{}^{127}I + H \cdot \qquad (9.21)$$

$$C_2H_5{}^{127}I + {}^{128}I \cdot \rightarrow CH_3CH{}^{127}I^{128}I + H \cdot \qquad (9.22)$$

Free iodine produced by reaction (9.18) can be extracted. ^{128}I-labelled ethyl iodide is formed by reactions (9.19) and (9.20). The substitution products can be separated by gas chromatography.

The chemical effects observed after neutron irradiation of ethyl iodide have found great practical interest, because they allow general application to various compounds and chemical separation of isotopic products of nuclear reactions. Above all, isotopic nuclides of high specific activity can be obtained by Szilard–Chalmers reactions (section 9.6).

For the investigation of "hot"-atom induced reactions the technique of scavenging is often applied. In this method, small amounts of reactive compounds (scavengers) are added which react preferentially with atoms or radicals produced by the chemical effects of nuclear reactions. Whereas "hot" reactions are not influenced by the presence of scavengers, scavengers are highly selective in the range of thermal reactions.

9.5 Solids

In general, atoms or ions in solids are not mobile. They are bound rather strongly to neighbouring atoms or ions, as indicated in Fig. 9.10. The strong embedding of atoms in crystalline solids leads to an apparently high mass m_1. Two possibilities may be distinguished:

(a) The momentum transmitted is high enough to break all bonds with neighbouring atoms and the atom involved in the reaction is pushed out from its position in the lattice.
(b) The momentum is too small for a rupture of chemical bonds, and the atom stays at its place without suffering a recoil ($m_1 \rightarrow \infty$, $E_1 \rightarrow 0$). Excitation energy transmitted to the atom will quickly be distributed among neighbouring atoms in the lattice within about 10^{-11} s.

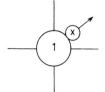

Figure 9.10. Recoil effect in a crystalline solid; x: particle or photon emitted; 1: atom bound in the solid.

At high recoil energies or high energies transmitted by incident particles or photons, the "hot" atoms collide with other atoms, pushing them from their lattice sites and producing a series of lattice defects (vacancies and atoms on interstitial sites), as indicated for a simple lattice in Fig. 9.11. In the course of these subsequent reactions the "hot" atom gives off its kinetic energy and stays on a lattice site or on an interstitial site. The range of a recoiling atom and the number of lattice defects produced depend on the recoil energy, the mass of the recoiling atom and the density of the solid. The range of recoil atoms in aluminium is plotted in Fig. 9.12 as a function of the recoil energy. This range is short after (n, γ) reactions $(0.5–5 \, nm)$ and long after (n, α), (n, p), (d, p) and (γ, n) reactions $(50–1000 \, nm)$. The lattice defects in the solid may cause remarkable changes of the properties (conductivity, volume, reactivity).

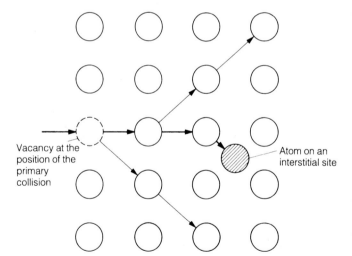

Vacancy at the position of the primary collision

Atom on an interstitial site

Figure 9.11. Generation of disorder in solids by a cascade of collisions (schematically).

After the recoil atoms have come to rest, subsequent reactions are stopped. Further reactions including recombination of reactive atoms or molecular fragments are possible after diffusion of the reactive species or after dissolution of the solid. Diffusion can be enhanced by increasing the temperature (thermal annealing) or by irradiation with γ rays or electrons (radiation annealing). Dissolution may lead to recombination, reaction with other species or reaction with the solvent. Due to these processes secondary retention may increase or decrease. As an example, the retention of ^{35}S in the form of sulfate after neutron irradiation of ammonium sulfate and dissolution is plotted in Fig. 9.13 as a function of the time of thermal annealing at

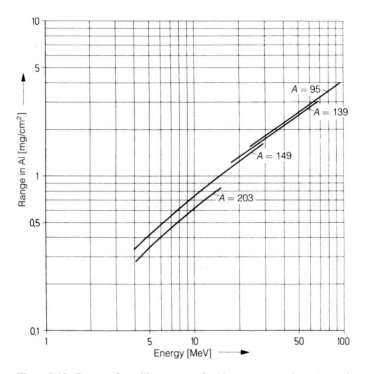

Figure 9.12. Range of recoiling atoms of various mass numbers A as a function of the recoil energy. (According to J. Alexander, M. F. Gazdik: Phys. Rev. **120**, 874 (1960); B. G. Harvey Annu. Rev. Nucl. Sci. **10**, 235 (1960).)

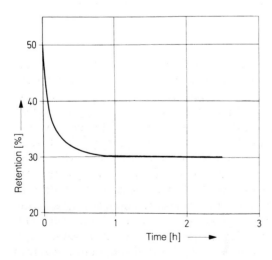

Figure 9.13. Thermal annealing of ammonium sulfate: relative activity of ^{35}S in the form of sulfate (retention) as a function of the time of annealing at $180\,^\circ C$.

180 °C. Major proportions of the ^{35}S atoms are found in the form of sulfur and of sulfite.

Because of the low concentrations of the products of primary and secondary reactions, their identification is only possible in some cases and by application of special techniques, such as Mössbauer spectroscopy.

With respect to chemical separation of isotopic nuclides from target nuclides after (n, γ) reactions, changes of the valence state and of complexation are of special interest. Some examples are listed in Table 9.3. All nuclides produced by (n, γ) reactions are found in appreciable amounts in lower valence states and free from complexing ligands, respectively.

Table 9.3. Radioactive products found after nuclear reactions.

Irradiated compound	Nuclear reaction	Reaction products
Perchlorates	$^{37}Cl(n, \gamma)^{38}Cl$	ClO_3^-, Cl^-
Periodates	$^{127}I(n, \gamma)^{128}I$	I^-, IO_3^-
Chlorates	$^{37}Cl(n, \gamma)^{38}Cl$	Cl^-
Bromates	$^{79}Br(n, \gamma)^{80m}Br$ $\Big\}$	Br^-, Br_2
	$^{81}Br(n, \gamma)^{82}Br$	
Iodates	$^{127}I(n, \gamma)^{128}I$	I^-
Sulfates	$^{34}S(n, \gamma)^{35}S$	S, SO_3^{2-}, S^{2-}
Phosphates	$^{31}P(n, \gamma)^{32}P$	PO_3^- and others
Permanganates	$^{55}Mn(n, \gamma)^{56}Mn$	$Mn^{2+}(MnO_2)$
Chromates	$^{50}Cr(n, \gamma)^{51}Cr$	Cr^{3+}(mono-, bi- und polynuclear)
Ferrocene	$^{58}Fe(n, \gamma)^{59}Fe$ $\Big\}$	Fe^{2+}
	$^{54}Fe(n, \gamma)^{55}Fe$	
Copper phthalocyanine	$^{63}Cu(n, \gamma)^{64}Cu$	Cu^{2+}

According to H. Müller; Angew. Chem. **79**, 132 (1967).

Several models have been brought forward to explain the chemical effects of nuclear reactions in solids. Elastic collisions of the recoiling atom with surrounding atoms have been assumed by Libby (1947) in his billiard ball model: the "hot" atom loses its energy in steps and after its last collision it is in a reaction cage, from which either it may escape as a free atom if its energy is high enough or it may recombine with the fragments in this cage with resulting secondary retention. However, some experimental results cannot be explained by this model. In the "hot zone" model (Harbottle, 1958), distribution of the kinetic energy of the recoiling atom over neighbouring atoms by collisions within about 10^{-11} s is assumed with the resultant formation of a molten hot zone in which chemical reactions (e.g. exchange and substitution reactions) take place. The molten zone cools down quickly and contains many dislocations. The disorder model (Müller, 1965) has been developed for crystalline solids. In this model formation of a great number of dislocations is assumed instead of formation of a hot zone.

The fact that atoms in crystalline solids do not suffer a recoil at low energies (<100 keV) of emitted or incident particles is made use of in Mössbauer spectroscopy (recoilless γ-ray resonance absorption; section 10.2).

9.6 Szilard–Chalmers Reactions

The background of Szilard–Chalmers reactions has already been mentioned in section 9.4. Isotopic nuclides produced by nuclear reactions can be separated by chemical methods from the target nuclides due to the chemical effects of the nuclear reactions, such as changes of the oxidation state or other changes of chemical bonds. The specific activity of the product nuclides may be high, but it depends on the degree of radiation decomposition of the compound containing the target nuclide. The possibility of separation of isotopic products from target nuclides by Szilard–Chalmers reactions is preferably used in the case of (n, γ) reactions, but it may also be applied for (γ,n), $(n, 2n)$ and (d, p) reactions. If the product nuclide does not contain inactive isotopes it is called "carrier-free". However, due to the ubiquity of stable elements and long-lived natural radioelements, the presence of small amounts of carriers must always be taken into account.

Szilard–Chalmers reactions are applicable to elements existing in different stable oxidation states or forming substitution-inert complexes. Exchange reactions between the oxidation states or with the complexes should not take place during irradiation and chemical separation, because they would cause a decrease of the specific activity. Therefore, substitution-labile complexes are not suitable.

Szilard–Chalmers reactions are characterized by the enrichment factor (i.e. the ratio of the specific activity of the radionuclide considered after separation to the average specific activity before separation), and by the yield (i.e. the ratio of the activity of the radionuclide obtained after separation to its total activity). Enrichment factors of up to about 1000 or more may be obtained, and yields of about 50 to 100% are of practical interest.

Examples of Szilard–Chalmers reactions are given in Table 9.4. Radionuclides of the halides may be obtained in high specific activities by neutron irradiation of alkyl or aryl halides or of the salts of the oxoacids. Radionuclides of other elements may also be produced in high specific activities by neutron irradiation of covalent compounds.

Szilard–Chalmers reactions are of special interest for the investigation of nuclear isomers, because they offer the possibility of separating isomeric nuclides.

Table 9.4. Yield and retention for some Szilard–Chalmers reactions.

Irradiated compound	Nuclear reaction	Yield [%]	Retention [%]
CH_3CH_2I	$^{127}I(n, \gamma)^{128}I$	54	46
$LiIO_3$		34	66
$NaIO_3$		33	67
KIO_3		33	67
NH_4IO_3		78	22
CH_2Br_2	$^{81}Br(n, \gamma)^{82}Br$	57	43
$CHBr_3$		53	47
$NaBrO_3$		90	10
$KBrO_3$		91	9
C_6H_5Cl	$^{37}Cl(n, \gamma)^{38}Cl$	65	35
$NaClO_3$		98	2
Li_2CrO_4	$^{50}Cr(n, \gamma)^{51}Cr$	34	66
Na_2CrO_4		26	74
$Na_2Cr_2O_7$		20	80
K_2CrO_4		39	61
$K_2Cr_2O_7$		10	90
$(NH_4)_2CrO_4$		82	18
$(NH_4)_2Cr_2O_7$		68	32
$LiMnO_4$	$^{55}Mn(n, \gamma)^{56}Mn$	91	9
$NaMnO_4$		91	9
$KMnO_4$		77	23
Na_3PO_4	$^{31}P(n, \gamma)^{32}P$	50	50
Na_2HPO_4		55	45
$Na_4P_2O_7$		42	58
Na_3AsO_3	$^{75}As(n, \gamma)^{76}As$	10	90
Na_2HAsO_4		40	60

(According to G. Narbottle N. Sutin, in: Advances in Inorganic Chemistry and Radiochemistry, Vol. 1 (Eds. H. J. Emelens, A. G. Sharpe), Academic Press, New York 1959)

9.7 Recoil Labelling and Self-Labelling

Chemical effects of nuclear reactions do not only cause rupture of chemical bonds, they also lead to formation of new chemical bonds, a result that may be used for preparation of labelled compounds. Recoil labelling and self-labelling both involve radiation-induced reactions and also belong to the field of radiation chemistry.

Recoil labelling was first observed by Reid in 1934. After neutron irradiation of a mixture of ethyl iodide and pentane, ^{128}I-labelled amyl iodide was found. It is produced by substitution of an H atom by an "hot" ^{128}I atom.

T- and ^{14}C-labelled compounds are of special interest in organic chemistry. T and ^{14}C can be produced with relatively high yields by the following reactions:

$$^6\text{Li}(n, \alpha)^3\text{H}; \quad \sigma = 940 \text{ b}; \quad E_1(\text{T}) = 2.74 \text{ MeV} \tag{9.23}$$

$$^3\text{He}(n, p)^3\text{H}; \quad \sigma = 5327 \text{ b}; \quad E_1(\text{T}) = 0.190 \text{ MeV} \tag{9.24}$$

$$^{14}\text{N}(n, p)^{14}\text{C}; \quad \sigma = 1.81 \text{ b}; \quad E_1(^{14}\text{C}) = 0.042 \text{ MeV} \tag{9.25}$$

(E_1 is the recoil energy.) The range of tritium atoms produced by reaction (9.23) is relatively long (about 40 μm in organic compounds), and a heterogeneous mixture of fine-grained organic substance and a lithium compound is suitable for labelling the compound. Reaction (9.24) may be used for T-labelling in the gaseous phase. For recoil labelling with ^{14}C by means of reaction (9.25) a homogeneous mixture is needed, because of the short range of the ^{14}C recoil atoms.

Substantial disadvantages of recoil labelling are the multitude of labelled compounds produced and the radiation decomposition due to the the "hot" recoil atoms and the incident radiation, which require careful chemical separation. Furthermore, the yield of a certain labelled compound is relatively small. For example, recoil labelling with ^{14}C leads to a great variety of substitution products in which any carbon atom in the molecule may be substituted by ^{14}C, and products containing additional carbon atoms. These products due to substitution and addition of carbon atoms contain about 0.1 to 10% of the ^{14}C produced. The rest of the ^{14}C is found in the form of degradation products and polymers. Recoil labelling with T leads to the formation of simple compounds such as HT and CH_3T. The specific activity is limited by the radiation decomposition. Values of the order of about 10^{11} Bq/g may be obtained in the case of recoil labelling with T and values of the order of about 10^8 Bq/g in the case of recoil labelling with ^{14}C.

Self-labelling of organic compounds with tritium has been described in detail by Wilzbach and is also called Wilzbach labelling. In this method the organic compound is stored together with tritium gas in a closed vial for about one week. Labelling proceeds by reactions with "hot" atoms and by radiation-induced reactions: on the one hand the ions $^3\text{HeT}^+$ and T^+ produced by the decay of T_2 react with the organic compound and on the other hand the β^- particles emitted by T cause ionization of the organic compounds and formation of radicals followed by reaction with T_2. The radiation-induced reactions can be enhanced by addition of a chemically inert β^- active radionuclide such as ^{85}Kr or by electric discharges.

For self-labelling with T, high partial pressures of tritium gas and gaseous or finely dispersed organic compounds are favourable. The products can be separated by gas chromatography or other chromatographic methods. Specific activities of the order of about 10^{12} Bq/g are obtained.

Self-labelling with ^{14}C by use of $^{14}\text{CO}_2$ or $^{14}\text{C}_2\text{H}_2$ is of little practical importance, because of the low specific activities of the products.

Literature

W. F. Libby, Chemistry of Energetic Atoms Produced in Nuclear Reactions, J. Amer. Chem. Soc. *69*, 2523 **(1947)**

H. A. C. McKay, The Szilard–Chalmers Process, in: Progress in Nuclear Physics (Ed. O. Frisch), Vol. I. Pergamon, Oxford, **1950**

E. H. S. Burhop, The Auger Effect, Cambridge University Press, Cambridge, **1952**

J. E. Willard, Chemical Effects of Nuclear Transformations, Annu. Rev. Nucl. Sci. *3*, 193 **(1953)**

G. Harbottle, N. Sutin, The Szilard–Chalmers Reaction in Solids, in: Advances in Inorganic Chemistry and Radiochemistry (Eds. H. J. Emeleus and A. G. Sharpe), Vol. 1, Academic Press, New York, **1959**

B. G. Harvey, Recoil Techniques in Nuclear Reaction and Fission Studies, Annu. Rev. Nucl. Sci. *10*, 235 **(1960)**

J. W. T. Spinks, R. J. Woods, An Introduction to Radiation Chemistry, Wiley, New York, **1964**

G. Harbottle, Chemical Effects of Nuclear Transformations in Inorganic Solids, Annu. Rev. Nucl. Sci. *15*, 89 **(1965)**

R. Wolfgang, Hot Atom Chemistry, Annu. Rev. Phys. Chem. *16*, 15 **(1965)**

H. Müller, Chemische Folgen von Kernumwandlungen in Festkörpern, Angew. Chem. *79*, 128 **(1967)**

A. G. Maddock, R. Wolfgang, The Chemical Effects of Nuclear Transformations, in: Nuclear Chemistry, Vol. II (Ed. L. Yaffe), Academic Press, New York, **1968**

G. Stöcklin, Chemie heißer Atome, Chemische Reaktionen als Folge von Kernprozessen, Chemische Taschenbücher Bd. 6, Verlag Chemie, Weinheim, **1969**

G. W. A. Newton, Chemical Effects of Nuclear Transformations, in: Radiochemistry, Vol. 1, Specialist Periodical Reports, The Chemical Society, London, **1972**

F. Baumgärtner, D. R. Wiles, Radiochemical Transformations and Rearrangements in Organometallic Compounds, Fortschr. Chem. Forsch. *32*, 63 **(1972)**

D. S. Urch, Nuclear Recoil Chemistry in Gases and Liquids, in: Radiochemistry, Vol. 2, Specialist Periodical Reports, The Chemical Society, London, **1975**

P. Glentworth, A. Nath, Recoil Chemistry of Solids, in: Radiochemistry, Vol. 2, Specialist Periodical Reports, The Chemical Society, London, **1975**

T. A. Carlson, Photoelectron and Auger Spectroscopy, Plenum Press, New York, **1975**

G. Harbottle, A. G. Maddock (Eds.), Chemical Effects of Nuclear Transformations in Inorganic Systems, North-Holland, Amsterdam, **1979**

T. Tominaga, E. Tachikawa, Modern Hot Atom Chemistry and its Application, Springer, Berlin, **1981**

T. Matsuura (Ed.), Hot Atom Chemistry, Kodansha, Tokyo, and Elsevier, Amsterdam, **1984**

K. H. Lieser, Einführung in die Kernchemie, 3rd ed., VCH Verlag, Weinheim, **1991**

J. P. Adloff, P. P. Gaspar, M. Imamura, A. G. Maddock, T. Matsuura, H. Sano, K. Yoshihara (Eds.), Handbook of Hot Atom Chemistry, Kodansha, Tokyo, and VCH Verlag, Weinheim, **1992**

J. P. Adloff, R. Guillaumont, Fundamentals of Radiochemistry, CRC Press, Boca Raton, FL, **1993**

A. G. Maddock, Radioactivity and the Nuclear Environment, Radiochim. Acta *70/71*, 323 **(1995)**

H. K. Yoshihara, X-Rays and Radiochemistry, Radiochim. Acta *70/71*, 333 **(1995)**

10 Influence of Chemical Bonding on Nuclear Properties

10.1 Survey

In the preceding chapter chemical effects of nuclear reactions have been discussed. On the other hand, the electronic structure, in particular chemical bonds, may affect nuclear properties. However, because the binding energies of the electrons are smaller by a factor of the order of 10^3 to 10^6 than the binding energies of the nucleons, the influence of chemical bonding on nuclear properties is, in general, relatively small.

The most drastic change of the properties of a nucleus is obtained if all the electrons are stripped off, and such a nucleus which is stable in the presence of its electron shell may become unstable. This has been discussed in section 5.5. Special conditions exist if the nuclides have only one electron in their electron shell, e.g. a nucleus of ^{238}U with one electron. The investigation of such "hydrogen-like" ions will give valuable information about their physical properties.

As far as the nuclei are concerned, low-energy excited states are most sensitive towards changes in chemical bonding. This is the field of Mössbauer spectroscopy, which has become a very important tool for the investigation of chemical bonding. It will be discussed in section 10.3.

Interaction between the magnetic field of the electrons and the nuclear spin is the basis for various techniques that are broadly applied in chemistry, atomic physics, nuclear physics and solid-state physics. The magnetic field of the electrons is due to their spin and orbital angular momentum and much larger than the magnetic field of the nucleus. Consequently, the nuclear spin is oriented in relation to the field produced by the electron shell. This leads to hyperfine spectra which can be resolved by means of optical spectrometers of very high resolution.

Nuclear magnetic resonance (NMR) is observed if an atom is placed in an external magnetic field of varying field strength B so that decoupling of J and I (the total angular momentum of the atom and the nuclear spin) is obtained. The vector $\vec{\mu}_I$ of the nuclear magnetic momentum precesses around the direction of the field in such a way that the components in the direction of the field are restricted to

$$\mu_I = g_I B_n m_I \tag{10.1}$$

where g_I is the nuclear g-factor, B_n is the nuclear magneton and m_I is the quantum number of the magnetic angular momentum. In the external field the states with different m_I have slightly different energies. The potential magnetic energy of the nucleus is given by

$$E_{\mathrm{magn}} = -\vec{\mu}_I \vec{B} = -g_I B_n m_I B \tag{10.2}$$

and the energy spacing between two adjacent levels with $m_I = \pm 1$ is

$$\Delta E = g_I B_n B \qquad (10.3)$$

These relations can be used to calculate the nuclear magnetic momentum if I is known, or vice versa. The nuclear field experienced by the nucleus is not exactly equal to the external field because of the shielding effect of the electron shell, which depends on the electron structure. Although this shielding effect is very small, it can easily be measured by use of NMR spectrometers.

Nuclear magnetic resonance (NMR) techniques are broadly applied in chemistry, but the interactions between nuclear spin and electronic structure including the NMR techniques are not discussed here in detail, because they are not considered to be part of nuclear chemistry.

10.2 Dependence of Half-lives on Chemical Bonding

As already mentioned, changes of transmutation properties are observed if electrons of the electron shell are involved in the transmutations, as in the case of electron capture (ε) or of emission of conversion electrons (e^-). The rate of both processes depends on the electron density at the nucleus. Consequently, the half-life of electron capture and the probability of emission of conversion electrons vary to a small degree with the number and the distribution of the electrons, in particular K electrons, in the electron shell.

An influence of chemical bonding on the half-life of electron capture has been measured for light nuclides such as ^7Be. In metallic ^7Be the density of the 1s electrons at the nucleus is somewhat higher than in $^7Be^{2+}$ ions. The relative changes of the decay constants of ^7Be in various compounds compared with that in metallic ^7Be are listed in Table 10.1.

Application of external pressure also leads to an increase of the decay constant of ^7Be: $\Delta\lambda/\lambda \approx 2.2 \cdot 10^{-5}$ per kilobar for ^7BeO.

An influence of chemical bonding on the emission of conversion electrons has been observed for ^{99m}Tc, ^{90m}Nb, ^{125m}Te and ^{235m}U. ^{99m}Tc ($t_{1/2} = 6.0\,h$) changes in

Table 10.1. Relative changes of the half-life of ^7Be ($\Delta\lambda/\lambda$) compared with that in ^7Be metal for various compounds of ^7Be.

Compound	$\Delta\lambda/\lambda \cdot 10^3$
BeS	+5.3
Be $(H_2O)_4^{2+}$	+2.3
Be metal	± 0
BeO	−1.4
Hexagonal BeF$_2$	−7.8
Be(C$_5$H$_5$)$_2$	−9.4
Amorphous BeF$_2$	−12.0
BeBr$_2$	−16.2

0.8% directly from the metastable state at 142.66 keV into the ground state ^{99}Tc ($t_{1/2} = 2.13 \cdot 10^5$ y) and in 99.2% via an excited state at 140.49 keV. The 2.17 keV transition is practically fully converted by emission of electrons from the M and N shells which are involved in chemical bonding. Correspondingly, changes $\Delta\lambda/\lambda$ of about 10^{-3} have been observed for Tc metal, Tc_2S_7 and $KTcO_4$.

Relatively great effects have been found for ^{235m}U ($t_{1/2} = 26.1$ min). The energy of the isomeric state is 68 eV higher than that of the ground state of ^{235}U. At these low energy differences only conversion of electrons in the 6s, 6p, 5f, 6d and 7s orbitals is possible, and because these electrons are engaged in chemical bonding, an appreciable influence of chemical bonding on the probability of isomeric transition is to be expected. Accordingly, $\Delta\lambda/\lambda \approx 0.3\%$ has been found for ^{235m}UC compared with ^{235m}U metal and $\approx 10\%$ for ^{235m}UO$_2$ ($t_{1/2} = 24.7$ min) compared with ^{235m}U implanted in Ag ($t_{1/2} = 27.1$ min).

An influence of temperature on the probability of transmutation of nuclear isomers is expected at temperatures of several hundred million Kelvin, but these influences have not yet been measured.

10.3 Dependence of Radiation Emission on Chemical Bonding

The emission of K_α- and K_β-X rays after electron capture depends on the electron density in the 2p and 3p orbitals, respectively. If these orbitals are affected by changes of chemical bonding, the ratio of the intensities of K_α- and K_β-X rays may also change. For instance, in the case of ^{51}Cr this ratio is about 10% higher for Cr(VI) compared with Cr(0).

Changes of the angular correlation between particles or γ rays emitted in immediate succession has been observed for nuclides with $I = 0$. These nuclides have a magnetic momentum and an electric quadrupole momentum which interact with the surrounding field. If this field is influenced by variations in chemical bonding, the angular correlation may also be affected.

10.4 Mössbauer Spectrometry

Mössbauer spectrometry has already been mentioned in discussing the chemical effects of nuclear reactions in solids (section 9.5).

Electrons in the inner orbitals of atoms have a finite probability of entering the nucleus, interacting with the nuclear charge distribution and thereby affecting the nuclear energy levels and γ transitions. The probability of the interaction of the electrons with the nucleus varies with the properties of the electron orbitals and consequently with chemical bonding. A γ-ray photon emitted from an isomeric state of an atom bound in a certain chemical compound will have a slightly different energy compared with that of a photon emitted by the same atom bound in another compound. This energy difference is called an isomer shift. It is extremely small – only about 10^{-10} of the energy of the γ ray emitted.

In order to be able to measure such small energy differences, several effects have to be taken into account.

The natural linewidth $\Gamma = \Delta E$ of γ rays is given by the Heisenberg uncertainty principle,

$$\Gamma \cdot \tau = h/2\pi \qquad (10.4)$$

where τ is the mean half-life of the excited state. Values of τ between about 10^{-9} and 10^{-7} are the most suitable for Mössbauer spectrometry, because they ensure sufficient resolution. At higher values of τ the linewidth is too small and at lower values of τ it is too large to be measured without experimental problems.

The most frequently used Mössbauer nuclide is ^{57}Fe, originating from the Mössbauer source ^{57}Co by electron capture (Fig. 10.1). Source and Mössbauer nuclide form a Mössbauer pair. The half-life of the first excited state of ^{57}Fe at $E^* = 14.4$ keV is 98 ns ($\tau = 1.4 \times 10^{-7}$ s) and the natural linewidth is $\Gamma = 4.6 \cdot 10^{-9}$ eV.

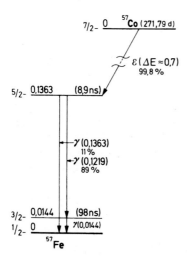

Figure 10.1. Transmutation of ^{57}Co (Mössbauer source) into ^{57}Fe (Mössbauer nuclide); Mössbauer level at 0.0144 MeV (excitation energy).

In the case of free atoms of ^{57}Fe the recoil energy is $E_1 = 1.95 \cdot 10^{-3}$ eV (eq. (9.7)), and the same kinetic energy is transferred to a free ^{57}Fe atom by an incident γ-ray photon of the energy E_1. Due to the energy E_1 transmitted to the nuclei by recoil or absorption, the emission line is shifted to lower energies, $E = E^* - E_1$, and the absorption line to higher energies, $E = E^* + E_1$, as shown in Fig. 10.2.

In crystalline solids, however, the atoms are firmly bound to neighbouring atoms, the whole system behaves like a rigid block of high mass, the ^{57}Fe atoms emitting the 14.4 keV γ-ray photons do not suffer a recoil, and photons take away the full energy of the γ transition, as explained in section 9.5. On the other hand, by absorption of the 14.4 keV γ-ray photon, an ^{57}Fe atom embedded in a crystalline solid does not receive measurable amounts of kinetic energy, and the energy of the photon is quickly distributed among the neighbouring atoms.

Another effect is due to the vibration of atoms in solids. At room temperature, this vibration leads to a Doppler effect and line broadening of the order of $D \approx 10^{-2}$ eV.

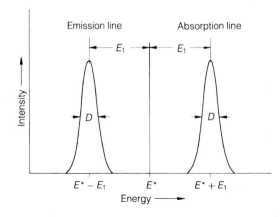

Figure 10.2. Absorption of γ-ray photons by free atoms; E^* = excitation energy; E_1 = recoil energy; D = line broadening by the Doppler effect.

To obtain sufficient resolution, this line broadening has to be suppressed by cooling to the temperature of liquid nitrogen.

Thus, the emitting and absorbing atoms have to be embedded in solids and both have to be kept at low temperatures, in order to measure recoilless resonance absorption of γ rays. These are essential conditions for Mössbauer spectrometry.

The original Mössbauer experiment has been carried out with ^{191}Os as source and an iridium foil as absorber in an arrangement shown schematically in Fig. 10.3. The Mössbauer nuclide is ^{191}Ir. Some of the 129 keV γ rays emitted after β^- decay of ^{191}Os from the first excited state of ^{191}Ir are absorbed by the atoms of ^{191}Ir in the foil, exciting these atoms from the ground state to the first excited state (resonance absorption). The latter changes with a half-life of 0.13 ns to the ground state, re-emitting 129 keV γ-ray photons at random. As a result, a decrease of the intensity is measured by the detector.

If source and absorber are in different chemical states, the nuclear energy levels of the atoms in the source and in the absorber differ by a small amount ΔE_γ, as mentioned above, and resonance absorption is obtained by moving the source with a velocity v, in order to change the kinetic energy of the photons by adding or subtracting this small amount ΔE_γ. In this way, a Mössbauer spectrum of a certain

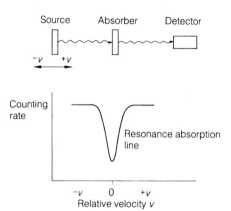

Figure 10.3. Mössbauer experiment (schematically).

compound relative to a reference compound is obtained. The location of the absorption maximum is the isomer shift δ (mm/s) which is characteristic for the compound.

The isomer shift δ is due to the electric monopole interaction between the electrons in the nucleus. Different oxidation states or different ligands lead to different values of δ. Low-spin and high-spin complexes can be distinguished. Electric quadrupole interaction and magnetic dipole interaction between the electrons and the nucleus cause electric quadrupole splitting ΔE_Q and magnetic splitting ΔE_M, respectively, of the resonance lines. The three Mössbauer parameters δ, ΔE_Q and ΔE_M give information about chemical bonding and electronic structure.

Electric quadrupole splitting ΔE_Q is only observed with nuclei exhibiting non-spherical charge distribution (nuclear spin $I > \frac{1}{2}$) and for compounds with aniso-tropic distribution of valence electrons or ligands, respectively. It gives information about the arrangement of ligands in complexes. Magnetic dipole interaction ΔE_M (also called the nuclear Zeeman effect) may be observed if a nucleus has a magnetic dipole momentum ($I \neq 0$). It causes splitting of the energy levels of nuclei into $2I + 1$ levels. For example, nuclei with $I = \frac{1}{2}$ show magnetic dipole splitting ΔE_M into two levels, but no electric quadrupole splitting ΔE_Q, and nuclei with $I = \frac{3}{2}$ exhibit electric quadrupole splitting and magnetic splitting into four levels.

Application of Mössbauer spectrometry depends on the availability of suitable sources with half-lives of excited states between about 10^{-9} and 10^{-7} s. The photon energy must not exceed 100 keV and conversion must not be too high to ensure recoilless emission and absorption. As already mentioned, ^{57}Fe, the daughter of ^{57}Co, is the most frequently used Mössbauer nuclide. ^{57}Co is used as Mössbauer source and iron of natural isotopic composition (2.17% ^{57}Fe) or enriched ^{57}Fe as absorber.

About 70 other Mössbauer pairs have also been applied, including Mössbauer nuclides such as ^{61}Ni, ^{67}Zn, ^{83}Kr, ^{99}Tc, ^{99}Ru, ^{101}Ru, ^{107}Ag, ^{117}Sn, ^{119}Sn, ^{123}Sb, ^{125}Te, ^{127}I, ^{129}Xe, ^{133}Cs, ^{139}La, ^{151}Eu, ^{161}Dy, ^{169}Tm, ^{181}Ta, ^{182}W, ^{187}Re, ^{189}Os, ^{190}Os, ^{191}Ir, ^{193}Ir, ^{195}Pt, ^{197}Au, ^{199}Hg and ^{237}Np.

^{237}Np is formed by α decay of ^{241}Am via an excited state at 59.6 keV (Mössbauer level) which changes with a half-life of 68 ns into the ground state. This transition exhibits an exceptionally wide range of isomer shifts for the various oxidation states of Np from about +70 mm/s for Li_5NpO_6 (Np(VII)) to about +47 mm/s for $K_3NpO_2F_5$ (Np(VI)), about +18 mm/s for $NpO_2(OH) \cdot H_2O$ (Np(V)), about +5 mm/s for NpF_4 and about −41 mm/s for NpF_3. This wide range allows unambiguous identification of the oxidation state of Np in solid compounds.

Mössbauer spectrometry gives information about the chemical environment of the Mössbauer nuclide in the excited state at the instant of emission of the photon. It does not necessarily reflect the normal chemical state of the daughter nuclide, because of the after-effects that follow the decay of the mother nuclide (recoil and excitation effects, including emission of Auger electrons). At very short lifetimes of the excited state, ionization and excitation effects may not have attained relaxation at the instant of emission of the γ-ray photon; this results in a time-dependent pattern of the Mössbauer spectrum.

Literature

Mössbauer Spectroscopy

R. L. Mössbauer, Kernresonanzfluoreszenz von Gammastrahlen in ^{191}Ir, Z. Physik *151*, 124 **(1958)**

R. L. Mössbauer, Recoilless Nuclear Resonance Absorption, Annu. Rev. Nucl. Sci. *12*, 1 **(1962)**

N. N. Greenwood, T. C. Gibb, Mössbauer Spectroscopy, Chapman and Hall, London, **1971**

G. M. Bancroft, Mössbauer Spectroscopy, Wiley, New York, **1973**

P. Gütlich, Mössbauer Spectroscopy in Chemistry, in: Topics in Applied Physics, Vol. 5 (Ed. U. Gonser), Springer, Berlin, **1975**

T. C. Gibb, Principles of Mössbauer Spectroscopy, Halsted, New York, **1976**

R. L. Cohen (Ed.), Applications of Mössbauer Spectroscopy, Academic Press, New York, **1976**

P. Gütlich, R. Link, A. Trautwein, Mössbauer Spectroscopy and Transition Metal Chemistry, Inorganic Chemistry Concepts 3, Springer, Berlin, **1978**

G. K. Shenoy, F. E. Wagner (Eds.), Mössbauer Isomer Shifts, North-Holland, Amsterdam, **1978**

Other Effects

J. I. Vargas, The Chemical Applications of Angular Correlation and Half-life Measurements, in: Radiochemistry, International Review of Sciences, Inorganic Chemistry, Series One, Vol. 8 (Ed. A. G. Maddock), Butterworths, London, **1971**

C. Keller, Zum Einfluss der chemischen Bindung auf die Halbwertzeit von Radionukliden, Chem. Ztg. *99*, 365 **(1975)**

J. P. Adloff, Application to Chemistry of Electric Quadrupole Perturbation of γ–γ Angular Correlations, Radiochim. Acta *25*, 57 **(1978)**

T. Matsuura (Ed.), Hot Atom Chemistry, Kodansha, Tokyo, **1984**

K. Yoshihara, Chemical Nuclear Probes Using Photon Intensity Ratios, in: Topics in Current Chemistry, Vol. 157, Springer, Berlin, **1990**

N. C. Pyper, M. R. Harston, Atomic Effects on β decay, Proc. Roy. Soc. London, A 420, 277 **(1988)**

J. F. Burke, C. M. Archer, K. Wei Chu, I. A. Latham, R. G. Egdell, A Study of Core Electron Binding Energies in Technetium-99m Complexes by Internal Conversion Spectroscopy, Appl. Rad. Isotop. *42*, 49 **(1991)**

11 Nuclear Energy, Nuclear Reactors, Nuclear Fuel and Fuel Cycles

11.1 Energy Production by Nuclear Fission

For the production of energy by nuclear fission the following features are decisive:

- The energy ΔE set free by fission of heavy nuclei is very high (section 8.9).
- As several neutrons are liberated by fission of heavy nuclei ($\nu = 2$–3, section 8.9), a chain reaction is possible if at least one of these neutrons triggers another fission reaction.

The energy ΔE can be assessed from the mean binding energy per nucleon (Fig. 2.6) to amount to ≈ 200 MeV. In case of fission of ^{235}U, ΔE is shared out in the following way:

- Kinetic energy of the fission products ≈ 167 MeV
- Kinetic energy of the neutrons $\approx$ 5 MeV
- Prompt γ rays $\approx$ 6 MeV
- β decay of the fission products $\approx$ 8 MeV
- γ transmutation of the fission products $\approx$ 6 MeV
- Neutrinos $\approx$ 12 MeV

The sum of the kinetic energy of the fission products and the energy of β decay can be determined calorimetrically. The energy of the neutrons and the γ rays is usable only inasmuch as neutrons and γ rays are absorbed in the medium considered. The energy of the neutrinos is lost, because of their small interaction with matter.

Disregarding the energy of the neutrinos, the sum of the energies listed above is ≈ 192 MeV. To this value the energy set free by absorption of the neutrons has to be added. This energy is mainly given off by emission of γ-ray photons. Setting $\bar{\nu} = 2.43$ for the fission of ^{235}U and assuming a mean value of 5 MeV for the energy set free by absorption of one neutron, the contribution of neutron absorption is about 12 MeV, and the usable energy amounts to about 204 MeV.

The greatest proportion of fissile nuclides loaded into a nuclear reactor undergoes fission, but another part of these nuclides is transformed by nuclear reactions, in particular (n, γ) reactions, such as ^{238}U(n, γ)^{239}U or ^{235}U(n, γ)^{236}U, into other nuclides that are, at least partly, less fissile. The fraction of nuclides suffering fission is given by Σ_f / Σ_a, the ratio of the macroscopic cross sections for fission and for neutron absorption, where Σ_a comprises (n, f) as well as (n, γ) reactions. For ^{235}U and thermal neutrons this ratio is 0.855. Consequently, from 1 kg of ^{235}U up to $1.98 \cdot 10^7$ kW h could be produced, provided that the energy of the γ rays could be used. In practice, however, about 60% of the energy of the γ rays is lost which reduces the usable energy per fission to about 190 MeV and the energy producible from 1 kg of ^{235}U to a maximum value of $1.85 \cdot 10^7$ kW h ($7.70 \cdot 10^5$ MW d per ton of ^{235}U).

Comparison between the energy production by nuclear fission and by burning of coal leads to the following result: by burning of 1 kg of carbon or coal an energy of 9.4 kW h is set free, and the ratio of the energies producible from 1 kg of ^{235}U and from 1 kg of coal is $\approx 2 \cdot 10^6$. This is the stimulus for the use of nuclear energy. Comparing the energies set free by explosives leads to the result that fission of 1 kg ^{235}U or ^{239}Pu gives the same energy as explosion of $20 \cdot 10^6$ kg of trinitrotoluene (TNT). This provided the impetus to develop nuclear weapons.

The possibility of operating nuclear fission in the form of a chain reaction is governed by the effective multiplication factor k_{eff}, the ratio of the number of neutrons in the second generation to that in the first generation:

$$k_{\text{eff}} = \bar{v}\varepsilon(1 - l_{\text{f}})p(1 - l_{\text{t}})f\frac{\Sigma_{\text{f}}}{\Sigma_{\text{a}}} = \eta\varepsilon(1 - l_{\text{f}})p(1 - l_{\text{t}})f \tag{11.1}$$

This is illustrated in Fig. 11.1. Fission of ^{235}U may be taken as an example: $\bar{v}$ is the average number of neutrons liberated in nuclear fission of ^{235}U by thermal neutrons and $\Sigma_{\text{f}}/\Sigma_{\text{a}}$ is the ratio of the macroscopic cross sections for fission and for absorption of neutrons. For pure nuclides (e.g. pure ^{235}U), $\Sigma_{\text{f}}/\Sigma_{\text{a}} = \sigma_{\text{f}}/\sigma_{\text{a}}$. The fission factor $\eta = \bar{v}\Sigma_{\text{f}}/\Sigma_{\text{a}}$ is the number of fission neutrons relative to the number of thermal neutrons absorbed in uranium; ε, also called the fast fission factor, takes into account the production of neutrons by fission of ^{238}U ($1.0 < \varepsilon < 1.1$). The fraction l_{f} of neutrons is lost, whereas the rest are slowed down to low energies in a moderator. Suitable moderators are materials with low absorption cross section for neutrons and low mass, such as H_2O, D_2O or graphite. The fraction p of neutrons escaping capture while slowing down is referred to as the resonance escape probability. The fraction $(1 - p)$ of neutrons is captured by ^{238}U, inducing the reaction ^{238}U(n, γ)^{239}U followed by β^- decay into ^{239}Np and ^{239}Pu. From the remaining neutrons, the fraction

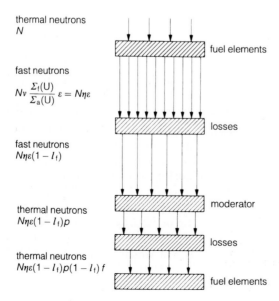

thermal neutrons
N

fast neutrons

$Nv\dfrac{\Sigma_{\text{f}}(U)}{\Sigma_{\text{a}}(U)}\varepsilon = N\eta\varepsilon$

fast neutrons
$N\eta\varepsilon(1 - l_{\text{f}})$

thermal neutrons
$N\eta\varepsilon(1 - l_{\text{f}})p$

thermal neutrons
$N\eta\varepsilon(1 - l_{\text{f}})p(1 - l_{\text{t}})f$

fuel elements

losses

moderator

losses

fuel elements

Figure 11.1. Effective multiplication factor k_{eff} for neutrons (schematically).

l_t escapes from the system, the fraction f is captured by uranium atoms and the fraction $(1 - f)$ by other atoms; f is called the thermal utilization. Both p and f depend on the nature and the arrangement of fuel and moderator.

To make the losses of fast and thermal neutrons (l_f and l_t, respectively) as small as possible, the reactor core is surrounded by a reflector, preferably made of graphite or beryllium.

The fission cross section σ_f for ^{235}U, ^{239}Pu and ^{238}U, the ratio σ_f/σ_a for ^{235}U and ^{238}U and the values of $\eta = \bar{\nu}\sigma_f/\sigma_a$ for ^{235}U and ^{238}U are plotted in Figs. 11.2, 11.3 and 11.4, respectively, as a function of the neutron energy.

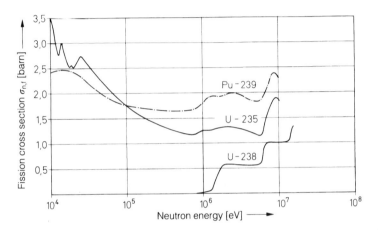

Figure 11.2. Fission cross section $\sigma_{n,f}$ for ^{235}U, ^{238}U and ^{239}Pu as a function of the neutron energy.

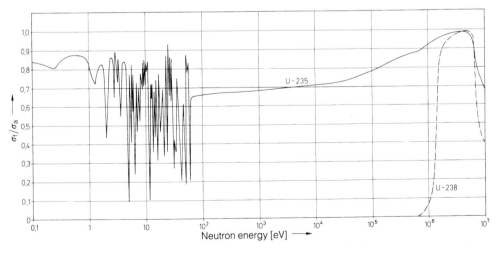

Figure 11.3. The ratio σ_f/σ_a for ^{235}U and ^{238}U as a function of the neutron energy.

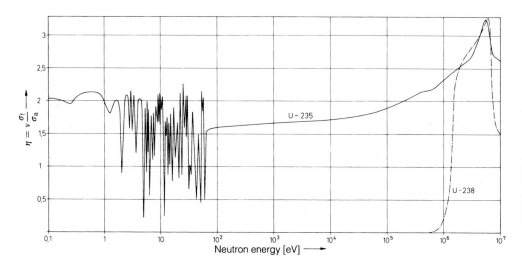

Figure 11.4. $\eta = v\dfrac{\sigma_f}{\sigma_a}$ as a function of the neutron energy for ^{235}U and ^{238}U.

The following operation conditions are distinguished:

– Thermal neutrons ($E_n \leq 1\,\text{eV}$): Fission of ^{235}U and ^{239}Pu prevails ($\sigma_{n,f} > \sigma_{n,\gamma}$). Although $\sigma_{n,\gamma}$ for ^{238}U is small, (n, γ) reactions play an important role if large amounts of ^{238}U are present.
– Epithermal neutrons ($1\,\text{eV} \leq E_n \leq 0.1\,\text{MeV}$): The influences of neutron capture and fission resonances increase. Heavier isotopes of U and Pu are formed by (n, γ) reactions, such as ^{235}U(n, γ)^{236}U, ^{238}U(n, γ)^{239}U, ^{239}Pu(n, γ)^{240}Pu.
– Fast neutrons ($E_n > 0.1\,\text{MeV}$): Fission prevails ($\sigma_{n,f} > \sigma_{n,\gamma}$). ^{238}U also becomes fissile at $E_n \approx 0.6\,\text{MeV}$, reaching a constant value of $\approx 0.5\,\text{b}$ at $E_n \geq 2\,\text{MeV}$.

The multiplication factor k_{eff} can be appreciably increased by heterogeneous arrangement of uranium and moderator, because resonance absorption of the neutrons by ^{238}U is low after the neutrons have been slowed down in the moderator. Then p becomes markedly higher, f somewhat lower and pf becomes higher than in the case of homogeneous arrangement of fissile material and moderator.

The following possibilities are distinguished:

– $k_{\text{eff}} < 1$: The reactor is subcritical, a chain reaction cannot occur.
– $k_{\text{eff}} = 1$: The reactor is critical, a chain reaction is possible.
– $k_{\text{eff}} > 1$: The reactor is supercritical.

The term $k_{\text{eff}} - 1$ is called excess reactivity, and $(k_{\text{eff}} - 1)/k_{\text{eff}}$ is called reactivity. Because the fissile material is continuously used up by fission and because the fission products absorb neutrons, a certain excess reactivity is necessary to operate a nuclear reactor. This excess reactivity is compensated by control rods that absorb the excess neutrons. These control rods contain materials of high neutron absorption cross section, such as boron, cadmium or rare-earth elements. The excess reactivity can also be balanced by addition to the coolant of neutron-absorbing substances such as boric acid.

In a medium of infinite extent the neutron losses l_f and l_t become negligible and the multiplication factor is given by

$$k_\infty = \eta \, \varepsilon \, p \, f \qquad (11.2)$$

While ε is somewhat greater than 1, both p and f are somewhat smaller than 1. Therefore, for approximate calculations, $\varepsilon \, p \, f$ can be set ≈ 1. The fission factor η varies appreciably with the energy of the neutrons, as shown in Fig. 11.4 for ^{235}U.

The neutron losses in a reactor of finite dimensions can be taken into account approximately by the sum $L_S^2 + L^2$, where L_S is the mean slowing-down length of the fission neutrons in the moderator and L the mean diffusion length in the fuel–moderator mixture. For a spherical reactor of radius R the approximate relation is

$$k_\infty - k_{\text{eff}} = \pi^2 \frac{L_S^2 + L^2}{R^2} \quad \text{or} \quad R = \pi \left(\frac{L_S^2 + L^2}{k_\infty - k_{\text{eff}}} \right)^{1/2} \qquad (11.3)$$

L_S is of the order of 10 cm (H_2O: 5.7 cm; D_2O: 11.0 cm; Be: 9.9 cm; C: 18.7 cm); furthermore, in most cases of practical interest, $L^2 \ll L_S^2$. By use of eqs. (11.2) and (11.3) the critical size of spherical nuclear reactors, given by $k_{\text{eff}} = 1$, can be assessed.

For operation of nuclear reactors, the delayed neutrons (section 8.9) play an important role, because they cause an increase in the time available for control. The multiplication factor due to the prompt neutrons alone is $k_{\text{eff}}(1 - \beta)$, β being the contribution of the delayed neutrons, and as long as $k_{\text{eff}}(1 - \beta) < 1$, the delayed neutrons are necessary to keep the chain reaction going. In the fission of ^{235}U, 0.65% of the fission neutrons are emitted as delayed neutrons from some neutron-rich fission fragments such as ^{87}Kr or ^{137}Xe.

The ratio Σ_f / Σ_a and correspondingly also the value of η change continuously with the consumption of nuclear fuel, and in practice it is not possible to use up the fuel quantitatively. The fraction that can be used depends on the composition of the fuel and the operation conditions of the reactor. The widely used boiling- or pressurized-water reactors contain enriched uranium with a content of about 3% of ^{235}U as fuel which can be burned up until a residual concentration of about 0.8% ^{235}U is obtained. About half of the energy gained during the time of operation is produced by fission of ^{235}U and the rest by fission of ^{238}U and of ^{239}Pu formed by neutron capture and β^- decay from ^{238}U.

11.2 Nuclear Fuel and Fuel Cycles

For the use of nuclides as nuclear fuel, their fissionability is the most important aspect. High fission yields by thermal neutrons are obtained if the binding energy of an additional neutron is higher than the fission barrier. Fission barriers, neutron binding energies and fission cross sections are listed for some nuclides in Table 11.1. The fission cross sections are high for ^{233}U, ^{235}U and ^{239}Pu, as already mentioned in section 8.9. These nuclides are fissile with high yields by thermal neutrons (thermal reactors). Fission of the even–even nuclides ^{232}Th and ^{238}U requires the use of high-energy (fast) neutrons (fast reactors).

Table 11.1. Fission barriers, binding energies of an additional neutron and fission cross sections for some heavy nuclides.

Nuclide	Fission barrier [MeV]	Binding energy of an additional neutron [MeV]	Fission cross section $\sigma_{n,f}$ [b]
^{232}Th	7.5	5.4	0.00004
^{233}U	6.0	7.0	531
^{235}U	6.5	6.8	582
^{238}U	7.0	5.5	<0.0005
^{239}Pu	5.0	6.6	743

^{233}U, ^{235}U and ^{239}Pu are most suitable as nuclear fuel in reactors operating with thermal neutrons. ^{235}U is present in natural uranium with an isotopic abundance of 0.72%. Because of this low concentration, use of natural uranium as a nuclear fuel is only possible if the neutron losses are kept as low as possible. For this purpose, D_2O or graphite may be used as moderators. Graphite was applied in the first nuclear reactors. D_2O is still used as moderator and coolant in heavy-water reactors (HWR).

^{239}Pu and ^{233}U are produced from ^{238}U and ^{232}Th, respectively, by the following reactions:

$$^{238}U(n,\gamma)^{239}U \xrightarrow[23.5\,m]{\beta^-} {}^{239}Np \xrightarrow[2.35\,d]{\beta^-} {}^{239}Pu \tag{11.4}$$

$$^{232}Th(n,\gamma)^{233}Th \xrightarrow[22.3\,m]{\beta^-} {}^{233}Pa \xrightarrow[27.0\,d]{\beta^-} {}^{233}U \tag{11.5}$$

By reaction (11.4) ^{239}Pu is produced in all reactors operated with uranium. Special types of reactors are designed with the aim of producing larger amounts of ^{239}Pu or ^{233}U, respectively. The concept of these reactors is to use one of the neutrons released by fission to initiate another fission, and a second one to produce another fissile atom. The ratio of the number of fissile atoms produced to the number of atoms used up by fission is called the conversion factor c. If $c > 1$, the reactor is called a breeder reactor; if $c < 1$, it is called a converter.

Table 11.2. Some data on nuclear fuel.

	^{233}U	^{235}U	^{239}Pu
Half-life [y]	$1.59 \cdot 10^5$	$7.038 \cdot 10^8$	$2.411 \cdot 10^4$
$\sigma_{n,\gamma}$ for thermal neutrons [b]	48	99	269
$\sigma_{n,f}$ for thermal neutrons [b]	531	582	743
Average number of neutrons liberated in thermal neutron fission	3.13	2.43	2.87

Some properties of ^{233}U, ^{235}U and ^{239}Pu are summarized in Table 11.2. The following kinds of fuel are distinguished:

– natural uranium;
– weakly enriched uranium ($\approx 3\%$ ^{235}U);
– highly enriched uranium ($>90\%$ ^{235}U);
– mixtures of uranium and plutonium;
– mixtures of uranium and thorium.

The energy output of a nuclear reactor is characterized by the "burn-up" which is usually given in megawatt-days (MW d) per ton of fuel. By use of natural uranium a burn-up of about 10^4 MW d per ton is achieved. This corresponds to the fission of about 13 kg of fissile nuclides per ton of fuel, the greatest part being ^{239}Pu produced by reaction (11.4) The fission of ^{235}U leads to a decrease of its concentration below the natural isotopic abundance of 0.72%. From the economic point of view, only the recovery of plutonium is of interest.

Weakly enriched uranium, containing about 3% ^{235}U, is most widely used in nuclear power stations. The usual burn-up is about $3.4 \cdot 10^4$ MW d per ton of fuel, corresponding to the fission of about 45 kg of the fuel. About half of this is ^{235}U, and the abundance of this nuclide in the fuel decreases from $\approx 3.0\%$ to $\approx 0.8\%$. The concentration of long-lived isotopes of plutonium increases to $\approx 0.9\%$ and that of fission products to $\approx 3.4\%$.

Highly enriched uranium containing $>90\%$ ^{235}U is, in general, only used in research reactors. The production of fissile nuclides is negligible, and the maximum burn-up is of the order of 10^5 MW d per ton of fuel, corresponding to the consumption of about 13% of the fuel. Recovery of the remaining ^{235}U is of economic interest.

Mixtures of uranium and plutonium may be used instead of weakly enriched uranium in thermal reactors and are applied in fast breeder reactors, which are operated with the aim of producing more fissile material than is consumed by fission. The main fissile nuclide is ^{239}Pu, which is continuously reproduced according to reaction (11.4) from ^{238}U. In fast breeder reactors operating with about 6 tons of Pu and about 100 tons of U a net gain of fissile ^{239}Pu may be obtained. The burn-up is about 10^5 MW d per ton of fuel, and reprocessing with the aim to recover the plutonium is expedient.

Mixtures of enriched uranium and thorium are preferably used in high-temperature reactors operating as thorium converters. This means that ^{233}U is produced according to reaction (11.5) and serves as nuclear fuel. The conversion factor, given by the ratio of ^{233}U produced by reaction (11.5) to the amount of nuclides used up by

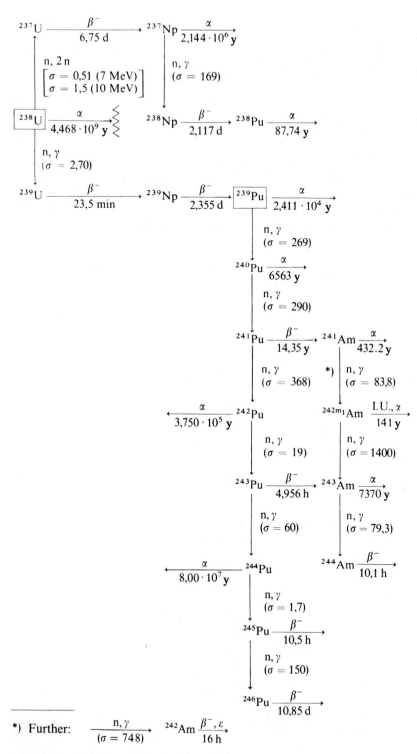

Figure 11.5. Nuclear reactions with ^{238}U (σ [barn] for thermal neutrons).

fission, varies between about 0.65 and 0.95. Mixtures of highly enriched uranium ($>90\%$ ^{235}U) and thorium in a ratio of the order of 1:10 are preferably applied. The burn-up is also of the order of 10^5 MW d per ton of fuel. Reprocessing is carried out with the aim of separating U, Th and the relatively small amounts of Pu. The uranium fraction contains ^{238}U, ^{235}U and ^{233}U.

High-temperature reactors may also be operated with enriched uranium containing about 10% ^{235}U. As in the case of weakly enriched uranium, ^{235}U is the only fissile material at the beginning, and is supplemented by the production of ^{239}Pu according to reaction (10.4).

The reactions taking place with ^{238}U in a nuclear reactor are summarized in Fig. 11.5, those occurring with ^{232}Th in Fig. 11.6. The main products are the long-lived nuclides ^{239}Pu and ^{233}U, respectively. However, other isotopes of Pu and isotopes of Am and Cm are also produced from ^{238}U and isotopes of U and Pa from ^{232}Th. The relative amounts of these radionuclides increase with the time of irradiation.

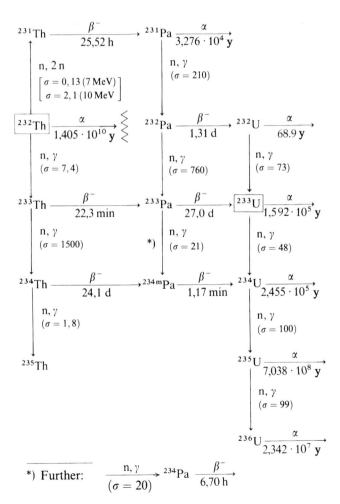

Figure 11.6. Nuclear reactions with ^{232}Th (σ [barn] for thermal neutrons).

The process route of uranium as nuclear fuel is shown in Fig. 11.7. It begins with processing of uranium ores, from which pure uranium compounds are produced. These may be used in the natural isotopic composition or transformed into other compounds suitable for isotope separation. The next step is the production of fuel elements for the special requirements of reactor operation. Solutions of uranium compounds are applied only in homogeneous reactors.

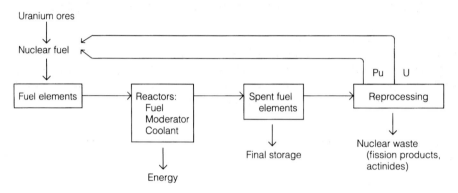

Figure 11.7. Process route of uranium as a nuclear fuel.

Besides the fuel elements, the other main components in a nuclear reactor are the moderator and the coolant. Moderator and coolant may be identical.

In general, the fuel elements remain in the reactor for several years. During this time the chemical composition and the properties of the fuel change markedly. By burning-up, a great variety of fission products is produced. The multiplication factor decreases due to the decrease of the concentration of fissile material, and the generation of fission products leads to an increasing absorption of neutrons. Absorption cross sections for thermal neutrons are exceptionally high for some fission products, such as ^{133}Xe (fission yield 6.4%, $\sigma_a = 2.7 \cdot 10^6$ b) and many lanthanides. This high neutron absorption must be compensated by an excess of fissile material. However, at a certain burn-up further use of the fuel elements becomes uneconomic, and the fuel elements are exchanged for new ones.

After burn-up, the fuel elements are stored under water for radiation protection and cooling, for at least several months and generally for about one year. Afterwards, they may be either disposed of or reprocessed in order to separate the fuel into three fractions: uranium, plutonium, and fission products including the rest of the actinides. Uranium and plutonium may be re-used as nuclear fuel, thus closing the U/Pu fuel cycle. The fission products and the rest of the actinides are converted into chemical forms that are suitable for long-term storage.

In the case of fuel elements containing ^{232}Th, uranium and thorium may be recycled (U/Th fuel cycle).

Handling of fissile material (plutonium and enriched uranium) requires strict observance of criticality conditions. As elucidated in section 11.1, criticality depends on the properties and the mass of fissile material in the system, its concentration (including local concentrations) and the presence of a moderator such as water. In all operations with Pu and enriched U only limited amounts are permitted to be handled, and samples of these materials have to be stored in portions of limited mass and at appropriate distances from each other.

The process route of nuclear fuel will be considered in more detail in sections 11.3 to 11.8, mainly for uranium, but also for other kinds of fuel.

11.3 Production of Uranium and Uranium Compounds

The route from uranium ores to uranium concentrates is summarized in Fig. 11.8. The mean concentration of U in the earth's crust is only about 0.0003%. Ores containing high percentages of U are rare. Many uranium ore deposits contain only about 0.1 to 1% U. Relatively high amounts of U are dissolved in the oceans (about $4 \cdot 10^9$ tons), but in rather low concentrations ($\approx 3 \, mg/m^3$).

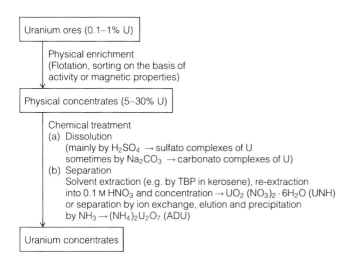

Figure 11.8. The route from uranium ores to uranium concentrates.

In the course of processing of uranium ores, appreciable amounts of long-lived radioactive decay products of uranium are obtained, as listed in Table 11.3. With respect to radiation hazards, they have to be handled carefully. Some of them, such as ^{230}Th, ^{231}Pa, ^{226}Ra or ^{210}Pb, may be isolated for practical use.

For production of uranium compounds suitable for use in nuclear reactors or for isotope separation, further chemical procedures are applied, as indicated in Fig. 11.9. Nuclear purity means that the compounds are free of nuclides with high neutron absorption cross section, i.e. free of boron, cadmium and rare-earth elements. Selective extraction procedures are most suitable for this purpose. Uranyl nitrate hexahydrate ($UO_2(NO_3)_2 \cdot 6H_2O$; UNH) is obtained by concentration of solutions of $UO_2(NO_3)_2$, and ammonium diuranate (($NH_4)_2U_2O_7$; ADU) by precipitation with ammonia.

Table 11.3. Long-lived members of the ^{238}U and ^{235}U decay series in secular radioactive equilibrium.

Atomic number Z	Element	Mass number A		Half-life
		Uranium family $4n + 2$	Actinium family $4n + 3$	
92	U	238		$4.468 \cdot 10^9$ y
			235	$7.038 \cdot 10^8$ y
		234		$2.455 \cdot 10^5$ y
91	Pa		231	$3.276 \cdot 10^4$ y
90	Th	234		24.10 d
		230 (Ionium)		$7.54 \cdot 10^4$ y
			227	18.72 d
89	Ac		227	21.77 y
88	Ra			1600 y
			223	11.43 d
86	Rn	222		3.825 d
84	Po	210		138.38 d
83	Bi	210		5.013 d
82	Pb	210		22.3 y

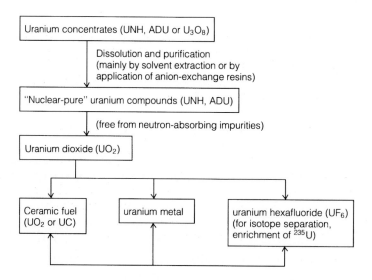

Figure 11.9. The route from uranium concentrates to nuclear fuel.

From UNH or ADU, UO_2 is obtained in two steps:

$$UNH \text{ or } ADU \xrightarrow{350\,°C} UO_3 \xrightarrow{H_2/600\,°C} UO_2 \tag{11.6}$$

UO_2 may be used as nuclear fuel or transformed into metallic uranium or into UF_6:

$$UO_2 \xrightarrow{HF/450\,°C} UF_4 \begin{array}{c} \xrightarrow{Ca \text{ (or Mg,Na)}} U \\ \xrightarrow{F_2 \text{ (or ClF}_3\text{)}} UF_6 \end{array} \tag{11.7}$$

Uranium metal was used as fuel in early types of reactors. UF_6, sublimating at 56 °C, is used to separate the isotopes ^{235}U and ^{238}U. After isotope separation, UO_2 may be obtained from UF_6 by hydrolytic decomposition, precipitation of U as ADU and heating, and uranium metal may be produced by reduction with hydrogen to UF_4 and further reduction with Ca. Handling of enriched uranium compounds requires small-scale operations with amounts of the order of 1 to 10 kg, depending on the conditions, to exclude criticality.

11.4 Fuel Elements

The design of fuel elements depends on the type of reactor and on the operating conditions. Fabrication of fuel elements does not apply for homogeneous reactors in which the fuel is used in the form of a solution of uranyl sulfate or uranyl [^{15}N] nitrate. In heterogeneous reactors, the fuel is applied in the form of metals or alloys or in the form of ceramic substances, such as UO_2, UC or mixtures with other components.

In order to prevent corrosion of the fuel and escape of fission products, the fuel is tightly enclosed in fuel rods. Good heat transfer and low neutron absorption are important properties of the cladding. Generally, the fuel rods are assembled to fuel elements to make their exchange easier.

If metallic uranium is used as fuel, the modifications of the metal and their properties have to be taken into account (Table 11.4). The anisotropic thermal expansion of α-U leads to plastic deformations which restrict the use of metallic uranium considerably. Furthermore, by the difference in the density of α-U and β-U the applica-

Table 11.4. Modifications of uranium metal.

	Temperature range [°C]	Crystal lattice	Density [$g\,cm^{-3}$]
α-U	Up to 668	Orthorhombic	19.04 (25 °C)
β-U	668–774	Tetragonal	18.11 (720 °C)
γ-U	774–1132	Cubic (b.c.)[a]	18.06 (805 °C)

[a] Body-centred.

tion of uranium metal is limited to temperatures up to about 660 °C. By addition of Mo the γ-phase can be stabilized down to room temperature, but the negative influence of the high neutron absorption of Mo must be compensated by a higher content of ^{235}U. For these reasons, metallic uranium is not used in modern reactors. It is applied in gas-cooled, graphite-moderated reactors operating with natural uranium (Calder Hall type). In some research reactors alloys of uranium and aluminium or zirconium are used, containing up to about 20% uranium.

The metallurgical properties of metallic plutonium are even more unfavourable than those of uranium. The melting point of Pu is 639 °C and six solid phases are known. Furthermore, the critical mass of a reactor operating with pure Pu as fuel is below 10 kg, and it would be very difficult to take away the heat from such a small amount of material. A great number of plutonium alloys have been investigated with respect to their possible use as nuclear fuel, but they have not found practical application.

Some properties of the ceramic fuels UO_2 and UC are summarized in Table 11.5. UO_2 is preferably used as nuclear fuel in all modern light-water reactors (LWR) of the boiling-water (BWR) as well as of the pressurized-water (PWR) type. The main advantages of UO_2 are the high melting point and the resistance to H_2, H_2O, CO_2 and radiation. The main disadvantage is the low thermal conductivity, which has to be compensated by application of thin fuel rods.

Table 11.5. Properties of uranium dioxide and uranium carbide.

	UO_2	UC
Density at 20 °C [g cm^{-3}]	10.96	13.63
Melting point [°C]	2750	2375
Thermal conductivity [J cm^{-1} s^{-1} K^{-1}]	0.036	0.213
Specific heat [J g^{-1} K^{-1}]	0.239	0.201
	(at 25 °C)	(at 100 °C)
Coefficient of thermal expansion [K^{-1}]	$9.1 \cdot 10^{-6}$	$10.4 \cdot 10^{-6}$
Crystal lattice type	Cubic (b.c.)[a]	Cubic (f.c.)[b]
	(CaF$_2$ type)	(NaCl type)

[a] Body-centred.
[b] Face-centred.

UO_2 is a non-stoichiometric compound. Freshly reduced with hydrogen, it has the composition $UO_{2.0}$, but in air it takes up oxygen and the composition varies with the partial pressure of O_2 between $UO_{2.0}$ and $UO_{2.25}$. For use as nuclear fuel, pellets of UO_2 about 1 cm in diameter and 1 cm in height are produced. By sintering at 1600–1700 °C in hydrogen, the content of excess oxygen in UO_{2+x} is reduced to x < 0.03 and about 98% of the theoretical density is obtained.

The behaviour of UO_2 pellets in a nuclear reactor is determined by the high temperature gradient in the pellets. Recrystallization takes place and hollow spaces are formed in the centre. However, up to a burn-up of about 20 000 MW d per ton these effects are of little importance, and UO_2 is the most favourable fuel for light-water reactors.

PuO$_2$ is also well suited as a nuclear fuel. It is often used in the form of a UO$_2$/PuO$_2$ mixture ("mixed oxides"; MOX) containing up to about 20% PuO$_2$. UO$_2$/PuO$_2$ mixtures may be applied in thermal reactors instead of enriched uranium, or in fast breeder reactors. Pellets of ThO$_2$ can be used in thermal converters for production of ^{233}U.

The main advantage of UC is the high thermal conductivity. On the other hand, the low chemical resistance is a major disadvantage: UC is decomposed by water below 100 °C, which is prohibitive for its use in water-cooled reactors. However, UC may be applied in gas-cooled reactors or in the form of UC/PuC mixtures in fast sodium-cooled breeder reactors.

The properties of some metals that are discussed as cladding materials for the manufacture of fuel rods are listed in Table 11.6. Al has many advantages, but it reacts with U at higher temperatures to intermetallic phases such as UAl$_3$. Mg was applied in the first reactor of the Calder Hall type, starting operation in 1956. However, the use of Mg limits the maximum temperature of operation to 400 °C. Be is not corrosion-resistant to water. Zr is very resistant to corrosion as well as to temperature. However, it must be carefully refined to separate it from Hf, which exhibits high neutron absorption. Zr and its alloys zircaloy-2 and zircaloy-4 are preferably used in modern nuclear reactors. Steel has favourable mechanical properties, but it can only be used in the form of thin sheets, because of its relatively high neutron absorption cross section. The other metals listed in Table 11.6 are also unfavourable, because of the relatively high values of σ_a.

Table 11.6. Properties of some metals considered as cladding materials for nuclear fuel.

Element	Atomic number Z	Absorption cross section for thermal neutrons σ_a [barn]	Melting point [°C]	Thermal conductivity at 20 °C [J cm^{-1} s^{-1} K^{-1}]	Specific heat [J g^{-1} K^{-1}]	Coefficient of thermal expansion [K^{-1}]	Density at 20° [g cm^{-3}]
Be	4	0.009	1285	1.591	1.800 (20 °C)	11.6 · 10^{-6}	1.848
Mg	12	0.063	650	1.574	1.047 (25 °C)	25.8 · 10^{-6}	1.74
Zr	40	0.185	1845	0.209	0.335 (200 °C)	6.11 · 10^{-6}	6.51
Al	13	0.232	660.2	2.106	0.871 (0 °C)	23.8 · 10^{-6}	2.699
Nb	41	1.15	2468	0.553	0.272 (0 °C)	7.2 · 10^{-6}	8.57
Fe	26	2.55	1539	0.754	0.473 (20 °C)	11.7 · 10^{-6}	7.866
Mo	42	2.65	2622	1.340 (0 °C)	0.247 (0 °C)	5.1 · 10^{-6}	10.22
Cr	24	3.1	1875	0.670	0.465 (25 °C)	6.2 · 10^{-6}	7.19
Ni	28	4.43	1455	0.670	0.448 (25 °C)	13.3 · 10^{-6}	8.90
V	23	5.04	1710	0.310 (100 °C)	0.502 (0 °C)	8.3 · 10^{-6}	6.11
W	74	18.5	3410	1.675 (0 °C)	0.137 (20 °C)	4.98 · 10^{-6}	19.30
Ta	73	21.1	2996	0.544	0.151 (20 °C)	6.5 · 10^{-6}	16.6

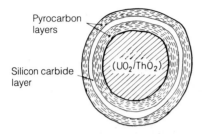

Pyrocarbon
layers

Silicon carbide
layer

(UO_2/ThO_2)

Figure 11.10. Coated particles.

Coated particles have been designed and developed as a special form of nuclear fuel for use in high-temperature gas-cooled reactors. These particles have a diameter of about 100 µm. They consist of a core of UO_2, UC or UO_2/ThO_2 coated with layers of graphite and silicium carbide (Fig. 11.10). Fuel elements are obtained by filling the coated particles into hollow spheres of graphite of about 6 cm outer diameter and 0.5 cm wall thickness, or by filling the fuel into graphite rods.

Another form of fuel element are the matrix elements in which the fuel is dispersed in a matrix of non-fissile material such as Al. Some combinations of ceramic fuel in a metallic matrix ("cermets") have found interest, because of the high thermal conductivity. On the other hand, the metallic matrix causes relatively high neutron absorption, and therefore matrix elements including cermets have only found very limited application.

As an example, a fuel element of the type used in a pressurized water reactor (PWR) is shown in Fig. 11.11. The fuel element has 16×16 positions for 236 fuel rods and 20 control rods.

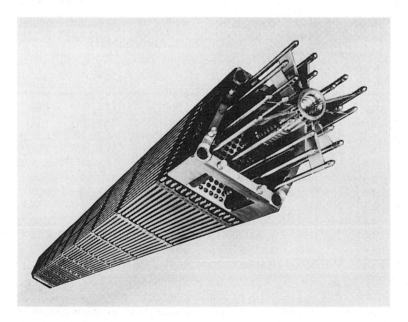

Figure 11.11. Fuel element used in pressurized-water reactors (16×16 positions for 236 fuel rods and 20 control rods).

11.5 Nuclear Reactors, Moderators and Coolants

Several types of nuclear reactors have already been mentioned in the previous section with respect to the use of nuclear fuel and the manufacture of fuel elements. The various types of nuclear reactors are distinguished on the basis of the following aspects:

– the kind of fuel used (e.g. natural U, enriched U, Pu, U/Pu mixtures);
– the energy of the neutrons used for fission (thermal or fast reactors);
– the kind of moderator (e.g. graphite, light water, heavy water);
– the combination of fuel and moderator (homogeneous or heterogeneous);
– the kind of coolant (e.g. gas, water, sodium, organic compounds, molten salts);
– the operation of the coolant (boiling water, pressurized water); and
– the application (e.g. research reactors, test reactors, power reactors, breeder reactors, converters, plutonium production, ship propulsion).

The first nuclear reactor was built up by Fermi and co-workers, beneath the stand in a football stadium in Chicago, by use of natural uranium and bars of graphite, and reached criticality in december 1942. It looked like a pile, the thermal power was 2 W, and cooling and radiation protection were not provided. The next nuclear reactor was operated from 1943 at Oak Ridge (USA) by use of 54 tons of uranium metal in the form of fuel rods inserted into a block of graphite 5.6 m long, shielded by concrete. Several reactors of a similar type (graphite-moderated, water-cooled, natural uranium reactors) were built in 1943 and in the following years at Hanford (USA) for production of plutonium to be used as a nuclear explosive.

The concept of energy production by nuclear reactors has found greater interest since about 1950. The first nuclear power station (graphite-moderated, gas-cooled (CO_2), natural uranium) began operation in 1956 at Calder Hall (UK). To-day, pressurized-water reactors (PWR) and boiling-water reactors (BWR) are the most widely used power reactors (Table 11.7). They contain about 100 tons of weakly enriched U (about 3.0 to 3.5% ^{235}U). Fast breeder reactors contain about 100 tons of natural U and about 6 tons of Pu, but no moderator. They exhibit several advantages: the high burn-up is due to the fact that large amounts of ^{238}U are transformed into the easily fissile ^{239}Pu. In this way, the energy production from U is increased by a factor of about 100, and enrichment of ^{235}U by isotope separation is not needed. Some problems are caused by use of liquid sodium as coolant due to its high reactivity. High-temperature gas-cooled reactors (HTGR) have some advantages also, because the high temperature of the coolant gives high efficiency, but they have not found broad application. The operation of gas-cooled reactors (GCR), boiling-water reactors (BWR), pressurized-water reactors (PWR) and high-temperature gas-cooled reactors (HTGR) is shown schematically in Figs. 11.12, 11.13, 11.14 and 11.15, respectively.

Power reactors have also been developed and installed for ship propulsion, for instance in submarines (e.g. "Nautilus", USA) or icebreakers (e.g. "Lenin", Russia).

World-wide the production of energy by nuclear power amounts to about 470 GW$_e$ (1995) and increases by about 4% per year, although the problems with respect to the storage of the radioactive waste (fission products and actinides) are not yet solved in a satisfactory way.

Table 11.7. Most widely used power reactors.

Reactor type	Percentage[a]	Fuel[b]	Canning	Moderator	Coolant	Coolant temp. [°C]	Coolant pressure [MPa]	Power [MWe]	Burn-up [MW d per kg]
Pressurized-water reactor (PWR)	≈64	UO_2 pellets (2.2–3.2%)	Zircaloy	H_2O	H_2O	300–320	14–16	1000–1300	30–35
Boiling-water reactor (BWR)	≈22	UO_2 pellets (2.2–3.2)	Zircaloy	H_2O	H_2O	280–290	≈7	1000–1300	30–35
Pressurized heavy-water reactor (PHWR)	≈5	UO_2 pellets (natural U)	Zircaloy	D_2O	D_2O	280–310	8–11	700–800	8–10
Pressurized-tube boiling-water reactor (RBMK) (Russian design)	≈4	UO_2 (≈2%)	Zr (1% Nb)	Graphite	H_2O	280–300	7–8	1000	18–19
Gas-cooled reactor (GCR)	≈1.5	U metal (natural U)	Magnox	Graphite	CO_2	340–420	2.8–3.5	600–700	3–4
Advanced gas-cooled reactor (AGR)	≈2.5	UO_2 pellets (≈2%)	Stainless steel	Graphite	CO_2	400–680	3–4	620–930	17–19
Liquid-metal fast breeder reactor (FBR)	≈1	UO_2/PuO_2 pellets (≈15% PuO_2)	Stainless steel	–	Na	530–560	≈0.1	1200–1300	70–100
High-temperature gas-cooled reactor (HTGR)	< 0.1	UO_2 (90–95%) + ThO_2 e.g. coated particles	Graphite	Graphite	He	750–850	1–4	≥300	70–100

[a] Percentage of the total nuclear power capacity in the world in 1994.
[b] In parentheses: enrichment.

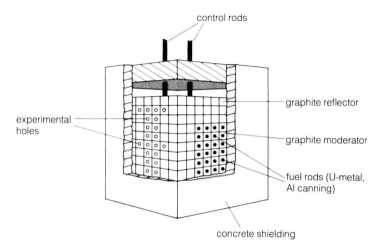

Figure 11.12. Gas-cooled reactor (GCR) operating with natural uranium (schematically).

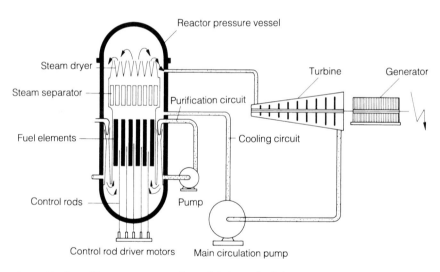

Figure 11.13. Boiling-water reactor (BWR) (schematically).

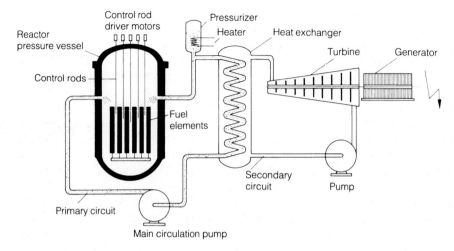

Figure 11.14. Pressurized-water reactor (PWR) (schematically).

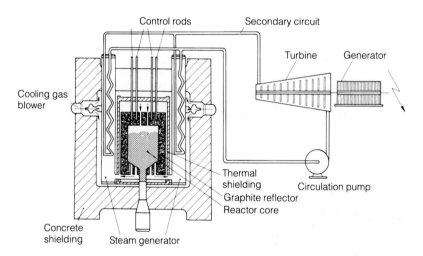

Figure 11.15. High-temperature gas-cooled reactor (HTGR) (schematically).

Research and test reactors are designed for special purposes, such as development of new reactor concepts, material testing or use as neutron sources. Generally, the energy production and the operating temperature are low. For material testing, high neutron flux densities are required. Various irradiation facilities are installed in research reactors for neutron irradiation, such as irradiation channels, pneumatic dispatch systems, neutron windows, thermal columns or uranium converters. The neutron fluxes vary between about 10^{11} and 10^{16} cm^{-2} s^{-1}, and the power varies between about 10 kW and 100 MW. Reactors of the swimming pool type (Fig. 11.16) are often used for research purposes. In the relatively small Triga reactors, mixtures of enriched uranium (e.g. 20% ^{235}U) and zirconium hydride serve as fuel and moderator, respectively. With increasing temperature the moderation properties of zirconium hydride decrease, with the result of a negative temperature coefficient of the reactivity which gives the reactor an inherent safety. Furthermore, taking out the fuel rods leads within about 0.1 s to a sudden increase of the neutron flux by a factor of about 1000 for a period of about 10 ms. These pulses can be used for the investigation of short-lived radionuclides. Homogeneous reactors containing solutions of enriched uranium (e.g. 20% ^{235}U) as sulfate, in a small tank, as fuel have also been designed for research purposes.

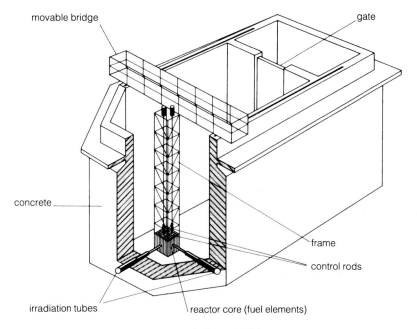

Figure 11.16. Swimming-pool reactor (schematically).

The properties of some moderators and coolants are listed in Table 11.8. As already mentioned, the purpose of the moderator is to take away the energy of the fission neutrons by collisions, without absorbing appreciable amounts of the neutrons. The dependence of the fission cross section σ_f of ^{235}U on the neutron energy is illustrated in Fig. 11.17. The absorption cross section σ_a is relatively low for graph-

Table 11.8. Properties of some moderators and coolants.

	Absorption cross section for thermal neutrons σ_a [barn]	Density at 20 °C [g cm^{-3}]	Melting point [°C]	Boiling point [°C]	Thermal conductivity at 20 °C [J cm^{-1} s^{-1} K^{-1}]	Specific heat at 20 °C [J g^{-1} K^{-1}]
graphite	0.0045	2.256	Sublimation	3650	1.674	0.720 (25 °C)
D_2O	0.0011	1.105	3.8	101.42	0.00586	4.212
H_2O	0.66	0.998	0	100.0	0.00586	4.183
CO_2	0.0038	$1.977 \cdot 10^{-3}$	Sublimation	−78.5	0.000184 (30 °C)	0.833 (15 °C)
He	0.007	$0.177 \cdot 10^{-3}$	−272.2	−268.6	0.000611 (50 °C)	5.200
Na	0.53	0.928 (100 °C)	97.7	883	0.863 (100 °C)	1.386 (100 °C)

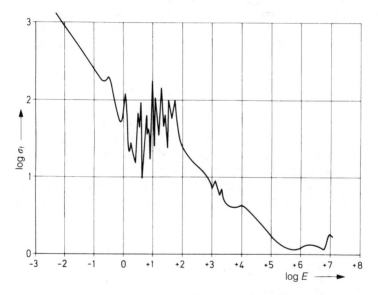

Figure 11.17. Fission cross section σ_f [b] for the fission of ^{235}U as a function of the neutron energy E [eV]. (According to D. J. Hughes, J. A. Harvey: BNL, Neutron Cross Sections. United States Atomic Energy Commision. McGraw-Hill Book Comp. New York.)

ite, D_2O, CO_2 and He (Table 11.8). The relatively high absorption cross sections of H_2O and liquid metals require use of enriched U or of Pu as fuel. Thermal conductivity is of special importance in power reactors. Application of the same materials as coolant and as moderator is desirable in the case of thermal reactors.

Disadvantages of water as a coolant are the low boiling temperature and the influence of corrosion. For operation at high temperatures, gases are preferable as coolants. As the ratio of heat transfer to pumping power is proportional to $M^2 c_p^2$ (M = molecular mass, c_p = specific heat), hydrogen would be the most favourable coolant at high temperatures. However, because of its reactivity, use of hydrogen is prohibitive. Helium is rather expensive; CO_2 is suitable as a coolant for graphite-moderated reactors, but at high temperatures transport of graphite due to the equilibrium $C(s) + CO_2(g) \rightleftharpoons 2\,CO(g)$ has to be taken into account. Liquid metals exhibit high thermal conductivity, but because of their reactivity special precautions are necessary.

All other materials used in nuclear reactors for construction or as tubes should exhibit low neutron absorption, low activation, no change in the properties under the influence of the high neutron and γ-ray fluxes and high corrosion resistance. These requirements are best met by zirconium which has found wide application in nuclear reactors. Al, Be and Mg have limited applicability. Steel and other heavy metals are only applicable if their relatively high neutron absorption is acceptable.

The range of the fission products is small (about 5–10 µm in solids and about 25 µm in water), but their specific ionization is high. This leads to high temperatures in solid fuel, in particular in UO_2 (up to several thousand degrees Celsius). Furthermore, lattice defects and deformations are produced in solids, and gaseous fission products migrate under the influence of the temperature gradient into hollow spaces formed in the central part of the solids. Volatile fission products may escape if there are leaks in the canning material. This makes continuous control of the activity in the coolant and purification of the latter by passage through ion exchangers necessary.

Neutrons also produce lattice defects in solid fuel, but in lower local concentrations. (n, γ) reactions lead to activation products and contribute to the secondary γ radiation in the reactor.

The intense primary γ radiation due to nuclear fission, the secondary γ radiation emitted by the fission and activation products and the β^- radiation from the fission products give rise to radiation-induced chemical reactions. The most important reaction is the radiation decomposition of water in water-cooled reactors, leading to the formation of H_2, H_2O_2 and O_2. Many substances dissolved in the water influence the formation of H_2 (Fig. 11.18). In most closed coolant systems equipment for

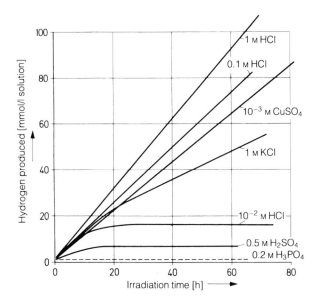

Figure 11.18. Influence of various components on the formation of hydrogen by radiolysis. (According to J. K. Dawson, R. G. Sowden: Chemical Aspects of Nuclear Reactors. Vol. 2. Water-Cooled Reactors. Butterworths. London 1963.)

catalytic recombination of H_2 and O_2 is installed. In pressurized-water reactors H_2 is added to suppress the radiolytic effects.

As long as the reaction $\cdot OH + H_2 \rightarrow H_2O + H\cdot$ is faster than the reaction $\cdot OH + \cdot HO_2 \rightarrow H_2O + H_2$, net decomposition of water does not occur, although all water molecules present in the cooling system are turned over at least once per day.

In order to compensate the excess reactivity (section 11.1), in water-cooled reactors boric acid is added to the coolant in concentrations up to about 0.2%. The concentration is reduced with increasing burn-up. The pH is adjusted to ≈ 9 by addition of 1 to 2 mg ^7LiOH per litre water to lower the solubility of the metal oxides and hydroxides, respectively, produced by corrosion on the walls of the cooling system.

11.6 Reprocessing

Mastery of the handling of spent nuclear fuel, including safe disposal of the radio-active waste, is a prerequisite for the use of nuclear energy.

The main effects of the fission products are:

– radiation defects in the fuel elements which may lead to damage;
– poisoning of the reactor by neutron-absorbing fission products.

An example of the effect of poisoning is the series of isobars with mass number $A = 135$:

$$U(n,f)^{135}Te \xrightarrow[18.6\,s]{\beta^-} {}^{135}I \xrightarrow[6.61\,h]{\beta^-} {}^{135}Xe \xrightarrow[9.10\,h]{\beta^-} {}^{135}Cs \xrightarrow[2.3\cdot10^6\,y]{\beta^-} {}^{135}Ba\ (\text{stable})$$

$$
\begin{array}{ccccc}
\downarrow (n,\gamma) & \downarrow (n,\gamma) & \downarrow \begin{array}{l}(n,\gamma)\\ \sigma=2.65\cdot10^6\,b\end{array} & \downarrow \begin{array}{l}(n,\gamma)\\ \sigma=8.7\,b\end{array} & \downarrow \begin{array}{l}(n,\gamma)\\ \sigma=5.8\,b\end{array}\\
{}^{136}Te & {}^{136}I & {}^{136}Xe & {}^{135}Cs & {}^{136}Ba
\end{array}
$$

$$(11.8)$$

The neutron absorption of ^{135}Xe is extremely high. The ratio of the number of atoms transmuting by β^- decay to the number transformed by (n,γ) reactions depends on the neutron flux density Φ:

$$\frac{(dN/dt)_\beta}{(dN/dt)_{n,\gamma}} = \frac{N\lambda}{N\sigma\Phi} = \frac{\lambda}{\sigma\phi\Phi} \tag{11.9}$$

At a neutron flux density $\Phi = 10^{14}\,cm^{-2}\,s^{-1}$, 93% of ^{135}Xe undergoes (n,γ) reactions.

The fission products that are mainly responsible for neutron absorption are listed in Table 11.9. With respect to the mass the lanthanides represent the greatest fraction, but with regard to neutron absorption the noble gases are most important due to the high value of σ_a for ^{135}Xe $(2.65\cdot10^6\,b)$.

The effect of poisoning can be compensated to a certain extent by an excess reactivity or by installation of a breeder blanket (an outer layer of ^{232}Th) in which new fissile material is produced. In fast reactors the effect of poisoning is less important,

Table 11.9. Most important long-lived fission products.

Element	Relative abundance in the fission products [%]	Relative neutron absorption [%]
Noble gases	7	72
Samarium	}70	14
Other lanthanides		11
Technetium	10	1
Caesium	4	0.5
Molybdenum	1	0.2
Other elements	8	1.3

because fast neutrons exhibit a lower absorption cross section. However, in all cases fuel elements have to be exchanged for new ones after 10 to 80% of the fissile nuclides have been used up.

Further handling of nuclear fuel is determined by its activity and the heat production due to radioactive decay. The relation between energy production by fission and the number of fissions per second is given by

$$1\,\text{MW} = 6.25 \cdot 10^{18}\,\text{MeV s}^{-1} \triangleq \approx 3.3 \cdot 10^{16}\,\text{fissions s}^{-1} \tag{11.10}$$

As the disintegration rate of the fission products with $t_{1/2} > 1\,\text{s}$ is about five times the rate of fission, the activity of the fuel several seconds after shutting off the reactor is $\approx 17 \cdot 10^{16}\,\text{Bq}$ ($\approx 5 \cdot 10^6\,\text{Ci}$) per MW of thermal energy produced. The β activity per MW and the heat production of the fission products are plotted in Fig. 11.19 as a function of the time after shutting off the reactor. The heat production requires cooling of the fuel elements, because melting of the fuel and volatilization of fission products may occur under unfavourable conditions. ^{237}U produced by the nuclear reactions ^{235}U(n, γ)^{236}U(n, γ)^{237}U and ^{238}U(n, 2n)^{237}U causes a relatively high initial activity of uranium. As ^{237}U decays with a half-life of 6.75 d:

$$^{237}\text{U} \xrightarrow[6.75\,\text{d}]{\beta^-,\gamma} {}^{237}\text{Np} \tag{11.11}$$

storage of the fuel elements for about 100 d is necessary before further handling.

The composition of spent nuclear fuel from light water reactors after storage of 1 y is given in Table 11.10. The following options are possible:

(a) final disposal of the spent fuel elements;
(b) long-term intermediate storage with the aim of later reprocessing;
(c) short term interim storage and reprocessing.

After interim storage for at least several months under water for cooling and radiation protection, the choice has to be made between options (a), (b), and (c).

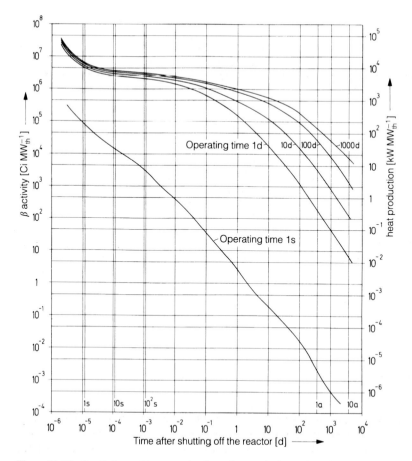

Figure 11.19. β activity and heat production of spent fuel as a function of the time after shutting off the reactor. (The γ activity amounts to about half of the β activity; the heat production is calculated for an average β energy of 0.4 MeV and quantitative absorption.)

The main aim of reprocessing is the recovery of fissile and fertile material. If U or U–Pu mixtures are used as fuel, the fissile nuclides are ^{235}U and ^{239}Pu, and the fertile nuclide is ^{238}U. Reprocessing of these kinds of fuel closes the U-Pu fuel cycle. The U-Th fuel cycle is closed by reprocessing of spent fuel containing mixtures of U and Th. In the case of final storage of the spent fuel elements, the fuel cycle is not closed; fissile and fertile nuclides are not retrieved for further use.

Reprocessing is started after about six months to several years of cooling time, depending on the aim of reprocessing (e.g. recovery of Pu), the storage costs of unprocessed fuel, the reprocessing capacity and the advantages of lower activites after longer interim storage.

In the case of the U-Pu fuel cycle, the main steps of reprocessing are separation of uranium, plutonium and fission products including the other actinides. U and Pu are to be recovered in high purity and free from fission products and other actinides, in order to make further use possible. Pu causes problems, because it may be used

Table 11.10. Composition of spent nuclear fuel from a light-water reactor at an initial enrichment of 3.3% ^{235}U, a burn-up of 34 000 MW d per ton and a storage time of 1 year.

Nuclide	Weight percent	
Uranium and transuranium elements		
^{235}U	0.756	
^{236}U	0.458	
^{237}U	$3 \cdot 10^{-9}$	95.4
^{238}U	94.2	
^{237}Np	0.05	
^{238}Pu	0.018	
^{239}Pu	0.527	
^{240}Pu	0.220	0.908
^{241}Pu	0.105	
^{242}Pu	0.038	
Americium isotopes	0.015	
Curium isotopes	0.007	
Fission products		
Stable fission products	3.00	
^{85}Kr	0.038	
^{90}Sr	0.028	3.62
^{129}I	0.09	
^{134}Cs $+^{137}$Cs	0.275	
Others	0.19	

for production of nuclear weapons. With respect to non-proliferation of nuclear weapons, strict controls of Pu input and output are necessary to avoid its misuse. Reprocessing of U–Th mixtures comprises separation of uranium, plutonium, thorium and the fission products, including the other actinides.

The details of reprocessing depend on the kind of fuel and of fuel elements. In the "head-end" process the fuel elements are taken apart and the fuel is chopped if necessary. Canning material and fuel may be separated mechanically by use of special machines or by chemical means. The chemical "head end" may consist in the dissolution of the canning, but more favourably fuel and canning are separated by dissolving the fuel after chopping the fuel elements ("chop–leach" process).

The result of the separation of U and Pu from the fission products and other actinides is characterized by the decontamination factor, given by the ratio of the activity of the fission products and actinides in the fuel to that in uranium and plutonium after separation. The decontamination factors should be in the order of 10^6 to 10^7, and the recoveries of U and Pu should be near to 100%. These requirements are best met by solvent extraction procedures. With respect to the high activity of the fuel, remote control of all operations is necessary.

In general, 11 M HNO_3 is used for dissolution of uranium metal and 7.5 M HNO_3 for dissolution of UO_2. After dissolution in HNO_3, U is present in the form of $UO_2(NO_3)_2$, and Pu mainly as $Pu(NO_3)_4$.

Gaseous or volatile fission products such as ^{85}Kr, ^{129}I (mainly as I_2, HI and HOI), ^{106}Ru (as RuO_4) and a part of T are liberated in the course of chopping and dissolution of the fuel and are found in the off-gas. Another part of T is oxidized to tritiated water and is found mainly in the dissolver. ^{129}I and ^{106}Ru are retained in the off-gas scrubbers and filters containing silver-impregnated zeolite or charcoal for the fixation of iodine. ^{85}Kr may be separated from the off-gas by low-temperature rectification. After dissolution, the solution is adjusted to 1–2 M HNO_3, and by addition of $NaNO_2$ quantitative transformation of Pu into Pu(IV) is secured.

In the first separation procedure operated in a technical scale, Pu was separated as Pu(IV) from U and fission products by coprecipitation with $BiPO_4$ (bismuth phosphate process). Today, solvent extraction is applied, because it leads to higher decontamination factors and can be operated as a continuous process.

Solvent extraction can be carried out in pulsated extraction columns, in mixer–settlers or in centrifuge extractors. Organic compounds such as esters of phosphoric acid, ketones, ethers or long-chain amines are applied as extractants for U and Pu. Some extraction procedures are listed in Table 11.11. The Purex process has found wide application because it may be applied for various kinds of fuel, including that from fast breeder reactors. The Thorex process is a modification of the Purex process and has been developed for reprocessing of fuel from thermal breeders.

Table 11.11. Extraction procedures used for reprocessing of spent nuclear fuel.

Process	Organic phase	Aqueous phase
Purex	$\approx 30\%$ TBP (tri-n-butyl phosphate) in kerosine or in dodecane	HNO_3 (0.1–3 M)
Redox	Hexon (methyl isobutyl ketone)	HNO_3 (0.1–3 M) containing $Al(NO_3)_3$
Butex	Dibutylcarbitol	HNO_3 (0.1–3 M)
Eurex	TLA (trilaurylamine)	HNO_3 (0.1–3 M)
Thorex	$\approx 40\%$ TBP (tri-n-butyl phosphate) in kerosine	HNO_3 (0.1–3 M) containing $Al(NO_3)_3$

In the first extraction cycle, U, Pu and fission products (including other actinides) are separated in the following steps:

– Extraction of U(VI) and Pu(IV) from 1–2 M HNO_3 into the organic phase (extractor 1). The major part of the fission products and other actinides remain in the aqueous phase.
– Reduction of Pu to Pu(III) by addition of Fe(II) or U(IV) and separation of Pu(III) from U by extraction into 1–2 M HNO_3 (extractor 2). U remains in the organic phase.
– Back-extraction of U into the aqueous phase (≈ 0.01 M HNO_3). The organic phase is refined and recycled.

In the following extraction cycles U and Pu, respectively, are refined.

– Uranium cycle: U is extracted from $1–2\,M\,HNO_3$ into the organic phase and back-extracted into $\approx 0.01\,M\,HNO_3$. The organic phase is refined and recycled.
– Plutonium cycle: Pu is oxidized to Pu(IV), extracted into the organic phase, reduced to Pu(III) and back-extracted into the aqueous phase. The organic phase is refined and recycled. Further purification of Pu is obtained by application of anion-exchange resins.

Refinement of the organic phase is necessary, because of the formation of radiolysis products. The main product of radiolysis of tributylphosphate (TBP) is dibutylphosphoric acid (DBP) which forms very stable complexes with Pu(IV) and prohibits quantitative reduction to Pu(III). DBP is separated by washing the organic phase with an aqueous solution of Na_2CO_3.

The last step of reprocessing ("tail end") is the production of compounds of U and Pu suitable for further use, e.g. uranyl nitrate hexahydrate (UNH) or a concentrated solution of $UO_2(NO_3)_2$, and PuO_2 or a solution of $Pu(NO_3)_4$, respectively.

Dry reprocessing procedures, such as volatilization of U as UF_6, fusion with $Na_2S_2O_7$ or $NaOH/Na_2O_2$, chemical reactions in molten salts or pyrometallurgical procedures, have also been proposed, but have not found practical application.

An additional advantage of reprocessing may be seen in the possibility of converting the radioactive waste into chemical forms that are most suitable for long-term disposal, thus minimizing possible hazards. In this context, separation of long-lived radionuclides from the rest of the waste and their transformation into chemical forms of high stability have also been discussed. The recovery of actinides other than U and Pu (e.g. ^{237}Np and transplutonium elements) and of fission products such as metals of the platinum group, ^{85}Kr, ^{90}Sr and ^{137}Cs is only of secondary interest, because of the limited practical use of most of these radionuclides. The elements of the platinum group have a certain value as noble metals. ^{85}Kr, ^{90}Sr, ^{137}Cs and long-lived isotopes of transplutonium elements may be applied as radiation sources, and ^{237}Np may be transformed by neutron irradiation into ^{238}Pu, which has found application as an energy source in radionuclide generators.

11.7 Radioactive Waste

The radioactive wastes originating from the operation of nuclear reactors and reprocessing plants is classified according to the activity level:

– low-level waste (LLW or LAW, low-active waste);
– medium-level waste (MLW or MAW, medium-active waste);
– high-level waste (HLW or HAW, high-active waste);

and according to the state of matter (gaseous, liquid or solid).

In nuclear reactors, radionuclides are produced by nuclear reactions in the coolant and in various solid materials in the reactor vessel. Furthermore, fission products or actinides may leak into the cooling system from faulty fuel elements. T, ^{14}C, ^{13}N, ^{16}N, ^{19}O, ^{18}F and ^{41}Ar are produced by a multitude of nuclear reactions with various nuclides in the coolant, e.g. $^2H(n, \gamma)^3H$, $^{10}B(n, 2\alpha)^3H$, $^{13}C(n, \gamma)^{14}C$, $^{14}N(n, p)^{14}C$, $^{13}C(n, p)^{13}N$, $^{14}N(n, 2n)^{13}N$, $^{16}O(n, p)^{16}N$, $^{18}O(n, \gamma)^{19}O$, $^{18}O(n, p)^{18}F$, $^{40}Ar(n, \gamma)^{41}Ar$. Other radionuclides are produced by nuclear reactions with metals and their corrosion products (^{51}Cr, ^{54}Mn, ^{59}Fe, ^{58}Co, ^{60}Co, ^{65}Zn, ^{124}Sb). Corrosion products may dissolve in ionic form or they may be carried in the form of fine particles with the coolant. The dominating radionuclides released from faulty fuel elements are ^3H, ^{85}Kr, ^{131}I, ^{133}I, ^{133}Xe, ^{135}Xe, ^{134}Cs and ^{137}Cs.

Off-gas and coolant are purified continuously in the nuclear power stations. The off-gas passes a catalytic hydrogen–oxygen combiner and a delay system allowing short-lived radionuclides to decay. The remaining radionuclides are retained in charcoal filters. Under normal operating conditions, only limited amounts of radionuclides are given off into the air. Part of the coolant is branched off from the primary circuit and passed through a filter and an ion exchanger to retain particulates and ionic species, respectively. Other radioactive wastes are produced by decontamination and various purification procedures.

The amounts of wastes produced from a 1000 MW$_e$ power station per year are of the order of 5000 m^3 LLW and 500 m^3 MLW. Liquid wastes may be reduced in volume by purification or concentration, combustible solid wastes by incineration. The general tendency is to transform larger volumes of LLM into smaller volumes of MLW. The latter are enclosed in blocks of concrete or mixed with bitumen and loaded into steel drums.

Dismantling of decommissioned nuclear reactors requires special procedures. The outer parts can be handled like normal industrial waste, whereas the inner parts, mainly the reactor vessel and some core components, exhibit high radioactivity due to activation. Radioactive deposits on the inner surface of the reactor vessel may be removed by chemical decontamination. Altogether, the relatively large volumes of LLM and MLW which are obtained by dismantling are further processed and then preferably enclosed in concrete or bitumen.

The largest amounts of radioactivity are in the spent fuel elements, which represent high-level waste (HLW). For final disposal, the spent fuel elements may be taken apart and the fuel rods may either be canned and put into casks or they may be cut into pieces and loaded into canisters and casks.

Reprocessing of nuclear fuel by the Purex process leads to the following amounts of waste per ton of U: ≈ 1 m^3 HLW (fission products and actinides in HNO$_3$ solution), ≈ 3 m^3 MLW as organic solution, ≈ 17 m^3 MLW as aqueous solution, ≈ 90 m^3 LLW (aqueous solution). By further processing a volume reduction is achieved: ≈ 0.1 m^3 HLW, ≈ 0.2 m^3 MLW (organic), ≈ 8 m^3 MLW (aqueous), ≈ 3 m^3 LLW (aqueous).

The HLW solutions obtained by reprocessing are collected in tanks near the reprocessing plants. They contain relatively high concentrations of HNO$_3$ which can be decomposed by addition of formaldehyde or formic acid. At low acid concentrations, however, hydrolysis and precipitation of elements of groups III and IV occurs. Concentration by heating may lead to volatilization of fission products, such as Ru isotopes in the form of RuO$_4$.

After one year of intermediate storage of spent nuclear fuel and reprocessing, the initial activity of the HLW solutions is of the order of 10^{14} Bq/l. The activity due to the presence of ^{90}Sr and ^{137}Cs is about 10^{13} Bq/l, and after 10 y the activity of the HLW solution decreases approximately with the half-life of these radionuclides

(28 y and 30 y, respectively). After about 1000 y of storage the activity is of the order of 10^4 Bq/l, determined by long-lived fission products (e.g. ^{99}Tc and ^{129}I) and actinides.

The high initial activity of the HLW solutions causes heating and radiation decomposition. If no countermeasures are taken, the solutions are self-concentrating by evaporation, strong corrosion occurs, nitrous gases and gaseous fission products are given off, and hydrogen and oxygen are formed by radiolysis. These effects require extensive precautions, such as cooling, continuous supervision of the tank, ventilation, filtering and control of the air.

After various times of storage, the HLW solutions are processed with the aim of transforming the waste into a stable form suitable for long-term disposal and to reduce the volume as far as possible. Most suitable chemical forms are ceramics or glasses obtained by calcination or vitrification, respectively, of the waste. Solutions containing large amounts of $Al(NO_3)_3$ due to dissolution of Al canning may be calcinated by spraying them into a furnace. By addition of silicates or SiO_2, leaching of the products by water can be suppressed. Glasses are obtained by addition of borax or phosphates. The liquid drops of glass formed in the furnace can be collected in steel containers.

Final disposal of HLW is planned and in some places performed in underground repositories. Several kinds of geological formations are discussed or used, respectively, e.g. tuff (USA), salt domes (Germany), granite (Switzerland) or clay (Belgium). The geological formations should remain unchanged for long periods of time ($>10^4$ y) and access of water should be excluded.

In safety assessments, the following scenario is discussed: access of water, corrosion of the containers, leaching of the waste and migration of the radionuclides to upper layers of the earth where they may enter the biosphere.

Besides the HLW solutions, various other kinds of waste are obtained by reprocessing. Dismantling and chopping of the fuel elements give rise to solid MLW and LLW. Finely dispersed particles of undissolved metals or metal oxides (e.g. Ru, Rh, Mo, Tc) are separated as MLW from the dissolver solution by filtering or centrifugation. Chopping and dissolution of the fuel leads to volatilization of T (as T_2 or HTO), ^{14}C (as CO_2), ^{85}Kr, ^{129}I and ^{106}Ru (as RuO_4). The amount of T is relatively high. To avoid release into the air, T may be oxidized to T_2O by treating the fuel after chopping and before dissolution at elevated temperature (500–700 °C) with oxygen ("voloxidation"). HTO formed by dissolution may be condensed and led back into the dissolver. ^{14}C formed by the reaction ^{14}N$(n, \gamma)^{14}$C from nitrogen impurities in the fuel is oxidized in the dissolver to $^{14}CO_2$.

The off-gas passes scrubber containing NaOH solution and filters with silver-impregnated zeolite or charcoal in which Ru and I are retained. Kr may be separated by condensation or adsorption on charcoal at liquid-nitrogen temperature.

The extraction cycles lead to the production of various aqueous and organic waste solutions, mainly from solvent clean-up and washing. Some of the aqueous LLW solutions may be released directly into the environment if their activity is low enough. Others are decontaminated by precipitation, coprecipitation, ion exchange or sorption procedures. The general tendency in handling liquid wastes is to reduce the volume as far as possible and to transform LLW into MLW, as already mentioned. Liquid organic wastes are either incinerated or the radionuclides contained therein are separated by precipitation or other procedures.

Solid wastes containing long-lived α emitters (^{239}Pu and other actinides), including wastes from handling these radionuclides in the laboratory, require application of special procedures, such as wet oxidation by a mixture of H_2SO_4 and HNO_3 followed by solvent extraction and isolation of the α emitters.

Several concepts have been developed to avoid or to minimize the risks of long-term storage of HLW. An interesting concept is that of "burning" long-lived waste by bombarding it with high-energy protons of about 800 MeV. This accelerator-driven transmutation technology (ADTT) project developed at Los Alamos, USA, is based on the liberation of neutrons by spallation of heavy nuclei. A proton beam from a linear accelerator is directed onto a target of molten Pb–Bi eutectic surrounded by a graphite moderator in which the neutrons produced by spallation of Pb and Bi nuclei are slowed down. The waste to be "burned" is dissolved in a molten salt and circulated within the moderator, where it is transmuted by fission (actinides) and/or neutron capture followed by β^- decay. In this way, a great part of the waste is transformed into stable or short-lived products. The heat generated by fission is converted into electricity by usual techniques. It is estimated that about 20% of this electricity is sufficient to power the accelerator. A fraction of the circulating molten salt is continuously diverted to a processing system in which the fission products are removed. Stable products and those with half-lives <30 y are sent for managed storage, whereas those with half-lives >30 y are returned to the furnace. Such a facility may also be used for "burning" of Th and provide an important energy supply.

11.8 The Natural Reactors at Oklo

In 1972 it was found that uranium in ore deposits at Oklo, Gabon, contains significantly smaller concentrations of ^{235}U than other deposits of natural uranium (<0.5% compared with 0.72%). At these places, the isotopic composition of other elements is also different from the mean composition in nature. For instance, natural Nd contains ≈ 27% ^{142}Nd and ≈ 12% ^{143}Nd, whereas Nd at Oklo contains <2% ^{142}Nd and up to 24% ^{143}Nd. Comparison with the yields of nuclear fission leads to the result that Nd produced by fission contains ≈ 29% ^{143}Nd, but practically no ^{142}Nd. From the high isotope ratio ^{143}Nd/^{142}Nd and the low isotope ratio ^{235}U/^{238}U, it must be concluded that chain reactions have occurred. The age of the Oklo deposits was found by ^{87}Rb/^{87}Sr analysis to be $\approx 1.7 \cdot 10^9$ y. At that time the concentration of ^{235}U in natural uranium was ≈ 3%. The presence of water in the sedimentary ore deposits led to high values of the resonance escape probability p (eq. (11.1)) and to criticality of the systems ($k_{eff} > 1$).

The natural reactors at Oklo have been in operation for about 10^6 y, probably with intermissions, depending on the presence of water. The neutron flux density, the power level and the temperature were relatively low ($\leq 10^9$ cm^{-2} s^{-1}, ≤ 10 kW and about 400–600 °C, respectively). About 6 tons of ^{235}U have been consumed and about 1 ton of ^{239}Pu has been produced. Since then, the latter decayed into ^{235}U.

If can be assumed that natural nuclear reactors have been in operation about (1 to 3) $\cdot 10^9$ y ago at many other places containing uranium-rich ore deposits in the presence of water.

Analysis of the natural reactors at Oklo gives valuable information about the migration behaviour of fission products and actinides in the geosphere. Uranium and the lanthanides have been redistributed locally. Plutonium produced in the Oklo reactors did not move during its lifetime from the site of its formation; 85–100% of the lanthanides, 75–90% of the Ru and 60–85% of the Tc were retained within the reactor zones. Small amounts of U, lanthanides, Ru and Tc moved with the water over distances of up to 20–50 m.

11.9 Controlled Thermonuclear Reactors

Controlled thermonuclear reactors (CTR), also referred to as fusion reactors, are in the development stage. Operation on an industrial scale before the year 2050 is unlikely.

The ignition temperature for the D–D reaction is $3 \cdot 10^8$ K, that for the D–T reaction $3 \cdot 10^7$ K. These reactions are most important for the operation of controlled thermonuclear reactors. Because of the lower ignition temperature, the D–T reaction is more attractive, at least for the first generation of these reactors, although it requires handling of large amounts of T. The ignition temperature for the D–^{3}He reaction is similar to that of the D–D reaction, whereas it is appreciably higher for the H–H reaction (about 10^{10} K).

The preferred concept is the magnetic confinement of the plasma of D^+ or T^+, respectively, and free electrons at temperatures of the order of 10^7 K. Because no material is able to withstand a plasma of this temperature, it must be kept away from the walls. This can be done by strong magnetic fields. The torus design is most promising, and the experimental reactors based on this design are called "Tokamaks". The JET machine (Joint European Torus) is operated at Abingdon, UK; another machine, ITER (International Thermonuclear Reactor Experiment), will have an appreciably higher power output.

For a controlled nuclear fusion reaction the following parameters are important:

– The fusion reaction rate parameter σv, where σ is the reaction cross section, which depends on the particle energy, and v is the velocity of the ions, averaged over the Maxwell velocity distribution and proportional to the temperature.
– The Lawson limit $n\tau$, where n is the particle density and τ the confinement time (confinement quality) indicating the ability of the plasma to retain the heat.
– The triple product $n\tau T$, where T is the average ion temperature.

The temperature must be $\geq 10^8$ K, and $n\tau$ must be $\geq 10^{21}$ s $\cdot$ m^{-3} for the D–D reaction and $\geq 10^{20}$ s $\cdot$ m^{-3} for the D–T reaction. The triple product $n\tau T$ must be $\geq 5 \cdot 10^4$ s $\cdot$ eV $\cdot$ m^{-3}. Since two particles are involved in each collision, the fusion power density increases with the square of the particle density. At 1 Pa ($n = 3 \cdot 10^{20}$ m^{-3}) the power density is several tenths of a megawatt per cubic metre, and the required confinement time would be 0.1–1 s. Best results have been obtained so far with central ion temperatures of the order of $\approx 4 \cdot 10^8$ K, particle densities of $\approx 5 \cdot 10^{19}$ m^{-3} and confinement times of ≈ 2 s.

Steady-state machines may be operated in a pulsed or continuous mode. In both cases, the fuel gases D and T, respectively, must be injected and the fusion products

^{4}He and ^{3}He must be withdrawn. Up to now, this is done by shutting down, empty-ing the torus and filling it with a fresh D–T mixture. Suitable technical solutions for continuous operation of the machines have to be developed.

To meet the Lawson criterion, the plasma density is increased by increasing the magnetic field. This also results in an increase of temperature. The ignition temper-ature may be obtained by injection of D^+ and T^+ of high kinetic energy or by high-frequency heating. After ignition, the high kinetic energy of ^{4}He^{2+} ions produced by fusion leads to a rapid rise of the plasma temperature. The high magnetic field strength necessary to confine and to compress the plasma requires high currents ($\geq 30\,000$ A) and the use of superconducting magnets.

A thermonuclear reactor operating on the basis of the D–T reaction will contain several kilograms of T. This requires a well-developed tritium technology compris-ing production and safe handling, taking into account diffusion through metals and chemical reactions, including exchange reactions.

In the D–T reaction about 80% of the energy is given off in the form of kinetic energy of the neutrons (14 MeV). In most steady-state concepts a blanket is provided containing lithium in the form of Li metal or a lithium salt (e.g. Li_2BeF_4) to capture these high-energy neutrons by the reactions:

$$^6\text{Li}(7.4\%) + \text{n} \rightarrow {}^3\text{H} + {}^4\text{He} \quad (\Delta E = 4.78\,\text{MeV}) \tag{11.12}$$

$$^7\text{Li}(92.6\%) + \text{n} \rightarrow 2\,{}^4\text{He} \quad (\Delta E = 18.13\,\text{MeV}) \tag{11.13}$$

By these reactions most of the energy of the fusion process is transferred to the blan-ket, which is heated to high temperatures. The heat is given off in a cooling cycle by pumping the molten Li metal or salt through a heat exchanger in which the heat is transferred to a secondary circuit for production of steam and operation of a turbine. By reaction (11.12) additional amounts of T are produced (breeding effect) which must be recovered and recycled. For this purpose, the Li metal or salt is treated con-tinuously or intermittently to recover the tritium.

Bremsstrahlung and synchrotron radiation emitted during operation are absorbed in the walls of the vessel, which must be cooled. Very good heat insulation is required between the hot walls of the vessel ($\approx 1000\,^\circ$C) and the superconducting coils of the magnets (≈ 5 K).

A major problem is the proper choice of construction materials, which must be heat- and radiation-resistant and exhibit low neutron absorption. The following components are under discussion:

(a) a vacuum chamber of 10 cm stainless steel, on the inside lined with reinforced carbon shielding to protect the steel, and on the outside water-cooled;
(b) a water-cooled blanket about 1.5 cm thick, made of a vanadium alloy and con-taining the lithium or lithium compound;
(c) a water-cooled shield to absorb neutrons and γ rays. The vacuum chamber and the blanket are exposed to high radiation intensities leading to radiation damage and activation. Hydrogen atoms escaping from the magnetic confinement react with the walls, which will be loaded with D and T. On the other hand, atoms are emitted from the walls, polluting the plasma.

With respect to the operation of thermonuclear reactors, laser-induced fusion and heavy-ion-induced fusion are also discussed. In these concepts compression of T or D–T mixtures to high density and heating to high temperatures are achieved by irradiation with a laser beam of very high intensity or with a beam of high-energy heavy ions.

Compared with fission reactors, operation of fusion reactors is more complicated because of the high ignition temperatures, the necessity to confine the plasma, and problems with the construction materials. On the other hand, the radioactive inventory of fusion reactors is appreciably smaller. Fission products are not formed and actinides are absent. The radioactivity in fission reactors is given by the tritium and the activation products produced in the construction materials. This simplifies the waste problems considerably. Development of thermonuclear reactors based on the D–D reaction would reduce the radioactive inventory even further, because T would not be needed. The fact that the energy produced by fusion of the D atoms contained in 1 litre of water corresponds to the energy obtained by burning 120 kg coal is very attractive.

11.10 Nuclear Explosives

The high amounts of energy liberated by nuclear fission and fusion led very early to the production of nuclear explosives, as already mentioned in section 11.1. ^{235}U, ^{239}Pu and ^{233}U can be used as nuclear explosives, because they have sufficiently high cross sections for fission by fast neutrons. By use of the equations in section 11.1, it can be assessed that, in the absence of a reflector, a sphere of about 50 kg uranium metal containing 94% ^{235}U or a sphere of about 16 kg plutonium metal (^{239}Pu) is needed to reach criticality. If a reflector is provided, the critical masses are about 20 kg for ^{235}U and about 6 kg for ^{239}Pu. The critical masses for ^{233}U are similar to those for ^{239}Pu.

For use in nuclear weapons, the concentration of ^{240}Pu in the plutonium should be low, because the presence of this nuclide leads to the production of appreciable amounts of neutrons by spontaneous fission; if the concentration of ^{240}Pu is too high the neutron multiplication would start too early with a relatively small multiplication factor, and the energy release would be relatively low. Higher concentrations of ^{241}Pu also interfere, because of its transmutation into ^{241}Am with a half-life of only 14.35 y. To minimize the formation of ^{240}Pu and ^{241}Pu, Pu for use in weapons is, in general, produced in special reactors by low burn-up ($<20\,000$ MW$_{th}$ d per ton).

Criticality can be reached by shooting one subcritical hemisphere onto another one by normal explosives (gun-type) or by compressing a subcritical spherical shell into a supercritical sphere (implosion-type). The bomb dropped over Hiroshima (energy release corresponding to ≈ 15 kilotons of TNT) was of the gun type and made from ^{235}U, whereas that dropped over Nagasaki was of the implosion type and made from ^{239}Pu (energy release corresponding to ≈ 22 kilotons of TNT). Nuclear weapons have been made with explosive forces corresponding to from 0.01 to about 500 kilotons of TNT, sometimes with very low efficiency.

Without application of special measures, it would take about 0.2 µs to increase the number of neutrons to that required to fission all fissile atoms. However, much earlier the energy released would blow apart the material, resulting in a very low efficiency. In order to reach a high neutron flux within the short duration of supercritical configuration, a few microseconds before maximum criticality a large number of neutrons are injected by a special neutron source, triggering a rapid neutron multiplication.

Generally, the fissile core is surrounded by a heavy material, in order to reflect the neutrons and to increase the inertia and consequently the time in which the supercritical configuration is held together.

The explosion of fissile material leads to temperatures of about 10^8 K which are sufficient to initiate D–T fusion. This is the basis of the development of hydrogen bombs, in which the energy of fission is used for ignition of fusion. LiD serves as a source of D and T, the latter being produced by thermal neutrons (^{6}Li(n, α)t) and by fast neutrons (^{7}Li(n, αn)t). If the temperature is high enough, the D–D reaction can contribute to the energy production. The fast neutrons released by the fusion reactions react very effectively with natural or depleted U initiating fission of ^{238}U. By these kinds of weapons large amounts of fission products are formed ("dirty weapons"). If a surrounding of non-fissile heavy material is used, fission products are released only by the ignition process ("clean weapons").

Between 1960 and 1963 hydrogen bombs with explosive forces of up to 60 megatons of TNT were discharged into the atmosphere and underground.

Literature

General

R. Stephenson, Introduction to Nuclear Engineering, McGraw-Hill, New York, **1958**
S. E. Liverhout, Elementary Introduction to Reactor Physics, Wiley, New York, **1960**
C. R. Tipton, Reactor Handbook, Vol. 1, Materials, Interscience, London, **1960**
H. Kouts, Nuclear Reactors, in: Methods of Experimental Physics, Vol. 3B, in: Nuclear Physics (Eds. L. C. L. Yuan and C. S. Wu), Academic Press, New York, **1963**
J. K. Dawson, R. D. Sowden, Chemical Aspects of Nuclear Reactors, 3 Vols., Butterworths, London, **1963**
S. Peterson, R. G. Wymer, Chemistry in Reactor Technology, Pergamon, Oxford, **1963**
Reactor Handbook, 2nd ed., Interscience, New York, **1960–1964**
S. Glasstone, Source Book on Nuclear Energy, 3rd ed., Van Nostrand, Princeton, NJ, **1967**
M. Benedict, T. Pigford, H. W. Levi, Nuclear Chemical Engineering, 2nd ed., McGraw-Hill, New York, **1981**
S. Glasstone, A. Sesonke, Nuclear Reactor Engineering, 3rd ed., Van Nostrand, London, **1981**
International Atomic Energy Agency, Nuclear Power, the Environment and Man, IAEA, Vienna, **1982**
A. McKay, The Making of the Atomic Age, Oxford University Press, Oxford, **1984**
W. C. Patterson, Nuclear Power, Penguin Books, Hardmondsworth, UK, **1985** (more popular)
K. H. Lieser, Einführung in die Kernchemie, 3rd ed., VCH, Weinheim, **1995**
G. Choppin, J. Rydberg, J. O. Liljenzin, Radiochemistry and Nuclear Chemistry, 2nd ed., Butterworths–Heinemann, Oxford, **1995**

More Special

UN Conferences on Peaceful Uses of Atomic Energy, Geneva, **1955, 1958, 1965, 1972**
International Atomic Energy Agency, Disposal of Radioactive Wastes, Vols. 1 and 2, IAEA, Vienna, **1960**
J. Flagg (Ed.), Chemical Processing of Reactor Fuels, Academic Press, New York, **1961**
S. M. Stoller, R. B. Richards, Reactor Handbook, Vol. II, Fuel Reprocessing, Interscience, London, **1961**

F. R. Bruce, J. M. Fletcher, H. H. Hyman, J. J. Katz, Process Chemistry, Vol. 1, 2, 3, in: Progress in Nuclear Energy, Series III. Pergamon, Oxford, **1961**

C. B. Amphlett, Treatment and Disposal of Radioactive Wastes, Pergamon, Oxford, **1961**

J. R. Lamarch, Introduction to Nuclear Reactor Theory, Addison–Wesley, Reading, MA, **1966**

J. G. Wills, Nuclear Power Plant Technology, Wiley, New York, **1967**

C. O. Smith, Nuclear Reactor Materials, Addison–Wesley, Reading, MA, **1967**

J. T. Long, Engineering for Nuclear Fuel Reprocessing, Gordon and Breach, New York, **1967**

G. I. Bell, S. Glasstone, Nuclear Reactor Theory, Van Nostrand, London, **1970**

W. Häfele, D. Faude, E. A. Fischer, H. J. Laue, Fast Breeder Reactors, Annu. Rev. Nucl. Sci. *20*, 393 **(1970)**

R. F. Pocock, Nuclear Ship Propulsion, Ian Allan, London, **1970**

J. F. Flagg, Chemical Processing of Reactor Fuels, Vol. 1, in: Nuclear Science and Technology Series, Academic Press, New York, **1971**

J. R. Lamarsh, Introduction to Nuclear Engineering, Addison–Wesley, Reading, MA, **1975**

A. M. Perry, A. M. Weinberg, Thermal Breeder Reactors, Annu. Rev. Nucl. Sci. *22*, 317 **(1972)**

T. H. Pigford, Environmental Aspects of Nuclear Energy Production, Annu. Rev. Nucl. Sci. *24*, 515 **(1974)**

S. Ahrland, J. O. Liljenzin, J. Rydberg, (Actinide) Solution Chemistry, in: Comprehensive Inorganic Chemistry, Vol. 5, Pergamon, Oxford, **1975**

International Atomic Energy Agency, Management of Radioactive Wastes from the Nuclear Fuel Cycle, Vols. 1 and 2, IAEA, Vienna, **1976**

International Atomic Energy Agency, Int. Conf. on Nuclear Power and its Fuel Cycle, Salzburg **1977**, IAEA, Vienna, **1977**

R. A. Rydin, Nuclear Reactor Theory and Design, PBS Publications, **1977**

J. Guéron, Les Materiaux Nucléaires, Presse Universitaire de France, Paris, **1977**

W. S. Lyon, Analytical Chemistry in Fuel Reprocessing, Science Press, Princeton, **1978**

F. Baumgärtner (Ed.), Chemie der Nuklearen Entsorgung, 2 Vols., Thieme, München, **1978**

H. W. Grawes, jr., Nuclear Fuel Management, Wiley, New York, **1979**

R. G. Wymer, B. L. Vondra, Technology of the Light Water Reactor Fuel Cycle, CRC Press, Boca Raton, FL, **1980**

Le Dossier Electronucléaire, Editions du Deuil, Paris, **1980** (data on nuclear energy production)

A. Judd, Fast Breeder Reactors, Pergamon, Oxford, **1981**

R. G. Wymer, B. L. Vondra (Eds.), Light Water Reactor Nuclear Fuel Cycle, CRC Press, Boca Raton, FL, **1981**

Z. Dlouhy, Disposal of Radioactive Wastes, Elsevier, Amsterdam, **1982**

J. Guéron, L'Energie Nucléaire, Presse Universitaire de France, Paris, **1982**

G. Kessler, Nuclear Fission Reactors, Potential Role and Risks of Converters and Breeders, Springer, Berlin, **1983**

F. Baumgärtner, K. Ebert, E. Gelfort, K. H. Lieser (Eds.), Nuclear Fuel Cycle, 4 Vols., VCH Weinheim, **1981**, **1983**, **1986**, **1988**

International Atomic Energy Agency, Uranium, Resources, Production and Demand, OECD/NEA and IEAE, Vienna, **1989**

E. P. Steinberg, Radiochemistry of the Fission Products, J. Chem. Educ. *66*, 367 **(1989)**

The Management of Radioactive Waste, The Uranium Institute, London, **1991**

J. Rydberg, C. Musika, G. Choppin (Eds.), Principles and Practices of Solvent Extraction, Marcel Dekker, New York, **1992**

World Nuclear Industry Handbook, Nucl. Eng. Spec. Publ. **1992**

W. F. Miller, Present and Future Nuclear Reactor Designs, Weighing the Advantages and Disadvantages of Nuclear Power with an Eye of Improving Safety and Meeting Future Needs, J. Chem. Educ. *70*, 109 **(1993)**

Proc. Fourth Int. Conf. Nuclear Fuel Reprocessing and Waste Management, April **1994**, London, British Nuclear Industry Forum, London, **1994**

H. Röthemeyer, E. Warnecke, Radioactive Waste Management – the International Approach, Kerntechnik *59*, 1, 7 **(1994)**

Kernkraftwerke **1994**, Weltübersicht, atw Internat. Zeitschr. Kernenergie *40*, No.3 **(1995)**

L. Koch, Radioactivity and Fission Energy, Radiochim. Acta *70/71*, 397 **(1995)**

C. D. Bowman, Basis and Objectives of the Los Alamos Accelerator-Driven Transmutation Technology Project, Los Alamos National Laboratory Report LA-UR-95-206

Natural Reactors

International Atomic Energy Agency, The Oklo Phenomenon, IAEA, Vienna, **1975**
International Atomic Energy Agency, The Natural Fission Reactors, IAEA, Vienna, **1978**
P. K. Kuroda, The Origin of the Chemical Elements and the Oklo Phenomenon, Springer, Berlin,
 1982

Fusion Technology

R. F. Post, Controlled Fusion Research and High Energy Plasmas, Annu. Rev. Nucl. Sci. *20*, 509
 (1970)
D. M. Gruen, The Chemistry of Fusion Technology, Plenum Press, New York, **1970**
H. J. Ache, Chemical Aspects of Fusion Technology, Angew. Chem., Int. Ed. *28*, 1 **(1989)**
R. W. Conn, V. A. Chuyanov, N. Inoue, D. R. Sweetman, The International Thermonuclear
 Experimental Reactor, Scientific American, April **1992**, 75
J. G. Cordey, R. J. Goldston, R. R. Parker, Progress Toward a Tokomak Fusion Reactor, Physics
 Today, Jan. **1992**, 22
T. H. Jensen, Fusion, A. Potential Power Source, J. Chem. Educ. *71*, 820 **(1994)**
H. Kudo, Radioactivity and Fusion Energy, Radiochim. Acta *70/71*, 403 **(1995)**

Nuclear Explosives

H. D. Smith, Atomic Energy for Military Purposes, Princeton University Press, Princeton, NY, **1946**
E. Teller, W. K. Talley, G. H. Higgins, G. W. Johnson, The Constructive Uses of Nuclear Explo-
 sives, McGraw-Hill, New York, **1968**
Nuclear Weapons, Report of the Secretary-General of the United Nations, Autumn Press, New
 York, **1981**
R. Rhodes, The Making of the Atomic Bomb, Simon and Schuster, New York, **1986**

12 Production of Radionuclides and Labelled Compounds

12.1 Production in Nuclear Reactors

Most radionuclides used in science and technology are produced in nuclear reactors, because neutrons are available with high flux densities $\Phi \approx 10^{10}$ to $10^{16}\,\mathrm{cm^{-2}\,s^{-1}}$ and because the cross sections of (n, γ) reactions are relatively high. In general, research reactors are applied for production of radionuclides, because they are mostly equipped with special installations for that purpose. Power reactors may also be used, preferably for the production of greater amounts of long-lived radionuclides, such as ^{60}Co as a radiation source. Larger reactors offer more space for introducing a greater number of samples without affecting the neutron flux too much. As the activity increases with Φ (eq. (8.27)), high neutron flux densities are required for the production of high activities, particularly in the case of long-lived radionuclides.

The samples used for irradiation must withstand the temperature and the radiation in the reactor without decomposition. In general, they are encapsulated in aluminium cans or in quartz ampoules which are introduced in irradiation tubes or directly into special positions within the reactor.

Some research reactors are equipped with a so-called thermal column of about $1\,\mathrm{m} \times 1\,\mathrm{m} \times 1\,\mathrm{m}$, consisting of blocks of graphite and installed near the reactor core. Due to the moderator properties of graphite, only thermal neutrons are present in such a column.

Fast pneumatic transport systems are needed for the investigation of short-lived radionuclides. They can be operated over distances of several metres and the samples can be brought, within 0.1 to 1 s, directly from the place of irradiation to the counter.

A survey of neutron-induced reactions used for the production of radionuclides is given in Table 12.1. (n, γ) reactions are preferably triggered by thermal neutrons giving isotopic products of limited specific activity.

As an example, irradiation of 1 g NaCl at a thermal neutron flux density $\Phi = 10^{13}\,\mathrm{cm^{-2}\,s^{-1}}$ may be considered:

$$
\begin{aligned}
&^{23}\mathrm{Na}(n, \gamma)^{24}\mathrm{Na} \quad (\sigma = 0.53\,\mathrm{b};\ t_{1/2} = 14.96\,\mathrm{h}) \\
&^{35}\mathrm{Cl}(n, \gamma)^{36}\mathrm{Cl} \quad (\sigma = 0.43\,\mathrm{b};\ t_{1/2} = 3.0 \cdot 10^5\,\mathrm{y}) \\
&^{35}\mathrm{Cl}(n, \gamma)^{35}\mathrm{S} \quad (\sigma = 0.49\,\mathrm{b};\ t_{1/2} = 87.5\,\mathrm{d}) \\
&^{37}\mathrm{Cl}(n, \gamma)^{38}\mathrm{Cl} \quad (\sigma = 0.43\,\mathrm{b};\ t_{1/2} = 37.18\,\mathrm{min})
\end{aligned}
\tag{12.1}
$$

The activities obtained after an irradiation time of 24 h are $3.7 \cdot 10^{10}\,\mathrm{Bq}$ ^{24}Na $(2.2 \cdot 10^{12}\,\mathrm{Bq/mol})$, $2.1 \cdot 10^4\,\mathrm{Bq}$ ^{36}Cl $(1.2 \cdot 10^6\,\mathrm{Bq/mol})$, $1.1 \cdot 10^{10}\,\mathrm{Bq}$ ^{38}Cl $(6.4 \cdot 10^{10}\,\mathrm{Bq/mol})$, $3.0 \cdot 10^8\,\mathrm{Bq}$ ^{35}S. The activities as well as the specific activities increase with the neutron flux density and the irradiation time. Whether the specific activity is sufficient or not depends on the purpose for which the radionuclides are used. For

Table 12.1. Neutron-induced reactions in nuclear reactors (survey).

Reaction	Decay modes of the products	Remarks
(n, γ)	Predominantly β^-; rarely β^+ or ε	Generally applicable; high yields with thermal neutrons
$(n, 2n)$	Predominantly β^+; sometimes β^-	Strongly endoergic; high-energy neutrons ($>10\,\mathrm{MeV}$) required
(n, p)	Nearly always β^-	Mostly endoergic (exceptions: $^{14}N(n, p)^{14}N$, $^{35}Cl(n, p)\,^{35}S$ and some others); frequently at small mass numbers ($A < 40$)
(n, α)	Predominantly β^-	Mostly endoergic (exceptions $^6Li(n, \alpha)^3H$ and $^{10}B(n, \alpha)^7Li$); predominantly at small mass numbers
(n, f)	Fission with thermal neutrons: always β^-; fission with high-energy neutrons: also β^+ or ε	Nuclides with atomic numbers $A > 90$; fission of ^{233}U, ^{235}U and ^{239}Pu with thermal neutrons

medical applications, high specific activities are mostly required. The specific activity of ^{35}S depends on the presence of sulfur impurities in the sample. In general, the specific activity of non-isotopic products is high provided that no carrier is added.

With respect to high specific activity, (n, γ) reactions followed by relatively quick β^- decay are favourable. This is illustrated by the production of ^{131}I by neutron irradiation of Te. The following reactions have to be considered:

$$
^{130}Te(n, \gamma)\,^{131m}Te \xrightarrow[30\,h]{\mathrm{IT}} {}^{131}Te \xrightarrow[25\,min]{\beta^-,\gamma} {}^{131}I \xrightarrow[8.02\,d]{\beta^-,\gamma}
$$
$$(H = 0.338)$$

$$
^{128}Te(n, \gamma)\,^{129m}Te \xrightarrow[33.6\,d]{\mathrm{IT}} {}^{129}Te \xrightarrow[69.6\,min]{\beta^-,\gamma} {}^{129}I \xrightarrow[1.57\cdot10^7\,y]{\beta^-,\gamma}
$$
$$(H = 0.317)$$

$$
^{126}Te(n, \gamma)\,^{127m}Te \xrightarrow[109\,d]{\mathrm{IT}} {}^{127}Te \xrightarrow[9.35\,h]{\beta^-} {}^{127}I(\text{stable})
$$
$$(H = 0.189)$$

$$(12.2)$$

The (n, γ) reactions lead partly to the isomeric states and partly to the ground states, and the isomeric states change partly first into the ground states and partly directly to the daughter nuclides. From eqs. (12.2) it follows that after irradiation a decay time of several days is favourable in order to increase the yield of ^{131}I. Longer decay times are unfavourable, because of the decay of ^{131}I and the formation of increasing amounts of ^{129}I. In general, I is separated from Te by distillation.

Szilard–Chalmers reactions may also be applied to obtain radionuclides of high specific activity after (n, γ) reactions (section 9.6.).

Some radionuclides can be produced by (n, p) and (n, α) reactions with thermal neutrons, for example

$$^3\mathrm{He}(n, p)^3\mathrm{H} \quad (\sigma = 5327\,\mathrm{b};\ t_{1/2} = 12.323\,\mathrm{y}) \tag{12.3}$$

$$^{14}\mathrm{N}(n, p)^{14}\mathrm{C} \quad (\sigma = 1.81\,\mathrm{b};\ t_{1/2} = 5730\,\mathrm{y}) \tag{12.4}$$

$$^6\mathrm{Li}(n, \alpha)^3\mathrm{H} \quad (\sigma = 940\,\mathrm{b};\ t_{1/2} = 12.323\,\mathrm{y}) \tag{12.5}$$

$$^{40}\mathrm{Ca}(n, \alpha)^{37}\mathrm{Ar} \quad (\sigma = 0.0025\,\mathrm{b};\ t_{1/2} = 35.0\,\mathrm{d}) \tag{12.6}$$

As the product nuclides are non-isotopic, their separation from the target nuclides does not cause problems. Production of $^{14}\mathrm{C}$ according to eq. (12.4) is of great practical interest. The long half-life of $^{14}\mathrm{C}$ requires a long irradiation time ($\geq 1\,\mathrm{y}$) to obtain sufficient yield. Therefore, the nitrogen compound used as target must be heat- and radiation-resistant. Nitrides, such as AlN or $\mathrm{Be_3N_2}$, are most suitable. After irradiation, they are heated in a stream of oxygen; $^{14}\mathrm{C}$ escapes as $^{14}\mathrm{CO_2}$ and is bound by $\mathrm{Ba(OH)_2}$ as $\mathrm{Ba}^{14}\mathrm{CO_3}$. The specific activity is of the order of $2\,\mathrm{GBq/mmol}$ ($2.3\,\mathrm{GBq/mmol}$ if free from impurities and up to $1.6\,\mathrm{GBq/mmol}$ in practice). Because of the long irradiation time, $\mathrm{Ba}^{14}\mathrm{CO_3}$ is rather expensive (about US\$ 10^4 per gram at the time of writing). $\mathrm{Ba}^{14}\mathrm{CO_3}$ and $^{14}\mathrm{CO_2}$, respectively, are used for the preparation of a great variety of $^{14}\mathrm{C}$-labelled organic compounds by chemical or biochemical methods. Short synthetic routes and high yields are preferred to keep the losses as small as possible.

Tritium is produced on a large scale according to reaction (12.5) by irradiation of Li or lithium compounds in nuclear reactors. Due to the high cross section of this reaction and the relatively short half-life of T compared with that of $^{14}\mathrm{C}$, the yield is relatively high and the costs of tritium production are relatively low. T is obtained as a gas containing about 90% T and can be oxidized to tritiated water ($\mathrm{T_2O}$) for further use. 1 Ci ($3.7 \cdot 10^{10}\,\mathrm{Bq}$) of T is about $0.104\,\mathrm{mg}$ and as $\mathrm{T_2}$ gas it has a volume of $0.38\,\mathrm{cm}^3$ under normal temperature and pressure. Tritium is mainly used for controlled thermonuclear fusion experiments. It has also been produced for application in nuclear weapons (section 11.10). Relatively small amounts of T are used for labelling of organic compounds.

Pulsed research reactors, such as reactors of the Triga type, are especially designed for production and investigation of short-lived radionuclides. In these reactors the neutron flux is increased for about 10 ms to about $10^{16}\,\mathrm{cm}^{-2}\,\mathrm{s}^{-1}$ by taking out the control rods (section 11.5). Due to the negative temperature coefficient of the zirconium hydride moderator, the outburst of power causes a sudden decrease of the moderator properties and shutting off of the reactor. After several minutes the effects have vanished and a new pulse can be started. The activities of radionuclides of various half-lives obtained with pulsed reactors are compared in Table 12.2 with those produced at constant neutron flux densities. The table shows that pulsed reactors are useful for production and investigation of radionuclides with half-lives $<10\,\mathrm{s}$.

A great number of radionuclides are produced in nuclear reactors by fission. Some of these fission products may be separated for practical applications. For this purpose, either special separation procedures may be included in the course of reprocessing, or samples of enriched uranium may be irradiated in nuclear reactors and

Table 12.2. Activities of radionuclides of various half-lives obtained in pulsed reactors of the Triga type compared with those obtained at constant flux densities (activities in MBq/g for $\sigma_{n,\gamma} = 1$ barn, $H = 1$ and $M = 100$)

Half-life	Pulsed reactor $(\Phi_{max} = 10^{16}\,\mathrm{cm^{-2}\,s^{-1}};$ 10 ms)	Activities at constant neutron flux (saturation activity)		
		$\Phi = 10^{11}\,\mathrm{cm^{-2}\,s^{-1}}$	$\Phi = 10^{12}\,\mathrm{cm^{-2}\,s^{-1}}$	$\Phi = 10^{13}\,\mathrm{cm^{-2}\,s^{-1}}$
0.01 s	$3.0 \cdot 10^7$			
0.1 s	$4.0 \cdot 10^6$			
1 s	$4.2 \cdot 10^5$	$6.0 \cdot 10^2$	$6.0 \cdot 10^3$	$6.0 \cdot 10^4$
10 s	$4.2 \cdot 10^4$			
1 min	$7.0 \cdot 10^3$			

reprocessed separately with the aim of isolating the desired radionuclides. Some radionuclides that may be obtained from fission products and methods of their isolation are listed in Table 12.3. The main problems with fission products are the high activities to be handled, and the fact that their application requires careful purification procedures to reach acceptable radionuclide purity. Long-lived radioactive impurities, in particular actinides, have to be separated effectively. Radionuclide purity is of special importance in the case of medical applications (e.g. the application of fission ^{99}Mo in radionuclide generators).

Reactions with fast neutrons, such as (n, 2n), (n, p) and (n, α) reactions, are only of minor importance for production of radionuclides in nuclear reactors. However, special measures may be taken for irradiation of samples with high-energy neutrons. For instance, the samples may be irradiated in special fuel elements of ring-like cross section as shown in Fig. 12.1, or they may be irradiated in a receptacle made of enriched uranium. In both cases, the fast neutrons originating from the fission of ^{235}U enter the samples directly and their flux density is higher by about one order of magnitude than that at other places in the reactor.

Another possibility is the generation of fast neutrons by mixing the sample with LiD or introducing LiD together with the sample into the reactor. Fast neutrons with energies of ≈ 14 MeV are produced in LiD by te successive reactions ^{6}Li (n, α)t and d(t, n)α. The flux density of 14 MeV neutrons is smaller by a factor of about 10^5 than that of thermal neutrons.

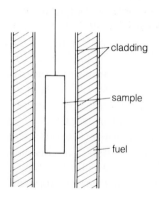

Figure 12.1. Irradiation in a ring-like fuel element.

Table 12.3. Radionuclides that may be obtained by separation from fission products.

Radionuclide	Separation procedure
^{85}Kr, ^{133}Xe	Driving out of the noble gases from boiling fission product solution. Adsorption on activated carbon. Purification by repeated de- and adsorption.
^{90}Sr	Precipitation of Sr and Ba as the nitrates from HNO_3 solution. Separation of Sr and Ba by oxalate precipitation.
^{95}Zr	(a) Adsorption on silica gel. Elution of other fission products with H_2SO_4/HNO_3. Elution of Zr with 0.5 M oxalic acid.
	(b) Extraction with thenoyltrifluoroacetone (TTA) in benzene. Back-extraction with 2 M HF and precipitation as $BaZrF_6$. Separation of Ba as $BaSO_4$.
^{99}Mo	(a) Adsorption of Mo from 2 M HNO_3 on Al_2O_3. Elution with 1 M NH_3. Separation from I by filtration through freshly precipitated AgCl.
	(b) Extraction with ether from 6 M HCl. Back-extraction into water. Separation of other fission products by coprecipitation with iron(III) hydroxide.
	(c) Extraction with organic complexing agents followed by purification.
^{99}Tc	(a) Precipitation as tetraphenylarsonium pertechnetate.
	(b) Extraction as tetraphenylarsonium pertechnetate into $CHCl_3$ and back-extraction with 0.2 M $HClO_4$.
^{103}Ru	(a) Oxidation in H_2SO_4 to RuO_4 and distillation.
	(b) Distillation from $HClO_4$ in the presence of $NaBiO_3$.
	(c) Extraction of RuO_4 from acid solution into CCl_4. Precipitation of RuO_2 with methanol.
^{131}I	(a) Reduction to I_2. Steam distillation. Extraction into CCl_4. Purification by repeated reduction and oxidation.
	(b) Separation of I^- or I_2 on AgCl. Elution with hypochlorite solution pH 9.
^{137}Cs	(a) Extraction as caesium tetraphenylborate into amyl acetate. Back-extraction with 3 M HCl.
	(b) After separation of alkali ions, Ru and rare-earth elements, precipitation of Cs as $CsAl(SO_4)_2$.
^{141}Ce, ^{144}Ce	(a) Extraction of Ce(IV) from HNO_3 solution with tributyl phosphate or di(2-ethylhexyl)phosphate.
	(b) Precipitation of Ce(IV) as $CeHIO_6$.
^{140}Ba	Precipitation as $BaCl_2 \cdot H_2O$ with a mixture of conc. HCl and ether (5:1). Purification by repeated precipitation.
^{147}Pm	Separation of the rare-earth fraction on a cation-exchange resin. Elution with lactic acid of increasing concentration (0.85 to 1.0 M) at pH 3 and about 80 °C.

If activation by epithermal or fast neutrons is to be favoured over activation by thermal neutrons, the flux of thermal neutrons can be suppressed by wrapping the samples in foils of cadmium, which has a high neutron absorption cross section for thermal neutrons ($\sigma_a = 2520\,$b). This possibility is also used in activation analysis.

The high intensity of γ radiation in nuclear reactors may be reduced by shielding the samples with Pb or Bi. The absorption coefficients of these metals are $\mu_\gamma\,(\text{Pb}) = 0.8\,\text{cm}^{-1}$ and $\mu_\gamma\,(\text{Bi}) = 0.7\,\text{cm}^{-1}$ for 1 MeV γ rays and $\mu_n\,(\text{Pb}) = 0.56 \cdot 10^{-2}\,\text{cm}^{-1}$ and $\mu_n\,(\text{Bi}) = 0.93 \cdot 10^{-3}\,\text{cm}^{-1}$ for thermal neutrons. By application of 2 cm Pb or Bi the intensity of γ radiation can be reduced by a factor of about five and four, respectively, whereas the neutron flux is not markedly diminished.

12.2 Production by Accelerators

The various types of accelerators offer the possibility of applying a great variety of projectiles of different energies. The most frequently used projectiles are protons, deuterons and α particles. Some features of the reactions induced by these particles are summarized in Table 12.4. Neutrons may be produced indirectly by nuclear reactions, γ rays are generated as bremsstrahlung in electron accelerators, and heavy ions are available in heavy-ion accelerators.

Table 12.4. Production of radionuclides by accelerators (survey).

Reaction	Decay modes of the products	Remarks
(p, γ)	β^+ or ε, rarely β^-	Sharp resonances with light nuclei; (p, n) reactions may compete
(p, n)	β^+ or ε, sometimes β^-	Mostly endoergic; threshold energy 2–4 MeV
(p, α)	–	Very seldom
(d, n)	β^+ or ε, sometimes β^-	Exoergic; high yields
$(d, 2n)$	β^+ or ε, rarely β^-	Threshold energy 5–10 MeV; at higher energies $(d, 3n)$ and $(d, 4n)$ reactions
(d, p)	mostly β^-	Generally relatively high yields; with 14 MeV deuterons practically applicable for all elements
(d, α)	β^+ or β^-	Mostly exoergic; frequently observed with light nuclei, e.g. ^{6}Li$(d, \alpha)\alpha$
(α, n)	β^+ or ε, rarely β^-	Applied for production of neutrons; yields decrease with increasing atomic number; at higher energies $(\alpha, 2n)$ and $(\alpha, 3n)$ reactions
(α, p)	mostly β^-, sometimes β^+ or ε	Relatively high threshold energies (like other reactions with α particles); reactions with nuclides of high atomic number require high energy of the α particles
(γ, n)	β^+, sometimes β^-	High-energy γ rays available as bremsstrahlung; always endoergic; threshold energy given by the binding energy of the neutron; at higher energies increasing contributions of $(\gamma, 2n)$ and $(\gamma, 3n)$ reactions

Deuterons are often preferred as projectiles, because of the relatively high cross sections obtained with them. Some examples of radionuclides produced by d-induced reactions are listed in Table 12.5.

Table 12.5. Reactions with deuterons applicable to production of radionuclides

Target	Nuclear reaction	Yield $\left[\dfrac{kBq}{\mu A\,h}\right]$ (thick targets)		
		8 MeV	14 MeV	19 MeV
Li, LiF, LiBO$_2$	^{6}Li(d, n)^{7}Be	–	Low	74
B$_2$O$_3$	^{10}Be(d, n)^{11}C	$1.85 \cdot 10^4$	$1.79 \cdot 10^4$	–
Mg, MgO	^{24}Mg(d, α)^{22}Na	–	39–67	–
Cr	^{50}Cr(d, α)^{48}V	–	Low	–
Cr	^{52}Cr(d, 2n)^{52}Mn	–	$2.96 \cdot 10^3$	–
Fe	^{56}Fe(d, α)^{54}Mn	–	37	–
Mn alloy	^{55}Mn(d, 2n)^{55}Fe	–	26	0.74
Fe	^{56}Fe(d, 2n)^{56}Co	–	High	–
Fe	^{56}Fe(d, n)^{57}Co	–	High	185
Fe	^{57}Fe(d, n)^{58}Co	–	High	–
Cu	^{65}Cu(d, 2n)^{65}Zn	–	130	–
Ge alloy	^{74}Ge(d, 2n)^{74}As	–	74	370
As alloy	^{75}As(d, 2n)^{75}Se	–	High	–
NaBr	^{79}Br(d, 2n)^{79}Kr	–	High	–
SrCO$_3$	^{88}Sr(d, 2n)^{88}Y	–	$1.41 \cdot 10^3$	–
Pd	^{107}Pd(d, n)^{106}Ag	–	Average	–
Te	^{130}Te(d, 2n)^{130}I	–	$3.3 \cdot 10^4$	–
NaI	^{127}I(d, 2n)^{127}Xe	–	Average	–
Au	^{197}Au(d, 2n)^{197}Hg	–	$3.0 \cdot 10^4$	–

Several positron emitters are of practical importance in nuclear medicine for positron emission tomography (PET). Some are listed in Table 12.6. Protons or deuterons with energies varying between about 10 and 40 MeV are available in small cyclotrons ("baby cyclotrons") and are applied for the production of suitable radionuclides.

The cross sections for production of ^{22}Na and ^{24}Na by the reactions ^{24}Mg(d, α) ^{22}Na and ^{26}Mg(d, α)^{24}Na, respectively, are plotted in Fig. 12.2 as a function of the deuteron energy. The cross sections for various nuclear reactions that are of interest with respect to the production of radioisotopes of Zn are plotted in Fig. 12.3, also as a function of the deuteron energy.

(d, p) reactions have, in general, relatively high cross sections, but they are not of practical interest because they are equivalent to (n, γ) reactions in nuclear reactors, which exhibit high yields.

Table 12.6. Production of positron emitters used in nuclear medicine.

Radionuclide	Half-life	Mode of decay $(E_{\beta^+(max)})$	Nuclear reactions	Yields (thick targets) [MBq/μA h] (Particle energies [MeV])
^{11}C	20.38 min	β^+ (99.8%) (0.96 MeV)	^{14}N(p, α)^{11}C ^{11}B(p, n)^{11}C ^{10}B(d, n)^{11}C	≈ 3800 (13) ≈ 3400 (10) ≈ 2500 (10)
^{13}N	9.96 min	β^+ (100%) (1.19 MeV)	^{12}C(d, n)^{13}N ^{16}O(p, α)^{13}N	≈ 2000 (8) ≈ 1700 (16)
^{15}O	2.03 min	β^+ (99.9%) (1.72 MeV)	^{14}N(d, n)^{15}O ^{16}O(p, pn)^{15}O	≈ 2400 (8) ≈ 3700 (25)
^{18}F	109.7 min	β^+ (97%) (0.635 MeV)	^{20}Ne(d, α)^{18}F ^{18}O(p, n)^{18}F ^{16}O(^{3}He, p)^{18}F	≈ 1100 (14) ≈ 3000 (16)[a] ≈ 500 (40)
^{38}K	7.6 min	β^+ (100%) (2.68 MeV)	^{35}Cl(α, n)^{38}K ^{38}Ar(p, n)^{38}K ^{40}Ca(d, α)^{38}K	≈ 260 (20) ≈ 800 (16)[a] ≈ 200 (12)
^{73}Se	7.1 h	β^+ (65%) (1.32 MeV)	^{75}As(p, 3n)^{73}Se ^{75}As(d, 4n)^{73}Se	≈ 1400 (40) ≈ 650 (45)
^{75}Br	1.6 h	β^+ (75.5%) (1.74 MeV)	^{76}Se(p, 2n)^{75}Br ^{76}Se(d, 3n)^{75}Br ^{75}As(α, 4n)^{75}Br	≈ 3700 (30) ≈ 3000 (35)[a] ≈ 250 (60)
^{76}Br	16.0 h	β^+ (57%) (3.90 MeV)	^{76}Se(p, n)^{76}Br ^{77}Se(p, 2n)^{76}Br	≈ 290 (16)[a] ≈ 250 (25)[a]

[a] Using highly enriched isotopes.
(From: G. Stöcklin, V. W. Pike (Eds.), *Radiopharmaceuticals for Positron Emission Tomography*, Kluwer Academic Publishers, Dordrecht, **1993**.)

Table 12.7. Examples of the production of radionuclides by (γ, n) reactions.

Target	Radionuclide produced by (γ, n) reaction	Half-life	Resonance energy [MeV]	Cross section at the resonance energy [barn]
TiO_2	^{45}Ti	3.08 h	–	–
$Co(NO_3)_2 \cdot 6 H_2O$	^{58}Co	70.86 d	16.9	0.13
$AgNO_3$	^{106m}Ag	8.3 d	14.0	0.24
KIO_3 or LiI	^{126}I	13.11 d	15.2	0.45
Cs_2CO_3	^{132}Cs	6.47 d	–	–
$TlNO_3$	^{202}Tl	12.23 d	≈17	0.092
$Pb(NO_3)_2$	^{203}Pb	51.9 h	–	–

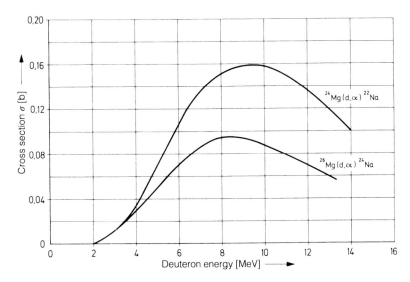

Figure 12.2. Cross sections of the reactions ^{24}Mg(d, α)^{22}Na and ^{26}Mg(d, α)^{24}Na as a function of the deuteron energy. (According to J. W. Irvine, E. T. Clarke: J. chem. Physics **16**, 686 (1948).)

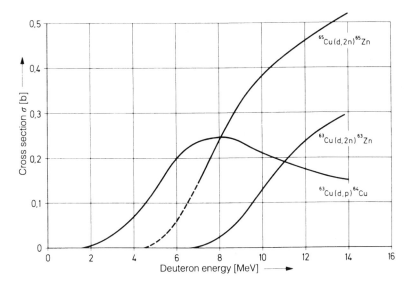

Figure 12.3. Cross sections for various nuclear reactions of deuterons with copper. (According to J. W. Irvine, jun.: J. chem. Soc. [London] **5**, 356 (1949).)

(γ, n) reactions are applied for the production of isotopic neutron-deficient nuclides. Some examples are listed in Table 12.7. The cross sections of these reactions vary between about 1 and 100 mb. They increase with E_γ, the energy of the γ-ray photons, and decrease again at higher values of E_γ, as illustrated in Fig. 12.4 for the reaction $^{141}Pr(\gamma, n)^{140}Pr$. The contribution of $(\gamma, 2n)$, $(\gamma, 3n)$ and $(\gamma, 4n)$ reactions increases with E_γ. For example, by irradiation of rubidium salts ^{83}Rb can be produced which changes by electron capture into ^{83m}Kr:

$$^{85}Rb(\gamma, 2n)^{83}Rb \xrightarrow[86.2d]{\varepsilon} {}^{83m}Kr \xrightarrow[1.83h]{IT} \tag{12.7}$$

^{83m}Kr is of interest for practical application and can be obtained from a $^{83}Rb/^{83m}Kr$ radionuclide generator (section 12.4).

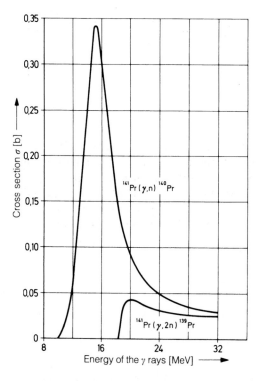

Figure 12.4. Cross sections of the reactions $^{141}Pr(\gamma, n)^{140}Pr$ and $^{141}Pr(\gamma, 2n)^{139}Pr$ as a function of the photon energy. (According to J. H. Carver, W. Turchinetz: Proc. physic. Soc. **73**, 110 (1959).)

The samples may be irradiated inside or outside the accelerator, as shown schematically in Fig. 12.5. In both cases heat is generated by absorption of radiation. For instance, a current of 100 µA protons at 10 MeV corresponds to a power of 1 kW which must be taken away by cooling to avoid overheating or melting of the samples. The thickness of the samples must also be taken into account. Because of the strong interaction of charged particles with matter, their flux density Φ decreases rapidly and their energy distribution changes with the penetration depth. Only in thin foils can Φ be considered to be constant.

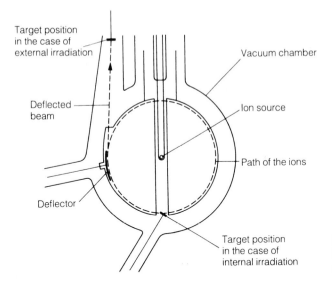

Target position
in the case of
external irradiation

Vacuum chamber

Deflected
beam

Ion source

Path of the ions

Deflector

Target position
in the case of
internal irradiation

Figure 12.5. External and
internal irradiation in a cyclotron (schematically).

12.3 Separation Techniques

The task of quantitative and effective separation of small amounts of radionuclides
has appreciably enhanced the development of modern separation techniques. High
radionuclide purity is of great importance for application in nuclear medicine as well
as for sensitive measurements. In this context, impurities of long-lived radionuclides
are of particular importance, because their relative activity increases with time. For
example, if the activity of ^{90}Sr is only 0.1% of that of ^{140}Ba after fresh separation, it
will increase to 11.5% in three months.

The most frequently used separation techniques are:

- crystallization, precipitation or coprecipitation;
- electrolysis;
- distillation;
- solvent extraction;
- ion exchange;
- chromatography.

Fractional crystallization was one of the first methods used in radiochemistry by
Marie and Pierre Curie to separate ^{226}RaCl$_2$ from BaCl$_2$. Precipitation is only applicable if the solubility product is exceeded, i.e. if the concentration of the radionuclide
to be separated is high enough. If the concentration is too low, coprecipitation may
be applied by addition of a suitable carrier, i.e. of compounds of identical or very
similar chemical properties. Application of isotopic carriers is very effective, but it
leads to a decrease in specific activity. Therefore, non-isotopic carriers with suitable
chemical properties are preferred. Hydroxides such as iron(III) hydroxide or other
sparingly soluble hydroxides give high coprecipitation yields, because of their high
sorption capacity. Coprecipitation of actinides with LaF$_3$ is also applied success-

fully. Coprecipitation by formation of anomalous solid solutions (anomalous mixed crystals), for example coprecipitation of actinides(III) and -(IV) with $BaSO_4$ or $SrSO_4$ in the presence of K^+ or Na^+ ions, respectively, is also possible.

If coprecipitation of radionuclides present in low concentrations is to be avoided, hold-back carriers are added. Isotopic carriers are most effective for this purpose, but they lead to low specific activity.

For filtration, Hahn suction filters are convenient (Fig. 12.6). They allow easy removal of the filter for further operations or subsequent measurements.

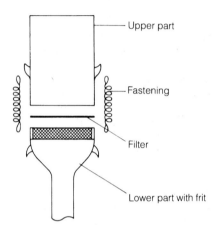

Figure 12.6. Hahn suction filter.

Electrolytic deposition of radionuclides is frequently applied. It gives thin samples and is well suited for preparation of standard samples. For instance, Po, Pb or Mn can be deposited with high yields on anodes of Cu, Pt or Ag, and by electrolysis of the nitrates or chlorides of Th and Ac in acetone or ethanol solutions these elements can be separated on cathodes. The preparation of thin samples by electrolytic deposition is of special interest for the measurement of α emitters, such as isotopes of Pu or other actinides.

Separation of radionuclides by distillation is applicable if volatile compounds are formed. Separation of ^{131}I from irradiated Te has already been mentioned in section 12.1. Other examples are separation of Ru as RuO_4 under oxidizing conditions, and volatilization of Tc as Tc_2O_7 from concentrated H_2SO_4 at 150–250 °C. ^{32}P may be purified by volatilization as PCl_5 in a stream of Cl_2.

Solvent extraction is widely used for separation of radionuclides, because this technique is simple, fast and applicable in the range of low concentrations. Addition of a carrier is not required. Some examples of separation of radionuclides by solvent extraction are given in Table 12.8. As already mentioned in section 11.6, solvent extraction plays an important role in reprocessing. Tributyl phosphate (TBP), methyl isobutyl ketone (Hexon) and trilaurylamine (TLA) are preferred complexing agents for separation and purification of U and Pu.

Ion-exchange procedures have also found broad application in radiochemistry. The selectivity S is measured by the difference in the logarithms of the distribution coefficients $K_d(1)$ and $K_d(2)$: $S = \log K_d(1) - \log K_d(2)$. Commercial ion-exchange resins exhibit relatively low selectivities. Higher values of S are obtained by use of

Table 12.8. Examples of the separation of radionuclides by extraction.

Element	Extractant	Remarks
Fe	(a) isopropylether	Extraction from HCl solutions as HFeCl$_4$
	(b) cupferron in CHCl$_3$	Quantitative extraction from HCl solution
Br	CCl$_4$	Extraction of Br$_2$ from HNO$_3$ solution
Sr	TTA[a]	At pH >10
Y	(a) D2EHPA[b] in toluene	Separation from Sr in HCl solution
	(b) TBP[c]	Effective separation from rare-earth elements at high HNO$_3$ concentration
Zr	(a) TBP[c]	Strong HCl or HNO$_3$ solution
	(b) DBP[d] in dibutyl ether	Quantitative extraction from acid solution
Tc	Tetraphenylarsonium chloride in CHCl$_3$	Extraction from neutral or alkaline solutions in the presence of H$_2$O$_2$ as (C$_6$H$_5$)$_4$AsTcO$_4$
I	CCl$_4$	Extraction from acid solution
Cs	Sodium tetraphenylborate in amyl acetate	Selective extraction as Cs(C$_6$H$_5$)$_4$B from buffered solution
Ce	(a) Ether	Separation of Ce(IV) from lanthanides(III)
	(b) TBP[c]	Extraction of Ce(IV); separation from lanthanides(III)
Ac	TTA[a] in benzene	At pH 6 from aqueous solution
Th	(a) Methyl isobutyl ketone	From HNO$_3$ solution
	(b) TTA[a] in benzene	From acid solution
	(c) TBP[c] in kerosine	From HNO$_3$/NaNO$_3$ solution
U	(a) Ether	Selective extraction of UO$_2^{2+}$ under certain conditions
	(b) TBP[c] in kerosine	Extraction of UO$_2^{2+}$ from HNO$_3$ solutions
Pu	TBP[c] in kerosine	Extraction of Pu(VI) and Pu(IV) from HNO$_3$ solutions
Am, Cm	TTA[a] in benzene	Quantitative extraction of Am(III) and Cm(III) at pH 4
Bk	D2EHPA[b] in heptane	Extraction of Bk(IV) from strong HNO$_3$ solution in presence of KBrO$_3$ – possibility of separating Bk from Cm and Cf

[a] α-thenoyltrifluoroacetone.
[b] Di(2-ethylhexyl)phosphoric acid.
[c] Tri-n-butyl phosphate.
[d] Di-n-butylphosphoric acid.

organic ion exchangers carrying chelating groups of high selectivity (preferably "tailor-made") as anchor groups or by application of inorganic ion exchangers. Highly selective organic ion exchangers are synthesized on the basis of polystyrene, cellulose or other substances as matrices. High selectivity with commercial ion-exchange resins is also obtained by addition of complexing agents, such as α-hydroxycarboxylic acids (lactic acid, α-hydroxyisobutyric acid and others) to the solution. Under these conditions, the selectivity is determined by the difference in the logarithms of the stability constants K_S of the complexes: $S = \log K_S(1) - \log K_S(2)$.

Inorganic ion exchangers comprise a great variety of compounds: hydrous oxides (e.g. Al$_2$O$_3 \cdot x$H$_2$O, SiO$_2 \cdot x$H$_2$O, TiO$_2 \cdot x$H$_2$O, ZrO$_2 \cdot x$H$_2$O, Sb$_2$O$_5 \cdot x$H$_2$O), acid salts (e.g. phosphates of Ti, Zr, hexacyanoferrates of Mo), salts of heteropoly acids (e.g. ammonium molybdophosphate), and clay minerals. Many ionic inorganic compounds also exhibit ion-exchange properties (e.g. BaSO$_4$, AgCl, CuS) and offer the possibility of highly selective separations (e.g. I$^-$ and I$_2$ on AgCl). The main dis-

Table 12.9. Examples of the separation of radionuclides by ion exchange.

Elements to be separated	Ion exchanger[a]	Eluant
Cs/Rb	zirconium tungstate	1 M NH_4Cl
Cs/Rb,K	Duolite C-3	0.3 M HCl (Rb, K,...), 3 M HCl (Cs)
Cs/Rb/K/Na	Titanium hexacyanoferrate	1 M HCl
alkaline earths	Dowex-50	1.5 M ammonium lactate, pH 7
Cs/Ba/rare earths	Dowex-50	$LiNO_3$ solution
Ba/Ra	Dowex-50	0.32 M ammonium citrate, pH 5.6
Rare earths	Dowex-50	5% ammonium citrate, pH 3–3.5, or 0.2–0.4 M α-hydroxyisobutyric acid, pH 4.0–4.6, or 0.025 M ethylene-diaminetetraacetic acid (EDTA)
Mn/Co/Cu/Fe/Zn	Dowex-1	HCl of decreasing concentration
Zr/Nb	Dowex-50	1 M HCl + H_2O_2 (Nb); 0.5% oxalic acid (Zr)
Zr/Pa/Nb/Ta	Dowex-1	HCl/HF solutions
Actinides/lanthanides	Dowex-50	20% ethanol, saturated with HCl (actinides are first eluted)
Actinides	Dowex-50	5% ammonium citrate, pH 3.5, or 0.4 M ammonium lactate, pH 4.0–4.5
Th/Pa/U	Dowex-1	10 M HCl or 9 M HCl/1 M HF

[a] Instead of the ion exchangers listed, others with similar properties may be applied.

advantage of inorganic ion exchangers is the relatively slow equilibration. Examples of separation of radionuclides by ion exchange are listed in Table 12.9.

Chromatographic separation techniques are based on adsorption, ion exchange or partition between a stationary and a mobile phase. Gas chromatography (GC) is applied for separation of volatile compounds. Thermochromatography (isothermal or in a temperature gradient) is frequently used for the study of the properties of small amounts of radionuclides. For the investigation of radionuclides of extremely short half-life the thermochromatographic column, usually a quartz tube, may be operated on-line with the production of the radionuclide by an accelerator, aerosols may be added as carriers and chemical reactions may be initiated by injection of reactive gases into the column.

For separation of radionuclides in solution, high-performance liquid chromatography (HPLC), paper chromatography or ion-exchange chromatography may be applied. Reversed-phase partition chromatography (RPC) offers the possibility of using organic extractants as the stationary phase in a multistage separation. On-line extraction techniques are also available.

12.4 Radionuclide Generators

Application of short-lived radionuclides has the advantage that the activity vanishes after relatively short periods of time. This aspect is of special importance in nuclear medicine. Short-lived radionuclides may be produced by irradiation in nuclear reactors or by accelerators, but their supply from irradiation facilities requires matching of production and demand, and fast transport. These problems are avoided by application of radionuclide generators containing a longer-lived mother nuclide from which the short-lived daughter nuclide can be separated.

The mother nuclide must be present in such a chemical form that the daughter nuclide can be separated repeatedly by a simple operation leaving the mother nuclide quantitatively in the generator. After each separation, the activity of the daughter nuclide increases again, and the daughter nuclide can be separated repeatedly at certain intervals of time. For example, after three half-lives of the daughter nuclide 87.5% of its saturation activity is reached and separation may be repeated with a relatively high yield. Fixation of the mother nuclide and simple separation of the daughter nuclide are achieved by use of separation columns containing a suitable ion exchanger or sorbent from which the daughter nuclide can be eluted.

Since the separation can be repeated many times, radionuclide generators are sometimes called "cows" and the separation procedure is called "milking". Some mother/daughter combinations used in radionuclide generators are listed in Table 12.10. They are very useful if the short-lived radionuclides are applied frequently, and have gained great practical importance in nuclear medicine, mainly for diagnostic purposes, where the main advantage of the application of short-lived radionuclides is the short radiation exposure. High radionuclide purity is an essential prerequisite and needs careful examination, in particular with respect to the presence of long-lived α emitters.

Table 12.10. Examples of mother and daughter nuclides suitable for use in radionuclide generators.

Mother nuclide			Daughter nuclide		
Nuclide	Half-life	Decay mode	Nuclide	Half-life	Decay mode
^{28}Mg	20.9 h	β^-, γ	^{28}Al	2.246 m	β^-, γ
^{42}Ar	33 y	β^-	^{42}K	12.36 h	β^-, γ
^{44}Ti	6.04 y	ε, γ	^{44}Sc	3.92 h	β^+, ε, γ
^{60}Fe	$1.5 \cdot 10^6$ y	β^-, γ	^{60m}Co	10.5 m	β^-, IT, γ
^{66}Ni	54.6 h	β^-	^{66}Cu	5.1 m	β^-, γ
^{62}Zn	9.13 h	ε, β^+, γ	^{62}Cu	9.74 m	β^+, ε, γ
^{72}Zn	46.5 h	β^-, γ	^{72}Ga	14.1 h	β^-, γ
^{68}Ge	270.8 d	ε	^{68}Ga	67.63 m	β^+, ε, γ
^{72}Se	8.5 d	ε, γ	^{72}As	26.0 h	β^+, ε, γ
^{83}Rb	86.2 d	ε, γ	^{83m}Kr	1.83 h	IT
^{82}Sr	25.34 d	ε	^{82}Rb	1.27 m	β^+, ε, γ
^{90}Sr	28.64 y	β^-	^{90}Y	64.1 h	β^-, γ
^{87}Y	80.3 h	ε, β^+, γ	^{87m}Sr	2.81 h	IT
^{99}Mo	66.0 h	β^-, γ	^{99m}Tc	6.0 h	IT
^{103}Ru	39.35 d	β^-, γ	^{103m}Rh	56.1 m	IT

Table 12.10. (continued)

Mother nuclide			Daughter nuclide		
Nuclide	Half-life	Decay mode	Nuclide	Half-life	Decay mode
^{100}Pd	3.7 d	ε, γ	^{100}Rh	20.8 h	α, γ
^{103}Pd	16.96 d	ε, γ	^{103m}Rh	56.1 m	β^-, γ
^{112}Pd	21.1 h	β^-, γ	^{112}Ag	3.12 h	α, γ
^{115}Cd	53.38 h	β^-, γ	^{115m}In	4.49 h	β^-, α, γ
^{115m}Cd	44.8 h	β^-, γ	^{115m}In	4.49 h	β^-
^{111}In	2.81 d	ε, γ	^{111m}Cd	49 m	α, γ
^{113}Sn	115.1 d	ε, γ	^{113m}In	99.49 m	α, γ
^{127}Sb	3.85 d	β^-, γ	^{127}Te	9.35 h	β^-, α, γ
^{118}Te	6.0 d	ε	^{118}Sb	3.5 m	α, γ
^{132}Te	76.3 h	β^-, γ	^{132}I	2.30 h	β^-, γ
^{137}Cs	30.17 y	β^-, γ	^{137m}Ba	2.55 m	β^-, γ (IT)
^{128}Ba	2.43 d	ε, γ	^{128}Cs	3.8 m	α, γ
^{140}Ba	12.75 d	β^-, γ	^{140}La	40.27 h	β^-, γ
^{134}Ce	75.9 h	ε	^{134}La	6.67 m	β^-, γ (IT)
^{144}Ce	284.8 d	β^-, γ	^{144}Pr	17.3 m	β^-, γ
^{178}W	22 d	ε	^{178}Ta	2.45 h	β^-, γ
^{188}W	69 d	β^-, γ	^{188}Re	16.98 h	β^-, γ
^{189}Re	24.3 h	β^-, γ	^{189m}Os	6.0 h	$\varepsilon, \beta^+, \gamma$
^{194}Os	6.0 y	β^-, γ	^{194}Ir	19.15 h	IT
^{189}Ir	13.3 d	ε, γ	^{189m}Os	6.0 h	β^-, γ
^{188}Pt	10.2 d	ε, γ	^{188}Ir	41.5 h	IT, β^-, γ
^{200}Pt	12.5 h	β^-	^{200}Au	48.4 m	IT, β^-, γ
^{194}Hg	520 y	ε	^{194}Au	38.0 h	IT
^{195m}Hg	40 h	IT, ε, γ	^{195m}Au	30.5 s	IT
^{210}Pb	22.3 y	β^-, γ	^{210}Bi	5.01 d	β^-, γ
^{204}Bi	11.22 h	ε, γ	^{204m}Pb	67.2 m	β^+, γ
^{211}Rn	14.6 h	$\varepsilon, \alpha, \gamma$	^{211}At	7.22 h	β^-, γ
^{224}Ra	3.66 d	α, γ	^{220}Rn	55.6 s	IT
^{224}Ra	3.66 d	α, γ	^{212}Pb	10.64 h	$\beta^+, \varepsilon, \gamma$
^{226}Ra	1600 y	α, γ	^{222}Rn	3.825 d	β^-, γ
^{228}Ra	5.75 y	β^-, γ	^{228}Ac	6.13 h	$\beta^+, \varepsilon, \gamma$
^{225}Ac	10.0 d	α, γ	^{221}Fr	4.9 m	β^-, γ
^{225}Ac	10.0 d	α, γ	^{213}Bi	45.59 m	$\varepsilon, \beta^+, \gamma$
^{225}Ac	10.0 d	α, γ	^{209}Pb	3.25 h	β^-, γ
^{226}Ac	29 h	$\beta^-, \varepsilon, \gamma$	^{226}Th	31 m	IT
^{227}Ac	21.77 y	β^-, α, γ	^{227}Th	18.72 d	β^-, γ
^{227}Ac	21.77 y	β^-, α, γ	^{223}Fr	21.8 m	IT
^{228}Th	1.913 y	α, γ	^{224}Ra	3.66 d	$\varepsilon, \beta^+, \gamma$
^{229}Th	7880 y	α, γ	^{225}Ra	14.8 d	β^-, γ
^{234}Th	24,10 d	β^-, γ	^{234m}Pa	1.17 m	$\varepsilon, \beta^+, \gamma$
^{230}U	20.8 d	α, γ	^{226}Th	31 m	IT
^{238}U	$4.468 \cdot 10^9$ y	α, γ	^{234}Th	24.10 d	β^-, γ
^{240}U	14.1 h	β^-, γ	^{240m}Np	7.22 m	IT
^{245}Pu	10.5 h	β^-, γ	^{245}Am	2.05 h	$\varepsilon, \alpha, \gamma$
^{246}Pu	10.85 d	β^-, γ	^{246m}Am	25 m	α, γ
^{254}Es	276 d	α, γ	^{250}Bk	3.22 h	β^-, γ

The ^{99}Mo/^{99m}Tc generator is the most frequently used radionuclide generator in medicine, due to the favourable properties of ^{99m}Tc (half-life 6.0 h; IT by emission of 141 keV γ rays). The activity of the ground state ^{99}Tc is negligible, because of its long half-life. Either ^{99}Mo is produced with relatively low specific activity by neutron irradiation of Mo or it is obtained with high specific activity as a fission product by neutron irradiation of ^{235}U followed by chemical separation. Usually, it is fixed as MoO_4^{2-} on hydrous Al_2O_3 where it is strongly bound, whereas ^{99m}TcO$_4^-$ can easily be eluted with water or physiological NaCl solution. Other radionuclide generators used in medicine are the ^{62}Zn/^{62}Cu, ^{68}Ge/^{68}Ga, ^{82}Sr/^{82}Rb and ^{113}Sn/^{113m}In generators. ^{62}Cu, ^{68}Ga and ^{82}Rb are of interest as positron emitters for positron emission tomography (PET). ^{113m}In changes with $t_{1/2} = 99.49$ min by emission of 392 keV γ rays into the stable ground state ^{113}In.

Some radionuclide generators have found application in chemical laboratories and in industry. An example is the ^{137}Cs/^{137m}Ba generator. ^{137m}Ba has a relatively short half-life (2.55 min) and changes by isomeric transition (IT) and emission of 662 keV γ rays into the stable ground state ^{137}Ba. The γ radiation can easily be detected and measured. The "milking" procedure can be repeated after several minutes. The long half-life of the mother nuclide ^{137}Cs (30.17 y) makes it possible to use this generator for long periods of time. However, ^{137}Cs must be bound so firmly that its migration within the time of use of the generator can be excluded. ^{137}Cs is produced in nuclear reactors as a fission product with a rather high yield (6.18%) and can be fixed quite firmly in thin layers of hexacyanoferrates generated on the surface of small pieces of metals (e.g. Fe, Mo).

12.5 Labelled Compounds

Labelled compounds have found broad application in various fields of science and technology. A great variety of labelled compounds are applied in nuclear medicine. The compounds are produced on a large scale as radiopharmaceuticals in cooperation with nuclear medicine, mainly for diagnostic purposes and sometimes also for therapeutic application. The study of metabolism by means of labelled compounds is of great importance in biology. More details on the application of radionuclides and labelled compounds in medicine and other areas of the life sciences will be given in chapter 19.

In chemistry, labelled compounds are used to elucidate reaction mechanisms and to investigate diffusion and transport processes. Other applications are the study of transport processes in the geosphere, the biosphere and in special ecological systems, and the investigation of corrosion processes and of transport processes in industrial plants, in pipes or in motors.

Organic or inorganic compounds may be labelled at various positions and by various nuclides. For that purpose, certain atoms in a molecule are substituted by isotopic radionuclides, by stable isotopes or even by non-isotopic radionuclides. To illustrate the great variety of possibilities, acetic acid is taken as an example: it may be labelled at the methyl or carboxyl group with ^{14}C, ^{13}C or ^{11}C, and the hydrogen atoms of the methyl group or the acidic hydrogen on the carboxyl group may be labelled with ^{2}H (D) or ^{3}H (T). Finally, the oxygen atoms in the carboxyl group may be labelled with ^{15}O, ^{17}O or ^{18}O. Sometimes double labelling with two different nuclides is of interest, in order to track the different pathways of certain groups. In the case of non-isotopic labelling it must be checked whether the different properties of the resulting compound are acceptable. The following discussion is restricted to labelling with radionuclides.

For preparation of labelled compounds the following parameters have to be considered:

- the kind of nuclide (isotopic or non-isotopic, half-life);
- the position of labelling (labelling at a certain position in a molecule or labelling of all atoms of a certain element at random);
- the specific activity (activity per unit mass of the element);
- the chemical purity (fraction of the compound in the desired chemical form);
- the radionuclide purity (fraction of the total radioactivity present as the specified radionuclide);
- the radiochemical purity (fraction of the radionuclide present in the desired chemical form and in the specified position in the molecule).

The choice of the parameters depends on the application of the labelled compound. Labelled compounds may be prepared in various ways:

- simple compounds are obtained by selection of suitable targets (e.g. ^{11}CO and ^{11}CO$_2$ by irradiation of nitrogen containing traces of oxygen with protons);
- chemical synthesis (most widely applied);
- biochemical methods (which allow labelling of complex organic compounds);
- exchange reactions (which offer the possibility of introducing radionuclides or stable nuclides into inactive compounds);
- recoil labelling and radiation-induced labelling (based on recoil and radiation-induced reactions (section 9.7); in general, a spectrum of labelled compounds is obtained).

Some points of view with respect to the synthesis of labelled compounds should be emphasized:

- in most cases the masses of the radionuclides to be handled in the synthesis are very small (of the order of a milligram or less), in particular if high specific activities are required;
- adequate methods are to be selected with regard to confinement of the radioactive substances (e.g. use of small closed apparatus);
- the problem of waste production, in particular in the case of long-lived radionuclides, or the cost of the radionuclides, e.g. in the case of ^{14}C, make it necessary to select simple chemical reactions with high yields. Reactions proceeding in one stage and in one receptacle are to be preferred to multistage reactions and use of different kinds of glassware.

In consideration of these points, known procedures of chemical synthesis have been modified. Special techniques have been developed for the preparation of a large number of ^{14}C-labelled organic compounds. Some reactions starting from Ba^{14}CO$_3$ are summarized in Fig. 12.7, in which, as an example, various preparation routes to

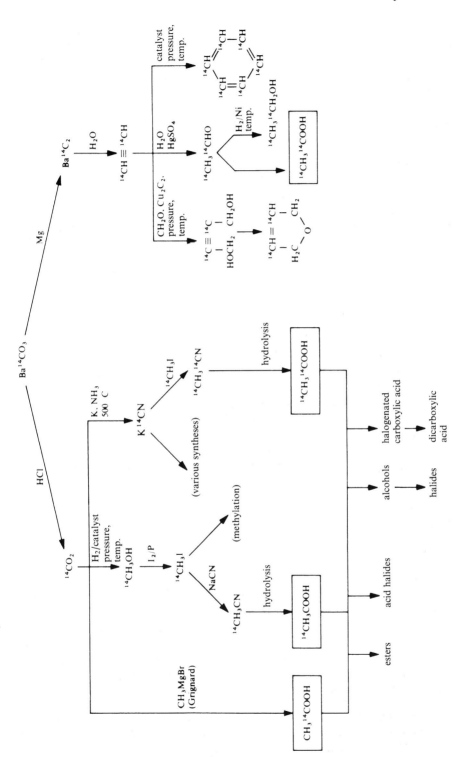

Figure 12.7. Preparation of ^{14}C-labelled compounds (example, acetic acid) from $Ba^{14}CO_3$.

labelled acetic acid are shown. Labelling of organic compounds with various other radionuclides is exemplified in the following paragraphs.

For labelling with ^{11}C the most important starting compound is $^{11}CO_2$. Irradiation of high-purity nitrogen containing traces of oxygen leads to the formation of ^{11}CO and $^{11}CO_2$, due to hot-atom reactions (recoil labelling). $^{11}CO_2$ may be separated in a cryogenic trap or molecular sieve trap. In the presence of small amounts of hydrogen ($\approx 5\%$) $^{11}CH_4$ and NH_3 are formed which react on heated Pt to give $H^{11}CN$. Syntheses with $^{11}CO_2$ are performed in a similar way to those with $^{14}CO_2$ (Fig. 12.7). Various methods are also available for introducing ^{11}C into organic compounds by means of $H^{11}CN$, e.g.

$$R-CH_2Cl \xrightarrow{Na^{11}CN} R-CH_2{}^{11}CN \xrightarrow{H_2/Pd} R-CH_2{}^{11}CH_2NH_2$$

In all syntheses with ^{11}C, the time needed is a decisive factor, because of the rather short half-life (20.38 min).

^{13}N ($t_{1/2} = 9.96$ min) is the longest-lived radioisotope of nitrogen. It has found limited application in biological studies, mainly as $^{13}NH_3$ or $^{13}NO_3^-$. Both compounds are obtained by use of a water target. If ethanol is added as a scavenger for oxidizing radicals, the main product is $^{13}NH_3$ which can be separated on a cation-exchange resin followed by elution with NaCl solution. Under oxidizing conditions, $^{13}NO_3^-$ and $^{13}NO_2^-$ are the predominant products. The by-product $^{13}NH_3$ can be separated on a cation exchanger and $^{13}NO_2^-$ can be decomposed by addition of acid. After expelling the nitrogen oxides hy heating, a solution of $^{13}NO_3^-$ is obtained. The mixture of $^{13}NO_3^-$ and $^{13}NO_2^-$ may also be reduced by Devarda's alloy to $^{13}NH_3$.

^{15}O ($t_{1/2} = 2.03$ min) is produced by proton irradiation of nitrogen containing small amounts (0.2–0.5%) of oxygen. From ^{15}O-labelled oxygen, $H_2{}^{15}O$ is obtained by reaction with H_2 on Pd, and $C^{15}O$ is produced by reaction with graphite at $1000\,°C$. ^{15}O-labelled organic compounds may be synthesized by rapid chemical reactions. An example is the reaction of ^{15}O-labelled O_2 with tri-n-butylborane to ^{15}O-labelled butanol for use as a lipophilic substance in nuclear medicine. If a mixture of nitrogen with about 2% CO_2 is irradiated with protons, ^{15}O-labelled CO_2 is obtained by reaction of the recoiling ^{15}O atoms with CO_2.

^{18}F ($t_{1/2} = 109.7$ min) is preferably produced by the nuclear reactions $^{18}O(p, n)^{18}F$ and $^{20}Ne(d, \alpha)^{18}F$, because of the relatively high yields at moderate projectile energies. If O_2 or Ne gas is used, the chemical form of ^{18}F obtained after irradiation depends on the impurities in the target gas and on the inner walls of the target. Application of H_2 or H_2O give $H^{18}F$. Proton irradiation of ^{18}O-enriched water is most effective for production of non-carrier-added (n.c.a.) ^{18}F. From aqueous solution, $^{18}F^-$ can be separated by sorption on an anion-exchange resin, and the ^{18}O-enriched water can be re-used. To facilitate recovery of ^{18}F, Ne gas to which 0.1–0.5% F_2 is added may be irradiated.

By nucleophilic substitution or electrophilic fluorination a great variety of ^{18}F-labelled compounds are synthesized for application in nuclear medicine, such as 2-[^{18}F]fluoro-2-deoxy-D-glucose, L-6-[^{18}F]fluoro-3,4-dihydroxyphenylalanine (L-6-[^{18}F]fluoro-DOPA), 3-N-[^{18}F]fluoroethylpiperone and many others. For labelling by nucleophilic substitution, ^{18}F must be applied free of water in a polar aprotic solvent. For that purpose, water is removed by distillation in the presence of large counter-ions (e.g. K^+, Rb^+, Cs^+, Bu_4N^+ or K^+-Kryptofix) which enable subsequent dissolution in an aprotic organic solvent.

Radioisotopes of iodine are used for some time in medicine, mainly for diagnosis of thyroid diseases. From the various radioisotopes available, ^{123}I has the most favourable properties. It changes by electron capture and emission of 159 keV γ rays into ^{123}Te. The radiation dose is relatively low and the γ radiation can easily be measured from outside. Direct production of ^{123}I is possible by irradiation of Te. However, because of the great number of stable isotopes of this element, many nuclear reactions have to be taken into account, leading to formation of ^{124}I and ^{125}I besides ^{123}I. Rather high yields of ^{123}I (≈ 1.5 GBq per μA h) are obtained by irradiation of highly enriched ^{124}Te with 25–30 MeV protons: $^{124}Te(p, 2n)^{123}I$. ^{123}I can be separated by heating, and ^{124}I can be used again. ^{123}I produced in this way contains about 1% of the unwanted ^{124}I, which emits β^+ radiation and has a longer half-life.

Indirect formation of ^{123}I via the production of ^{123}Xe leads to higher radionuclide purity. Several reactions are feasible:

$$
\left.
\begin{array}{l}
^{122}Te(\alpha, 3n)^{123}Xe \\[4pt]
^{123}Te(\alpha, 4n)^{123}Xe \\[4pt]
^{124}Te(\alpha, 5n)^{123}Xe \\[4pt]
^{122}Te(^3He, 2n)^{123}Xe \\[4pt]
^{123}Te(^3He, 3n)^{123}Xe \\[4pt]
^{124}Te(^3H, 4n)^{123}Xe \\[4pt]
^{127}I(p, 5n)^{123}Xe \\[4pt]
^{127}I(d, 6n)^{123}Xe
\end{array}
\right\}
\quad
\xrightarrow[\substack{2.08\,h}]{\substack{\beta^+(1.505;...) \\ \varepsilon(0.149;...)}}
\quad ^{123}I
\qquad (12.8)
$$

If 50 MeV protons are available, the (p, 5n) reaction with ^{127}I is the most favourable, because iodine has only one stable nuclide. However, measurable amounts of ^{125}Xe are produced by the (p, 3n) reaction. The volatile products formed by irradiation of I_2, LiI, NaI, KI or CH_2I_2 are transferred by He. In a first trap, cooled to $-79°C$, directly formed iodine and iodine compounds are separated and in a second trap, cooled to $-196°C$, ^{123}Xe and ^{125}Xe are collected. The second trap is taken off and after about 5 h the iodine formed by decay of ^{123}Xe is dissolved in dilute NaOH. During this time ^{123}Xe has largely decayed, but ^{125}Xe ($t_{1/2} = 16.9$ h) only to some extent. The yields are of the order of 100 to 500 MBq per μA h, and the relative activity of ^{125}I is of the order of 0.1%.

Labelling of organic compounds with ^{123}I is performed by exchange of halogen atoms or by iodination. Decay of ^{123}Xe in the presence of organic compounds also leads to formation of labelled compounds (recoil labelling).

^{75}Br ($t_{1/2} = 1.6$ h) is applied as a positron emitter in nuclear medicine. The most suitable reactions for the production of this radionuclide are $^{76}Se(p, 2n)^{75}Br$ and $^{75}As(^3He, 3n)^{75}Br$. Irradiation of highly enriched ^{76}Se with 30 MeV protons leads to yields of about 4 GBq per μA h. The ^{76}Br impurity is about 0.9%. Elemental Se or selenides such as Ag_2Se or Cu_2Se may be used as targets. After irradiation, ^{75}Br can be separated from elemental Se by thermochromatography at 300 °C and taken up in a small volume of water. In case of the reaction with ^{75}As enrichment is not necessary, because ^{75}As is the only stable nuclide of As. Cu_3As may be used as the target, and after irradiation ^{75}Br can be separated by thermochromatography at 950 °C.

α-emitting radionuclides may be used for therapeutic purposes if they can be transported in the form of labelled compounds to those places in the body where their action is desired. With respect to this application, ^{211}At has found interest, because it can be introduced into organic compounds and has a half-life suitable for medical application ($t_{1/2} = 7.22$ h). It can be produced by the nuclear reaction ^{209}Bi$(\alpha, 2n)^{211}$At (threshold energy 22 MeV) and separated from Bi in a gas stream at elevated temperature or by wet chemical procedures. Labelled compounds are obtained by methods similar to those used for iodine.

Generator-produced radionuclides are also introduced into compounds suitable for specific applications, in particular in medicine. For Instance, $^{99m}TcO_4^-$ eluted from a ^{99}Mo/^{99m}Tc radionuclide generator can be introduced into organic compounds by various chemical procedures that can be performed by use of special "kits" which allow easy handling.

For routine syntheses of labelled compounds, automated procedures have been developed which enable fast, safe, reproducible and reliable production. Automation has found broad application for the synthesis of radiopharmaceuticals. All steps must be as efficient as possible. For that purpose, target positioning and cooling, irradiation, removal of the target after irradiation, addition of chemicals, temperature and reaction time, purification of the product and dispensing are remotely controlled. Automation may be aided by computers and robotics may be applied.

Most biochemical methods are based on the assimilation of $^{14}CO_2$ by plants and the feeding of various kinds of microorganisms or animals with labelled compounds. Afterwards, the compounds synthesized in the plants, microorganisms or animals are isolated. In this way, glucose, amino acids, adenosine triphosphate, proteins, alkaloids, antibiotics, vitamins and hormones can be obtained that are labelled with ^{14}C, ^{35}S or ^{32}P. Cultures of various microorganisms such as clorella vulgaris may be applied and operated as "radionuclide farms". The labelled compounds produced by biochemical methods are, in general, labelled at random, i.e. not at special positions.

Exchange reactions have the advantage that long routes of synthesis with radionuclides or radioactive compounds are avoided. This is of particular interest if the chemical yields of the syntheses are low. Homogeneous exchange reactions may be applied for labelling of compounds with halogens, e.g.

$$RCl + Li^{36}Cl \rightleftharpoons R^{36}Cl + LiCl \tag{12.9}$$
$$(1) \quad (2) \qquad (1) \quad (2)$$

RCl and LiCl are dissolved in polar organic solvents, and exchange proceeds fast at elevated temperature. Instead of isotopic exchange according to eq. (12.9), non-isotopic exchange may also be applied, e.g.

$$RCl + Li^{131}I \rightleftharpoons R^{131}I + LiCl \tag{12.10}$$

The compounds can be separated by distillation or by chromatography. Aromatic compounds RX (X = halogen) may be labelled in the presence of catalysts such as

$AlCl_3$. In the isotope exchange equilibrium (12.9), the specific activity of (1) is given by

$$A_s(1) = \frac{n_1}{n_1 + n_2} A_s(2),$$ (12.11)

where n_1 and n_2 are the mole numbers of (1) and (2) and $A_s(2)$ is the initial specific activity of (2). In the non-isotopic exchange reaction (12.10), the equilibrium constant has to be taken into account.

Heterogeneous exchange reactions are applicable for labelling in the batch mode or in a continuous mode. Labelling in a continuous mode may be carried out by use of exchange columns loaded with radionuclides. For example, volatile compounds containing acidic hydrogen atoms may be passed through a gas chromatography column loaded with tritium in the form of T-labelled sorbite as stationary phase, where they are labelled by multistage isotope exchange and may approach the initial specific activity of the sorbite. In a similar way, volatile halides can be labelled with radioactive halogens.

Radiation-induced exchange may be applied instead of thermal exchange, if thermal exchange does not occur below the decomposition temperature whereas the chemical bonds involved in the exchange reaction are easily split under the influence of radiation. An example is the labelling of aromatic halogen compounds.

Recoil labelling and radiation-induced labelling (self-labelling) have been described in section 9.7. Examples of recoil-induced labelling of simple compounds such as CO, CO_2 or HCN with ^{11}C, NH_3 with ^{13}N, and CO, CO_2 or H_2O with ^{15}O have been described in the previous paragraphs. For the production of larger compounds recoil labelling and radiation-induced labelling are relatively seldom applied, because generally a greater number of different labelled compounds are produced, which must be separated from each other. Radiation decomposition and the number of degradation products increase with the size of the molecules, whereas the yields and the specific activities of individual labelled compounds decrease markedly.

Literature

General

A. C. Wahl, N. A. Bonner, Radioactivity Applied to Chemistry, Wiley, New York, **1951**

W. M. Garrison, J. G. Hamilton, Production and Isolation of Carrier-Free Radioisotopes, Chem. Rev. 49, 237 **(1951)**

C. B. Cook, J. F. Duncan, Modern Radiochemical Practice, Clarendon Press, Oxford, **1952**

W. J. Whitehouse, J. L. Putman, Radioactive Isotopes – An Introduction to their Preparation, Measurement and Use, Clarendon Press, Oxford, **1953**

H. A. C. McKay, Principles of Radiochemistry, Butterworths, London, **1971**

International Atomic Energy Agency, Radioisotope Production and Quality Control, IAEA Technical Reports Series No. 128, Vienna, **1971**

A. G. Maddock (Ed.), Radiochemistry, in: International Review of Science, Inorganic Chemistry, Butterworths, London, Vol. 7 **(1972)**, Vol. 8 **(1975)**

R. A. Faires, B. H. Parks, Radioisotope Laboratory Techniques, 3rd ed., Butterworths, London, **1973**

D. J. Malcolme-Lawes, Introduction to Radiochemistry, MacMillan, London, **1978**

W. Geary, Radiochemical Methods, Wiley, New York, **1986**

W. W. Meinke (Ed.), Monographs on the Radiochemistry of the Elements, Subcommittee on Radio-chemistry, National Acedemy of Sciences, National Research Council, Nuclear Science Series, NAS-NS 3001–3058, Washington, DC, **1959–1962**

W. W. Meinke (Ed.), Monographs on Radiochemical Techniques, Subcommittee on Radio-chemistry, National Academy of Sciences, National Research Council, Nuclear Science Series NAS-NS 3101–3120, Washington, DC, **1060–1965**

Labelled Compounds

A. Murray, D. L. Williams, Organic Synthesis with Isotopes, 2 Vols., Interscience, New York, **1958**

A. P. Wolf, Labeling of Organic Compounds by Recoil Methods, Annu, Rev. Nucl. Sci. *10*, 259 **(1960)**

J. R. Catch, Carbon-14 Compounds, Butterworths, London, **1961**

R. H. Herber (Ed.), Inorganic Isotopic Synthesis, Benjamin, New York, **1962**

L. Yaffe, Preparation of Thin Films, Sources and Targets, Annu. Rev. Nucl. Sci. *12*, 153 **(1962)**

M. Wenzel, P. E. Schulze, Tritium-Markierung, Darstellung, Messung und Anwendung nach Wilz-bach ^{3}H-markierter Verbindungen, Walter de Gruyter, Berlin, **1962**

V. F. Rasen, G. A. Ropp, H. P. Raaen, Carbon-14, McGraw-Hill, New York, **1968**

A. F. Thomas, Deuterium Labelling in Organic Chemistry, Appleton–Century–Crofts, New York, **1971**

E. A. Evans, Tritium and ist Compounds, Butterworths, London, **1974**

D. J. Silvester, Preparation of Radiopharmaceuticals and Labelled Compounds Using Short-lived Radionuclides, in: Radiochemistry Vol. 3, Specialist Periodical Reports, The Chemical Society, London, **1976**

W. M. Welch (Ed.), Radiopharmaceuticals and other Compounds Labelled with Short-lived Nuclides, Spec. Issue, Int. J. Appl. Rad. Isot. *28*, 1 **(1977)**

D. G. Ott, Synthesis with Stable Isotopes of Carbon, Nitrogen and Oxygen, Wiley, New York, **1981**

S. L. Waters, D. J. Silvester, Inorganic Cyclotron Radionuclides, Radiochim. Acta *30*, 163 **(1982)**

R. R. Muccino, Organic Syntheses with Carbon-14, Wiley, New York, **1983**

H. Deckart, P. H. Cox (Eds.), Principles of Radiopharmacology, Developments in Nuclear Medicine, Vol. 11, Kluwer Academic Publ., Dordrecht, Boston, London, **1987**

G. Stöcklin, V. W. Pike (Eds.), Radiopharmaceuticals for Positron Emission Tomography, Meth-odological Aspects, Developments in Nuclear Medicine, Vol. 24, Kluwer Academic Publ., Dordrecht, Boston, London, **1993**

Generators

V. I. Spitsyn, N. B. Mikheev, Generators for the Production of Short-lived Radioisotopes, Atomic Energy Rev. *9*, 787 **(1971)**

K. H. Lieser, Chemische Gesichtspunkte für die Entwicklung von Radionuklidgeneratoren, Radio-chim. Acta *23*, 57 **(1976)**

R. M. Lambrecht, Radionuclide Generators, Radiochim. Acta *34*, 9 **(1983)**

F. F. Knapp, jr., T. A. Butler (Eds.), Radionuclide Generators, ACS Symposium Series, Wash-ington, DC, **1984**

R. E. Boyd, Technetium Generators; Status and Prospects, Radiochim. Acta *41*, 59 **(1987)**

Separation Procedures

K. A. Kraus, F. Nelson, Radiochemical Separations by Ion Exchange, Annu. Rev. Nucl. Sci. *7*, 31 **(1957)**

H. Freiser, G. H. Morrison, Solvent Extraction in Radiochemical Separations, Annu. Rev. Nucl. Sci. *9*, 221 **(1959)**

J. Stary, The Solvent Extraction of Metal Chelates, Pergamon, Oxford, **1964**

Y. Marcus, A. S. Kertes, Ion Exchange and Solvent Extraction of Metal Complexes, Wiley, New York, **1969**

J. Korkisch, Modern Methods for the Separation of Rarer Metal Ions, Pergamon, Oxford, **1969**

A. K. De, S. M. Khopar, R. Chalmers, Solvent Extraction of Metals, Van Nostrand, London, **1970**

T. Braun, Radiochemical Separation Methods, Elsevier, Amsterdam, **1975**

N. Trautmann, G. Herrmann, Rapid Chemical Separation Procedures, J. Radioanal. Chem. *32*, 533 **(1976)**

J. A. Marinsky, Y. Marcus (Eds.), Ion Exchange and Solvent Extraction, A Series of Advances, Vols. 1–8, Marcel Dekker, New York, **1981**

J. Korkisch, Handbook of Ion Exchange Resins, their Application to Inorganic Analytical Chemistry, Vols. I–VI, CRC Press, Boca Raton, FL, **1989**

Accelerators

E. M. McMillan, Particle Accelerators, In: Experimental Nuclear Physics (Ed. E. Segrè), Vol. III, Wiley, New York, **1959**

M. S. Livingstone, J. P. Blewett, Particle Accelerators, McGraw-Hill, New York, **1962**

M. H. Blewett, The Electrostatic (Van de Graaff) Generator, in: Methods of Experimental Physics (Eds. L. C. L. Yuan, C. S. Wu), Vol. 5B, Academic Press, New York, **1963**

P. M. Lapostolle, L. Septier, Linear Accelerators, North-Holland, Amsterdam, **1970**

A. P. Wolf, W. B. Jones, Cyclotrons for Biomedical Radioisotope Production, Radiochim. Acta *34*, 1 **(1983)**

S. Humphries, Principles of Charged Particle Accelerators, Wiley, New York, **1986**

W. Scharf, Particle Accelerators and their Uses, Harwood, New York, **1986**

13 Special Aspects of the Chemistry of Radionuclides

13.1 Short-Lived Radionuclides and the Role of Carriers

The most important aspects of the chemistry of short-lived radionuclides are that

- the mass of the radionuclides is small and
- chemical procedures have to be fast.

The mass of a radionuclide is proportional to its half-life:

$$m = A \frac{M}{\ln 2 \ N_{Av}} t_{1/2} \tag{13.1}$$

where A is the activity in Bq (s^{-1}), M is the mass of the nuclide in atomic mass units (u), and N_{Av} is Avogadro's number. This relation is illustrated in Table 13.1 for the activity of 10 Bq and radionuclides of various half-lives. At short half-lives, the masses of the radionuclides are considerably smaller than the masses usually handled in chemical operations. Traces of the order of 10^{-10} g or less are, in general, only detectable on the basis of their radioactivity, and handling of those traces requires special attention. The following aspects have to be considered:

- the mass of the radionuclide to be handled,
- the presence of isotopic carriers,
- the presence of non-isotopic carriers.

Table 13.1. Number of atoms and mass of various radionuclides corresponding to 10 Bq.

Radionuclide	Half-life	Number of atoms	Mass [g]	Concentration if dissolved in 10 ml [mol/l]
^{238}U	$4.468 \cdot 10^9$ y	$2.0 \cdot 10^{18}$	$8.0 \cdot 10^{-4}$	$3.4 \cdot 10^{-4}$
^{226}Ra	1600 y	$7.3 \cdot 10^{11}$	$2.7 \cdot 10^{-10}$	$1.2 \cdot 10^{-10}$
^{227}Ac	21.77 y	$9.9 \cdot 10^9$	$3.7 \cdot 10^{-12}$	$1.6 \cdot 10^{-12}$
^{60}Co	5.272 y	$2.4 \cdot 10^9$	$2.4 \cdot 10^{-13}$	$4.0 \cdot 10^{-13}$
^{210}Po	138.38 d	$1.7 \cdot 10^8$	$6.0 \cdot 10^{-14}$	$2.9 \cdot 10^{-14}$
^{32}P	14.26 d	$1.8 \cdot 10^7$	$9.5 \cdot 10^{-16}$	$3.0 \cdot 10^{-15}$
^{24}Na	14.96 h	$7.7 \cdot 10^5$	$3.1 \cdot 10^{-17}$	$1.3 \cdot 10^{-16}$
^{251}Md	4.0 m	$3.5 \cdot 10^3$	$1.4 \cdot 10^{-18}$	$5.5 \cdot 10^{-19}$
^{258}Lr	3.9 s	$5.6 \cdot 10$	$2.4 \cdot 10^{-20}$	$0.9 \cdot 10^{-20}$

As already mentioned in section 12.3, carriers are elements or compounds, respectively, with identical or very similar chemical properties to the radionuclide. With respect to the suitability as carrier, the chemical state is decisive. Carriers are often added to ensure normal chemical behaviour of radionuclides. For that purpose, they must be in the same chemical state as the radionuclide considered.

In the case of radioisotopes of stable elements, such as ^{60}Co, ^{32}P and ^{24}Na, small amounts (traces) of these elements are always present, due to their ubiquity. The masses are, in general, higher than the masses of short-lived radioisotopes, and the omnipresent traces act as isotopic carriers of the short-lived radionuclides, provided that they are in the same chemical state. In the case of isotopes of radioelements, such as ^{226}Ra, ^{227}Ac, ^{210}Po, ^{251}Md or ^{258}Lr, however, stable nuclides are absent and the masses of the radionuclides are identical with the masses of the elements, provided that the presence of longer-lived radioisotopes can be excluded.

Traces of elements or compounds, respectively, of other elements with similar properties may serve as non-isotopic carriers for radioisotopes of stable elements as well as for isotopes of radioelements. The influence of non-isotopic carriers depends on the nature of the compounds and the chemical operation. For example, in precipitation reactions, non-isotopic carriers or hold-back carriers, respectively, may play a major role.

Separation of short-lived radionuclides requires application of fast methods. The time needed for the separation procedure should not exceed the half-life. Precipitation, including filtration or the usual ion-exchange methods followed by elution, and chromatographic methods require up to several minutes. Solvent extraction involving slow complexation reactions may also take too much time.

For fast filtration, the device used by Hahn (Fig. 12.6) is well suited. The filter with the precipitate can easily be removed and measured. The same device may be used for fast separation by ion exchange on thin layers. For example, short-lived radioisotopes of iodine may be separated rather effectively on a thin layer of AgI (or AgCl) by means of the ion-exchange reaction $AgI(AgCl) + {}^*I^- \rightleftharpoons Ag^*I + I^-(Cl^-)$ (Fig. 13.1).

Thermochromatography using a quartz column is applied for separation of volatile compounds (section 12.3). Aerosols may be injected for transportation of the radionuclides and a reactive gas (e.g. Cl_2) may be added to form special compounds or to investigate the chemical behaviour of the radionuclides, in particular radionuclides of heavy elements.

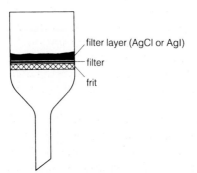

filter layer (AgCl or AgI)

filter

frit

Figure 13.1. Filter layer for separation of carrier-free iodine (I_2 or I^-) by exchange.

13.2 Radionuclides of High Specific Activity

In the absence of stable isotopes, the specific activity of radionuclides is given by

$$\frac{A}{m} = \frac{\ln 2}{t_{1/2}} \frac{N_{Av}}{M} \tag{13.2}$$

where A, m, $t_{1/2}$ and M are the activity, the mass, the half-life and the atomic mass of the radionuclide, respectively, and N_{Av} is Avogadro's number. Even for longer-lived radionuclides such as ^{14}C ($t_{1/2} = 5730$ y) the specific activity is rather high and the mass of the substance is rather small, if stable isotopes are absent. For example, 1 GBq ^{14}C (≈ 27 mCi) corresponds to a mass of only 6.06 mg ^{14}C and a specific activity of 165 GBq per g of ^{14}C. Due to the presence of small amounts of stable isotopes of carbon, specific activites of up to about 100 GBq ^{14}C per gram of carbon are obtained in practice. Synthesis of ^{14}C-labelled carbon compounds of this specific activity means handling of milligram amounts and requires small pieces of equipment and special precautions. Reactions in closed systems, use of vacuum lines and cooling traps are favourable.

The term "carrier-free" is often used to indicate the absence of stable isotopes or longer-lived radioisotopes of the radionuclide considered. However, due to the omnipresence of most stable elements, carrier-free radioisotopes of stable elements are, in general, not available. The presence of stable isotopes or longer-lived radioisotopes has to be taken into account, and the specific activity is smaller than calculated by eq. (13.2). As long as the presence of such other isotopes cannot be excluded, it is more correct to distinguish no-carrier-added (n.c.a.) and carrier-added radionuclides. On the other hand, radioisotopes of radioelements are carrier-free if longer-lived radioisotopes are absent.

13.3 Microamounts of Radioactive Substances

From the previous sections it is evident that radionuclides of high specific activity often represent very small amounts (microamounts, non-weighable amounts <1 µg) of matter, especially if the half-lives are short. Handling of such microamounts requires special precautions, because in the absence of measurable amounts of carriers the radionuclides are microcomponents and their chemical behaviour may be quite different from that observed for macrocomponents. This aspect is of special importance if the system contains liquid/solid, gas/solid or liquid/liquid interfaces. The percentage of radionuclides sorbed on the walls of a container depends on the chemical form (species) of the radionuclide, its concentration and specific activity, and on the properties of the container material. At high specific activity of a radionuclide in solution, the surface of a glass beaker generally offers an excess of surface sorption sites.

Glass surfaces have an ion-exchange capacity of the order of 10^{-10} mol/cm^2 ($\approx 10^{14}$ ions/cm^2) and a similar number of sorption sites is available for chemisorption. Thus, 100 ml glass beakers have an ion-exchange capacity of about 10^{-8} mol coresponding to a concentration of 10^{-7} mol/l in 100 ml. Therefore, sorption on glass walls may be marked at concentrations $<10^{-6}$ mol/l. By comparison with Table 13.1 it is obvious that in the absence of carriers even long-lived radionuclides may easily be sorbed on glass walls by ion exchange or chemisorption.

Chemisorption is very pronounced if the radionuclide is able to react with the surface. An example is the sorption of hydroxo complexes of tri- and tetravalent elements by the silanol groups on glass surfaces. Sorption increases with the formation of mononuclear hydroxo complexes in solution and decreases with the condensation to polynuclear complexes at higher pH values. Several measures may be taken to suppress ion exchange or chemisorption of traces of radionuclides on glass surfaces:

- high concentration of H^+ to suppress ion exchange and hydrolysis;
- high concentration of non-isotopic cations, anions or other substances to suppress ion exchange and adsorption of the radionuclides considered;
- hydrophobization of the glass surface (e.g. by treating with silicones) to prevent ion exchange and chemisorption.

The surfaces of plastic materials, such as polyethylene, polypropylene or perspex, do not exhibit ion exchange, but adsorption may be pronounced, in particular adsorption of organic compounds including organic complexes of radionuclides.

Sorption of radionuclides on particulates in solution is frequently observed. The particles may be coarsely or finely dispersed. Their surface properties (surface layer, charge, ion-exchange and sorption properties) play a major role. In general, they offer a great number of sorption sites on the surface, and microamounts of radionuclides may be found on the surface of these particles instead of in solution. Sorption of radionuclides on colloidal particles leads to formation of radiocolloids (carrier colloids, section 13.4).

Reactions at gas/solid interfaces may lead to sorption of radionuclides from the gas phase or they may cause loss of activity of solid samples by interaction with the gas phase. For example, $Ba^{14}CO_3$ exhibits loss of $^{14}CO_2$ to moist air due to the isotope exchange $^{14}CO_3^{2-}(s)/^{12}CO_2(g)$ in the presence of water vapour. The specific activity of solid samples may be reduced appreciably by such reactions.

Separation of non-weighable amounts of radioactive substances requires application of suitable techniques. Precipitation is, in general, not possible, because the solubility product cannot be exceeded. Coprecipitation may be used after addition of a suitable carrier. An isotopic carrier must be in the same chemical state as the radionuclide and leads to marked lowering of the specific activity. A non-isotopic carrier must coprecipitate the radionuclide efficiently and it must be separable from the radionuclide after coprecipitation.

According to Hahn, two possibilities are distinguished in the case of coprecipitation or cocrystallization:

- coprecipitation (cocrystallization) by isomorphous substitution and
- coprecipitation (cocrystallization) by adsorption.

Radionuclides that are able to form normal or anomalous mixed crystals with the macrocomponent are incorporated at lattice sites. In most cases the distribution in the lattice is heterogeneous, i.e. the concentration of the microcomponent varies with the depth. If the solubility of the microcomponent is lower than that of the macrocomponent, it is enriched in the inner parts of the crystals. Heterogeneous distribution may even out over longer periods of time by diffusion or recrystallization.

For homogeneous distribution of a microcomponent (1) and a macrocomponent (2) in a solid phase (s) and a liquid phase (l), Nernst's distribution law is valid and the following relation is obtained:

$$\frac{(c_1/c_2)_s}{(c_1/c_2)_l} = \frac{(n_1/n_2)_s}{(n_1/n_2)_l} = K_h \tag{13.3}$$

where c_1 and c_2 are the concentrations and n_1 and n_2 the mole numbers. K_h is called the homogeneous distribution coefficient.

In the case of heterogeneous distribution, Nernst's distribution law holds only for the respective surface layer in equilibrium with the solution and the relation is

$$\frac{\log n_1^o/(n_1^o - n_1)}{\log n_2^o/(n_2^o - n_2)} = K_l \tag{13.4}$$

where n_1 and n_2 are the mole numbers of the microcomponent and the macrocomponent in the solid and n_1^o and n_2^o are the total numbers of moles present. K_l is called the logarithmic distribution coefficient.

These relations have been verified for the distribution of ^{226}Ra as a microcomponent in BaSO$_4$. At elevated temperature, homogeneous distribution is obtained after some hours, whereas heterogeneous distribution is observed at low temperature and fast separation of solid and solution. Heterogeneous distribution is also found after crystallization by slow evaporation of the solvent.

The fraction of a radionuclide present as the microcomponent that is separated in the cases of homogeneous and heterogeneous distribution by coprecipitation or cocrystallization is plotted in Fig. 13.2 as a function of the precipitation or crystallization of the macrocomponent for various values of homogeneous and logarithmic distribution coefficients. Knowledge of these values is important for separation by fractional precipitation or crystallization. The figure shows that separation is more effective in the case of heterogeneous distribution, provided that $K_l > 1$. For example, at a distribution coefficient of 6, precipitation of 50% of the macrocomponent leads to coprecipitation of 98.4% of the microcomponent in the case of heterogeneous distribution and only to coprecipitation of 86% in the case of homogeneous distribution. Therefore, fast precipitation or slow evaporation at low temperature is

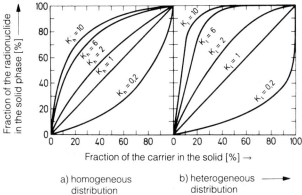

Figure 13.2. Fraction of the radionuclide in the solid phase as a function of fractional precipitation or crystallization.

most effective for separation of radionuclides by fractional precipitation or crystallization, respectively.

Coprecipitation of microamounts of radioactive substances by adsorption depends on the surface properties such as the surface charge and the specific surface area of the solid. For instance, cationic species are preferably sorbed on surfaces carrying negative surface charges, and hydroxides are very effective sorbents, because of their high specific surface area. Microcomponents adsorbed on particles formed in the early stages of precipitation or crystallization are partly occluded in the course of crystal growth (inner adsorption). As an example, coprecipitation of no-carrier-added ^{140}La with $BaSO_4$ by adsorption is shown in Fig. 13.3 as a function of the time after the beginning of precipitation. The amount of ^{140}La occluded in the $BaSO_4$ crystals decreases with time, mainly due to Ostwald ripening, whereas the amount of ^{140}La adsorbed at the surface of the $BaSO_4$ crystals increases.

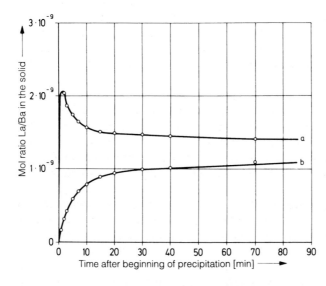

Figure 13.3. Coprecipitation of carrier-free La with $BaSO_4$: (a) La occluded; (b) La adsorbed at the surface. (According to K. H. Lieser, H. Wertenbach: Z. physik. Chem., Neue Folge **34**, 1 (1962).)

If macroamounts of other elements are to be separated from microamounts of radioactive substances by precipitation, isotopic or non-isotopic hold-back carriers may be added to suppress coprecipitation of the radioactive substances.

In ion-exchange and chromatographic procedures microamounts of radioactive sustances may be lost by sorption on the ion exchangers or sorbents or on the walls of the columns. Small amounts of impurities in the materials used may be responsible for unexpected reactions and losses.

The most favourable separation method for microamounts is solvent extraction, because the number of possible ion-exchange and sorption sites on solid surfaces is relatively small. However, small amounts of impurities in the organic or aqueous phases may also lead to unexpected behaviour of microcomponents.

Many effects described in this section increase with decreasing amounts of the radioactive substance. If the number of atoms or molecules becomes very small ($\ll 100$), the usual thermodynamic descriptions are no longer applicable. The distribution of one or several atoms of an element or of several molecules in a system is not defined, because the law of mass action, partition functions, and thermodynamic functions, such as chemical potential, Gibbs free energy and entropy, are based on the properties of a multitude of atoms or molecules of the same kind.

The rate of reaction between two microcomponents (1) and (2) is $R = k \cdot c_1 \cdot c_2$, and if both concentrations c_1 and c_2 are extremely small, R is also extremely small, and the microcomponents (1) and (2) cannot be assumed to be in equilibrium.

Single atom chemistry is of particular importance if only single atoms are available for chemical studies, as in the case of the heaviest elements. The short-lived isotopes of these elements can only be produced at a rate of one atom at a time, and the investigation of their chemical properties requires special considerations.

The equilibrium constant K of a chemical reaction $A + B \rightleftharpoons C + D$ is given by $K = (a_C \cdot a_D)/(a_A \cdot a_B)$, where a_A, a_B, a_C, and a_D are the chemical activities. However, if only one atom is present, it cannot exist at the same time in form of A and C and a_A or a_C must be zero. Consequently, in single atom chemistry K is no longer defined. The same holds for $\Delta G = -RT\ln K$, the free enthalpy of the reaction. In order to overcome this problem, in single atom chemistry the probabilities of finding the single atom in form of A or C are introduced instead of the chemical activities or concentrations, respectively, and by use of these probabilities equilibrium constants and single particle free enthalpies can be defined.

An example is the distribution equilibrium of a single atom between two phases, where the distribution coefficient K_d is defined by the probabilities of finding the single atom in one phase or in the other. If a one-step partition method is applied, K_d must be measured repeatedly many times, in order to obtain a statistically relevant result. Much more favourable is the use of a multi-stage method, particularly a chromatographic method, in which the partition of the single atom is taking place many times successively.

In chemical experiments with short-lived single atoms the time must be taken into account that is necessary to obtain chemical equilibrium. In this respect, chromatographic methods comprising fast adsorption and desorption processes are also very favourable.

Moreover, the half-life of a single radioactive atom cannot be measured, and even if the half-life is knwon, the instant at which the atom will undergo disintegration cannot be predicted. Radioactive decay, chemical kinetics and chemical equilibria are governed by the laws of probability and many measurements with single atoms are necessary to establish the statistics and to obtain relevant results. With regard to chemical properties, it is important that all the atoms are present in the same chemical form (same species).

The chemical behaviour of small numbers of atoms in the interstellar space, i.e. in an environment that is virtually free of matter, also exhibits some special features. Because energies cannot be transmitted to or from other atoms or molecules, exothermic reactions lead to immediate decomposition and endothermic reactions are not possible.

13.4 Radiocolloids

Radiocolloids are colloidal forms of microamounts of radioactive substances. Their formation was first observed by Paneth (1913) in his research on the separation of ^{210}Bi and ^{218}Po. Radiocolloids can be separated from aqueous solutions by ultra-filtration, centrifugation, dialysis and electrophoresis. They can be detected with high sensitivity by autoradiography. As an example, the autoradiography of a radio-colloid of ^{234}Th is shown in Fig. 13.4.

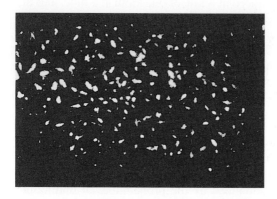

Figure 13.4. Autoradiograph of a radio-colloid (^{234}Th, pH $\approx$ 3).

In order to understand the nature of radiocolloids, knowledge of the general properties of colloids is needed. Colloids are finely dispersed particles in a liquid phase, a gas phase or a solid. The size of colloidal particles is in the range between that of molecules or ions and that of particles visible by means of a light microscope, i.e. between about 1 nm and about 0.45 µm. The upper value corresponds to the mean wavelength of visible light. Large molecules, in particular polymers and bio-molecules, approach or exceed the upper value and may also form colloids.

Like ions and small molecules, colloids are considered to be components of the phase in which they are suspended. In general, the metastable colloidal state exists for longer periods of time (up to several months) and colloidal particles may be transported with water or air over long distances.

Small particles have a large specific surface area and a relatively large specific surface energy. Therefore, they have the tendency to form particles of lower specific surface energy, i.e. they are metastable with respect to larger particles. The main feature of colloids is that aggregation to larger particles is prevented by mutual repulsion, for example in water by electric charges of the same sign in the case of hydrophobic colloids or by shells of water molecules in the case of hydrophilic colloids.

A colloidal solution is also called a sol, in contrast to a gel, which exhibits the properties of a solid but contains large amounts of solvent and is amorphous. Typical examples are a sol and a gel of silicic acid (silica gel). Colloidal particles of inorganic hydroxides ($Fe_2O_3 \cdot xH_2O$, $Al_2O_3 \cdot xH_2O$, $TiO_2 \cdot xH_2O$, $ThO_2 \cdot xH_2O$ and others) often carry positive charges. By addition of ions with opposite charge, neutralization and coagulation may occur. Therefore, colloids of this kind are sensitive towards electrolytes. Other hydrophilic colloids, such as colloidal polysilicic acid, are stabilized by a shell of water molecules and less sensitive towards the presence of electrolytes. Hydrophobic colloids may be stabilized by surface-active substances in such a way that the hydrophobic part of these substances is bound on the surface, whereas the hydrophilic part is directed to the outside and leads to mutual repulsion of the particles.

Many organic molecules are large enough to form colloids, and organic colloids are frequently encountered in the life sciences. Humic substances are found in natural waters and may form complexes with radionuclides.

Generally, radiocolloids may be generated in two ways:

(a) The radionuclide or the labelled compound may form an intrinsic colloid ("Eigenkolloid", sometimes also called a "real colloid").
(b) The radionuclide or the labelled compound may be sorbed on an already-existing colloid which serves as carrier (carrier colloid, "Fremdkolloid", sometimes also called a "pseudocolloid").

Formation of an intrinsic colloid is only possible if the solubility of the radionuclide or the labelled compound is exceeded. If an isotopic or a non-isotopic carrier is present, the radionuclide or the labelled compound will be incorporated if the solubility of the carrier is exceeded. In this case, a carrier colloid is formed containing the radionuclide or the labelled compound in homogeneous or heterogeneous distribution. A carrier colloid is also formed if the radionuclide or the labelled compound is sorbed on an already-existing colloid. This may be an inorganic colloid such as polysilicic acid or colloidal iron(III) hydroxide or an organic colloid such as humic acid. Because colloids have a high specific surface area, the probability of formation of carrier colloids by sorption of radionuclides or labelled compounds is high if colloids are present. This probability increases with increasing concentration of colloids and with decreasing concentration of radionuclides or labelled compounds and their carriers. Both kinds of carrier colloids are real colloids, not "pseudocolloids".

It is obvious that the formation of radiocolloids will be observed

(a) if colloids of other origins are present or
(b) if the solubility of the radionuclide, the labelled compound or a suitable carrier is exceeded.

Colloids are found in many systems, e.g. in natural waters and in the air. Traces of colloids formed by dust particles or by particles given off from the walls of containers are practically omnipresent. They can only be removed by careful ultrafiltration. If a radionuclide or a labelled compound enters such a system, there is a high probability that it will be sorbed on the colloids, provided that the competition of other ions or molecules is not too strong. Only if the presence of colloids from other origins can be excluded and if the solubilities of relevant species are not exceeded, formation of radiocolloids by microamounts of radionuclides can be neglected.

In aqueous solutions, generation of radiocolloids is favoured by hydrolysis of the radionuclide considered. Mononuclear hydroxo complexes may form colloids by condensation to polynuclear complexes if their concentration is high enough or they may be sorbed on the surface of colloids from other origins. Formation of radiocolloids of these kinds can be prevented by a low pH or addition of complexing agents. For instance, addition of F^- ions prevents formation of radiocolloids of ^{95}Zr.

Commonly, a distinction is made between genuine solutions containing only matter in molecular dispersion (ions, molecules) and colloidal solutions containing colloidal particles. As already mentioned, colloids are also considered to be components of the phase in which they are suspended.

Like normal colloids, radiocolloids show different behaviour from that of ions or molecules. For instance, they are generally not sorbed on ion exchangers or chromatographic columns.

It should be emphasized that the (metastable) colloidal state cannot be described by thermodynamic functions. Consequently, thermodynamic description of a system fails in the presence of colloids. For example, the solubility product is well defined for certain (in general crystalline) solids and takes into account the ions in equilibrium with these solids. Non-ionic and colloidal forms, however, are not taken into account.

13.5 Tracer Techniques

Tracer techniques comprise all methods in which microamounts (traces) of radionuclides or labelled compounds are added to a system, in order to pursue (trace) the fate, transport or chemical reaction of a certain element or compound in that system. Radioactive tracers are preferabley used, because they can be detected and measured in very low concentrations and with high sensitivity, as is evident from Table 13.1. With a measuring period of 10 min and an overall counting efficiency of 20%, 10 Bq can be determined with a statistical error of about 3%.

The prerequisite of tracer techniques is that the radionuclide is in the same chemical form as the species to be investigated. At least it must exhibit the same behaviour, for instance in the study of transport processes. The same chemical behaviour can be assumed in the case of isotopic tracers, provided that isotope effects can be neglected. However, isotope effects are only marked for light atoms, in particular hydrogen, for which kinetic and equilibrium isotope effects have to be taken into account in the case of substitution of H by D or T.

Tracer techniques offer the unique possibility of studying the kinetics of chemical reactions in chemical equilibria in which one isotope is exchanged for another (isotopic exchange reactions, reaction enthalpy $\Delta H \approx 0$, reaction entropy $\Delta S \neq 0$). Isotopic exchange reactions have found broad application for kinetic studies in homogeneous and heterogeneous systems.

Besides the investigation of isotopic exchange reactions, tracer techniques are applied in various fields of science and technology:

– radioanalysis;
– investigation of bonding and reaction mechanisms in chemistry and biochemistry;
– measurement of diffusion and self-diffusion;
– study of pathways of elements or compounds in biological systems, in the human body and in the environment;
– application for diagnostic purposes in nuclear medicine;
– investigation of transport processes in industrial equipment;
– study of corrosion and wear.

The various applications of tracer techniques will be discussed in more detail in chapters 17 to 20.

Literature

General

O. Hahn, Applied Radiochemistrry, Cornell University Press, Ithaca, NY, **1936**
A. C. Wahl, N. A. Bonner, Radioactivity Applied to Chemistry, Wiley, New York, **1951**
M. Haissinsky, Nuclear Chemistry and its Applications, Addison–Wesley, Reading, MA, **1964**
E. A. Evans, M. Muramatsu (Eds.), Radiotracer Techniques and Applications, Marcel Dekker, New York, **1977**
J. P. Adloff, R. Guillaumont, Fundamentals of Radiochemistry, CRC Press, Boca Raton, FL, **1993**

More Special

W. W. Meinke (Ed.), Monographs on the Radiochemistry of the Elements, Subcommittee on Radiochemistry, National Academy of Sciences, National Research Council, Nuclear Science Series, NAS-NS 3001–3058, Washington, DC, **1959–1962**
W. W. Meinke (Ed.), Monographs on Radiochemical Techniques, Subcommittee on Radiochemistry, National Academy of Sciences, National Research Council, Nuclear Science Series NAS-NS 3101–3120, Washington, DC, **1960–1965**
I. Zvara, One Atom at a Time Chemical Studies of Transactinide Elements, in: Transplutonium Elements (Eds. W. Muller, R. Lindner), North-Holland, Amsterdam, **1976**
G. Herrmann, N. Trautmann, Rapid Chemical Separation Procedures, J. Radioanal. Chem. *32*, 533 **(1976)**
P. Beneš, V. Majer, Trace Chemistry of Aqueous Solutions, Elsevier, Amsterdam, **1980**
R. Guillaumont, P. Chevallier, J. P. Adloff, Identification of Oxidation States of Ultra-trace Elements by Radiation Detection, Radiochim. Acta *40*, 191 **(1987)**
N. Trautmann, G. Herrmann, M. Mang, C. Mühlbeck, H. Rimke, P. Sattelberger, F. Ames, H. J. Kluge, E. W. Otten, D. Rehklau, W. Ruster, Low Level Detection of Actinides by Laser Resonance Spectroscopy, Radiochim. Acta *44/45*, 107 **(1988)**
P. Sattelberger, M. Mang, G. Herrmann, J. Riegel, H. Rimke, N. Trautmann, F. Ames, H. J. Kluge, Separation and Detection of Trace Amounts of Technetium, Radiochim. Acta *48*, 165 **(1989)**
R. Guillaumont, J. P. Adloff, A. Peneloux, Kinetic and Thermodynamic Aspects of Tracer Scale and Single Atom Chemistry, Radiochim. Acta *46*, 169 **(1989)**
R. Guillaumont, J. P. Adloff, A. Peneloux, P. Delamoye, Sub-tracer Scale Behaviour of Radionuclides, Application to Actinide Chemistry, Radiochim. Acta *54*, 1 **(1991)**
R. Guillaumont, J. P. Adloff, Behaviour of Environmental Plutonium at very Low Concentration, Radiochim. Acta *58*, 53 **(1992)**

14 Radioelements

14.1 Natural and Artificial Radioelements

Radioelements are elements existing only in the form of radionuclides, but not as stable isotopes, as already mentioned in section 2.1.

The natural radioelements are listed in Table 14.1. Isotopes of these elements are members of the uranium, actinium and thorium families (Table 1.2, and Tables 4.1 to 4.3). In the ores of U and Th the concentrations of natural radioelements are relatively high and proportional to the half-life. The average concentration of U in the earth's crust is about 2.9 mg/kg (ppm) and that of Th about 11 mg/kg (ppm). The

Table 14.1. The natural radioelements.

Atomic number Z	Name of the element (Symbol)	Longest-lived nuclide (Half-life)	Discovery	Remarks
84	Polonium (Po)	^{209}Po (102 y)	1898 P. and M. Curie	Similar to Te
85	Astatine (At)	^{210}At (8.3 h)	1940 Corson, McKenzie and Segrè	Halogen; volatile
86	Radon (Rn)	^{222}Rn (3.825 d)	1900 Rutherford and Soddy	Noble gas
87	Francium (Fr)	^{223}Fr (21.8 m)	1939 Perey	Alkali metal; similar to Cs
88	Radium (Ra)	^{226}Ra (1600 y)	1898 P. and M. Curie	Alkaline-earth metal; similar to Ba
89	Actinium (Ac)	^{227}Ac (21.773 y)	1899 Debierne	Similar to La; more basic
90	Thorium (Th)	^{232}Th ($1.405 \cdot 10^{10}$ y)	1828 Berzelius	Only in the oxidation state IV; similar to Ce(IV), Zr(IV) and Hf(IV); strongly hydrolysing; many complexes
91	Protactinium (Pa)	^{231}Pa ($3.276 \cdot 10^4$ y)	1917 Hahn and Meitner	Preferably in the oxidation state V; very strongly hydrolysing; many complexes
92	Uranium (U)	^{238}U ($4.468 \cdot 10^9$ y)	1789 Klaproth	Preferably in the oxidaton states IV and VI; in solution UO_2^{2+} ions; many complexes

concentration of U in seawater is $\approx 3 \, \mathrm{mg/m^3}$, corresponding to a total amount of about $4 \cdot 10^9$ tons of U in the oceans.

All natural radioelements with atomic numbers $Z = 84$ to 89 and $Z = 91$ have been identified as decay products of U and Th, but the first isotope of astatine (from Greek "unstable"; $Z = 85$) was obtained in 1940 by the nuclear reaction

$$^{209}\mathrm{Bi}(\alpha, 2n)^{211}\mathrm{At} \xrightarrow[7.22 \, \mathrm{h}]{\varepsilon, \alpha, \gamma} \tag{14.1}$$

The half-life is somewhat shorter than that of $^{210}\mathrm{At}$ ($t_{1/2} = 8.3 \, \mathrm{h}$), which has the longest half-life of all astatine isotopes. Three years later, At was also identified in nature. $^{219}\mathrm{At}$ produced in the α branching of $^{223}\mathrm{Fr}$ is the longest-lived natural isotope of At ($t_{1/2} = 5.4 \, \mathrm{s}$). The very short-lived isotopes $^{218}\mathrm{At}$ and $^{215}\mathrm{At}$ are present in small concentrations as members of the uranium and actinium families (Tables 4.2 and 4.3). Because the main importance of At is as a natural radioelement, it is included in this group.

Many natural radioelements are of great practical importance, in particular U with respect to its application as nuclear fuel, but also Th, Ra and Rn. $^{226}\mathrm{Ra}$ ($t_{1/2} = 1600 \, \mathrm{y}$) is found in many springs and the noble gas Rn is the main source of natural radioactivity in the air.

The artificial radioelements are listed in Table 14.2. Their number is now 25 and will probably continue to increase in the future. At present, the ratio of radioelements to the total number of known elements is about 30%.

The discovery of technetium ($Z = 43$) in 1937 and of promethium ($Z = 61$) in 1947 filled the two gaps in the Periodic Table of the elements. These gaps had been the reason for many investigations. Application of Mattauch's rule (section 2.3) leads to the conclusion that stable isotopes of element 43 cannot exist. Neighbouring stable isotopes could only be expected for mass numbers $A = 93$, $A \leq 91$, $A = 103$ and $A \geq 105$. However, these nuclides are relatively far away from the line of β stability. The report by Noddak and Tacke concerning the discovery of the elements rhenium and "masurium" (1925) was only correct with respect to Re ($Z = 75$). The concentration of element 43 (Tc) in nature due to spontaneous or neutron-induced fission of uranium is several orders of magnitude too low to be detectable by emission of characteristic X rays of element 43, as had been claimed in the report.

Many artificial radioelements have gained great practical importance. Production of $^{99}\mathrm{Tc}$ in nuclear reactors has already been mentioned. It is found in all steps of reprocessing of nuclear fuel and in all kinds of nuclear waste. In the environment it is very mobile under oxidizing conditions, as $\mathrm{TcO_4^-}$. The short-lived isomer $^{99m}\mathrm{Tc}$ is one of the most widely used radionuclides in nuclear medicine (chapter 19). $^{237}\mathrm{Np}$ is produced in nuclear reactors in amounts of about $0.5 \, \mathrm{kg}$ per ton of spent fuel after burn-up of $35\,000 \, \mathrm{MW \, d}$ per ton. In general, it goes with the high-level waste (HLW). Special steps have been developed to separate $^{237}\mathrm{Np}$ in the course of reprocessing, for instance for the production of $^{238}\mathrm{Pu}$ by the reaction

$$^{237}\mathrm{Np}(n, \gamma)^{238}\mathrm{Np} \xrightarrow[2.117 \, \mathrm{d}]{\beta^-} {}^{238}\mathrm{Pu} \tag{14.2}$$

Table 14.2. Artificial radioelements

Atomic number Z	Name of the element (Symbol)	Longest-lived nuclide (Half-life)	Discovery	Remarks
43	Technetium (Tc)	^{98}Tc (4.2 · 10^6 y)	1937; Perrier and Segrè	Similar to Re; preferred oxidation states IV and VII
61	Promethium (Pm)	^{145}Pm (17.7 y)	1947; Marinsky, Glendenin and Coryell	Only in the oxidation state III
93	Neptunium (Np)	^{237}Np (2.144 · 10^6 y)	1940; McMillan and Abelson	Oxidation states III to VII; Np(V) in aqueous soln.
94	Plutonium (Pu)	^{244}Pu (8.00 · 10^7 y)	1940; Seaborg et al.	Oxidation states III to VIII
95	Americium (Am)	^{243}Am (7370 y)	1944; Seaborg et al.	Oxidation states III to VII
96	Curium (Cm)	^{247}Cm (1.56 · 10^7 y)	1944; Seaborg et al.	Analogy to Gd; can be oxidized to Cm(IV)
97	Berkelium (Bk)	^{247}Bk (1380 y)	1949; Thompson, Ghiorso et al.	Analogy to Tb
98	Californium (Cf)	^{251}Cf (898 y)	1950; Thompson, Ghiorso et al.	Analogy to Dy
99	Einsteinium (Es)	^{252}Es (471,7 d)	1952; Thompson, Ghiorso et al.	Analogy to Ho
100	Fermium (Fm)	^{257}Fm (100.5 d)	1953; Thompson, Ghiorso et al.	Analogy to Er
101	Mendelevium (Md)	^{258}Md (51.5 d)	1955; Ghiorso et al.	Analogy to Tm
102	Nobelium (No)	^{259}No (58 m)	1958; Ghiorso et al.	Analogy to Yb; oxidation state II preferred
103	Lawrencium (Lr)	^{262}Lr (3.6 h)	1961; Ghiorso et al.	Analogy to Lu
104	Rutherfordium (Rf)	[b]	1969; Ghiorso et al.[a]	Similar to Zr and Hf
105	Dubnium (Db)	[b]	1970; Ghiorso et al.[a]	Similar to Nb and Ta
106	Seaborgium (Sg)	[b]	1974; Ghiorso et al., Flerov et al.	Similar to Mo and W
107	Bohrium (Bh)	[b]	1981; Münzenberg, Armbruster et al.	Homologue of Re
108	Hassium (Hs)	[b]	1984; Münzenberg, Armbruster et al.	Homologue of Os
109	Meitnerium (Mt)	[b]	1982; Münzenberg, Armbruster et al.	Homologue of Ir
110	–	[b]	1994; Hofmann et al.	Homologue of Pt
111	–	[b]	1994; Hofmann et al.	Homologue of Au
112	–	[b]	1996; Hofmann et al.	Homologue of Hg
114	–	[b]	1999; Oganessian et al.	Homologue of Pb
116	–	[b]	1999; Ninov et al.	Homologue of Po
118	–	[b]	1999; Ninov et al.	Homologue of Rn

[a] First reports from Dubna by Flerov et al.: 1964: element 104; 1968: element 105
[b] Not yet known

The latter is used as energy source in radionuclide batteries, for instance in satellites. Pu is the most important radioelement produced in nuclear reactors. About 9 kg Pu is generated per ton of spent fuel after a burn-up of 35 000 MW d per ton. The main part is ^{239}Pu (about 5.3 kg) produced via

$$^{238}U(n,\gamma)^{239}U \xrightarrow[23.5\,m]{\beta^-} {}^{239}Np \xrightarrow[2.355\,d]{\beta^-} {}^{239}Pu \xrightarrow[2.411\cdot10^4\,y]{\alpha} \qquad (14.3)$$

^{240}Pu (about 2.2 kg), ^{241}Pu (about 1.1 kg) and ^{242}Pu (about 0.4 kg) are generated by (n,γ) reactions from ^{239}Pu. ^{239}Pu is a valuable nuclear fuel and may also be used for production of nuclear weapons. The global production rate of ^{239}Pu in nuclear power reactors is of the order of 100 tons per year contained in spent fuel elements. Non-proliferation agreements should prevent uncontrolled distribution of Pu. Moreover, Pu is highly toxic. Am and Cm are generated in smaller amounts in nuclear reactors by (n,γ) reactions (about 0.15 kg Am and about 0.07 kg Cm per ton of spent fuel after a burn-up of 35 000 MW d per ton).

14.2 Technetium and Promethium

^{99}Tc was isolated as the first isotope of technetium (the "artificial" element) in 1937 from neutron-irradiated Mo where it was formed by the reactions

$$^{98}Mo(n,\gamma)^{99}Mo \xrightarrow[66\,h]{\beta^-} {}^{99m}Tc \xrightarrow[6\,h]{IT} {}^{99}Tc \xrightarrow[2.1\cdot10^5\,y]{\beta^-} \qquad (14.4)$$

Because of the small cross section (0.13 b) the yield of the (n,γ) reaction is low. Relatively large amounts of ^{99}Tc are produced by nuclear fission of ^{235}U (fission yield 6.2%). After a burn-up of 35 000 MW d per ton of uranium with an initial enrichment of 3% ^{235}U, the spent fuel contains about 1 kg ^{99}Tc per ton. The longest-lived isotope of Tc is ^{98}Tc ($t_{1/2} = 4.2 \cdot 10^6$ y); in contrast to ^{99}Tc, it has no practical significance.

Although ^{99}Tc is present in uranium ores in extremely small concentrations, the main importance of technetium is that of a man-made element and technetium is counted as an artificial radioelement, due to its production in nuclear reactors and by nuclear explosions.

With respect to chemical behaviour, Tc is more closely related to Re than to Mn and exhibits some similarities to the neighbouring elements Mo and Ru. Under oxidizing conditions the stable oxidation state is Tc(VII) as TcO_4^- in aqueous solution or as volatile Tc_2O_7 in the absence of water. Under reducing conditions, Tc(IV) is the most stable oxidation state, strongly hydrolyzing in aqueous solutions and very stable as TcO_2 in the absence of water. Relatively stable complexes of Tc(V) are formed in the presence of suitable complexing agents. Fluorination leads to the volatile TcF_6. For the separation of Tc various methods may be used, for example distillation of Tc_2O_7 from concentrated H_2SO_4 or coprecipitation with Re_2S_7 or with CuS, followed by dissolution in HCl and selective separation of TcO_4^-.

Element 61 (Pm) could not be found in nature, and the gap in the Periodic Table of the elements remained until 1947 when the element was discovered by Marinsky,

Glendenin and Coryell in the fission products of uranium after separating the rare-earth fraction by oxalate precipitation. ^{147}Pm is produced with a fission yield of 2.27%. The longest-lived isotope of Pm is ^{145}Pm ($t_{1/2} = 17.7$ y), followed by ^{146}Pm ($t_{1/2} = 5.53$ y). Isotopes of Pm can also be produced by neutron irradiation of Nd

$$
^{146}\text{Nd}(n,\gamma)^{147}\text{Nd} \xrightarrow[10.98\,\text{d}]{\beta^-} {}^{147}\text{Pm} \xrightarrow[2.62\,\text{y}]{\beta^-}
$$

$$
^{148}\text{Nd}(n,\gamma)^{149}\text{Nd} \xrightarrow[1.73\,\text{h}]{\beta^-} {}^{149}\text{Pm} \xrightarrow[53.1\,\text{h}]{\beta^-} \tag{14.5}
$$

$$
^{150}\text{Nd}(n,\gamma)^{151}\text{Nd} \xrightarrow[12.4\,\text{m}]{\beta^-} {}^{151}\text{Pm} \xrightarrow[28.4\,\text{h}]{\beta^-}
$$

The element was named promethium im memory of Prometheus who, according to Greek mythology, brought fire to mankind.

Promethium is a typical element of the lanthanide series. The relative abundances of the lanthanides are plotted in Fig. 14.1 as a function of the atomic number. This figure illustrates Harkin's rule: the abundance of elements with even atomic numbers is appreciably higher than that of elements with odd atomic numbers. For element 61 the natural abundance is zero.

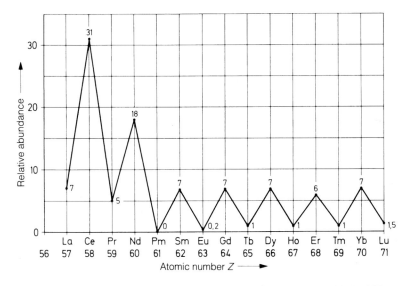

Figure 14.1. Relative abundances of the lanthanides (after V. M. Goldschmidt).

Usually, Pm is separated from other lanthanides by ion exchange in the presence of complexing agents in solution. This method was also applied by Marinsky, Glendenin and Coryell in 1947: after oxalate precipitation of the rare-earth fraction, the precipitate was treated with carbonate solution to remove the main part of Y, dissolved and passed as 5% citrate solution (pH 2.5) through a cation-exchange column. The result is shown in Fig. 14.2.

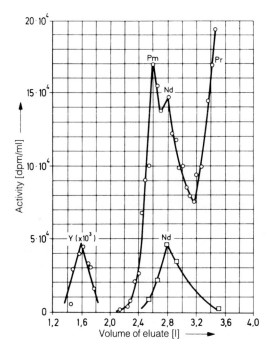

Figure 14.2. Separation of Pm from fission products of uranium on a cation-exchange column by elution with citrate solution ($\bigcirc$, β activity; $\square$, γ activity). (According to J. A. Marinsky, L. E. Glendenin, C. D. Coryell: J. Amer. chem. Soc. **69**, 2781 (1947).)

The valencies of the lanthanides are plotted in Fig. 14.3 as a function of the atomic number. The most stable electron configurations are $4f^0$ (La^{3+}), $4f^7$ (Gd^{3+}) and $4f^{14}$ (Lu^{3+}). These configurations are also favoured by neighbouring elements. The colours of the lanthanide ions show a similar variation with the electron configuration: whereas La^{3+}, Gd^{3+} and Lu^{3+} are colourless, the colour intensity of the other Ln^{3+} ions (Ln stands for lanthanides) increases with increasing distance from the atomic numbers of these three elements, with maxima halfway between La^{3+} and Gd^{3+} and between Gd^{3+} and Lu^{3+}. The ionic radii of the lanthanide ions decrease continuously with increasing atomic numbers, as shown in Fig. 14.4, due to the increasing charge on the nucleus at constant outer electron shell. This effect is known as lanthanide contraction.

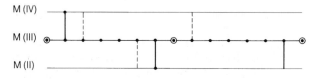

Figure 14.3. Oxidation states of the lanthanides.

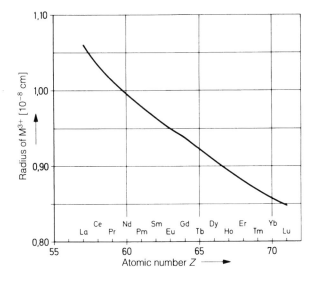

Figure 14.4. Ionic radii of lanthanide ions in the oxidation state +3. (According to D. H. Templeton, C. H. Paulen: J. Amer. chem. Soc. **76**, 5237 (1954).)

14.3 Production of Transuranium Elements

The various methods of production of transuranium elements are:

– irradiation with neutrons,
– irradiation with deuterons or α particles and
– irradiation with heavy ions.

The most important method of production of the first transuranium elements is neutron irradiation of uranium. After the discovery of the neutron by Chadwick in 1932, this method was applied since 1934 by Fermi in Italy and by Hahn in Berlin. The method is based on the concept that absorption of neutrons by nuclides with atomic number Z leads to formation of neutron-rich nuclides that change by β^- decay into nuclides with atomic numbers $Z + 1$. Unexpectedly, the experiments carried out by Hahn and Strassmann led to the discovery of nuclear fission in 1938.

The production of transuranium elements by neutron irradiation can be described by

$$^A Z(n, \gamma)^{A+1} Z \xrightarrow{\beta^-} {}^{A+1}(Z + 1) \tag{14.6}$$

After long irradiation times, elements with atomic numbers $Z + 2$, $Z + 3$ etc. are generated in amounts that increase with irradiation time. The formation of transuranium elements by neutron irradiation of ^{238}U is illustrated in Fig. 14.5. (n, γ) reactions and radioactive decay compete with each other. Formation of heavier nuclides is favoured if

$$\sigma_{n,\gamma} \Phi_n > \Lambda \tag{14.7}$$

where $\sigma_{n,\gamma}$ is the (n, γ) cross section and Φ_n is the neutron flux density. $\Lambda = \lambda + \Sigma \sigma_i \Phi_i$ is the sum of the decay constant λ and all products $\sigma_i \Phi_i$ of binuclear reactions

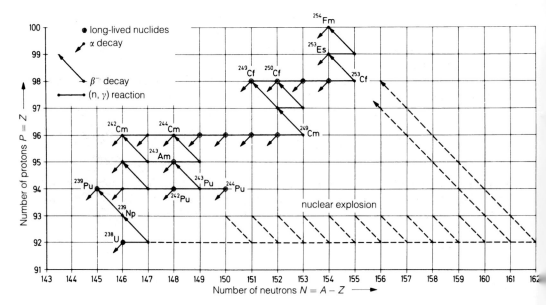

Figure 14.5. Production of transuranium elements by neutron irradiation of ^{238}U.

(e.g. nuclear fission) also leading to a decrease of the radionuclide considered. Condition (14.7) is fulfilled in a nuclear reactor at $\sigma_{n,\gamma} = 1\,\mathrm{b}$ and $\Phi_n = 10^{14}\,\mathrm{cm}^{-2}\,\mathrm{s}^{-1}$, if $\Lambda < 10^{-10}\,\mathrm{s}^{-1}$ or, neglecting the contribution of $\Sigma\sigma_i\Phi_i$, if $t_{1/2} > 200\,\mathrm{y}$. That means that under these conditions heavier nuclides are only produced in greater amounts if the half-lives of all intermediate nuclides are longer than $\approx 100\,\mathrm{y}$.

At the extremely high fluxes of a nuclear explosion, fast multiple neutron capture leads to very neutron-rich isotopes of U or Pu, respectively, changing rapidly into elements of appreciably higher atomic numbers by a quick succession of β^- transmutations. This method of formation of heavier elements is also indicated in Fig. 14.5. The elements can be found in the debris of nuclear underground explosions.

By irradiation with deuterons, one proton is introduced into the nucleus and (d n), or (d, 2n) reactions lead to the production of elements with $Z + 1$, for instance

$$^{A}Z(\mathrm{d},\mathrm{n})^{A+1}(Z+1) \tag{14.8}$$

For formation of heavier elements irradiation with α particles is preferable, because by (α,n) or $(\alpha,2\mathrm{n})$ reactions the atomic number increases by two units:

$$^{A}Z(\alpha,\mathrm{n})^{A+3}(Z+2) \tag{14.9}$$

The most important method of production of elements with $Z > 100$, however, is the irradiation, with heavy ions, of elements that are available in sufficient amounts. Fusion of projectiles (atomic number Z') with target nuclei (atomic number Z) may lead to elements with atomic number $Z + Z'$:

$$^{A}Z + {^{A'}}Z' \rightarrow {^{A+A'-x}}(Z+Z') + x\mathrm{n} \tag{14.10}$$

The energy needed to surmount the Coulomb barrier increases with Z and Z', whereas the cross section decreases. That is why, in general, only small amounts of heavier elements can be produced by heavy-ion reactions. Elements with $Z > 106$ are often obtained with a yield of only one atom at a time.

The first transuranium element, neptunium $(Z = 93)$, was discovered by McMillan and Abelson in 1940 while investigating the fission products of uranium. The neptunium isotope first to be identified was ^{239}Np produced by the reaction

$$^{238}\text{U}(n, \gamma)^{239}\text{U} \xrightarrow[23.5\,\text{m}]{\beta^-} {}^{239}\text{Np} \xrightarrow[2.355\,\text{d}]{\beta^-} \tag{14.11}$$

It was named in analogy to uranium after the planet Neptune. The Np isotope with the longest half-life $(t_{1/2} = 2.144 \cdot 10^6 \text{ y})$ is ^{237}Np, the mother nuclide of the (artificial) decay series with $A = 4n + 1$ (section 4.1). It is produced in nuclear reactors:

$$^{238}\text{U}(n, 2n)^{237}\text{U} \xrightarrow[6.75\,\text{d}]{\beta^-} {}^{237}\text{Np} \quad (\approx 70\%)$$

and $\left.\begin{array}{c} \\ \\ \\ \end{array}\right\}$ (14.12)

$$^{235}\text{U}(n, \gamma)^{236}\text{U}(n, \gamma)^{237}\text{U} \xrightarrow[6.75\,\text{d}]{\beta^-} {}^{237}\text{Np} \quad (\approx 30\%)$$

Very small amounts of ^{237}Np are present in uranium ores, where this nuclide is produced by neutrons from cosmic radiation. The ratio ^{237}Np/^{238}U in uranium ores is of the order of 10^{-12}.

Plutonium $(Z = 94)$ was discovered in 1940 by Seaborg and co-workers. It was also named in analogy to uranium, after the planet Pluto. The first isotope of Pu was produced by cyclotron irradiation of uranium with 16 MeV deuterons:

$$^{238}\text{U}(d, 2n)^{238}\text{Np} \xrightarrow[2.117\,\text{d}]{\beta^-} {}^{238}\text{Pu} \xrightarrow[87.74\,\text{y}]{\alpha} \tag{14.13}$$

The discovery of Pu has been described in detail by Seaborg in his "Plutonium Story" (chapter 1 of the book "The Transuranium Elements"; 1958). First, the separation of Pu from Th caused some difficulties, because both elements were in the oxidation state +4. After oxidation of Pu(IV) by persulfate to Pu(VI), separation became possible. ^{239}Pu is produced in appreciable amounts in nuclear reactors (section 14.1), but it has not immediately been detected, due to its low specific activity caused by its long half-life. After the discovery of ^{239}Pu, plutonium gained great practical importance, because of the high fission cross section of ^{239}Pu by thermal neutrons. Very small amounts of ^{239}Pu are present in uranium ores, due to (n, γ) reaction of neutrons from cosmic radiation with ^{238}U. The ratio ^{239}Pu/^{238}U is of the order of 10^{-11}. In 1971, the longest-lived isotope of plutonium, ^{244}Pu $(t_{1/2} = 8.00 \cdot 10^7 \text{ y})$ was found by Hoffman in the Ce-rich rare-earth mineral bastnaesite, in concentrations of the order of 10^{-15} g/kg.

Americium ($Z = 95$) was discovered by Seaborg and co-workers in 1944 after longer neutron irradiation of ^{239}Pu:

$$^{239}\text{Pu}(n, \gamma)^{240}\text{Pu}(n, \gamma)^{241}\text{Pu} \xrightarrow[14.35 \text{ y}]{\beta^-} {}^{241}\text{Am} \tag{14.14}$$

It was named in analogy to the element europium, which has the same number of f electrons.

Curium ($Z = 96$) was also discovered in 1944 by Seaborg and co-workers. At first, ^{242}Cm was produced by irradiation of ^{239}Pu with α particles:

$$^{239}\text{Pu}(\alpha, n)^{242}\text{Cm} \tag{14.15}$$

Later it was obtained by neutron irradiation of ^{241}Am:

$$^{241}\text{Am}(n, \gamma)^{242}\text{Am} \xrightarrow[16 \text{ h}]{\beta^-} {}^{242}\text{Cm} \tag{14.16}$$

Curium was named in analogy to the element gadolinium (having the same number of f electrons) in memory of research scientists – in this case in honour of Marie and Pierre Curie.

The elements with the atomic numbers 97 and 98 (berkelium and californium) at first could not be produced by irradiation with neutrons, because isotopes of Cm exhibiting β^- transmutation were not known. After milligram amounts of ^{241}Am had been produced by reaction (14.14), ^{243}Bk was obtained in 1949 by Thompson, Ghiorso and others by irradiation with α particles:

$$^{241}\text{Am}(\alpha, 2n)^{243}\text{Bk} \tag{14.17}$$

Berkelium was named in analogy to terbium after a city (Berkeley).

Californium ($Z = 98$) was discovered in 1950, also by Thompson, Ghiorso and others, by irradiation of ^{242}Cm with α particles:

$$^{242}\text{Cm}(\alpha, n)^{245}\text{Cf} \tag{14.18}$$

It was named after the state of its discovery (California).

Einsteinium ($Z = 99$) and fermium ($Z = 100$) were identified in 1952 and 1953, respectively, by Ghiorso and others in the radioactive debris of the first thermonuclear explosion. Hints of the formation of these elements were found in dust samples from the remotely controlled aircrafts used in this test. Then, the elements were isolated by processing larger amounts of the radioactive coral material from the test site and named in honour of Einstein and Fermi. The elements had been formed by multineutron capture,

$$^{238}\text{U} \xrightarrow{15 \times (n,\gamma)} {}^{253}\text{U} \xrightarrow{7 \times \beta^-} {}^{253}\text{Es} \tag{14.19}$$

$$^{238}\text{U} \xrightarrow{17 \times (n,\gamma)} {}^{255}\text{U} \xrightarrow{8 \times \beta^-} {}^{255}\text{Fm} \tag{14.20}$$

Later, einsteinium was synthesized by bombarding U with ^{14}N:

$$^{238}\text{U}(^{14}\text{N, xn})^{252-x}\text{Es} \tag{14.21}$$

and fermium was isolated as a product of reactor irradiation of Es.

Mendelevium ($Z = 101$) was produced in 1955 by Ghiorso and others by irradiation of ^{253}Es with α particles:

$$^{253}\text{Es}(\alpha, \text{n})^{256}\text{Md} \tag{14.22}$$

The amount of ^{253}Es available at this time was very small: about $N = 10^9$ atoms ($\approx 4 \cdot 10^{-13}$ g). At a flux density of α particles $\Phi_\alpha = 10^{14}\,\text{cm}^{-2}\,\text{s}^{-1}$, a cross section $\sigma_{\alpha,\text{n}} = 1$ mb and an irradiation time of 10^4 s a yield $N\Phi_\alpha\sigma_{\text{n},\alpha}t$ of about one atom per experiment was expected. In order to detect these single atoms, the recoil technique was applied (Fig. 14.6). Es was electrolytically deposited on a thin gold foil. The recoiling atoms of ^{256}Md were sampled on a catcher foil. After irradiation, the catcher foil was dissolved and Md was separated on a cation-exchange resin. In 8 experiments 17 atoms of ^{256}Md were detected and identified by their transmutation into the spontaneously fissioning ^{256}Fm, the properties of which were known:

$$^{256}\text{Md} \xrightarrow[\text{1.3 h}]{\varepsilon} {}^{256}\text{Fm} \xrightarrow[\text{2.63 h}]{\text{sf}} \tag{14.23}$$

Mendelevium was named in honour of Mendeleyev.

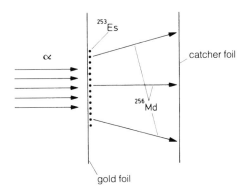

Figure 14.6. Separation of ^{256}Md by the recoil technique (schematically).

For the production of elements with atomic numbers $Z > 101$ irradiation with ions of atomic numbers $Z > 2$ is necessary, because for irradiation with α particles actinides with adequate half-lives which can be used as targets are not available. Two concepts for the synthesis of new, heavy nuclides can be distinguished:

- Irradiation of actinides with ions of relatively low atomic numbers (e.g. $Z = 5$ to 16). In general, these reactions lead to high excitation energies of the compound nuclei ("hot fusion").
- Irradiation of spherical closed-shell nuclei, like ^{208}Pb, with ions of medium atomic numbers (e.g. $Z = 18$ to 36). In these reactions, the excitation energy of the compound nuclei is relatively low ("cold fusion").

The first reports of the discovery of element 102 came from Stockholm and Dubna in 1957. The element was named nobelium after Alfred Nobel. However, the results could not be confirmed, and the new element was identified in 1958 by Ghiorso and others by the reaction

$$^{246}\text{Cm}(^{12}\text{C}, 4\text{n})^{254}\text{No} \xrightarrow[55\,\text{s}]{\alpha} {}^{250}\text{Fm} \tag{14.24}$$

In these experiments, the recoil technique was modified into a double recoil technique by application of a moving belt (Fig. 14.7). The recoiling atoms generated by the heavy-ion reaction (first recoil) are deposited on the belt and transported along a catcher foil on which the recoiling atoms from α decay (second recoil) are collected. From the activity recorded as a function of the distance, the half-life can be calculated.

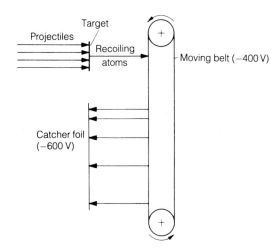

Figure 14.7. Separation of transuranium elements by means of the double-recoil technique (schematically).

By use of the same technique, lawrencium ($Z = 103$) was discovered and identified in 1961 by Ghiorso, Seaborg and others. ^{258}Lr was produced by bombarding Cf with B

$$^{250,251,252}\text{Cf}(^{10,11}\text{B}, (2\text{–}5)\text{n})^{258}\text{Lr} \xrightarrow[3.9\,\text{s}]{\alpha} \tag{14.25}$$

In this case, the recoiling products of the heavy-ion reaction were transported on the moving belt to an array of energy-sensitive solid-state detectors. The half-life of ^{258}Lr was too short to allow chemical separation, and Lr was the first element to be identified by purely instrumental methods. It was named in honour of Lawrence, the inventor of the cyclotron.

The applicability of heavy-ion reactions to the production of heavy elements increased with the development of efficient heavy-ion accelerators at Berkeley, Dubna and Darmstadt. On the other hand, the importance of instrumental methods for the identification of new elements or new radionuclides increased, because of the short half-lives of the order of seconds or less.

The first report concerning the discovery of element 104 (rutherfordium) came from Dubna (Flerov et al., 1964). By irradiation of ^{242}Pu with ^{22}Ne a radionuclide exhibiting a spontaneous fission half-life of ≈ 0.3 s was found and attributed to ^{260}104. Because only few atoms were obtained, the details of the report were rather inaccurate. Further investigations of the Dubna group revealed that two isotopes of element 104 were formed

$$^{242}\text{Pu}(^{22}\text{Ne}, 5n)^{259}\text{Rf}; t_{1/2}(\text{sf}) = 3.0 \text{ s} \tag{14.26}$$

$$^{242}\text{Pu}(^{22}\text{Ne}, 4n)^{260}\text{Rf}; t_{1/2}(\text{sf}) = 21 \text{ ms} \tag{14.27}$$

In the meantime, the Berkeley group (Ghiorso et al., 1969) was able to produce element 104 by other reactions and to measure the half-lives of two isotopes of this element and the energy of their α decay:

$$^{249}\text{Cf}(^{14}\text{C}, 4n)^{257}\text{Rf} \xrightarrow[4.0 \text{ s}]{\alpha} {}^{253}\text{No} \tag{14.28}$$

$$^{249}\text{Cf}(^{13}\text{C}, 3n)^{259}\text{Rf} \xrightarrow[3.0 \text{ s}]{\alpha} {}^{255}\text{No} \tag{14.29}$$

By repeating the experiments many times, results of relatively high statistical reliability were obtained. Afterwards, the same group applied other reactions for the production of element 104:

$$^{248}\text{Cm}(^{18}\text{O}, 5n)^{261}\text{Rf} \xrightarrow[78 \text{ s}]{\alpha} {}^{257}\text{No} \tag{14.30}$$

$$^{248}\text{Cm}(^{16}\text{O}, 5n)^{259}\text{Rf} \xrightarrow[3.0 \text{ s}]{\alpha} {}^{255}\text{No} \tag{14.31}$$

Two names were proposed for element 104, kurtchatovium (in honour of the Russian physicist Kurtchatov) by the Dubna group and rutherfordium (in honour of Rutherford) by the Berkeley group.

The Dubna group (Flerov et al., 1968) was also the first to announce the discovery of element 105 (dubnium) by bombardment of ^{243}Am with ^{22}Ne. However, the assignment of the mass numbers 260 or 261 was not possible unambiguously. In Berkeley, element 105 was produced by the reactions (Ghiorso et al., 1970)

$$^{249}\text{Cf}(^{15}\text{N}, 4n)^{260}\text{Db} \xrightarrow[1.5 \text{ s}]{\alpha} {}^{256}\text{Lr} \tag{14.32}$$

$$^{249}\text{Bk}(^{16}\text{O}, 4n)^{261}\text{Db} \xrightarrow[1.5 \text{ s}]{\alpha} {}^{257}\text{Lr} \tag{14.33}$$

$$^{249}\text{Bk}(^{18}\text{O}, 5n)^{262}\text{Db} \xrightarrow[34 \text{ s}]{\alpha} {}^{258}\text{Lr} \tag{14.34}$$

Again, two names were proposed, nielsbohrium (in honour of Niels Bohr) by the Dubna group and hahnium (in honour of Otto Hahn) by the Berkeley group.

Element 106 (seaborgium) was produced and identified in 1974 at Berkeley (Ghiorso et al.) and at Dubna (Flerov et al.) by the reactions

$$^{249}\text{Cf}(^{18}\text{O}, 4\text{n})^{263}\text{Sg} \xrightarrow[0.9\,\text{s}]{\alpha} {}^{259}\text{Rf} \xrightarrow[3.0\,\text{s}]{\alpha} {}^{255}\text{No} \tag{14.35}$$

and

$$^{207,208}\text{Pb}(^{54}\text{Cr}, 2\text{n})^{259,260}\text{Sg} \tag{14.36}$$

respectively. By the Berkeley group, the recoiling atoms of ^{263}Sg were transported by a helium jet to a turning wheel and passed to an array of solid-state detectors by which α decay was measured. At Dubna the products were also isolated by the recoil technique. The name seaborgium was proposed by the Berkeley group in honour of Glenn T. Seaborg.

Synthesis of heavy elements by use of nuclei with closed shells, like ^{208}Pb or ^{209}Bi, as target nuclei was proposed by Oganessian with the argument that an appreciable amount of the energy of the projectiles will be used to bring additional nucleons into the empty shells of the nucleus and therefore not appear in form of excitation energy. This "cold fusion" was demonstrated at Dubna 1974 by the reaction of ^{40}Ar with ^{208}Pb and then by reaction (14.36): The small number of two neutrons emitted was a proof of the low excitation energy of the compound nucleus. The concept of cold fusion was applied at GSI (Darmstadt) for the synthesis of elements 107 to 112.

The main features of cold fusion reactions with the spherical nuclei of ^{208}Pb or ^{209}Bi as targets are: low excitation energies of the compound nuclei ($E_x \approx 15$ to 20 MeV) with the consequence of emission of only one or two neutrons, low probability of fission, and relatively high fusion cross sections σ_{fus}. On the other hand, the reaction products have relatively small neutron numbers and short half-lives. Suitable projectiles are neutron-rich stable nuclei, such as ^{48}Ca, ^{50}Ti, ^{54}Cr, ^{58}Fe, ^{64}Ni, ^{70}Zn, and ^{86}Kr.

Hot fusion reactions with the deformed nuclei of the actinides (e.g. ^{238}U or ^{244}Pu) as targets lead to high excitation energies of the compound nuclei ($E_x \approx 50$ MeV), emission of about 4 to 5 neutrons, high probability of fission and relatively low values of σ_{fus}. On the other hand, the neutron numbers of the reaction products are relatively high and the half-lives relatively long.

At Darmstadt the velocity filter SHIP (Separator for Heavy-Ion reaction Products) was developed in order to separate the products of the heavy-ion reactions, which were then identified by α spectrometry of the radionuclides and their decay products. The heavy ions are hitting the target, which is put on a rotating wheel to avoid overheating. The reaction products enter the velocity filter, which consists of an arrangement of focusing devices and electric and magnetic fields, in which the heavy nuclei are separated in-flight. The velocity of the separated products and their activities are measured, the α activities by means of solid-state Si detectors and the γ activities by Ge detectors.

Two periods of detection of new elements at GSI can be distinguished: elements 107, 108 and 109 from 1981 to 1984 and, after new target and detector arrangements, elements 110, 111 and 112 from 1994 to 1996.

Element 107 (bohrium) was synthesized by the GSI group (Münzenberg, Armbruster et al.) in 1981 by the reactions

$$^{209}\text{Bi}(^{54}\text{Cr}, \text{n})^{262}\text{Bh} \xrightarrow[102\text{ ms}/8.0\text{ ms}]{\alpha} {}^{258}\text{Db} \xrightarrow[4.4\text{ s}]{\alpha} {}^{254}\text{Lr} \tag{14.37}$$

and

$$^{209}\text{Bi}(^{54}\text{Cr}, 2\text{n})^{261}\text{Bh} \xrightarrow[11.8\text{ ms}]{\alpha} {}^{257}\text{Db} \xrightarrow[\approx 1.3\text{ s}]{\alpha} {}^{253}\text{Lr} \tag{14.38}$$

The name bohrium was proposed in honour of Niels Bohr.

Element 108 (hassium) was produced by the same group in 1984 by the reactions

$$^{208}\text{Pb}(^{58}\text{Fe}, 2\text{n})^{264}\text{Hs} \xrightarrow[0.45\text{ ms}]{\alpha} {}^{260}\text{Sg} \xrightarrow[3.6\text{ ms}]{\alpha} {}^{256}\text{Rf} \tag{14.39}$$

The element is named after "Hassia" for Hessen, the state in which Darmstadt is situated.

Element 109 (meitnerium) was synthesized already in 1982 by the GSI group by the reaction

$$^{209}\text{Bi}(^{58}\text{Fe}, \text{n})^{266}\text{Mt} \xrightarrow[1.7\text{ ms}]{\alpha} {}^{262}\text{Bh} \xrightarrow[102\text{ ms}/8.0\text{ ms}]{\alpha} {}^{258}\text{Db} \tag{14.40}$$

The name meitnerium was given to element 109 in honour of Lise Meitner.

Elements 110 and 111 were synthesized and identified in November and December 1994 by the GSI group (Hofmann et al.) by the reactions

$$^{208}\text{Pb}(^{62}\text{Ni}, \text{n})^{269}110 \xrightarrow[0.17\text{ ms}]{\alpha} {}^{265}\text{Hs} \xrightarrow[2.0\text{ ms}]{\alpha} {}^{261}\text{Sg} \xrightarrow[0.23\text{ s}]{\alpha} {}^{257}\text{Rf} \tag{14.41}$$

$$^{208}\text{Pb}(^{64}\text{Ni}, \text{n})^{271}110 \xrightarrow[1.1\text{ ms}]{\alpha} {}^{267}\text{Hs} \xrightarrow[59\text{ ms}]{\alpha} {}^{263}\text{Sg} \xrightarrow[0.9\text{ s}]{\alpha} {}^{259}\text{Rf} \tag{14.42}$$

and

$$^{209}\text{Bi}(^{64}\text{Ni}, \text{n})^{272}111 \xrightarrow[1.5\text{ ms}]{\alpha} {}^{268}\text{Mt} \xrightarrow[70\text{ ms}]{\alpha} {}^{264}\text{Bh} \xrightarrow[\approx 0.44\text{ s}]{\alpha} {}^{260}\text{Db}$$

$$\xrightarrow[1.5\text{ s}]{\alpha} {}^{256}\text{Lr} \xrightarrow[25.9\text{ s}]{\alpha} {}^{252}\text{Md} \tag{14.43}$$

About one year later (February 1996) the same group announced the discovery of element 112 by the reaction

$$^{208}\text{Pb}(^{70}\text{Zn}, \text{n})^{277}112 \xrightarrow[0.24\text{ ms}]{\alpha} {}^{273}110 \xrightarrow[118\text{ ms}]{\alpha} {}^{269}\text{Hs} \xrightarrow[9.3\text{ s}]{\alpha} {}^{265}\text{Sg} \xrightarrow[7.4\text{ s}]{\alpha} {}^{261}\text{Rf}$$

$$\xrightarrow[78\text{ s}]{\alpha} {}^{257}\text{No} \xrightarrow[26\text{ s}]{\alpha} {}^{253}\text{Fm} \tag{14.44}$$

Most of the new nuclides have been identified by their α-decay chains, as indicated in eqs. (14.39) to (14.44). Formation of $^{277}112$ has been proved by six successive α decays leading to ^{253}Fm (eq. (14.44)). Names for elements 110, 111 and 112 have not yet been proposed.

The cross sections of the reactions ^{208}Pb + ^{64}Ni, ^{209}Bi + ^{64}Ni and ^{208}Pb + ^{70}Zn decrease from ≈ 15 pb to ≈ 1 pb (1 pb = 10^{-36} cm^2). As the production of one atom per day can be expected at a cross section of about 100 pb, irradiation for weeks is necessary to record one atom.

Recently, three more elements have been detected. Synthesis of element 114 was reported 1999 at Dubna by Oganessian et al. who irradiated plutonium with ^{48}Ca. Formation of $^{289}114$ was inferred from the following decay chain:

$$^{244}\text{Pu}(^{48}\text{Ca}, 3\text{n})^{289}114 \xrightarrow[\approx 30\,\text{s}]{\alpha} {}^{285}112 \xrightarrow[\approx 15\,\text{min}]{\alpha} {}^{281}110 \xrightarrow[\approx 2\,\text{min}]{\alpha} {}^{277}108 \xrightarrow{\text{sf}} \qquad (14.45)$$

The remarkably long half-life indicates the predicted relatively high stability of element 114.

Continuing the application of cold fusion, elements 118 and 116 were synthesized simultaneously 1999 at Berkeley by Ninov et al.

$$^{208}\text{Pb}(^{86}\text{Kr}, \text{n})^{293}118 \xrightarrow{\alpha} {}^{289}116 \xrightarrow{\alpha} {}^{285}114 \xrightarrow{6\times\alpha} {}^{269}\text{Sg} \qquad (14.46)$$

Both elements were identified by the α-decay chain (14.46).

Synthesis of further heavy nuclei is in progress at Berkeley, Darmstadt and Dubna. With regard to the search for the island of relatively high nuclide stability, target nuclei and projectile nuclei with closed shells and high neutron numbers are required.

14.4 Further Extension of the Periodic Table of the Elements

The half-lives of the longest-lived isotopes of transuranium elements (Fig. 14.8) show a continuous exponential decrease with increasing atomic number Z. Whereas up to element 103 the half-life is mainly determined by α decay, the influence of spontaneous fission seems to become predominant for elements with $Z \geq 106$. The drop model of nuclei predicts a continuous decrease of the fission barrier from about 6 MeV for uranium to about zero for element 110. That means that according to the drop model, elements with $Z > 110$ are not expected to exist, because normal vibrations of the nuclei should lead to fission.

On the other hand, closed nucleon shells stabilize the spherical form of nuclei, and according to the shell model appreciable fission barriers are expected in the region of closed nucleon shells (magic numbers). This led to the question of the next closed proton shell. First, $Z = 126$ was assumed to be the next "magic number" of protons, in analogy to $N = 126$ (section 2.3). However, theoretical calculations revealed that the next closed proton shell is to be expected for $Z = 114$, whereas the next "magic number" of neutrons should be $N = 184$. Accordingly, an island of relative stability is expected at $Z = 114$ and $N = 184$. Other calculations lead to the prediction of an

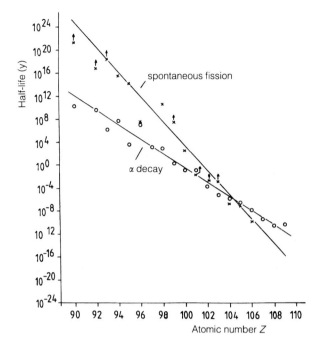

Figure 14.8. α-decay (○) and spontaneous fission (×) half-lives of the longest-lived isotopes of heavy elements.

island of relative nuclear stability around $Z = 108$ and $N = 162$ with spontaneous fission half-lives of the order of 1 s. Recent relativistic mean field calculations predict major closed shells for $Z = 120$ and $N = 172$ or 184 instead or in addition to $Z = 114$ and $N = 184$. On the other hand, extrapolations of the systematics of nuclear properties of the elements support $Z = 126$ and $N = 184$ as the next magic numbers. Further shell effects are predicted for $Z = 164$ and $N = 308$.

The islands of relative stability are shown in Fig. 14.9. The stability gap around mass number $A = 216$ is evident from this figure: nuclides with half-lives ≥ 1 s do not exist for $A = 216$. The search for superheavy elements concentrates on the islands around $Z = 108$ and $Z = 114$. At neutron numbers $N = 162$ the nuclei should exhibit a high degree of deformation, whereas spheric nuclei are expected for $N = 184$.

Theoretical calculations of the stability of superheavy elements have to take into account all possible modes of decay – spontaneous fission, α decay and β decay. The energy barriers for spontaneous fission are assessed for nuclei with closed nucleon shells to be about 10 to 13 MeV with an error of 2 to 3 MeV which causes an uncertainty in the half-life of about 10 orders of magnitude. Calculations of the half-lives of α and β decay are less problematic, but they have also an uncertainty of about 3 orders of magnitude. Predictions of half-lives of $>10^5$ y for even–even nuclei in the region of $^{298}114$ and of $\approx 10^9$ y for $^{294}110$ led to an intense search for superheavy elements in nature, in particular by the group in Dubna (Flerov et al.). However, this search was not successful.

The expected island of superheavy elements near $Z = 114$ and $N = 184$ has now been reached with respect to the atomic number Z, whereas there is still a gap in the neutron number. This gap can only be surmounted by fusion of neutron-rich nuclei.

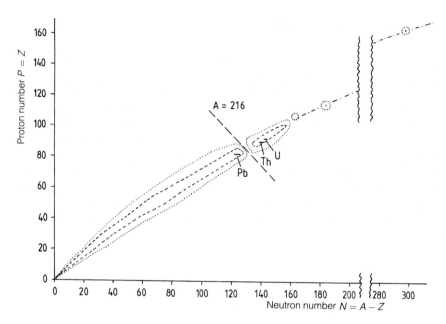

Figure 14.9. Regions of nuclide stability and predicted islands of relatively high stability (······, contours of regions with $t_{1/2} > 1$ s; - - - -, contours of regions with $t_{1/2} > 1$ h).

As projectile, the neutron-rich nucleus of ^{48}Ca, containing closed proton and neutron shells, is very favourable, but suitable target nuclei with high neutron numbers are not available. On the other hand, there is some hope of synthesizing in the near future nuclides with half-lives of the order of minutes that may be used for chemical experiments in order to reveal the chemical properties of the elements with atomic numbers up to $Z = 114$.

The general aspects of the synthesis of superheavy elements can be summarized as follows:

- High excitation energies (hot fusion) lead with high probability to immediate fission.
- Cold fusion by use of target nuclei with closed nucleon shells is most promising, because of the low excitation energies (emission of only 1 or 2 neutrons). However, the stabilizing effect of closed shells fades at excitation energies of the order of 50 MeV.
- Synthesis of new elements up to $Z \approx 120$ can be expected by application of cold fusion and refined techniques.
- Highly sophisticated methods are required to identify one atom of a new element or a new radionuclide at a time.
- The production cross sections of heavy elements by cold fusion decrease with increasing number of protons, because higher projectile energies are necessary to surmount the Coulomb barrier.
- Spontaneous fission does not prevail as assumed by extrapolation of the curve in Fig. 14.8. The main decay mode of the new elements with atomic numbers 108 to 118 is α decay.

14.5 Properties of the Actinides

All actinides are radioelements and only Th and U have half-lives long enough to justify neglecting their radioactivity in some special chemical or technical operations. Ac and Pa are present in small amounts as decay products of U and Th (Table 11.3). Extremely small amounts of Np and Pu are produced in U by neutrons from cosmic radiation: $^{237}Np/^{238}U \approx 10^{-12}:1$, $^{239}Pu/^{238}U \approx 10^{-14}:1$. Harkin's rule is also observed with the actinides, if the half-lives of the longest-lived isotopes of the elements are plotted as a function of the atomic number Z (Fig. 14.10). A characteristic feature of the heavier actinides is the tendency to decay by spontaneous fission. The ratio of the cross sections of thermal neutron fission $(\sigma_{n,f})$ and of (n, γ) reactions $(\sigma_{n,\gamma})$ is plotted in Fig. 14.11 for various nuclides as as function of the difference between the binding energy of an additional neutron and the threshold energy of fission.

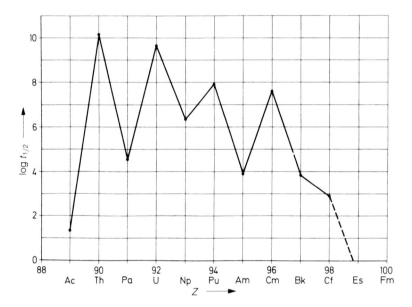

Figure 14.10. Logarithm of the half-life [y] of the longest-lived isotopes of the actinides as a function of the atomic number Z.

The electron configurations of the actinides in the gas phase are listed in Table 14.3. Whereas in the case of the lanthanides only up to two f electrons are available for chemical bonding, in the case of the actinides more than two f electrons may be engaged in chemical bonds (e.g. all the electrons in compounds of U(VI) and Np(VII)). This is due to the relatively low differences in the energy levels of the 5f and 6d electrons up to $Z \approx 95$ (Am). However, these differences increase with Z and the chemistry of elements with $Z \geq 96$ becomes similar to that of the lanthanides. The special properties of the actinides are evident from their oxidation states, plotted in Fig. 14.12 as a function of the atomic number. In contrast to the lanthanides, a tendency to form lower oxidation states is observed with the heavier actinides. The

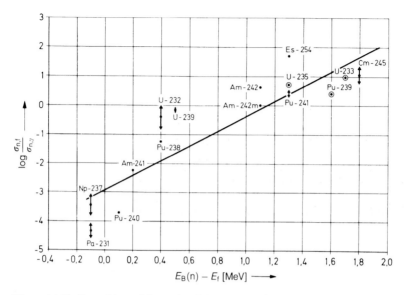

Figure 14.11. Logarithm of the ratio of the cross sections $\sigma_{n,f}$ and $\sigma_{n,\gamma}$ for various nuclides of the actinides as a function of the difference between the neutron binding energy $E_B(n)$ and the energy barrier of fission E_f. (According to G. T. Seaborg: The Transuranium Elements. Yale University Press 1958; Addison-Wesley Publ. Comp., Reading, Mass., S. 166/167; S. 240/241.)

Table 14.3. Electron configuration of the actinides in the gaseous state.

Atomic number	Symbol	Name of the element	Electron configuration
89	Ac	Actinium	$6d\ 7s^2$
90	Th	Thorium	$6d^2\ 7s^2$
91	Pa	Protactinium	$5f^2\ 6d\ 7s^2$ (or $5f^1\ 6d^2\ 7s^2$)
92	U	Uranium	$5f^3\ 6d\ 7s^2$
93	Np	Neptunium	$5f^5\ 7s^2$ (or $5f^4\ 6d\ 7s^2$)
94	Pu	Plutonium	$5f^6\ 7s^2$
95	Am	Americium	$5f^7\ 7s^2$
96	Cm	Curium	$5f^7\ 6d\ 7s^2$
97	Bk	Berkelium	$5f^8\ 6d\ 7s^2$ (or $5f^9\ 7s^2$)
98	Cf	Californium	$5f^{10}\ 7s^2$
99	Es	Einsteinium	$5f^{11}\ 7s^2$
100	Fm	Fermium	$5f^{12}\ 7s^2$
101	Md	Mendelevium	$5f^{13}\ 7s^2$
102	No	Nobelium	$5f^{14}\ 7s^2$
103	Lr	Lawrencium	$5f^{14}\ 6d\ 7s^2$

analogy between the 4f and 5f elements is also obvious from Fig. 14.13, in which the ionic radii of actinide and lanthanide ions are plotted as a function of the atomic number: the contraction of the actinide ions runs parallel to that of the lanthanide ions. In view of this analogy, the name actinides was introduced in 1925 by Goldschmidt.

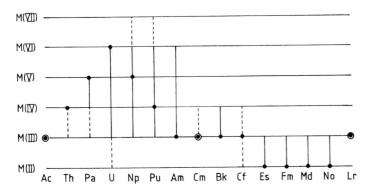

Figure 14.12. Oxidation states of the actinides.

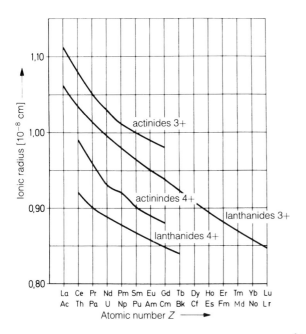

Figure 14.13. Ionic radii of actinide and lanthanide ions M^{3+} and M^{4+} as a function of the atomic number Z. (According to G. T. Seaborg: The Transuranium Elements. Yale University Press 1958; Addison-Wesley Publ. Comp., S. 137.)

With respect to the chemical properties of the actinides, a new effect becomes noticeable: with increasing atomic number Z, the influence of the positive nuclear charge on the electrons increases in such a way that their velocity approaches the velocity of light, which leads to relativistic effects. The valence electrons are more effectively screened from the nuclear charge, with the result of stabilization of the spherical 7s and $7p_{1/2}$ orbitals and destabilization of the 6d and 5f orbitals.

Relativistic effects on the valence electrons are already evident by comparing the electropositive character of Fr and Ra with that of their preceding homologues. The ionization potentials of both elements are not lower than those of their homologues Cs and Ba, respectively, as expected by extrapolation, but the ionization potential of Fr is about the same as that of Cs and the ionization potential of Ra is somewhat higher than that of Ba. The influence of relativistic effects on the properties of the actinides is evident also from the tendency of the heavier actinides to form lower oxidation states. For example, Es already prefers the oxidation state Es^{2+}.

The colours of the various oxidation states of the actinides Ac to Cm are listed in Table 14.4. Similarly to the lanthanides, the colour intensity of the M^{3+} ions increases with the distance from the electron configurations f^0 and f^7. The analogy between actinides and lanthanides is also valid for the magnetic susceptibilities, e.g. $U(VI) \triangleq Pa(V) \triangleq Th(IV) \triangleq Ac(III) \triangleq La(III)$ or $Pu(VI) \triangleq Np(V) \triangleq U(IV) \triangleq Pa(III) \triangleq Pr(III)$.

Table 14.4. Colours of actinide ions.

	M^{3+}	M^{4+}	MO_2^+	MO_2^{2+}
Actinium	Colourless	–	–	–
Thorium	–	Colourless	–	–
Protactinium	–	Colourless	–	–
Uranium	Reddish-brown	Green	–	Yellow
Neptunium	Blue to crimson	Yellowish green	Green	Pink
Plutonium	Violet	Orange	Reddish	Orange
Americium	Pink	Pink[a]	Yellow	Yellowish
Curium	Colourless	Pale yellow[a]	–	–

[a] Fluoro complexes.

Similarly to the lanthanides, actinides in the elemental state are reactive electropositive metals and pyrophoric in finely dispersed form. Strong reducing agents are necessary to prepare the metals from their compounds, for instance reduction of the halides by Ca or Ba at 1200 °C (e.g. $PuF_4 + 2Ca \rightarrow Pu + 2CaF_2$). Some properties of the actinides in the metallic state are listed in Table 14.5. The number of metallic modifications and the densities are remarkably high for U, Np and Pu. Some modifications of these elements are of low symmetry; this is an exception for metals that is explained by the influence of the f electrons. The properties of Am and the following elements correspond to those of the lanthanides.

Actinide(III) compounds are similar to lanthanide(III) compounds. Th(III) is unstable. In aqueous solution U(III) liberates hydrogen, whereas Np(III) and Pu(III) are oxidized in the presence of air.

The properties of actinides(IV) are similar to those of Ce(IV) and Zr(IV). Th(IV) is very stable; U(IV) and Np(IV) are stable in aqueous solution, but are oxidized slowly in the presence of air to U(VI) and Np(V). Pu(IV) is stable at high concentrations of acid or of complexing agents; otherwise it disproportionates to Pu(III) and Pu(VI). In moist air and at temperatures up to 350 °C, PuO_2 is slowly oxidized

Table 14.5. Properties of the actinide metals.

Element	Melting point (°C)	Phase	Structure	Density [g cm^{-3}]
Ac	1100 ± 50	–	Face-centred cubic	–
Th	1750	α (up to 1400 °C)	Face-centred cubic	11.72 (25 °C)
		β (1400–1750 °C)	Body-centred cubic	–
Pa	1873	–	Tetragonal	15.37
U	1132	α (up to 668 °C)	Orthorhombic	19.04 (25 °C)
		β (668–774 °C)	Tetragonal	18.11 (720 °C)
		γ (774–1132 °C)	Body-centred cubic	18.06 (805 °C)
Np	637	α (up to 280 °C)	Orthorhombic	20.45 (25 °C)
		β (280–577 °C)	Tetragonal	19.36 (313 °C)
		γ (577–637 °C)	Cubic	18.00 (600 °C)
Pu	639.5	α (up to 122 °C)	Monoclinic	19.74 (25 °C)
		β (122–203 °C)	Body-centred monoclinic	17.77 (150 °C)
		γ (203–317 °C)	Orthorhombic	17.19 (210 °C)
		δ (317–453 °C)	Face-centred cubic	15.92 (320 °C)
		δ' (453–477 °C)	Tetragonal	15.99 (465 °C)
		ε (477–640 °C)	Body-centred cubic	16.48 (500 °C)
Am	995	α (up to ≈600 °C)	Hexagonal	13.67 (20 °C)
Cm	1340	α (up to ≈150 °C)	Hexagonal	13.51 (25 °C)
		β (>150 °C)	Face-centred cubic	12.66 (150 °C)
Bk	986	α	Hexagonal	14.78 (25 °C)
		β	Face-centred cubic	13.25 (25 °C)
Cf	900	up to 600 °C	Hexagonal	15.1
		600–725 °C	Face-centred cubic	13.7
		>725 °C	Face-centred cubic	8.70

to PuO_{2+x} and hydrogen is set free. PuO_{2+x} contains Pu(VI) on the cationic sites of the fluorite structure and additional oxygen ions on octahedral interstices. In solution, Am(IV) and Cm(IV) are only known as fluoro complexes. Bk(IV) behaves similarly to Ce(IV). All actinides form sparingly soluble iodates and arsenates. The basicity decreases in the order $Th^{4+} > U^{4+} \approx Pu^{4+} > Ce^{4+} > Zr^{4+}$. Hydrolysis of actinides(IV) is very pronounced, but it can be prevented by complexation with ligands which form complexes that are more stable than the hydroxo complexes. Hydrolysis decreases in the order $M^{4+} > MO_2^{2+} > M^{3+} > MO_2^+$. Complexation of actinides(IV) with inorganic ligands decreases in the order $F^- > NO_3^- > Cl^- > ClO_4^-$ and $CO_3^{2-} > C_2O_4^{2-} > SO_4^{2-}$.

In the oxidation state V, Pa shows distinct differences from U and the following elements. Hydrolysis of Pa(V) in aqueous solutions can only be prevented by the presence of concentrated acids (e.g. 8 M HCl) or of complexing agents such as F^-. In contrast to Pa(V), U(V), Np(V) and Pu(V) form dioxocations MO_2^+ in which oxygen is strongly bound by the metal. Obviously, the formation of these dioxocations depends on the availability of a sufficient number of f electrons. In aqueous solution, UO_2^+ exists in small amounts in equilibrium with U^{4+} and UO_2^{2+}. NpO_2^+ is quite stable, whereas PuO_2^+ and AmO_2^+ disproportionate easily.

The oxidation state VI is preferred by U, but it is also found with Np, Pu and Am. In aqueous solution, this oxidation state always exists in the form of "yl" ions MO_2^{2+}. These ions are not formed by hydrolysis and are also stable at high acid concentrations. In a linear arrangement, the oxygen atoms are firmly bound by the metals and the interatomic distance is in agreement with the formation of double bonds. Four to six ligands are coordinated in the equatorial plane perpendicular to the linear MO_2^{2+} group.

By application of strong oxidizing agents, e.g. by melting with alkali peroxides, Np and Pu can be transformed into the oxidation state VII.

Actinide metals react with hydrogen at about 300 °C to form non-stoichiometric metallic hydrides of composition MH_2 to MH_3. The formation of uranium hydride is reversible at higher temperatures and can be used to store tritium.

The formation of non-stoichiometric compounds is very pronounced in the case of the oxides (Fig. 14.14). In thermal equilibrium, the composition of the oxides of Pa, U, Np and Pu varies with the partial pressure of oxygen and reflects the different oxidation states of the actinides.

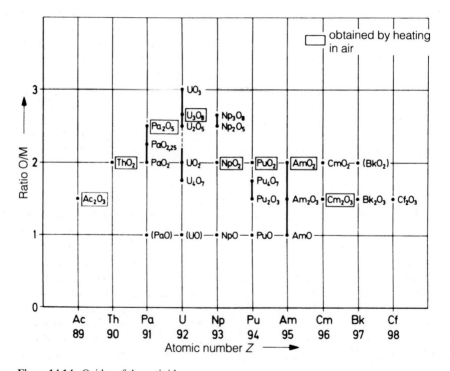

Figure 14.14. Oxides of the actinides.

Complexation of the actinides by inorganic and organic ligands is very important for their chemical separation and their behaviour in the environment. Examples are the separation of trivalent actinides by complexation with α-hydroxy acids and the high solubility of U(VI) due to complexation with carbonate ions to give the tris-carbonato complex $[UO_2(CO_3)_3]^{4-}$ found in seawater.

14.6 Properties of the Transactinides

Three ways can be distinguished to elucidate the properties of the transactinides:

– Extrapolation of the properties on the basis of the tendencies in the Periodic Table of the elements.
– Calculation of the electronic structure of the elements and the energy levels of the electrons in the atoms and their compounds.
– Elaborate chemical investigations with the microamounts of the elements available (few atoms and down to one atom at a time).

Straight-forward extrapolation allows prediction of the place of a new element with atomic number Z in the Periodic Table and assessment of its chemical properties. However, precise information about the properties can not be obtained in this way, mainly because of the relativistic effects in the electron shells, mentioned in the previous section, particularly as these effects increase with Z^2.

Relativistic calculations provide information about expected electron configurations in the ground state and energy levels of excited states. The relativistic effects lead to stabilization of the outer s and p orbitals and to their spatial contraction, and indirectly to destabilization and expansion of the outer d and f orbitals. An appreciable influence on the 7s, 7p and 6d orbitals of the transactinides and on the properties of these elements, such as electron configuration, ionization potentials and ionic radii, is expected. The properties of various compounds of the transactinides are predicted by quantum mechanical calculations. The most important result of these calculations is that an inversion of the trend in the properties is to be expected when going from the 5d to the 6d elements. Accordingly, the sequence of the properties should be Zr–Rf–Hf in the IVb group of the elements, and Nb–Db–Ta in the Vb group.

Experimental investigation of the chemical properties of the transactinides is only possible by use of nuclides with half-lives of at least 10 s. With respect to the short half-lives, these investigations require fast transport systems, preferably on-line gas-phase techniques, and fast chemical procedures, e.g. on-line separations in the gas phase or automated experiments in the aqueous phase. Furthermore, the experiments have to be repeated several times in order to obtain results that are statistically significant. Mainly α spectrometry and fission track counting are used for detection and counting of the radionuclides. Knowledge of the α-decay chains makes it possible to identify short-lived mother nuclides by measuring the daughter activities.

By application of the gas-phase technique, element 104 was identified by Zvara in Dubna as the first transactinide element. The arrangement is shown schematically in Fig. 14.15. Element 104 was produced by reaction (14.26), and the experiment was

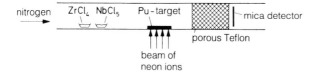

Figure 14.15. Gas-phase technique for investigating the chemical properties of element 104 (schematically).

based on the difference in volatility of the chlorides MCl_4 ($M = Zr$ and element 104) and MCl_3 ($M =$ lanthanides). $NbCl_5$ served as chlorinating agent. Element 104 was transported together with $ZrCl_4$ and detected by use of a mica fission track detector, proving that it is a homologue of Zr and the first transactinide element, whereas the lanthanides including element 103 were not able to pass the porous Teflon, because of the lower volatility of the chlorides. Application of cation-exchange techniques with the longer-lived $^{261}104$ performed later by the Berkeley group led to the same conclusion. Element 104 behaved like Zr and Hf, whereas actinides(III) were not eluted under the given conditions.

Recent chemical experiments with transactinides have been carried out by application of refined methods. Fast transport is achieved by thermalizing the products of nuclear reactions recoiling out of the target in helium gas loaded with aerosol particles (e.g. KCl, MoO_3, carbon clusters) of 10 to 200 nm on which the reaction products are adsorbed. Within about 2 to 5 s the aerosols are transported with the gas through capillary tubes over distances of several tens of metres with yields of about 50%.

In gas-phase experiments with single atoms adsorption enthalpies on the surface of the chromatographic column can be determined. These are correlated with the standard sublimation enthalpies and used as a measure of the volatility of the compounds. Two methods are applied, thermochromatography and isothermal gas-chromatography. Thermochromatography operates with a negative temperature gradient in the flow direction of the carrier gas. At the entrance of the chromatographic column halogenating and/or oxidizing gases are introduced, in order to produce the chemical species to be studied. Volatile species are deposited in the column according to their volatility, and mica sheets inserted into the column serve as fission track detectors. The advantage of the method is the high speed of production and separation of the volatile species. Disadvantages are the corrections necessary to take into account the influence of the half-lives on the observed deposition temperatures and the fact that real-time detection of the decay, determination of the half-life and identification of the decaying nuclide are not possible.

The on-line isothermal gas-chromatographic apparatus (OLGA) was developed to avoid the disadvantages of thermochromatography. In the first section, heated to 900-1000 °C, the aerosols are stopped in a quartz wool plug, and reactive gases (e.g. HCl, HBr, Cl_2, $SOCl_2$, BBr_3, O_2) are added. The second part of the column is the isothermal section, in which the volatile species experience numerous sorption/desorption steps with retention times that are characteristic of the volatility at the given temperature. The chemical yield of the volatile species is measured as a function of the temperature in the isothermal section. Above a certain temperature the yield rises steeply and reaches a plateau at higher temperatures. At 50% of the plateau, the retention time is equal to the half-life. At the exit of the chromatographic column the separated species enter a water-cooled chamber and are adsorbed on new aerosols. From this chamber they are transported through a capillary to a detection system. A rotating wheel or a moving tape bring the deposited nuclides in front of a series of detectors which register α particles and spontaneous fission. By evaluation of the registered data time-correlated parent-daughter particles are identified.

In contrast to gas-phase chemistry, aqueous chemistry is performed batch-wise, and it is necessary to repeat the same experiment many times with a cycle of about 1 min, in order to obtain statistically significant results. The experiments are carried out either manually (liquid/liquid extraction or chemisorption) or with an automated apparatus (e.g. ARCA = automated rapid chemistry apparatus). Automated separations are preferably carried out by computer controlled fast repetitive high performance liquid chromatography (HPLC). The apparatus comprises chromatographic columns, reservoirs for solutions, pumps and valves. The time necessary for one separation procedure varies between about 1/2 and 1 min. Complexation can be studied and distribution coefficients can be determined by use of such an apparatus.

For chemical experiments with rutherfordium, ^{261}Rf is preferably used, because of its relatively long half-life ($t_{1/2} \approx 78$ s). It is produced by the reaction ^{248}Cm(^{18}O, 5n) ^{261}Rf (production cross section $\sigma \approx 5$nb). Volatilities were found to be RfCl$_4 \approx$ ZrCl$_4 >$ HfCl$_4$, and RfBr$_4 >$ HfBr$_4$. Experiments with the automatic HPLC apparatus ARCA indicated similar behaviour of Rf and Th in HF solutions. The extraction sequence is Zr > Rf > Hf, and the distribution coefficients are approximately the same for Rf and Hf. This proves the predicted inversion of the trend in the properties from the 5d to the 6d elements. Further experiments with Rf are in progress by use of ^{263}Rf, produced by the reaction ^{248}Cm(^{22}Ne, α3n)^{263}Rf.

Gas-chromatographic experiments with dubnium using the isotopes ^{262}Db ($t_{1/2} \approx 34$ s) and ^{263}Db ($t_{1/2} \approx 27$ s) showed that the volatility of the oxychlorides is NbOCl$_3 >$ DbOCl$_3$. ^{262}Db and ^{263}Db are produced by the reactions ^{249}Bk(^{18}O, 5n)^{262}Db ($\sigma \approx 6$nb) and ^{249}Bk(^{18}O, 4n)^{263}Db, respectively. The sequence of complex formation and extraction with various extractants, studied with the ARCA apparatus, is Pa > Nb $\geq$ Db > Ta, and the tendency to hydrolyse is Ta > Db > Nb.

Isotopes of seaborgium suitable for chemical experimentss are ^{266}Sg ($t_{1/2} \approx 21$ s) and ^{265}Sg ($t_{1/2} \approx 7$ s), which are produced by the reactions ^{248}Cm(^{22}Ne, 4n)^{266}Sg ($\sigma \approx 0.03$nb) and ^{248}Cm(^{22}Ne, 5n)^{265}Sg ($\sigma \approx 0.2$nb), respectively. By gas-chromatographic experiments the volatility of the oxychlorides was found to be MoO$_2$Cl$_2 >$ WO$_2$Cl$_2 \geq$ SgO$_2$Cl$_2$. At higher temperatures, Sg is volatile in oxygen as SgO$_2$(OH)$_2$ which is adsorbed reversibly on the surface of quartz due to the reaction MO$_2$(OH)$_2$(g) $\rightleftharpoons$ MO$_3$(ads) + H$_2$O(g). Sg and W can be separated by use of cation exchangers, on which Sg is sorbed, whereas W is eluted. This is explained by a smaller tendency of Sg(VI) to hydrolyse.

For chemical experiments with bohrium the isotopes ^{267}Bh ($t_{1/2} \approx 17$ s) and ^{266}Bh are produced simultaneously by the reactions ^{249}Bk(^{22}Ne, 4–5n)^{267,266}Bh ($\sigma \approx$ 50pb). Bh is a typical member of group VIIb of the Periodic Table. Similar to TcO$_3$Cl and ReO$_3$Cl, BhO$_3$Cl is rather volatile: TcO$_3$Cl (b.p.25 °C) > ReO$_3$Cl (b.p.131 °C) $\geq$ BhO$_3$Cl.

Chemical experiments with hassium are in progress by use of ^{269}Hs, the isotope with the longest half-life known until now ($t_{1/2} \approx 10$ s), which can be produced by the reaction ^{248}Cm(^{26}Mg, 5n)^{269}Hs.

The extension of the Periodic Table of the elements is shown in Fig. 14.16. The transactinides with atomic numbers 104 to 121 are expected to be homologues of the elements 72 (Hf) to 89 (Ac). As already mentioned, this has been proved for elements 104 to 107 and can also be assumed for the following elements.

The relativistic effects explained in section 14.5 are more pronounced for the transactinides. The modified electronic structure may strongly influence their chemical properties and appreciably limits straightforward extrapolation. On the other hand, comparison of empirical extrapolations and results of careful measurements makes it possible to assess the influence of relativistic effects.

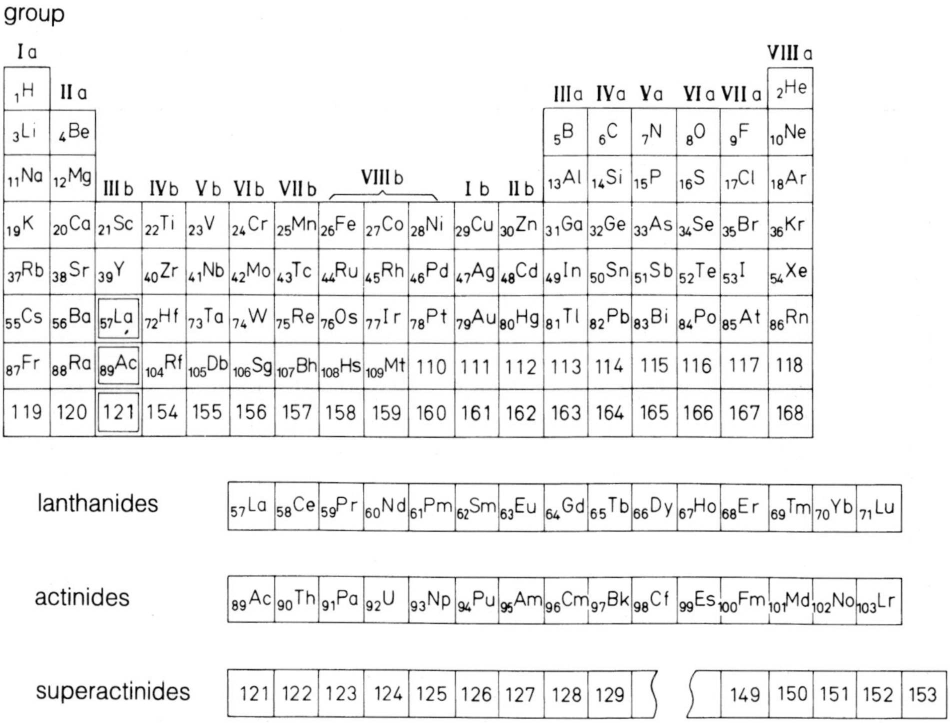

group

I a — VIII a

I a	II a	III b	IV b	V b	VI b	VII b	VIII b			I b	II b	III a	IV a	V a	VI a	VII a	VIII a
$_1$H	II a																$_2$He
$_3$Li	$_4$Be											$_5$B	$_6$C	$_7$N	$_8$O	$_9$F	$_{10}$Ne
$_{11}$Na	$_{12}$Mg											$_{13}$Al	$_{14}$Si	$_{15}$P	$_{16}$S	$_{17}$Cl	$_{18}$Ar
$_{19}$K	$_{20}$Ca	$_{21}$Sc	$_{22}$Ti	$_{23}$V	$_{24}$Cr	$_{25}$Mn	$_{26}$Fe	$_{27}$Co	$_{28}$Ni	$_{29}$Cu	$_{30}$Zn	$_{31}$Ga	$_{32}$Ge	$_{33}$As	$_{34}$Se	$_{35}$Br	$_{36}$Kr
$_{37}$Rb	$_{38}$Sr	$_{39}$Y	$_{40}$Zr	$_{41}$Nb	$_{42}$Mo	$_{43}$Tc	$_{44}$Ru	$_{45}$Rh	$_{46}$Pd	$_{47}$Ag	$_{48}$Cd	$_{49}$In	$_{50}$Sn	$_{51}$Sb	$_{52}$Te	$_{53}$I	$_{54}$Xe
$_{55}$Cs	$_{56}$Ba	$_{57}$La	$_{72}$Hf	$_{73}$Ta	$_{74}$W	$_{75}$Re	$_{76}$Os	$_{77}$Ir	$_{78}$Pt	$_{79}$Au	$_{80}$Hg	$_{81}$Tl	$_{82}$Pb	$_{83}$Bi	$_{84}$Po	$_{85}$At	$_{86}$Rn
$_{87}$Fr	$_{88}$Ra	$_{89}$Ac	$_{104}$Rf	$_{105}$Db	$_{106}$Sg	$_{107}$Bh	$_{108}$Hs	$_{109}$Mt	110	111	112	113	114	115	116	117	118
119	120	121	154	155	156	157	158	159	160	161	162	163	164	165	166	167	168

lanthanides: $_{57}$La $_{58}$Ce $_{59}$Pr $_{60}$Nd $_{61}$Pm $_{62}$Sm $_{63}$Eu $_{64}$Gd $_{65}$Tb $_{66}$Dy $_{67}$Ho $_{68}$Er $_{69}$Tm $_{70}$Yb $_{71}$Lu

actinides: $_{89}$Ac $_{90}$Th $_{91}$Pa $_{92}$U $_{93}$Np $_{94}$Pu $_{95}$Am $_{96}$Cm $_{97}$Bk $_{98}$Cf $_{99}$Es $_{100}$Fm $_{101}$Md $_{102}$No $_{103}$Lr

superactinides: 121 122 123 124 125 126 127 128 129 ... 149 150 151 152 153

Figure 14.16. Extended Periodic Table of the elements.

Relativistic calculations allow more detailed predictions of the chemical properties of transactinides compared with those of their lighter homologues. Electronic configurations and oxidation states predicted for the transactinide elements 104 to 120 on the basis of relativistic Hartree–Fock calculations are listed in Table 14.6. An important result of these calculations is the splitting of the p levels into a $p_{1/2}$ sublevel for 2 electrons and a $p_{3/2}$ sublevel for 4 electrons.

Elements 104 to 112 are transition elements ($6d^2s^2$ to $6d^{10}s^2$). For the first half of these elements high oxidation states are predicted. Elements 112 and 114 are of special interest, because of the relativistic effects of the filled $7s^2$ level of 112 and the filled $7p_{1/2}^2$ sublevel of 114, which give these elements a noble character. The formation of the $7p_{1/2}$ sublevel is also expected to influence the oxidation states of elements 115 to 117. With increasing atomic number, the energy difference between the $p_{1/2}$ and $p_{3/2}$ sublevels increases with the result that only the $p_{3/2}$ electrons will be available as valence electrons. Element 118 should be a noble gas but, due to its low ionization energy, compounds should easily be formed in which this element has the oxidation state IV or VI. Some chemical properties predicted for elements 104 to 121 are summarized in Table 14.7.

Prediction of the chemical and physical properties of the transactinides is of great value with respect to their detection and identification. On the other hand, correspondence between predictions and experimental results would confirm the extendibility of the Periodic Table of the elements.

Table 14.6. Predicted electron configurations and some other predicted properties of the elements 104 to 120.

Element	Atomic mass	Electrons in the outer orbitals	Preferred oxidation state	Ionization potential [eV]	Ionic radius in the metallic state [10^{-8} cm]	Density [g/cm^3]
104	278	$6d^2 7s^2$	IV	5.1	1.66	17.0
105	281	$6d^3 7s^2$	V	6.2	1.53	21.6
106	283	$6d^4 7s^2$	VI	7.1	1.47	23.2
107	286	$6d^5 7s^2$	VII	6.5	1.45	27.2
108	289	$6d^6 7s^2$	VIII	7.4	1.43	28.6
109	292	$6d^7 7s^2$	VI	8.2	1.44	28.2
110	295	$6d^8 7s^2$	IV	9.4	1.46	27.4
111	298	$6d^9 7s^2$	III	10.3	1.52	24.4
112	301	$6d^{10} 7s^2$	II	11.1	1.60	16.8
113	304	$7s^2 7p^1$	I	7.5	1.69	14.7
114	307	$7s^2 7p^2$	II	8.5	1.76	15.1
115	310	$7s^2 7p^3$	I	5.9	1.78	14.7
116	313	$7s^2 7p^4$	II	6.8	1.77	13.6
117	316	$7s^2 7p^5$	III	8.2	–	–
118	319	$7s^2 7p^6$	0	9.0	–	–
119	322	$8s^1$	I	4.1	2.6	4.6
120	325	$8s^2$	II	5.3	2.0	7.2

Table 14.7. Predicted chemical properties of the elements 104 to 120.

Element	Predicted oxidation states	Chemical properties of elements and compounds
104	(II, III) <u>IV</u>	similar to Zr and Hf, M^{4+} ions in aqueous solution and in solids
105	III, IV, <u>V</u>	similar to Nb and Ta
106	IV, <u>VI</u>	similar to Mo and W, MF_6 volatile, oxyanions MO_4^{2-} in aqueous solution
107	III, V, <u>VII</u>	similar to Re, MF_6 volatile, oxyanions MO_4^- in aqueous solution
108	III, IV, VI, <u>VIII</u>	similar to Os, MO_4 volatile
109	I, III, <u>VI</u>, (VIII)	similar to Ir, but nobler
110	0, I, II, II, <u>IV</u>, VI	similar to Pt, but nobler
111	<u>III</u> (I unstable)	similar to Au, M^- questionable
112	(0) I, <u>II</u>, (>II?)	noble metal, very volatile or gaseous oxides, chlorides and bromides unstable due to relativistic effect on $7s^2$ electrons, MF_2 stable, stable complexes MX_4^{2-} (X = I, Br, Cl) in aqueous solutions, many complexes properties of M^+ between those of Tl^+ and Ag^+, MCl soluble in NH_3 and HCl solutions
113	<u>I</u>, III	similar to Pb, but nobler, also similar to 112 due to the filled $p_{1/2}$ sublevel, element volatile or gaseous, MF_2 and MCl_2 stable
114	<u>III</u>	similar to Tl^+
115	<u>I</u>, III	similar to Po
116	<u>II</u>, IV	semimetallic, low electron affinity
117	I, <u>III</u>, V (-I questionable)	noble gas, but more reactive than Xe
118	<u>0</u>, II, IV, VI	similar to Rb, Cs and Fr
119	<u>I</u> (III)	similar to Sr, Ba and Ra
120	<u>II</u> (IV)	

The term superactinides (Fig. 14.16) was introduced in 1968 for the elements with $Z \geq 121$. In these elements, the 6f and 5g levels are expected to be filled as indicated in Table 14.8. Four sublevels ($8p_{1/2}$, $7d_{3/2}$, $6f_{5/2}$ and $5g_{7/2}$) are assumed to compete with each other and to determine, together with the 8s electrons, the properties and the chemistry of these elements.

Table 14.8. Predicted electron configurations of superactinides.

Element	5g	6f	7s	7p	7d	8s	8p	Element	6f	7s	7p	7d	8s	8p	9s	9p
121			2	6		2	1	145	3	2	6	2	2	2		
122			2	6	1	2	1	146	4	2	6	2	2	2		
123		1	2	6	1	2	1	147	5	2	6	2	2	2		
124		3	2	6		2	1	148	6	2	6	2	2	2		
125	1	3	2	6		2	1	149	6	2	6	3	2	2		
126	2	2	2	6	1	2	1	150	6	2	6	4	2	2		
127	3	2	2	6		2	2	151	8	2	6	3	2	2		
128	4	2	2	6		2	2	152	9	2	6	3	2	2		
129	5	2	2	6		2	2	153	11	2	6	2	2	2		
130	6	2	2	6		2	2	154	12	2	6	2	2	2		
131	7	2	2	6		2	2	155	13	2	6	2	2	2		
132	8	2	2	6		2	2	156	14	2	6	2	2	2		
133	8	3	2	6		2	2	157	14	2	6	3	2	2		
134	8	4	2	6		2	2	158	14	2	6	4	2	2		
135	9	4	2	6		2	2	159	14	2	6	4	2	2	1	
136	10	4	2	6		2	2	160	14	2	6	5	2	2	1	
137	11	3	2	6	1	2	2	161	14	2	6	6	2	2	1	
138	12	3	2	6	1	2	2	162	14	2	6	8	2	2		
139	13	2	2	6	2	2	2	163	14	2	6	9	2	2		
140	14	3	2	6	1	2	2	164	14	2	6	10	2	2		
141	15	2	2	6	2	2	2	165	14	2	6	10	2	2	1	
142	16	2	2	6	2	2	2	166	14	2	6	10	2	2	2	
143	17	2	2	6	2	2	2	167	14	2	6	10	2	2	2	1
144	18	1	2	6	3	2	2	168	14	2	6	10	2	2	2	2

Literature

General

O. Hahn, Künstliche Neue Elemente, Verlag Chemie, Weinheim, **1948**

K. W. Bagnall, Chemistry of the Rare Radioelements (Polonium, Actinium), Butterworths, London, **1957**

E. K. Hyde, G. T. Seaborg, The Transuranium Elements, Handbuch der Physik, Vol. XLII, Springer, Berlin, **1957**

G. T. Seaborg, The Transuranium Elements, Yale University Press, New Haven, **1958**

G. T. Seaborg, Man-made Transuranium Elements, Prentice-Hall, Englewood Cliffs, NJ, **1963**

M. Haissinsky, J. P. Adloff, Radiochemical Survey of the Elements, Elsevier, Amsterdam, **1965**

C. Keller, The Chemistry of the Transuranium Elements (Ed. K. H. Lieser), Verlag Chemie, Weinheim, **1971**

K. W. Bagnall (Ed.), International Review of Science, Inorganic Chemistry, Vol. 7, Lanthanides and Actinides, Series One and Two, Butterworths, London, **1972** and **1975**

A. J. Freeman, J. B. Darby, jr. (Eds.), The Actinides: Electronic Structure and Related Properties, Academic Press, New York, **1974**

G. T. Seaborg (Ed.), Transuranium Elements – Products of Modern Alchemy, Dowden, Hutchinson and Ross, Stroudsburg, **1978**

N. M. Edelstein (Ed.), Lanthanide and Actinide Chemistry and Spectroscopy, ACS Symp. Ser. 131, Washington, D. C., **1980**

N. M. Edelstein (Ed.), Actinides in Perspective, Pergamon, Oxford, **1982**

J. J. Katz, G. T. Seaborg, L. R. Morse (Eds.), The Chemistry of the Actinide Elements, 2nd ed., 2 Vols., Chapman and Hall, London, **1986**

G. T. Seaborg, W. D. Loveland, The Elements Beyond Uranium, Wiley, New York, **1990**

L. R. Morss, J. Fuger (Eds.), Transuranium Elements, A Half Century, Amer. Chem. Soc., **1992**

G. T. Seaborg, Transuranium Elements: the Synthetic Actinides, Radiochim. Acta *70/71*, 69 **(1995)**

G. Münzenberg, M. Schädel, Die Jagd nach den schwersten Elementen, Vieweg, Braunschweig **1996**

Handbooks

Gemlins Handbook of Inorganic Chemistry, Transuranium Elements, 8th ed., Vols. 7a, 7b, 8, 31, Verlag Chemie, Weinheim **1973**, **1974**, **1973**, **1976**

Handbook on the Physics and Chemistry of the Actinides, Vols. I, II, V (Eds. A. J. Freeman, G. H. Lander), Vols. III, IV (Eds. A. J. Freeman, C. Keller), North-Holland, Amsterdam, **1984** onward

More Special

E. H. P. Cordfunke, The Chemistry of Uranium, Elsevier, Amsterdam, **1969**

D. I. Ryabchikov, E. K. Golbraikh, Analytical Chemistry of Thorium, Ann Arbor–Humphrey, Ann Arbor, **1969**

M. S. Milyukova, N. I. Gusev, I. G. Sentyurin, I. S. Sklyarenko, Analytical Chemistry of Plutonium, Ann Arbor–Humphrey, Ann Arbor, **1969**

J. M. Cleveland, The Chemistry of Plutonum, Gordon and Breach, New York, **1970**

A. K. Lavrukhina, A. A. Podznyakov, Analytical Chemistry of Technetium, Astatine and Francium, Ann Arbor–Humphrey, Ann Arbor, **1970**

D. I. Ryabchikov, V. A. Ryabukhin, Analytical Chemistry of Yttrium and the Lanthanide Elements, Ann Arbor–Humphrey, Ann Arbor, **1970**

E. S. Palshin, B. F. Myasoedov, Analytical Chemistry of Protactinium, Ann Arbor–Humphrey, Ann Arbor, **1970**

P. N. Palei, Analytical Chemistry of Uranium, Ann Arbor–Humphrey, Ann Arbor, **1970**

D. C. Hoffman, F. U. Lawrence, J. L. Mewherter, F. M. Rourke, Detecton of Plutonium-244 in Nature, Nature (London) *234*, 132 **(1971)**

R. J. Silva, Transcurium Elements, in: International Review of Science, Inorganic Chemistry Series One, Vol. 8, Radiochemistry (Ed. A. G. Maddock), Butterworths, London, **1972**

V. A. Mikhailov, Analytical Chemistry of Neptunium, Halsted Pres, New York, **1973**

B. F. Myasoedov, L. I. Guseva, I. A. Lebedev, M. S. Milyukova, M. K. Chmutova, Analytical Chemistry of Transplutonium Elements, Wiley, New York, **1974**

W. Muller, R. Lindner (Eds.), Transplutonium Elements, North-Holland, Amsterdam, **1976**

W. Muller, H. Blank (Eds.), Heavy Element Properties, North-Holland, Amsterdam, **1976**

H. Blank, R. Lindner (Eds.), Plutonium and other Actinides, North-Holland, Amsterdam, **1976**

F. David, Radiochemical Studies of Transplutonium Elements, Pure Appl. Chem. *53*, 997 **(1981)**

S. P. Sinha (Ed.), Systematics and the Properties of the Lanthanides, Kluwer, Dordrecht, **1983**

L. Stein, The Chemistry of Radon, Radiochim. Acta *32*, 163 **(1983)**

W. T. Carnall, G. R. Choppin (Eds.), Plutonium Chemistry, ACS Symposium Series 216, Amer. Chem. Soc., Washington, DC, **1983**

E. K. Hulet, Chemistry of the Elements Einsteinium through 105, Radiochim. Acta *32*, 7 **(1983)**

R. K. Narayama, H. J. Arnikar (Eds.), Artificial Radioactivity, Tata McGraw-Hill, New Delhi, **1985**

G. R. Choppin, J. D. Navratil, W. W. Schulz (Eds.), Actinides–Lanthanides Separation, World Scientific, Singapore, **1985**

P. K. Hopke (Ed.), Radon and its Decay Products. Occurrence, Properties and Health Effects, Amer. Chem. Soc., Washington, DC, **1987**

I. Zvara, Thermochromatographic Method of Separation of Chemical Elements in Nuclear and Radiochemistry, Isotopenpraxis *26*, 251 **(1990)**

G. R. Choppin, K. L. Nash, Actinide Separation Science, Radiochim. Acta *70/71*, 225 **(1995)**

N. Trautmann, Fast Radiochemical Separations for Heavy Elements, Radiochim. Acta *70/71*, 237 **(1995)**

P. K. Kuroda, Formation of Heavy Elements in Nature, Radiochim. Acta *70/71*, 229 **(1995)**

J. M. Haschke, Th. H. Allen, L. A. Morales, Reaction of Plutonium Dioxide with Water: Formation and Properties of PuO_{2+x}, Science *287*, 285 **(2000)**

Transactinides

J. D. Hemingway, Superheavy Elements, in: Radiochemistry, Vol. 1, Specialist Periodical Reports, The Chemical Society, London, **1972**

J. D. Hemingway, Transactinide Elements, in: Radiochemistry Vol. 2, Specialist Periodical Reports, The Chemical Society, London, **1975**

G. Herrmann, Superheavy Elements, in: International Review of Science, Inorganic Chemistry Series Two, Vol. 8, Radiochemistry (Ed. A. G. Maddock), Butterworths, London, **1975**

B. Fricke, Superheavy Elements, in: Structure and Bonding, (Eds. W. L. Jørgensen et al.), Vol. 21, Springer, Berlin, **1975**

O. L. Keller, jr., G. T. Seaborg, Chemistry of the Transactinide Elements, Annu. Rev. Nucl. Sci. *27*, 139 **(1977)**

G. Herrmann, Superheavy Element Research, Nature (London) *280*, 543 **(1979)**

Y. T. Oganessian, M. Hussonnois, A. G. Demin, Y. P. Khaitonov, H. Bruchertseifer, O. Constantinescu, Y. Korotkin, S. P. Tretyakova, V. K. Utyonkov, I. V. Shirokovsky, J. Estevez, Experimental Studies on the Formation and Radioactive Decay of Isotopes with $Z = 104–109$, Radiochim. Acta *37*, 113 **(1984)**

P. Armbruster, On the Production of Heavy Elements by Cold Fusion; The Elements 106 to 109, Annu. Rev. Nucl. Part. Sci. *35*, 113 **(1985)**

G. T. Seaborg, Superheavy Elements, Contemp. Phys. *28*, 33 **(1987)**

E. K. Hyde, D. C. Hoffman, O. R. Keller, jr., A History and Analysis of the Discovery of Elements 104 and 105, Radiochim. Acta *42*, 57 **(1987)**

G. Herrmann, Synthesis of Heaviest Elements, Angew. Chem., Int. Ed. *27*, 1417 **(1988)**

K. Kumar, Superheavy Elements, Hilger, London, **1989**

Z. Szeglowski, H. Bruchertseifer, V. P. Domanov, B. Gleisberg, L. J. Guseva, M. Hussonnois, G. S. Tikhomirova, I. Zvara, Y. T. Oganessian, Study of the Solution Chemistry of Element 104, Kurtchatovium, Radiochim. Acta *51*, 71 **(1990)**

R. Bock, G. Herrmann, G. Siegert, Schwerionenforschung, Wiss. Buchgesellschaft, Darmstadt, **1993**

G. Münzenberg, Discovery, Synthesis and Nuclear Properties of the Heaviest Elements, Radiochim. Acta *70/71*, 193 **(1995)**

M. Schädel, Chemistry of the Transactinide Elements, Radiochim. Acta *70/71*, 207 **(1995)**

W. Greiner, R. K. Gupta (Eds.), Heavy Elements and Related Phenomena, World Scientific, Singapore **1998**, Vol. I and II.

J. V. Kratz, Chemical Properties of the Transactinide Elements, in: Heavy Elements and Related Phenomena (Eds. W. Greiner, R. K. Gupta), World Scientific, Singapore **1998**, Vol. I.

15 Radionuclides in Geo- and Cosmochemistry

15.1 Natural Abundances of the Elements and Isotope Variations

A main concern of geochemistry is the investigation of the abundance and the distribution of the elements on the surface and in deeper layers of the earth, and of transport processes. The components of the geosphere are the lithosphere, the hydrosphere and the atmosphere. The relative abundance of the elements on the surface of the earth is plotted in Fig. 15.1 as a function of the atomic number. This relative abundance is similar within the solar system. The elements H, O, Si, Ca and Fe exhibit the highest abundances and maxima are observed at the magic numbers $Z = 8, 20, 50$ and 82. The abundances of the elements and their isotopes are determined by the nuclear reactions by which they have been produced and by their nuclear properties, whereas the chemical properties of the elements are only responsible for distribution and fractionation processes.

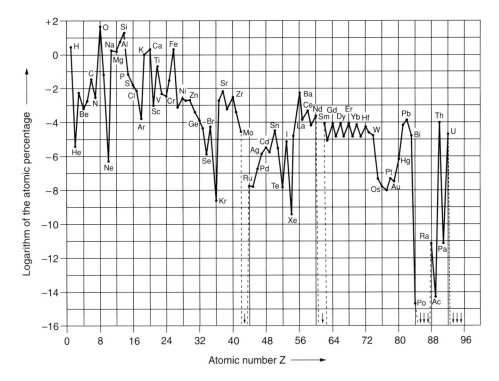

Figure 15.1. Abundance of the elements on the surface of the earth (lithosphere, hydrosphere and atmosphere).

The abundances of the isotopes of the elements in the geosphere show some variations caused by their formation, by isotope effects or by transport processes. A certain isotope ratio IR is taken as standard and the relative deviation from this standard is expressed as the δ-value:

$$\delta = \frac{\text{IR(sample)} - \text{IR(standard)}}{\text{IR(standard)}} \cdot 1000 \tag{15.1}$$

High variations of the isotope ratio are observed for the isotopes of hydrogen, H and D, because of their high relative mass difference. The average isotope ratio D:H in the oceans (IR $= 1.56 \cdot 10^{-4}$) is taken as standard. In natural waters $\delta(D)$ varies between $+6$ and -200. Evaporation of water leads to appreciable isotope effects and high isotope effects are also found in the water of hydrates.

Some values of the isotope ratios $^{16}O:^{18}O$, $^{12}C:^{13}C$ and $^{32}S:^{34}S$ are listed in Table 15.1. Because oxygen is the most abundant element on the earth, determination of

Table 15.1. Isotope ratios IR of oxygen, carbon and sulfur isotopes in various samples.

Sample	$^{16}O/^{18}O$
Fresh water	488.95
Ocean water	484.1
Water from the Dead Sea	479.37
Oxygen in the air	474.72
Oxygen from photosynthesis	486.04
CO_2 in the air	470.15
Carbonates	470.61

	$^{12}C/^{13}C$
CO_2 in the air	91.5
Limestone	88.8–89.4
Shells of sea animals	89.5
Ocean water	89.3
Meteorites	89.8–92.0
Coal, wood	91.3–92.2
Petroleum, pitch	91.3–92.8
Algae, spores	92.8–93.1

	$^{32}S/^{34}S$
Sulfates in the oceans	21.5–22.0
Volcanic sulfur	21.9–22.2
Magmatic rocks	22.1–22.2
Meteorites	21.9–22.3
Living things	22.3
Petroleum, coal	21.9–22.6

(According to S. R. Silverman: Geochim. Acta **2**, 26 (1951); B. F. Murphey, A. O. Nier: Physic Rev. **59**, 772 (1941); E. K. Gerling, K. G. Rik, cited by A. P. Vinogradov: Bull Acad. Sci. USSR. Serie Geol., Nr. **3**, 3 (1954).)

the $^{18}O:^{16}O$ ratio is of great interest in geochemistry. For instance, ^{18}O is strongly enriched in silicates containing only Si–O–Si bonds, whereas the $^{18}O:^{16}O$ ratio is lower in Si–O–Al bonds and particularly low in Si–OH groups. As the isotope exchange equilibria between minerals and water depend on temperature, information about the temperature of formation of the minerals is obtained by measuring the isotope ratio $^{18}O:^{16}O$ (geochemical isotope thermometry). The same holds for the exchange equilibrium between carbonates and water,

$$CaC^{16}O_3 + H_2{}^{18}O \rightleftharpoons CaC^{16}O_2{}^{18}O + H_2{}^{16}O \tag{15.2}$$

From the isotope ratio $^{18}O:^{16}O$ in carbonates the temperature of their formation can be obtained.

In the case of the stable isotopes of carbon, ^{13}C and ^{12}C, two isotope effects are noticeable: the kinetic isotope effect in photosynthesis, leading to an enrichment of ^{12}C in plants, and the equilibrium isotope effect in the exchange reaction

$$^{13}CO_2 + H^{12}CO_3^- \rightleftharpoons {}^{12}CO_2 + H^{13}CO_3^- \tag{15.3}$$

causing an enrichment of ^{13}C in hydrogencarbonate. The equilibrium constant K of reaction (15.3) depends on the temperature ($K = 1.009$ at $10\,°C$ and 1.007 at $30\,°C$) and the ratio $CO_2:HCO_3^-$ depends strongly on pH. In seawater of pH 8.2, 99% of the dissolved CO_2 is present in the form of HCO_3^-, whereas at lower pH CO_2 prevails. From the measurement of $\delta(^{12}C)$ values, conclusions can be drawn with respect to the conditions of formation of carbonates.

The isotope ratios of the sulfur isotopes are also affected by kinetic and equilibrium isotope effects. Kinetic isotope effects are marked in the reduction of sulfates to hydrogen sulfide by bacteria (enrichment of the lighter isotopes in H_2S). The equilibrium isotope effect in the reaction

$$^{32}SO_4^{2-} + H_2{}^{34}S \rightleftharpoons {}^{34}SO_4^{2-} + H_2{}^{32}S \tag{15.4}$$

leads also to an enrichment of ^{32}S in H_2S (equilibrium constant $K = 1.075$ at $25\,°C$). Isotope exchange reactions between sulfidic minerals also lead to a shift of the isotope ratios. For example, ^{34}S is enriched in relation to ^{32}S in the order pyrite > sphalerite > galerite. Measurement of the $^{34}S:^{32}S$ ratio gives information about the conditions of formation of sulfidic minerals (e.g. magmatic or hydrothermal, and temperature). Sulfates precipitated in seawater exhibit $^{34}S:^{32}S$ ratios that are characteristic for the geological era in which they have been formed.

For special geochemical investigations, isotope ratios of other elements, such as B, N, Si, K and Se are also determined. The measurement of the distribution of the natural radioelements U and Th and their daughter nuclides in minerals, sediments, oil, water and the air gives information about the genesis of the minerals, sediments and oils, and about the processes taking place in the lithosphere, the hydrosphere and the atmosphere. The nuclear methods of dating will be discussed in chapter 16.

15.2 General Aspects of Cosmochemistry

The concern of cosmochemistry is the investigation of extraterrestrial matter (sun, moon, planets, stars and interstellar matter) and their chemical changes. Meteorites are an object of special interest in cosmochemistry, because of the nuclear reactions induced by high-energy protons in cosmic radiation ($E(p)$ up to about 10^9 GeV) and by other particles, such as α particles and various heavy ions. Measurement of the radionuclides produced in meteorites by cosmic radiation gives information about the intensity of this radiation in interstellar space and about the age and the history of meteorites.

With the exception of a few special cases, the isotope ratios in meteorites are the same as on the earth, which means that during the formation of the various parts of the solar system only some fractionation of the elements occurred, but no isotope fractionation. Differences in the isotope ratios in meteorites and on the earth can be explained by radioactive decay, nuclear reactions triggered by cosmic radiation and some isotope fractionation of light elements.

For the investigation of meteorites various experimental methods are applied, in particular mass spectrometry, neutron activation analysis, measurement of natural radioactivity by low-level counting and track analysis. The tracks can be caused by heavy ions in cosmic radiation, by fission products from spontaneous or neutron-induced fission and by recoil due to α decay. Etching techniques and measurement of the tracks give information about the time during which the meteorites have been in interstellar space as individual particles (irradiation age).

Lunar samples have been investigated by similar methods, with the result that many details have been learned about the chemical composition of the surface of the moon and the nuclear reactions occurring there under the influence of cosmic radiation that hits the moon surface without hindrance by an atmosphere.

The first elements were formed in the early stage of the universe and their production continued by subsequent nuclear reactions, in particular nuclear fusion (thermonuclear reactions), (n,γ) reactions and radioactive transmutations. The most probable and generally accepted concept of the beginning of the universe is that of a single primordial event, called the "big bang", at the time zero. Since then the universe has been continuously expanding and the galaxies are moving away from each other. If this expansion is extrapolated back to the stage when the galaxies were close to each other, a value of $(15 \pm 3) \cdot 10^9$ y is obtained for the age of the universe. This is called the Hubble time.

15.3 Early Stages of the Universe

According to the concept of the big bang, a mixture of fundamental particles and energy existed at time zero at an extremely high temperature (of the order of about 10^{30} K) and an extremely high density (of the order of about 10^{50} g cm^{-3}) in an extremely small volume. The unified theory of fundamental particles and forces assumes that only one elementary constituent of matter and only one force existed at that time, but the origin of the big bang is not clear.

Early stages of the universe are listed in Table 15.2. Primeval matter was merging into elementary particles, huge amounts of energy were released and the big bang immediately caused a rapid expansion of the universe. Within about 1 s the temperature decreased markedly, matter and antimatter annihilated each other, quarks combined into mesons and baryons and enormous amounts of energy were liberated causing further expansion. Formation of the first protons and leptons is assumed after about 1 s, when the temperature of the early universe was about 10^{10} K.

Synthesis of nuclei heavier than protons started after about 100 s at a temperature of about 10^9 K by the merging of a proton with a neutron into a deuteron:

$$p + n \rightarrow d + \gamma \tag{15.5}$$

Table 15.2. Evolution of the universe according to the concept of the big bang.

Time	Universe
0	$T \approx 10^{30}$ K, density $\approx 10^{50}$ g cm^{-3}; primordial matter (quark-gluon plasma)
	Merging of primordial matter into elementary particles, such as protons and neutrons; release of huge amounts of energy, beginning of rapid expansion
1 s	$T \approx 10^{10}$ K
	Further formation of protons and other elementary particles
100 s	$T \approx 10^9$ K
	Synthesis of d and ^{4}He
$3 \cdot 10^5$ y	$T \approx 3 \cdot 10^3$ K; about 75% of the matter consists of p and about 25% of ^{4}He
	Formation of atoms (H, He) by combination of nuclei and electrons, formation of first molecules
$1 \cdot 10^9$ y	–
	Beginning of formation of galaxies and stars, gravitational contraction of the stars → increase of temperature, thermonuclear reactions (hydrogen burning, helium burning)
$5 \cdot 10^9$ y	–
	Further gravitational contraction (depending on the mass of the stars) → further increase of temperature → carbon burning, oxygen burning, liberation of neutrons by photons → (n,γ) reactions
$15 \cdot 10^9$ y	Present

At this temperature, the energy of most of the protons was not high enough to split deuterons into nucleons, and the deuterons were able to combine with one or two more nucleons to give nuclei with mass numbers 3 and 4

$$\left.\begin{array}{l} d + p \rightarrow {}^3He + \gamma \\ d + n \rightarrow t + \gamma \\ d + d \rightarrow t + p \\ d + d \rightarrow {}^3He + n \\ t + p \rightarrow {}^4He + \gamma \\ {}^3He + n \rightarrow {}^4He + \gamma \\ t + d \rightarrow {}^4He + n \\ {}^3He + d \rightarrow {}^4He + p \end{array}\right\} \tag{15.6}$$

Complete conversion of protons into ^4He was not possible, because of the lack of neutrons which in the free state decay to protons. After about 250 s, the mass of the universe consisted of about 75% hydrogen (protons and deuterons in a ratio of about $10^3 : 1$), about 25% helium (^4He and ^3He in a ratio of about $10^4 : 1$) and traces of ^7Li formed by reactions such as

$$^4He + t \rightarrow {}^7Li + \gamma \tag{15.7}$$

These light nuclei existed in an environment of photons, electrons and neutrinos for several thousand years. The ambient temperature decreased slowly to the order of about 10^4 K, but it was too low for further nucleosynthesis and too high for the formation of atoms and molecules.

After about $3 \cdot 10^5$ y the temperature was about $3 \cdot 10^3$ K and by combination of nuclei and electrons the first atoms of hydrogen and helium were formed. At further stages of expansion and cooling the first H_2 molecules became stable. Due to gravitation, the matter began to cluster and the formation of galaxies and stars began after about 10^9 y. The further individual development of the stars depended and still depends mainly on their mass.

The density of the photons in the universe was always appreciably higher than that of the nucleons. At present, the average densities are about $5 \cdot 10^8$ photons and only about 0.05 to 5 nucleons per m^3. However, because of the low mass equivalent of the photons, their mass is small compared with that of the nucleons.

15.4 Synthesis of the Elements in the Stars

With the aggregation of matter in the stars under the influence of gravitation, new processes begin to dominate:

- Gravitational contraction of the stars causes increase of temperature associated with the emission of light and other kinds of electromagnetic radiation.
- If the temperature in the core of the stars becomes sufficiently high, thermonuclear reactions give rise to new phases of nucleogenesis, and the energy produced by these reactions leads to further emission of radiation, including visible light.

The influences of gravitational contraction and thermonuclear reactions depend primarily on the mass of the stars. The phases of contraction and thermonuclear reactions overlap and determine the nucleogenesis as well as the fate of the stars.

The first stage of nucleogenesis in the stars is the fusion of protons into ^{4}He by the net reaction (section 8.12, deuterium cycle, carbon–nitrogen cycle or others)

$$4\,\mathrm{p} \rightarrow {}^4\mathrm{He} + 2\,\mathrm{e}^+ + 2\,\nu_\mathrm{e} + \Delta E \quad (\Delta E = 26.7\,\mathrm{MeV}) \tag{15.8}$$

This process is also called hydrogen burning. The temperature in the core of the stars must be $\geq 10^7$ K, in order to overcome the Coulomb repulsion of the protons. The neutrinos escape into outer space due to their small interaction with matter. Hydrogen burning is the longest stage of the stars.

It has been calculated that in the sun at a core temperature of $1.6 \cdot 10^7$ K about half of the energy is produced by the deuterium cycle, about half by the reaction sequence

$$\left.\begin{aligned}
\mathrm{p} + \mathrm{p} &\rightarrow \mathrm{d} + \mathrm{e}^+ + \nu_\mathrm{e} \\
\mathrm{d} + \mathrm{p} &\rightarrow {}^3\mathrm{He} + \gamma \\
{}^3\mathrm{He} + {}^4\mathrm{He} &\rightarrow {}^7\mathrm{Be} + \gamma \\
{}^7\mathrm{Be} + \mathrm{e}^- &\rightarrow {}^7\mathrm{Li} + \nu_\mathrm{e} \\
{}^7\mathrm{Li} + \mathrm{p} &\rightarrow {}^8\mathrm{Be} + \gamma \\
{}^8\mathrm{Be} &\rightarrow 2\,{}^4\mathrm{He}
\end{aligned}\right\} \tag{15.9}$$

and about 4% by the carbon–nitrogen cycle. The contribution of the latter increases with increasing temperature. In the sun, hydrogen burning is lasting for about 10^{10} y.

Towards the end of the stage of hydrogen burning, the concentration of hydrogen in the core of the stars and the energy production decrease, and the helium-rich core continues to contract, until the densities and the temperature in the core increase to about 10^4 g cm^{-3} and 10^8 K, respectively. Under these conditions, the ^{4}He nuclei accumulated by hydrogen burning are able to undergo further thermonuclear reactions,

$$\left.\begin{aligned}
{}^4\mathrm{He} + {}^4\mathrm{He} &\rightarrow {}^8\mathrm{Be} + \gamma \\
{}^8\mathrm{Be} + {}^4\mathrm{He} &\rightarrow {}^{12}\mathrm{C} + \gamma
\end{aligned}\right\} \tag{15.9}$$

This process is called helium burning. Although the lifetime of ^{8}Be is extremely short, its concentration is high enough to bridge the unstable region at $A = 8$ and to allow production of ^{12}C. The stage of helium burning is reached earlier at higher mass. Hydrogen burning continues in the outer zones of these stars, which become brighter and red. If the mass is similar to or greater than that of the sun, the stars expand in the outer zones to red giants with diameters that are greater than before by a factor of 10^2 to 10^3. On the cosmic scale, helium burning is a relatively short stage.

At increasing concentration of ^{12}C in the core, the synthesis of heavier nuclei up to ^{20}Ne begins by successive fusion with ^{4}He:

$$\left.\begin{array}{l} ^{12}\text{C} + {}^4\text{He} \rightarrow {}^{16}\text{O} + \gamma \\ ^{16}\text{O} + {}^4\text{He} \rightarrow {}^{20}\text{Ne} + \gamma \end{array}\right\} \quad (15.10)$$

Furthermore, gravitational contraction continues with increasing concentration of ^{12}C and ^{16}O, and in stars with masses >3 times that of the sun the temperature and the density rise to about $6 \cdot 10^8$ K and about $5 \cdot 10^4$ g cm^{-3}, respectively. Under these conditions, fusion of two ^{12}C becomes possible and reactions such as

$$\left.\begin{array}{l} ^{12}\text{C} + {}^{12}\text{C} \rightarrow {}^{20}\text{Ne} + {}^4\text{He} \\ ^{12}\text{C} + {}^{12}\text{C} \rightarrow {}^{23}\text{Na} + \text{p} \\ ^{12}\text{C} + {}^{12}\text{C} \rightarrow {}^{23}\text{Mg} + \text{n} \\ ^{12}\text{C} + {}^{12}\text{C} \rightarrow {}^{24}\text{Mg} + \gamma \end{array}\right\} \quad (15.11)$$

occur. At high concentrations of ^{12}C, these reactions may take place in the form of an explosion (carbon flash). They may also lead to a supernova explosion, i.e. to splitting of the star into a great number of parts.

At still higher temperatures, fusion of ^{12}C and ^{16}O or of two oxygen nuclei leads to the synthesis of a variety of isotopes of Mg, Si and S, for example

$$\left.\begin{array}{l} ^{16}\text{O} + {}^{16}\text{O} \rightarrow {}^{32}\text{S} + \gamma \\ ^{16}\text{O} + {}^{16}\text{O} \rightarrow {}^{31}\text{P} + \text{p} \\ ^{16}\text{O} + {}^{16}\text{O} \rightarrow {}^{31}\text{S} + \text{n} \\ ^{16}\text{O} + {}^{16}\text{O} \rightarrow {}^{28}\text{Si} + {}^4\text{He} \end{array}\right\} \quad (15.12)$$

Fusion of ^{12}C and ^{16}O is less probable, because most of the ^{12}C nuclei are used up before the temperature necessary for fusion of two ^{16}O nuclei is reached. At the end of the carbon and oxygen burning stages the most abundant nuclei are ^{28}Si and ^{32}S.

Fusion of ^{28}Si and synthesis of heavier nuclei up to $A = 56$ begin at temperatures $\geq 10^9$ K. The nuclei produced by these thermonuclear reactions are in the region of $Z = 26$ (Fe) and $A = 56$, which is the range of highest stability (highest binding energy per nucleon; section 2.4). With these reactions, nucleosynthesis finds a temporary end. The abundances calculated on the basis of the above-mentioned nuclear reactions are in good agreement with the abundances found in the solar system.

With increasing temperature caused by gravitational contraction, more and more excited states of the nuclei are populated, the photon intensities increase and neutrons are liberated by (γ,n) reactions induced by high-energy photons. These neutrons trigger further nucleosynthesis by (n,γ) reactions. Three processes are distinguished:

(a) s(slow)-process: At low neutron flux densities, neutron absorption is slower than β^- decay, and unstable nuclides formed by (n,γ) reactions have enough time to change by β^- decay into stable nuclides. The whole process may last for more than 10^7 y. The abundance of the nuclides produced by the s-process can be calculated by use of the cross sections of the (n,γ) reactions. However, the formation of neutron-rich nuclides next to other β^--unstable nuclides and the nucleosynthesis of heavy nuclides such as ^{232}Th and ^{238}U cannot be explained by the s-process.
(b) r(rapid)-process: At high neutron flux densities, as in supernova explosions, neutron absorption becomes faster than β^- decay, and many successive (n,γ) reactions may occur before the nuclides undergo β^- decay (multineutron capture; section 14.3). Formation of ^{232}Th and ^{238}U can only be explained by the r-process. On the other hand, synthesis of heavier elements is limited by fission. If (n,f) reactions prevail over (n,γ) reactions, no more heavy nuclides are formed.
(c) p(proton)-process: Synthesis of proton-rich nuclides such as ^{78}Kr, ^{84}Sr, ^{92}Mo, ^{96}Ru, ^{102}Pd ^{106}Cd, ^{112}Sn and others cannot be explained by the s- or the r-process. It is assumed that these nuclides are formed by proton capture.

15.5 Evolution of Stars

Nucleogenesis and evolution of stars are strongly correlated. The evolution of the stars comprises different stages and depends mainly on their mass, as already mentioned. In all cases, stars of high density are formed at the end of the evolution.

In stars of small mass (<0.1 times the mass of the sun) the energy liberated by gravitational contraction is not sufficient to reach the temperature necessary to start thermonuclear reactions. These stars are directly entering the stage of black dwarfs (black holes).

Stars with masses like that of the sun change into red giants after hydrogen burning has come to an end and helium burning prevails. In this stage light atoms and molecules are able to escape the field of gravity, the stars are continuously losing matter and decreasing in volume. The next stage is that of a white dwarf, a star with a size similar to that of the earth, but with a density of the order of 10^6 to 10^8 g cm^{-3} and a temperature of about 10^7 K. Within a period of the order of about 10^9 y the white dwarf cools down and finally it becomes a black dwarf.

Stars containing more mass than the sun may explode under the influence of their gravitational contraction, generating a supernova. In this case, the contraction leads to an extremely high pressure in the core and bursts of neutrons may occur due to transformation of protons in the nuclei into free neutrons. Furthermore, appreciable amounts of matter are given off in supernova explosions. Two types of supernova explosions are distinguished: type I is expected to occur at relatively moderate mass (about 1.2 to 1.5 solar masses) and represents the final stage of an old star which fully disintegrates in a huge thermonuclear explosion within a few seconds. The temperatures in the various layers of the star range from about 10^9 to 10^{10} K. Type II will occur at relatively high mass (≥ 10 solar masses). The core of these stars consists of layers in which different kinds of thermonuclear reactions take place, hydrogen burning in the outer layers and oxygen burning in the inner layers. When nuclear fusion comes to an end because the matter has largely been transformed into nuclei of mass numbers $A \approx 56$, gravitational contraction of these stars leads to temperatures of about 10^9 K and densities of about 10^{14} g cm^{-3} in the core, giving rise to the birth of a neutron star in which heavier nuclides are disintegrated by high-energy photons, and β^- transmutation is reversed:

$$\text{p (in the nucleus)} + e^- \rightarrow n + \nu_e \tag{15.13}$$

Within a time of the order of 1 s a great amount of matter is converted into neutrons and the star collapses into an extraordinarily compact mass of neutrons with a density of the order of 10^{14} g cm^{-3}. Formation of neutron stars represents the reversal of nucleogenesis. Due to the explosion, the outer layers of the star are ejected into interstellar space.

Supernova explosions are rather rare events. The flash of the explosion is brighter than the sun by several orders of magnitude, fading away within a few days or weeks, while the cloud of dust ejected by the star in the form of a nebula expands continuously. The remnants of supernova are pulsating stars, also called pulsars. They are rapidly rotating, and emit electromagnetic radiation in the region of radiowaves at certain intervals of time, like a lighthouse. The final stage of a supernova is assumed to be a black hole, i.e. a region with an intense gravitational field that prevents escape of matter and radiation and makes the black hole invisible.

Formation and decay of stars are continuous processes in the cosmic time scale. Stars of the first generation are formed as a direct consequence of the big bang, whereas those of the second and of following generations are formed at a later stage by aggregation of matter from the debris of burned-out stars in the interstellar space.

With regard to the future development of the universe, the most important aspect is that of the mass of the neutrinos. According to the big bang theory, the number of neutrinos and photons remaining after the big bang should be similar, and the number of neutrinos should be about 10^8 times that of other particles. If the mass of the neutrino were $>10^{-8}$ u (corresponding to $>10\,\text{eV}$), the neutrinos would represent a dominant mass in the universe and appreciably contribute to gravitational attraction, which may eventually overcome the present expansion. This could result in a closed or possibly a pulsating universe.

15.6 Evolution of the Earth

Stages of the evolution of the earth are listed in Table 15.3. The evolution of the sun and the planets from solar nebula began about $4.6 \cdot 10^9$ y ago. Materials of this age are not found on the earth, because most primordial solids on the earth went through one or several metamorphoses. However, the material of meteorites which have been formed simultaneously with the earth make it possible to date the age of the earth at $4.5 \cdot 10^9$ y. The oldest minerals on the earth have an age of the order of $4.3 \cdot 10^9$ y and suffered metamorphoses about $3.8 \cdot 10^9$ y ago. The age of the oldest rock for-

Table 15.3. Stages of the evolution of the earth.

Time before present	Stage
$5 \cdot 10^9$ y	Solar nebula
$4.6 \cdot 10^9$ y	Formation of the solar system
$4.5 \cdot 10^9$ y	Formation of the earth, the moon and of meteorites
$4.3 \cdot 10^9$ y	First stages of the earth's crust, formation of the oldest minerals found on the earth, formation of hydrosphere and atmosphere
$3.9 \cdot 10^9$ y	End of major meteoritic impacts
$3.8 \cdot 10^9$ y	Beginning of formation of rocks
$(3.8–3.5) \cdot 10^9$ y	Formation of oldest rocks
$3.5 \cdot 10^9$ y	First traces of life (stromatolites)

mations is in the range $(3.8–3.5) \cdot 10^9$ y. The first indications of life are dated back to about $3.5 \cdot 10^9$ y ago.

Several proposals have been made about the origin of the matter from which the earth and the solar system have been formed. These proposals are mainly based on the isotopic composition. The supernova hypothesis explains the presence of heavy nuclei in the solar system by a supernova explosion some time before the evolution of the solar system. This hypothesis is supported by the isotopic analysis of meteorites which shows an anomaly in the ^{129}Xe content. This anomaly is attributed to the decay of ^{129}I ($t_{1/2} = 1.57 \cdot 10^7$ y) which must have been present during the evolution of the solar system.

In any case, the primordial radioactivity on the earth was appreciably higher than at present. The ratios of the activities at the time of the birth of the earth to those at present are listed in Table 15.4 for some long-lived radionuclides that represent the main radioactive inventory on the earth. The relatively high activity of ^{235}U about $2 \cdot 10^9$ y ago is the reason for the operation of the natural nuclear reactors at Oklo at that time (section 11.8).

Table 15.4. Ratio of the activities of some long-lived radionuclides at the time of the birth of the earth to those present.

Radionuclide	Activity ratio A/A_0
^{40}K	11.4
^{87}Rb	1.07
^{232}Th	1.02
^{235}U	84.1
^{238}U	2.01

15.7 Interstellar Matter and Cosmic Radiation

In interstellar space, matter is distributed very unevenly. As already mentioned in section 14.4, some stars are ejecting their matter in form of nebulas of dust and gas. These nebulas contain various elements (mainly H, C, O, Si and others) at temperatures between about 10^2 and 10^3 K. Far away from the stars, the density of interstellar matter is of the order of 0.1 atom per cm^3, mainly H and C. In some regions, however, matter is condensed in the form of big interstellar clouds, the mass of which may exceed the mass of the sun by a factor of 10^3 or more. Two types of interstellar clouds are distinguished: optically transparent, diffuse clouds containing $<10^3$ atoms per cm^3 (mainly H, but also some compounds such as CO or HCHO) at temperatures of the order of 100 K, and opaque, dense clouds containing 10^4 to 10^6 molecules per cm^3 (mainly H_2, but also a variety of compounds) at temperatures varying between about 10 and 10^3 K. Densities and temperatures increase from the outer parts to the core of the clouds.

Dense interstellar clouds have been investigated by microwave spectroscopy, and many compounds have been identified, comprising simple molecules such as H_2, H_2O, CO, NH_3, HCHO, HCN, SO_2, CH_3OH and C_2H_5OH, radicals such as $\cdot OH$ and $\cdot CN$, and more complex compounds containing many carbon atoms. The interstellar space can be looked upon as a big chemical laboratory, where chemical reactions take place at extremely low pressures and temperatures under the influence of electromagnetic and cosmic radiation, including light and charged particles. Radiation-induced reactions play a predominant role.

Altogether, the mass of the interstellar matter consists of $\approx 70\%$ H_2, $\approx 28\%$ He and $\approx 2\%$ other elements. The heavier elements such as C and S are assumed to be incorporated in the interstellar dust and the mass of this dust is estimated to amount to about 1% of the total interstellar matter or about 10^8 to 10^{10} tons. The dust grains may adsorb molecules from the gas phase and catalyse chemical reactions. The average residence time of dust grains in interstellar space is assumed to be of the order of 10^8 y. By aggregation of interstellar matter under the influence of gravitation, dust grains may be incorporated into a new star. The probability of the presence of biologically important compounds in the dust grains is also being discussed. However, knowledge about interstellar dust is very limited, because no samples are available.

The present state of knowledge about invisible dark matter indicates that it represents the predominant part (about 90%) of the total mass of the universe. It includes the so-called missing mass. Dark matter is not discernible by any kind of electromagnetic radiation in the region from γ rays to radiowaves, but its gravitational effects on other kinds of matter are observable. For example, the rotation of spiral galaxies such as the Milky Way can only be explained if 90% of the matter is invisible in the sense mentioned above. The question of the nature of the dark mass is still open. Various possibilities are discussed: matter different from that on the earth (no protons, neutrons and electrons), remnants of the big bang, neutrinos. However, the actual mass of neutrinos is still uncertain.

An important proportion of the matter in the interstellar space appears in the form of cosmic rays consisting of high-energy particles and photons. The intensity of cosmic radiation (about 10 particles per cm^2 per s) is relatively weak compared with the intensity of radioactive sources, but the energy of the particles in cosmic radiation is extremely high (from about 10^2 to about 10^{14} MeV). Potential sources of primary cosmic radiation are supernova explosions and pulsars. Furthermore, particles may be accelerated in the interstellar space by shock waves and cosmic magnetic fields. It is assumed that most of the particles are not able to escape from the galaxy in which they are produced.

Primary cosmic rays consist mainly of protons ($\approx 90\%$). Helium nuclei and electrons ($\approx 5\%$ each), ions heavier than ^{4}He ($\approx 1\%$) and γ-ray photons are less abundant. Most of the protons have energies of the order of 10^3 MeV.

The cosmic radiation incident on the earth is generated in our galaxy. It is effectively absorbed in the atmosphere, and the flux density is reduced from about $20\,\mathrm{cm}^{-2}\,\mathrm{s}^{-1}$ to about $1\,\mathrm{cm}^{-2}\,\mathrm{s}^{-1}$ at the surface of the earth. By interaction with the atoms and the molecules in the atmosphere showers of elementary particles are produced, making up the secondary cosmic radiation. Positrons, muons, several kinds of mesons and baryons were first detected in the secondary cosmic radiation. Furthermore, nuclear reactions induced by secondary cosmic radiation lead to the production of cosmogenic radionuclides, such as T and ^{14}C (section 1.2).

At the surface of the sun, high-temperature bursts are observed as particularly bright areas (flares) at certain intervals (≈ 11 y, corresponding to the intervals between sunspot activity). With these flares the sun ejects great amounts of protons, called solar wind. The flux of these protons may be several orders of magnitude higher than the flux of cosmic rays, but the energy of the solar protons is lower than in the cosmic rays (on average 10–100 keV, at maximum 10^2 MeV). These protons are absorbed in the atmosphere and do not reach the surface of the earth.

Part of the solar wind is trapped in the magnetic field of the earth, forming the outer van Allen radiation belt, whereas more energetic solar protons are accumulated in narrow zones above the magnetic field of the earth, producing the auroral displays.

Literature

General

O. Hahn, Applied Radiochemistry, Cornell University Press, Ithaca, NY, **1936**

M. Haissinsky, Nuclear Chemistry and its Applications, Addison–Wesley, Reading, MA, **1964**

E. A. Evans, M. Muramatsu, Radiotracer Techniques and Applications, Marcel Dekker, New York, **1977**

I. G. Draganic, Z. D. Draganic, J. P. Adloff, Radiation and Radioactivity on the Earth and Beyond, CRC Pres, Boca Raton, FL, **1990**

Geochemistry

H. Faul, Nuclear Geology, Wiley, New York, **1954**

T. P. Kohman, N. Saito, Radioactivity in Geology and Cosmology, Annu. Rev. Nucl. Sci. *4*, 401 **(1954)**

H. E. Suess, in: Abundances of the Elements in the Universe, Landolt-Börnstein, Physical Tables, New Series VI/1, Springer, Berlin, **1965**, p. 83

M. G. Rutten, The Origins of Life by Natural Causes, Elsevier, Amsterdam, **1971**

J. Hoefs, Stable Isotope Geochemistry, Springer, Berlin, **1973**

S. L. Miller, L. E. Orgel, The Origins of Life on the Earth, Prentice Hall, Englewood Cliffs, NJ, **1974**

V. Trimble, The Origin and Abundances of the Chemical Elements, Rev. Mod. Phys. *47*, 877 **(1975)**

B. F. Windley (Ed.), The Early History of Earth, Wiley, New York, **1976**

G. Faure, Principles of Isotope Geology, Wiley, New York, **1977**

J. Rydberg, G. Choppin, Elemental Evolution and Isotopic Composition, J. Chem. Ed. *54*, 742 **(1977)**

C. Ponnamperuma (Ed.), Chemical Evolution of the Early Earth, Academic Press, New York, **1977**

M. Ozima, The Earth, Its Birth and Growth, Cambridge University Press, Cambridge, **1981**

B. Mason, C. B. Moore, Principles of Geochemistry, Wiley, New York, **1982**

P. K. Kuroda, The Origin of the Chemical Elements and the Oklo Phenomenon, Springer, Berlin, **1982**

E. M. Durrance, Radioactivity in Geology, Principles and Applications, Ellis Horwood, Chichester, **1986**

P. A. Cox, The Elements, Oxford University Press, Oxford, **1989**

Cosmochemistry

G. Gamow, Expanding Universe and the Origin of the Elements, Phys. Rev. *70*, 572 **(1946)**

O. A. Schaeffer, Radiochemistry of Meteorites, Annu. Rev. Phys. Chem. *13*, 151 **(1962)**

H. Craig, S. L. Miller, G. J. Wassenburg (Eds.), Isotopic and Cosmic Chemistry, North-Holland, Amsterdam, **1963**

W. A. Fowler, F. Hoyle, Nucleosyntheis in Massive Stars and Supernovae, University of Chicago Press, Chicago, IL, **1965**

D. D. Clayton, Principles of Stellar Evolution and Nuclear Synthesis, McGraw-Hill, New York, **1968**

J. R. Arnold, H. E. Suess, Cosmochemistry, Annu. Rev. Phys. Chem. *20*, 293 **(1969)**

J. F. Fruchter, Chemistry of the Noon, Annu. Rev. Phys. Chem. *23*, 485 **(1972)**

O. Manolescu, Geochemistry and Cosmochemistry, Rev. Fiz. Chim. A **1974**, 226

B. Nagy, Carbonaceous Meteorites, Elsevier, Amsterdam, **1975**

J. E. Ross, L. H. Aller, The Chemical Composition of the Sun, Science *191*, 1223 **(1976)**

S. Weinberg, The First Three Minutes, Bantam, New York, **1977**

D. Schramen, R. V. Wagoner, Element Production in the Early Universe, Annu. Rev. Nucl. Sci. *27*, 1 **(1977)**

S. F. Dermott (Ed.), Origin of the Solar System, Wiley, New York, **1978**

C. Rolfs, H. P. Trautvetter, Experimental Nuclear Astrophysics, Annu. Rev. Nucl. Sci. *28*, 115 **(1978)**

R. N. Clayton, Isotopic Anomalies in the Early Solar System, Annu. Rev. Nucl. Sci. *28*, 501 **(1978)**

C. Sagan, Cosmos, Random House, New York, **1980**

R. A. Galant, Our Universe, Nat. Geographic Soc., Washington, DC, **1980**

A. W. Wolfendale (Ed.), Progress in Cosmology, Reidel, Dordrecht, **1982**

C. A. Barnes, D. D. Clayton, D. N. Schramm (Eds.), Essays in Nuclear Astrophysics, Cambridge University Press, Cambridge, **1982**

D. D. Clayton, Principles of Stellar Evolution and Nuclear Synthesis, University of Chicago Press, Chicago, **1983**

W. W. Dullet, D. A. Williams, Interstellar Chemistry, Academic Press, New York, **1984**

M. Taube, Evolution of Matter and Energy on a Cosmic and Planetary Scale, Springer, Berlin, **1985**

J. A. Wood, S. Chang (Eds.), The Cosmic History of the Biogenic Elements and Compounds, NASA Scientific and Technical Information Branch, Washington, DC, **1985**

V. E. Viola, Formation of the Chemical Elements and the Evolution of Our Universe, J. Chem. Educ. *67*, 723 **(1990)**

E. B. Norman, Stellar Alchemy: The Origin of the Chemical Elements, J. Chem. Educ. *71*, 813 **(1994)**

16 Dating by Nuclear Methods

16.1 General Aspects

The laws of radioactive decay are the basis of chronology by nuclear methods. From the variation of the number of atoms with time due to radioactive decay, time differences can be calculated rather exactly. This possibility was realized quite soon after the elucidation of the natural decay series of uranium and thorium. Rutherford was the first to stress the possibility of determining the age of uranium minerals from the amount of helium formed by radioactive decay. Dating by nuclear methods is applied with great success in many fields of science, but mainly in archaeology, geology and mineralogy, and various kinds of "chronometers" are available.

Two kinds of dating by nuclear methods can be distinguished:

- dating by measuring the radioactive decay of cosmogenic radionuclides, such as T or ^{14}C;
- dating by measuring the daughter nuclides formed by decay of primordial mother nuclides (various methods, e.g. K/Ar, Rb/Sr, U/Pb, Th/Pb, Pb/Pb).

All naturally occurring radionuclides can be used for dating. The time scale of applicability depends on the half-life. With respect to the accuracy of the results, it is most favourable if the age to be determined and the half-life $t_{1/2}$ of the radionuclide are of the same order. In general, the lower limit is about $0.1 \cdot t_{1/2}$ and the upper limit about $10 \cdot t_{1/2}$. Therefore the long-lived mother nuclides of the uranium, thorium and actinium decay series and other long-lived naturally occurring radionuclides such as ^{40}K and ^{87}Rb are most important for application in geology and mineralogy.

The attainment of radioactive equilibria in the decay chains depends on the longest-lived daughter nuclides in the series. These are ^{234}U $(t_{1/2} = 2.455 \cdot 10^5 \text{ y})$, ^{228}Ra $(t_{1/2} = 5.75 \text{ y})$ and ^{231}Pa $(t_{1/2} = 3.276 \cdot 10^4 \text{ y})$ in the uranium $(4n + 2)$, thorium $(4n)$ and actinium $(4n + 3)$ series, respectively. After about 10 half-lives of these radionuclides, equilibrium is practically established (sections 4.3 and 4.4) and dating on the basis of radioactive equilibrium is possible.

On the other hand, the concentration of stable decay products, such as ^{4}He, ^{206}Pb, ^{208}Pb, ^{207}Pb, ^{40}Ar (daughter of ^{40}K) and ^{87}Sr (daughter of ^{87}Rb), increases continuously with time. If one stable atom (subscript 2) is formed per radioactive decay of the mother nuclide (subscript 1), the number of stable radiogenic atoms is

$$N_2 = N_1^0 - N_1 = N_1^0(1 - e^{-\lambda t}) = N_1(e^{\lambda t} - 1) \tag{16.1}$$

(N_1^0 is the number of atoms of the mother nuclide at $t = 0$). For dating, N_2 and N_1 have to be determined. If several stable atoms are formed per radioactive decay of the mother nuclide, as in the case of ^{4}He formed by radioactive decay of ^{238}U, ^{232}Th, ^{235}U and their daughter nuclides, the number of stable radiogenic atoms is

$$N_2(\text{He}) = n(N_1^0 - N_1) = nN_1^0(1 - e^{-\lambda t}) = nN_1(e^{\lambda t} - 1) \qquad (16.2)$$

where n is the number of ^4He atoms produced in the decay series.

The following methods of dating by nuclear methods can be distinguished:

- Measurement of cosmogenic radionuclides
- Measurement of terrestrial mother/daughter nuclide pairs
- Measurement of members of the natural decay series
- Measurement of isotope ratios of stable radiogenic isotopes
- Measurement of radioactive disequilibria
- Measurement of fission tracks

The main problems with application of cosmogenic radionuclides are knowledge of the production rate during the time span of interest and the possibility of interferences (e.g. by nuclear explosions). For the other methods it is important whether the systems are closed or open, i.e. whether nuclides involved in the decay processes (mother nuclides, daughter nuclides or α particles in the case of measurement of He) are lost or entering the system during the time period of interest.

The methods will be discussed in more detail in the following sections. Non-nuclear methods of dating such as thermoluminescence and electron spin resonance (ESR) will not be dealt with.

16.2 Cosmogenic Radionuclides

Cosmogenic radionuclides applicable for dating are listed in Table 16.1. The radionuclides are produced at a certain rate by the interaction of cosmic rays with the components of the atmosphere, mainly in the stratosphere. If the intensity of cosmic rays (protons and neutrons) can be assumed to be constant, the production rate of the radionuclides listed in Table 16.1 is also constant. The cosmogenic radionuclides take part in the various natural cycles on the surface of the earth (water cycle, CO_2 cycle) and they are incorporated in various organic and inorganic products of these cycles, such as plants, sediments and glacial ice. If no exchange takes place, the activity of the radionuclides is a measure of the age.

The tritium atoms formed in the stratosphere are transformed into HTO and enter the water cycle as well as the various water reservoirs, such as surface waters, groundwaters and polar ice. This offers, in principle, the possibility of determining the age of samples of various kinds, for example groundwaters, polar ice or old wine. However, by the thermonuclear explosions carried out in the atmosphere, mainly in 1958 and in 1961/62, large quantities of T have been set free into the atmosphere causing an increase in the T:H ratio in the atmosphere from the natural value of several 10^{-18} up to about 1000 times this natural value at some places in the northern hemisphere. In the meantime, the concentration of T in the atmosphere has considerably decreased due to radioactive decay and transfer into the water reservoirs, but it is still higher than the natural value. This restricts reliable T dating appreciably. But the tritium method is successfully applied for dating of glacier and polar ice in layers in which the influence of nuclear explosions is negligible. Tritium may be

Table 16.1. Cosmogenic radionuclides applicable for dating.

Radio-nuclide	Production	Decay mode and half-life [y]	Production rate [atoms per m² per y]	Range of dating [y]	Application
³H (T)	¹⁴N(n, t) ¹²C	β^-, 12.323	$\approx 1.3 \cdot 10^{11}$	0.5–80	Water, ice
¹⁴C	¹⁴N(n, p) ¹⁴C	β^-, 5730	$\approx 7 \cdot 10^{11}$	$2.5 \cdot 10^2$–$4 \cdot 10^4$	Archaeology, climatology, geology (carbon, wood, tissue, bones, carbonates)
¹⁰Be	Interaction of p and n with ¹⁴N and ¹⁶O	β^-, $1.6 \cdot 10^6$	$\approx 1.3 \cdot 10^{10}$	$7 \cdot 10^4$–10^7	Sediments, glacial ice, meteorites
²⁶Al	Interaction of cosmic rays with ⁴⁰Ar	β^+, $7.16 \cdot 10^5$	$\approx 4.8 \cdot 10^7$	$5 \cdot 10^4$–$5 \cdot 10^6$	Sediments, meteorites
³²Si	Interaction of cosmic rays with ⁴⁰Ar	β^-, 172	$\approx 5.10^7$	10–10^3	Hydrology, ice
³⁶Cl	Interaction of cosmic rays with ⁴⁰Ar	β^-, $3.0 \cdot 10^5$	$(4.5$–$6.5) \cdot 10^8$	$3 \cdot 10^4$–$2 \cdot 10^6$	Hydrology, water, glacial ice
³⁹Ar	Interaction of cosmic rays with ⁴⁰Ar	β^-, 269	$\approx 4.2 \cdot 10^{11}$	10^2–10^4	–

measured either directly, preferably after preconcentration by multistage electrolytic decomposition of water samples, or by sampling the ³He formed by the decay of T.

Formation of ¹⁴C by the interaction of cosmic rays with the nitrogen in the atmosphere was proved in 1947 by Libby, who also demonstrated the applicability of ¹⁴C dating in the following years. The half-life of ¹⁴C is very favourable for dating of archaeological samples in the range of about 250 to 40 000 y. The ¹⁴C atoms are rather quickly oxidized in the atmosphere to CO_2, which is incorporated by the process of assimilation into plants and via the food chain into animals and man. With the death of living things the uptake of ¹⁴C ends, and its activity decreases with the half-life, provided that no exchange of carbon atoms with the environment takes place. ¹⁴C dating is based on the assumption that the ¹⁴C:¹²C ratio in living things is identical with that in the atmosphere and that this ratio has been constant in the atmosphere during the period of time considered. The second assumption, however, is not strictly correct. A periodic variation of the ¹⁴C:¹²C ratio with a period of about $9 \cdot 10^3$ y and an amplitude of about $\pm 5\%$ is correlated with the variation of the magnetic field of the earth causing changes in the intensity and composition of cosmic radiation and consequently in the production rate of ¹⁴C. Furthermore, the ¹⁴CO₂ concentration in the atmosphere is also changing with temperature and climate on the earth, because the exchange equilibria between the atmosphere and the various CO_2 reservoirs in the biosphere and the oceans vary with temperature. Of the total amount of exchangeable ^{14}C, $\approx 1.5\%$ is in the atmosphere, $\approx 2\%$ in the terrestrial biosphere, $\approx 3\%$ in humus, $\approx 4\%$ in the form of dissolved organic matter and the rest ($\approx 90\%$) in the form of carbonate and hydrogencarbonate in the oceans. Whereas the equilibration of ^{14}C between the atmosphere and the terrestrial components takes about 30 to 100 y, equilibration with the deep layers of the oceans takes up to about 2000 y. The determination of the $^{14}C:^{12}C$ ratio allows the calculation of the residence time of ^{14}C in the various reservoirs.

Drastic changes in the $^{14}C:^{12}C$ ratio have been caused by the activities of man since the beginning of the industrial age. On the one hand, by combustion of fossil fuels, such as carbon, oil, petrol and natural gas, large amounts of ^{14}C-free CO_2 are set free leading to a dilution of $^{14}CO_2$ in the atmosphere and a decrease of the $^{14}C:^{12}C$ ratio by about 2%. On the other hand, an opposite and much greater change in the $^{14}C:^{12}C$ ratio in the atmosphere was caused by nuclear explosions. The neutrons liberated by these explosions led to a sharp increase of the ^{14}C production in the upper layers of the atmosphere, and the $^{14}C:^{12}C$ ratio increased by a factor of about 2 in the northern hemisphere in 1962/63. With respect to these effects, the relatively small amounts of $^{14}CO_2$ liberated from nuclear power stations and reprocessing plants can be neglected. In the meantime, the $^{14}C:^{12}C$ ratio decreased to a value which is about 20% higher than the original value. Great amounts of the $^{14}CO_2$ produced by nuclear explosions have been transferred to the reservoirs, with the result that the $^{14}C:^{12}C$ ratio in the surface layers of the oceans has increased by about 20%. These changes, however, are without influence on dating by the ^{14}C method, because samples suitable for application of this method are more than 100 y old.

For the measurement of ^{14}C, proportional counters and liquid scintillation techniques are most suitable. The samples are oxidized in a stream of air or oxygen to CO_2, which is carefully purified, stored to allow decay of ^{222}Rn and introduced into a proportional counter. CO_2 may also be converted into CH_4 by reaction with hydrogen at 475 °C or into C_2H_2 by reaction with Li to Li_2C_2 at 600 °C followed by decomposition of Li_2C_2 by water to C_2H_2. By transformation into CH_4 or C_2H_2 the purification procedures necessary in the case of CO_2 are avoided. For measurement in a liquid scintillation counter, CO_2 is usually transformed into an organic compound, for instance by reaction to C_2H_2 as described above and catalytic trimerization of C_2H_2 to C_6H_6. The measuring times are usually rather long, of the order of many hours, in order to keep the statistical errors low. Various kinds of samples can be dated by the ^{14}C method (Table 16.1). Depending on the carbon content, samples weighing from about ten to several hundreds of grams are needed. In order to obtain reliable results, it is very important that the samples are free from impurities containing carbon of recent origin.

The ratio of the carbon isotopes $^{14}C:^{13}C:^{12}C$ in samples of recent origin is about $1:0.9\cdot10^{10}:0.8\cdot10^{12}$. This ratio cannot be measured by classical mass spectrometry (MS), because by this method ions of the same mass such as $^{14}C^+$, $^{14}N^+$, $^{13}CH^+$ and $^{12}CH_2^+$ are found at practically the same position. This restricts the measurement of isotope ratios in the case of ^{14}C to about 10^{-9}. However, during recent years new developments in mass spectrometry have made it possible to measure isotope ratios down to the order of about 10^{-15}. In accelerator mass spectrometry (AMS) the samples are bombarded with ions (e.g. Cs^+) from an ion source, negative ions such as $^{14}C^-$ and $^{12}C^-$ are formed and pass an accelerator (usually of the tandem type), and several separation steps in which ions of the same mass containing different elements are sorted. By application of AMS, ^{26}Al, ^{32}Si, ^{36}Cl, ^{41}Ca and ^{129}I have been identified in 1979/80 as cosmogenic radionuclides.

^{10}Be is found in concentrations of about $(3–7) \cdot 10^4$ atoms per gram of antarctic ice. Because of the longer half-life compared with that of ^{14}C, dating for a longer period of time (up to about 10^7 y) is possible. On the other hand, determination of ^{10}Be is more difficult, because its production rate is lower (Table 16.1).

The rate of cosmogenic production of ^{26}Al is still lower than that of ^{10}Be, and in fresh sediments the ratio $^{26}Al : ^{27}Al$ is of the order of 10^{-14}, whereas the ratio $^{10}Be : ^9Be$ is of the order of 10^{-8}. This makes the determination of ^{26}Al in terrestrial samples very difficult. On the other hand, the production rate of ^{26}Al in meteorites and samples from the surface of the moon is comparable with that of ^{10}Be, because in these samples low-energy protons from the sun contribute appreciably to the production of ^{26}Al. Measurement of the $^{26}Al : ^{10}Be$ ratio in extraterrestrial samples provides information about their history.

^{32}Si is produced in the atmosphere by the reaction $^{40}Ar(p,p2\alpha)^{32}Si$. It is oxidized to SiO_2 and stays in the atmosphere for some time in the form of fine particles, until it comes down with the precipitations. It is found in surface waters, glaciers, polar ice and marine sediments. Due to the low production rate of ^{32}Si (Table 16.1), the measurement of its radioactivity in water samples requires processing of large amounts of water or ice of the order of tons, from which hydrous $SiO_2 \cdot xH_2O$ has to be separated. ^{32}Si changes into stable ^{30}S by a sequence of two transformations,

$$^{32}Si \xrightarrow[\text{172 y}]{\beta^- (0.2\,\text{MeV})} {}^{32}P \xrightarrow[\text{14.26 d}]{\beta^- (1.7\,\text{MeV})} {}^{32}S$$

^{32}P is most suitable for activity measurements, because of the relatively high energy of the β^- particles. However, application of AMS is much more favourable for the determination of ^{32}Si, because smaller samples can be used, but elaborate separation techniques are also required. ^{32}Si is applicable for dating of groundwater, ocean water, glacier ice, polar ice and sediments. Its half-life of 172 y offers some advantages over other radionuclides with respect to dating of ages up to about 10^3 y.

^{36}Cl is produced in the atmosphere by (n,p) reaction with ^{36}Ar and (p,nα) reaction with ^{40}Ar. It also comes down with the precipitations. Appreciable amounts of ^{36}Cl have been formed in the atmosphere by the neutrons liberated by nuclear explosions. Consequently, the ^{36}Cl deposition by precipitations increased from about 20 to about 5000 atoms per m^2 per s in the years 1955 to 1962, and since then it has decreased slowly to the original value.

Because ^{36}Cl stays predominantly in the aqueous phase, it is mainly applied for hydrological studies, e.g. on the time of transport of water within deep layers, the rate of erosion processes and the age of deep groundwaters. In the case of groundwaters without access of cosmogenic ^{36}Cl, the production of ^{36}Cl by the reaction $^{35}Cl(n,\gamma)^{36}Cl$ induced by neutrons from spontaneous fission of uranium contained in granite has to be taken into account.

^{39}Ar is produced in the atmosphere by the nuclear reaction ^{40}Ar(n,2n) ^{39}Ar. Due to its properties as a noble gas, about 99% of the cosmogenic ^{39}Ar stays in the atmosphere and the applicability of this radionuclide for dating purposes is very limited.

The amounts of ^{41}Ca ($t_{1/2} \approx 10^5$ y) produced in the atmosphere are extremely small. Only at the surface of the earth some ^{41}Ca is formed by the reaction ^{40}Ca(n,γ) ^{41}Ca, but the ratio ^{41}Ca:^{40}Ca is only of the order of $0.8 \cdot 10^{-14}$. Because of this low concentration, measurement of ^{41}Ca is very difficult. It is of some interest for dating of bones in the interval between about $5 \cdot 10^4$ and 10^6 y.

16.3 Terrestrial Mother/Daughter Nuclide Pairs

Terrestrial mother/daughter nuclide pairs suitable for dating are listed in Table 16.2. Dating by means of these nuclide pairs requires evaluation of eq. (16.1). In doing this, it has to be taken into account that, in general, at time $t = 0$ stable nuclides identical with the radiogenic nuclides are already present. This leads to the equation

$$N_2 = N_2^0 + N_1(e^{\lambda t} - 1) \qquad (16.3)$$

where N_2 is the total number of atoms of the stable nuclide (2), N_2^0 is the number of atoms of this nuclide present at $t = 0$ and $N_1(e^{\lambda t} - 1)$ is the number of radiogenic atoms formed by decay of the mother nuclide (1).

Table 16.2. Terrestrial nuclide pairs applicable for dating.

Nuclide pair	Decay mode of the mother nuclide	Half-life of the mother nuclide [y]	Range of dating [y]	Application
^{40}K/^{40}Ar	β^- (89%) $\varepsilon + \beta^+$ (11%)	$1.28 \cdot 10^9$	10^3–10^{10}	Minerals
^{87}Rb/^{87}Sr	β^-	$4.8 \cdot 10^{10}$	$8 \cdot 10^6$–$3 \cdot 10^9$	Minerals, geochronology, geochemistry
^{147}Sm/^{143}Nd	α	$1.06 \cdot 10^{11}$	10^8–10^{10}	Minerals, geochronology, geochemistry
^{176}Lu/^{176}Hf	β^- (97%) ε (3%)	$3.8 \cdot 10^{10}$	10^7–10^9	Geochemistry
^{187}Re/^{187}Os	β^-	$5 \cdot 10^{10}$	10^6–10^{10}	Minerals

In practice, two approaches are used – independent determination of N_2 and N_1 or simultaneous determination of N_2 and N_1 by mass spectrometry (MS). The second approach is not applicable if the properties of the mother nuclide and the daughter nuclide are very different, e.g. in the case of dating by the ^{40}K/^{40}Ar method or by measuring ^{4}He formed by radioactive decay. Both methods require additional determination of the unknown number N_2^0, but in special cases N_2^0 can be neglected.

N_2 and N_1 can be determined independently by various analytical methods. Isotopic dilution (addition of a known amount of an isotope followed by MS) is often applied.

In the $^{40}K/^{40}Ar$ method, the samples are heated to melting in order to drive out all Ar and a measured quantity of ^{38}Ar (>99% enriched) is added (method of isotope dilution). After purification, the isotope ratios $^{40}Ar:^{38}Ar$ and $^{38}Ar:^{36}Ar$ are measured by means of MS. From the isotope ratios $^{40}Ar:^{38}Ar$ and $^{38}Ar:^{36}Ar$ in the added sample and in the air, the number of radiogenic ^{40}Ar atoms is calculated. The concentration of K in the sample is determined independently by the usual analytical methods. The main problem with the $^{40}K/^{40}Ar$ method is the possibility that some ^{40}Ar may have escaped or that additional amounts of ^{40}Ar may have entered the system in the course of time.

Simultaneous determination of N_2 and N_1 is conveniently performed by use of a stable non-radiogenic nuclide as reference nuclide (r) and measurement of the ratios N_2/N_r and N_1/N_r. Division of eq. (16.3) by N_r gives

$$\frac{N_2}{N_r} = \frac{N_2^0}{N_r} + \frac{N_1}{N_r}(e^{\lambda t} - 1) \tag{16.4}$$

The value of N_2^0/N_r can be assessed or found by iterative application of eq. (16.4). For constant values of t (the same age of samples), the plot of N_2/N_r as a function of N_1/N_r gives a straight line with slope $(e^{\lambda t} - 1)$ intersecting the ordinate at N_2/N_r. Such a plot is called an isochrone and is used for evaluation, in particular for identification of samples of the same age.

According to eq.(16.4) the age of the sample is

$$t = \frac{t_{1/2}}{\ln 2} \ln\left[1 + \frac{N_2/N_r - N_2^0/N_r}{N_1/N_r}\right] \tag{16.5}$$

where $t_{1/2}$ is the half-life of the radioactive mother nuclide.

Simultaneous determination of mother and daughter nuclide by MS is applied in the $^{87}Rb/^{87}Sr$ method. The stable Sr isotope ^{86}Sr serves as reference nuclide. $^{87}Rb/^{87}Sr$ dating has been used for many kinds of minerals, magmatic rocks and sedimentary rocks of various origins.

In the $^{147}Sm/^{143}Nd$ method, mother and daughter nuclides are also determined simultaneously by MS. ^{144}Nd serves as reference nuclide. The method has also found application in geochronology, mainly for dating of very old minerals.

Applications of the $^{176}Lu/^{176}Hf$ method and of the $^{187}Re/^{187}Os$ method have no advantages over the methods mentioned previously. For simultaneous determination of the mother and daughter nuclide by MS, suitable non-radiogenic nuclides have to be selected as reference nuclides. In the case of Hf, the use of ^{177}Hf is preferred for this purpose. However, the main drawbacks of the $^{176}Lu/^{176}Hf$ method are the low concentration of Lu in minerals (<1 mg/kg) and the difficulties in measuring Hf by MS due to the low ionization yield. In the $^{187}Re/^{187}Os$ method, ^{186}Os is used as reference nuclide. As in the case of dating by the $^{176}Lu/^{176}Hf$ method, the low concentration of Re in minerals (on the average ≈ 1 ng/kg) is a basic drawback with respect to broad application. The method has been used for dating of meteorites and

of minerals containing higher amounts of Re. For both systems independent determination of mother and daughter nuclides is also applied. Because of the low concentrations, elaborate instrumental techniques are required.

16.4 Natural Decay Series

The atomic ratios ^{206}Pb:^{238}U, ^{207}Pb:^{235}U and ^{208}Pb:^{232}Th due to radioactive decay
of the mother nuclides are plotted in Fig. 16.1 as a function of the age, provided that
no losses have occurred. Application of the natural decay series for dating is summarized in Table 16.3. The ^{238}U/^{206}Pb method and the ^{232}Th/^{208}Pb method offer the
possibility of determining the ages of many minerals in the range of 10^6 to 10^{10} y
with rather high precision. Taking into account the long-lived radionuclides, radioactive equilibrium is established after about 10^6 y in the case of the uranium and
actinium series and after about 10 y in the case of the thorium series. ^{235}U decays
faster than ^{238}U, and the ratio of the production rates of ^{207}Pb and ^{206}Pb decreases
appreciably with time. Therefore, variations in the ratio ^{207}Pb:^{206}Pb indicate geological processes. In contrast to ^{208}Pb, ^{207}Pb and ^{206}Pb, ^{204}Pb is not radiogenic. This
is the reason why Pb isotope ratios are usually related to ^{204}Pb as reference nuclide.

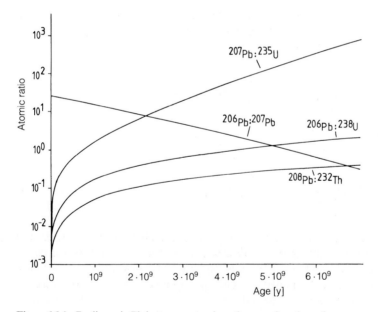

Figure 16.1. Radiogenic Pb isotopes: atomic ratios as a function of age.

Three kinds of systems can be distinguished:

(a) Systems losing parts of the members of the decay chains or the radiogenic Pb by
    diffusion or recrystallization processes (open systems). These losses may be pro-
    nounced if the half-lives of the intermediates are relatively long, as in the case of
    the uranium decay series $(t_{1/2}(^{226}\text{Ra}) = 1600$ y, $t_{1/2}(^{222}\text{Rn}) = 3.825$ d) or if the

Table 16.3. Natural decay series applicable for dating.

Decay series	Decay mode of the mother nuclide	Half-life of the mother nuclide [y]	Range of dating [y]	Application
$^{238}U \ldots ^{226}Ra \ldots ^{206}Pb$	α	$4.468 \cdot 10^9$	$10^6 - 10^{10}$	Minerals, geology, geochemistry
$^{235}U \ldots ^{207}Pb$	α (sf: $3.7 \cdot 10^{-7}\%$)	$7.038 \cdot 10^8$	$10^6 - 10^{10}$	Minerals, geology, geochemistry
$^{232}Th \ldots ^{208}Pb$	α	$1.405 \cdot 10^{10}$	$10^6 - 10^{10}$	Minerals, geology, geochemistry
$^{210}Pb \ldots ^{206}Pb$	β^-	22.3	$20 - 150$	Ice, exchange with the atmosphere

incorporation of Pb into the lattice of the mineral is hindered with the consequence of migration of Pb out of the crystals. An example is zircon, in which U^{4+} ions form solid solutions by substituting Zr^{4+}, whereas Pb^{2+} ions leave the lattice. Dating of those open systems encounters severe difficulties. Special correction methods have been proposed.

(b) Systems for which the loss of members of the decay chains can be neglected and in which the concentration of the mother nuclide can be taken as a measure of the age. For these systems eq. (16.4) can be applied in the forms

$$\left(\frac{^{206}Pb}{^{204}Pb}\right) = \left(\frac{^{206}Pb}{^{204}Pb}\right)_0 + \left(\frac{^{238}U}{^{204}Pb}\right)(e^{\lambda(238)t} - 1) \tag{16.6}$$

$$\left(\frac{^{207}Pb}{^{204}Pb}\right) = \left(\frac{^{207}Pb}{^{204}Pb}\right)_0 + \left(\frac{^{235}U}{^{204}Pb}\right)(e^{\lambda(235)t} - 1) \tag{16.7}$$

$$\left(\frac{^{208}Pb}{^{204}Pb}\right) = \left(\frac{^{208}Pb}{^{204}Pb}\right)_0 + \left(\frac{^{208}U}{^{204}Pb}\right)(e^{\lambda(232)t} - 1) \tag{16.8}$$

(c) Systems for which the loss of members of the decay chains can be neglected, but in which the concentration of the mother nuclide cannot be taken as a measure of the age, because of loss of uranium due to oxidation to UO_2^{2+} and dissolution. For these systems the Pb/Pb method of dating is applied; this will be discussed in more detail in section 16.5.

A practical application of eqs. (16.6), (16.7) and (16.8) is the calculation of the age of the solar system. MS analysis of meteorites containing negligible amounts of U gives the following values for the isotope ratios of the Pb isotopes $^{206}Pb : ^{204}Pb = 9.4$ and $^{207}Pb : ^{204}Pb = 10.3$. If these values are assumed to be the initial isotope ratios at the time of formation of the solar system, the age is obtained from the present isotope ratios of the Pb isotopes in the solar system and the ratios of the present abundances of U and Pb, for example by application of eq. (16.6):

$$t = \frac{1}{\lambda(238)} \ln\left[1 + \frac{(^{206}Pb/^{204}Pb) - (^{206}Pb/^{204}Pb)_0}{(^{238}U/^{204}Pb)}\right] \qquad (16.9)$$

Dating by means of ^{210}Pb is of special interest with respect to ages in the range between about 20 and 150 y, in particular for dating of glacier and polar ice, and climatology. The activity of ^{210}Pb in fresh snow is of the order of 10^{-2} Bq/kg, and the detection limit is of the order of 10^{-4} Bq/kg. The source of ^{210}Pb is ^{222}Rn emitted into the air in amounts of about $2 \cdot 10^3$ to $2 \cdot 10^4$ atoms per m^2 per s. Some ^{222}Rn is also emitted from volcanos. In the air, ^{210}Pb is sorbed on aerosols and found in the precipitations (rain or snow), wherein the average activity of ^{210}Pb is about 0.08 Bq/kg. The concentration of ^{210}Pb in the air and in the precipitations varies considerably within short periods of time and with the seasons, but the annual amount brought down with the precipitations is relatively constant. It has been found that the concentration of ^{210}Po, the daughter of ^{210}Pb, in the air is appreciably higher than expected for radioactive equilibrium with ^{210}Pb. This is explained by emission of relatively large amounts of ^{210}Po, together with sulfur, by volcanos.

Measurement of the β^- radiation of ^{210}Pb is difficult, because of its low energy, but the α activity of ^{210}Po is easily measurable by α spectrometry (detection limit $\approx 10^{-4}$ Bq) after attainment of radioactive equilibrium and chemical separation.

In the early stages of dating by nuclear methods, the measurement of ^4He formed by α decay in the natural decay series (9, 6 and 7 ^4He atoms in the uranium series, the thorium series and the actinium series, respectively) has been applied. The preferred method was the U/He method which allows dating of samples with very low concentrations of U of the order of 1 mg/kg. Helium produced by α decay is driven out by heating and measured by sensitive methods, e.g. by MS. However, it is difficult to ensure the prerequisites of dating by the U/He method: neither ^4He nor α-emitting members of the decay series must be lost and no ^4He atoms must be produced by other processes such as decay of ^{232}Th and spallation processes in meteorites.

16.5 Ratios of Stable Isotopes

Lead has four stable isotopes, ^{204}Pb, ^{206}Pb, ^{207}Pb and ^{208}Pb. The stability of ^{204}Pb has been debated, but the half-life is $>10^{17}$ y and, with respect to the age of the earth and the universe, ^{204}Pb can be considered to be stable. In the course of the genesis of the elements certain amounts of the above-mentioned four stable isotopes of Pb have been formed (primordial Pb). Additional amounts of ^{206}Pb, ^{207}Pb and ^{208}Pb (radiogenic Pb) have been produced by the decay of ^{238}U, ^{235}U and ^{232}Th, respectively.

On average, the relative percentages of the Pb isotopes in the earth's crust are: $^{204}Pb \approx 1.4\%$, $^{206}Pb \approx 24.1\%$, $^{207}Pb \approx 22.1\%$, $^{208}Pb \approx 52.4\%$. These ratios vary considerably with the concentrations of U and Th in the samples and with the age of the samples. For the dating of minerals, ^{204}Pb is taken as reference nuclide, and the isotope ratios $^{206}Pb:^{204}Pb$, $^{207}Pb:^{204}Pb$ and $^{208}Pb:^{204}Pb$ are determined by MS. If the contents of U or Th are known and losses can be neglected, eqs. (16.6), (16.7) and (16.8) can be applied.

However, measurement of the ratios of the Pb isotopes alone (Pb/Pb method) offers the possibilities of dating without knowledge of the contents of U and Th and of taking into account possible losses of U due to oxidation of U^{4+} to UO_2^{2+} and leaching of the latter, by simultaneous application of two or even three of the chronometers $^{238}U/^{206}Pb$, $^{235}U/^{207}Pb$, $^{232}Th/^{208}Pb$. The basis for the Pb/Pb method is given by eqs. (16.6), (16.7) and (16.8), knowledge of the ratio $^{235}U:^{238}U$ as a function of time and the fact that the ratio Th:U is practically constant for minerals of the same genesis. By this method, additional information about the history of the sample, in particular about losses of U can also be obtained.

The $^{39}Ar/^{40}Ar$ method is a variant on the $^{40}K/^{40}Ar$ method. For determination of the amount of K present in the sample, neutron activation is applied. The sample and a standard of known age (i.e. containing a known ratio $^{40}Ar:^{40}K$) are irradiated under the same conditions at a neutron flux density of about $10^{14}\,cm^{-2}\,s^{-1}$ for about 1 day. ^{39}Ar is produced by the nuclear reaction $^{39}K(n,p)\,^{39}Ar$ and the ratio $^{39}Ar:^{40}Ar$ is measured by MS. Because the half-life of ^{39}Ar is rather long ($t_{1/2} = 269\,y$), its decay after the end of irradiation can be neglected. From the relation

$$\left(\frac{N(39)}{N(40)}\right)_x = \left(\frac{N(39)}{N(40)}\right)_s \frac{e^{\lambda t_s} - 1}{e^{\lambda t_x} - 1} \tag{16.10}$$

the age of the sample can be calculated. $N(39)/N(40)$ are the measured isotope ratios $^{39}Ar:^{40}Ar$ of the sample (x) and the standard (s), λ is the decay constant of ^{39}Ar and t_s is the known age of the standard. Corrections for the presence of atmospheric ^{40}Ar, for production of ^{39}Ar by the reaction $^{42}Ca(n,\alpha)\,^{39}Ar$, for the production of ^{40}Ar by the reaction $^{40}K(n,p)\,^{40}Ar$ and for losses of ^{39}Ar may be necessary.

16.6 Radioactive Disequilibria

Mother and daughter nuclides of decay series are often separated by natural processes. Such separations are very common if the mother nuclide is dissolved in water (e.g. in the oceans), but they may also occur in solids. By measuring the decay of the separated daughter nuclide or the growth of the daughter nuclide in the phase containing the mother nuclide, the time can be determined at which the separation took place. This provides information about separation processes in minerals and ores and about formation of sediments in oceans or lakes. The prerequisite of application of radioactive disequilibria is that mother and daughter nuclide exhibit different chemical behaviour under the given conditions.

Radioactive disequilibria may be caused by different solubilities of mother and daughter nuclide, by different probabilities of escape or by different leaching rates due to recoil effects. Examples of different solubilities are U/Th and U/Pa. The probability of escape may be very high in radioactive equilibria with Rn. Recoil effects due to α decay lead to displacement and local lattice defects and may cause higher leaching rates. For instance, a change in the ratio $^{234}U:^{238}U$ caused by leaching is often observed.

An example of the application of radioactive disequilibria for dating is the ^{234}U/^{230}Th method. The chemical properties of U(VI) differ appreciably from those of Th(IV). Whereas UO_2^{2+} ions are found in natural waters, in particular in the oceans, in the form of $[UO_2(CO_3)_3]^{4-}$ ions, Th^{4+} ions are completely hydrolyscd and casily sorbed on particulates. With these they settle down in the sediments. On the other hand, corals and other inhabitants of the oceans form their skeletons by uptake of elements dissolved in the sea, e.g. U together with Ca, but no Th. In this way, ^{234}Th (daughter of ^{238}U) and the long-lived ^{230}Th are separated, and if the skeletons can be considered to be closed systems, the ingrowth of ^{230}Th is a measure of the age.

The ^{235}U/^{231}Pa method is based on the same principles. The chemical properties of U(VI) and Pa(V) differ even more than those of U(VI) and Th(IV).

Radioactive disequilibria are applied in geochemistry for dating of crystallization processes by measuring the ratio ^{238}U : ^{230}Th. Excess of ^{230}Th or ^{231}Pa found in marine sediments allows dating of these sediments and determination of the sedimentation rate. In archaeology, the ^{234}U/^{230}Th method is applied for dating of carbonates used by man or for dating of bones or teeth. U is taken up from natural waters during formation of the bones, and the content of ^{230}Th gives information about the age.

16.7 Fission Tracks

Fission tracks are observed in solids due to spontaneous or neutron-induced fission of heavy nuclei. The primary tracks can be made visible under an optical microscope by treatment with chemicals, by which track diameters of the order of 0.1 to 0.5 µm are obtained. The method is the same as that used with track detectors (section 7.12). The length of the fission tracks depends on the nature of the minerals and varies between about 10 and 20 µm. With respect to dating, the only important source of fission tracks is spontaneous fission of ^{238}U. Spontaneous fission of other naturally occurring heavy nuclides gives no measurable density of fission tracks; neutron-induced fission is, in general, negligible; and the tracks of recoiling atoms due to α decay are very short (of the order of 0.01 µm).

The track density (number of fission tracks per cm^2) in a mineral is a function of the concentration of U and the age of the mineral. For the purpose of dating, a sufficient number of tracks must be counted, which means that the concentration of U or the age (or both) should be relatively high. Usually, first the fission track density due to spontaneous fission of ^{238}U is counted, then the sample is irradiated at a thermal neutron flux density Φ_n in order to determine the concentration of U in the sample by counting the fission track density due to neutron-induced fission of ^{235}U. The age t of the mineral is calculated by the formula

$$t = \frac{1}{\lambda(238)} \ln\left[1 + 7.252 \cdot 10^{-3} \frac{D(\text{sf})}{D(\text{n,f})} \frac{\lambda(238)}{\lambda(\text{sf},238)} \sigma_{n,f} \Phi_n t_i\right] \tag{16.11}$$

where $\lambda(238)$ is the decay constant of ^{238}U, $7.252 \cdot 10^{-3}$ is the isotope ratio ^{235}U : ^{238}U, $D(\text{sf})$ and $D(\text{n,f})$ are the fission track densities due to spontaneous fission of ^{238}U and due to neutron-induced fission of ^{235}U, respectively, $\sigma_{n,f}$ is the cross section of fission of ^{235}U by thermal neutrons, and t_i is the irradiation time. In the case of homogeneous distribution of U in the sample, the values of $D(sf)$ and $D(n,f)$ can be determined in different aliquots of the sample. In the case of heterogeneous distribution of U, $D(sf)$ and the sum $D(sf) + D(n,f)$ must be counted in the same sample. $D(n,f)$ may also be determined by an external detector, e.g. a plastic foil, that is firmly pressed on a polished surface of the sample. After irradiation of the combination of sample plus detector, the fission tracks are counted in the detector and multiplied by the factor 2 to obtain $D(n,f)$. The factor 2 takes into account the different geometrical conditions, 2π for the detector and 4π for the sample.

Fission tracks are influenced by recrystallization processes in the solids (ageing), the rate of which increases appreciably with temperature. At higher temperatures, fission tracks may disappear quantitatively, but recrystallization processes may also be effective at normal temperatures, e.g. under the influence of pressure or of water. By comparing the ages calculated on the basis of fission track densities with those obtained by other methods, information can be obtained about the temperature to which the minerals have been exposed in the course of time.

Unusually high fission track densities are found in the vicinity of nuclear explosions and at the natural reactors at Oklo.

Literature

General

O. Hahn, Applied Radiochemistry, Cornell University Press, Ithaca, NY, **1936**
A. C. Wahl, N. A. Bonner, Radioactivity Applied to Chemistry, Wiley, New York, **1951**
H. Faul, Nuclear Geology, Wiley, New York, **1954**
L. T. Aldrich, and G. W. Wetherill, Geochronolgy by Radioactive Decay, Annu. Rev. Nucl. Sci. *8*, 257 **(1958)**
International Atomic Energy Agency, Symposium Radioactive Dating, IAEA, Vienna, **1963**
M. Haissinsky, Nuclear Chemistry and its Applications, Addison–Wesley, Reading, MA, **1964**
E. I. Hamilton, L. H. Ahrens, Applied Geochronology, Academic Press, New York, **1965**
E. Roth, Chimie Nucléaire Appliquée, Masson, Paris, **1968**
D. York, R. M. Farquhar, The Earth's Age and Geochronology, Pergamon, Oxford, **1972**
C. T. Harper, Geochronology, Dowden, Hutchinson and Ross, Stroudsburg, PA, **1973**
G. Faure, Principles of Isotope Geology, Wiley, New York, **1977**
E. Roth, B. Poty (Eds.), Méthodes de Datation par les Phénomènes Nucléaires Naturels. Applications, Masson, Paris, **1985**
E. M. Durrance, Radioactivity in Geology. Principles and Applications, Ellis Horwood, Chichester, **1986**
M. A. Geyh, H. Schleicher, Absolute Age Determination, Springer, Berlin, **1990**
H. R. von Gunten, Radioactivity: A Tool to Explore the Past, Radiochim. Acta *70/71*, 305 **(1995)**

More Special

W. F. Libby, Radiocarbon Dating, University of Chicago Press, Chigaco, IL, **1955**
A. G. Maddock, E. H. Willis, Atmospheric Activities and Dating Procedures, Adv. Inorg. Chem. Radiochem. *3*, 287 **(1961)**
I. Perlam, I. Asaro, H. V. Michel, Nuclear Applications in Art and Archaeology, Annu. Rev. Nucl. Sci. *22*, 383 **(1972)**
L. A. Currie, Nuclear and Chemical Dating Techniques. Interpreting the Environmental Record, ACS Symp. Series 176, Washington DC, **1982**
J. Aitken, Physics and Archaeology, 2nd ed., Clarendon Press, Oxford, **1984**

G. Furlan, P. Cassola Guida, C. Tuniz (Eds.), New Paths in the Use of Nuclear Techniques for Art and Archaeology, World Scientific, Singapore, **1986**

M. Ivanovich, R. S. Harmon, Uranium-Series Disequilibrium: Applications to Earth, Marine and Environmental Sciences, 2nd ed., Clarendon Press, Oxford, **1992**

Encyclopedia of Earth System Science, Academic Press, New York, **1992**

17 Radioanalysis

17.1 General Aspects

The low detection limits of radioactive substances are very attractive for use in analytical chemistry. In principle, a single radioactive atom can be detected provided that it is measured at the moment of its decay. In practice, however, a greater number of radioactive atoms is necessary to measure their radioactivity with a sufficiently low statistical error. The mass m of a radionuclide and its activity A are correlated by the half-life $t_{1/2}$:

$$m = A \frac{M}{\ln 2 \ N_{Av}} \ t_{1/2} \tag{17.1}$$

where M is the mass of the radionuclide in atomic mass units u and N_{Av} is Avogadro's number. On the assumption that 1 Bq can be measured with sufficient accuracy, the amounts of radionuclides listed in Table 17.1 can be determined quantitatively. From this table it is evident that short-lived radionuclides can be measured with extremely high sensitivity (extremely low detection limits). Because radioactivity is a property of atoms, radioanalytical methods are primarily applicable to the determination of elements.

Table 17.1. Detection limits of radionuclides (the amounts correspond to 1 Bq).

| $t_{1/2}$ | Detection limit | |
	Number of atoms N	mol
1 h	5 200	$8.64 \cdot 10^{-21}$
1 d	125 000	$2.08 \cdot 10^{-19}$
1 y	$4.55 \cdot 10^7$	$7.55 \cdot 10^{-17}$
10^5 y	$4.55 \cdot 10^{12}$	$7.55 \cdot 10^{-12}$
10^9 y	$4.55 \cdot 10^{16}$	$7.55 \cdot 10^{-8}$

The following applications of radionuclides in analytical chemistry can be distinguished:

- analysis on the basis of the inherent radioactivity of the elements to be determined;
- activation analysis, i.e. activity measurement after activation by nuclear reactions of the elements to be determined;
- analysis after addition of radionuclides as tracers (isotopic dilution and radiometric methods).

Inherent radioactivity is a property of elements containing radioisotopes, such as K, and of all radioelements, for example U, Ra, Th and others. If the daughter nuclides are also radioactive, they can be measured instead of the mother nuclides, provided that radioactive equilibrium is established.

Activation analysis may be applied in many variants. Neutron activation analysis (NAA) is the most widely used, but often charged particle activation or photon activation are more advantageous. If the energy of the projectiles can be varied, many variations are possible. The application of the manifold methods of activation depends on the availability of research reactors and accelerators. In addition, purely instrumental or radiochemical methods may be used. In instrumental activation analysis, the samples are measured after irradiation without chemical separation, whereas radiochemical activation analysis includes chemical separation.

If a microcomponent is to be determined in the presence of a macrocomponent, the conditions of irradiation are chosen in such a way that the microcomponent is highly activated, whereas the activation of the macrocomponent is as low as possible.

Addition of radioactive tracers for analytical purposes offers additional possibilities of radioanalysis. By isotopic dilution not only elements, but also compounds, can be determined quantitatively, provided that these compounds are available in labelled form. Radiometric methods comprise application of isotopic exchange, release of radionuclides and radiometric titration.

The various radioanalytical methods will be discussed in more detail in the following sections.

17.2 Analysis on the Basis of Inherent Radioactivity

The activity of naturally radioactive elements is a measure of their mass. Prerequisites of application of the correlation between mass and activity according to eq. (17.1) are that the isotopic composition of the element to be determined is constant and that interfering radioactive impurities are absent. If the daughter nuclides are also radioactive, radioactive equilibrium must be established or the daughter nuclides must be separated off quantitatively. Interference of radioactive impurities may be avoided by measuring the α or γ spectrum of the radionuclide considered.

Analytical determination on the basis of natural radioactivity is often used for K, U, Ra and Th. Potassium contains 0.0117% ^{40}K ($t_{1/2} = 1.28 \cdot 10^9$ y). The isotopic composition is practically constant and ^{40}K is easily measurable due to the emission of relatively high-energy β^-, β^+ and γ radiation. The natural radioactivity of 1 kg K is $3.13 \cdot 10^4$ Bq. On the assumption that 0.1 Bq can be measured with an acceptable error, the limit of quantitative determination is 3 mg. For measurement of the potassium concentration in salts, the arrangement shown in Fig. 17.1 is used. High counting efficiency is provided by a counter with a thin wall and a large volume, and by filling the salt into the space surrounding the counter. In this way, the concentration of K in salt mines can be determined directly. Higher concentrations of Rb interfere, because of the natural radioactivity of this element.

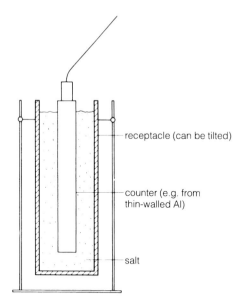

receptacle (can be tilted)

counter (e.g. from
thin-walled Al)

salt

Figure 17.1. Set-up for determination of K in
salts.

In natural U, the radionuclides of the uranium family and the actinium family are
present, and sometimes also radionuclides of the thorium family. Therefore, direct
determination of U in ores without chemical separation is difficult, especially since
the absorption of the radiation depends on the nature of the minerals. Generally, the
samples are dissolved and ^{234}Th is separated, e.g. by coprecipitation or by extraction
with thenoyltrifluoroacetone (TTA). Radioactive equilibrium between ^{234}Th and the
daughter nuclide ^{234m}Pa is rather quickly attained, and the high-energy β^- radiation
of the latter can easily be measured. A prerequisite of the determination of U by
measuring the activity of either ^{234}Th or ^{234m}Pa is the establishment of radioactive
equilibrium. This means that the uranium compound must not have been treated
chemically for about 8 months.

^{226}Ra can be determined with high sensitivity by measuring ^{222}Rn in radiochemical equilibrium with ^{226}Ra. Ra is loaded into a closed receptacle, and after
about 6 weeks Rn is transferred into an ionization chamber where its α activity is
measured as a function of time. By separating the radon, the influence of radioactive
impurities is excluded. In this way the very low ^{226}Ra content in human bones, of the
order of 10^{-12} g, can be determined.

In the case of ^{232}Th, the attainment of radioactive equilibrium with the daughter
nuclides is very slow, because of the long half-life of ^{228}Ra ($t_{1/2} = 5.75$ y). ^{232}Th can
be determined directly by measuring its α radiation, but the measurement of ^{212}Po is
more sensitive (about 10^{-6} g Th can be determined in this way in 1 g of rock material). Other methods are based on the separation and measurement of ^{228}Ra or
^{220}Rn. In all determinations of Th, the possibility of the presence of radioactive
impurities, mainly of members of the uranium and actinium families, has to be taken
into account.

17.3 Neutron Activation Analysis (NAA)

Activation analysis is based on the production of radionuclides by nuclear reactions. The specific activity is given by the equation of activation,

$$A_s = \sigma \Phi \frac{N_{Av}}{M} H \left(1 - \left(\frac{1}{2} \right)^{t/t_{1/2}} \right) \tag{17.2}$$

where σ is the cross section of the nuclear reaction in cm^2, Φ is the flux density of projectiles in $cm^{-2} s^{-1}$, N_{Av} is Avogadro's number, M is the atomic mass of the element (in atomic mass units u) and H is the relative abundance of the nuclide undergoing the nuclear reaction. Introducing $1\ b = 10^{-24}\ cm^2$ for σ and the value $N_{Av} = 6.022 \cdot 10^{23}$ gives

$$A_s = 0.602\ \sigma[b] \cdot \Phi \frac{H}{M} \left(1 - \left(\frac{1}{2} \right)^{t/t_{1/2}} \right) \tag{17.3}$$

From eqs. (15.2) and (15.3) it is evident that the limits of quantitative determination depend on the cross section σ of the nuclear reaction, the flux density Φ of the projectiles and the ratio $t/t_{1/2}$ (time of irradiation to half-life).

The following possible modes of neutron activation can be distinguished:

– activation by reactor neutrons (mainly (n,γ) reactions induced by thermal neutrons);
– activation by the neutrons from a spontaneously fissioning radionuclide (mainly (n,γ) reactions);
– activation by high-energy neutrons such as 14 MeV neutrons from a neutron generator (mainly (n,2n) reactions).

Reactor neutrons are most frequently used for activation analysis, because they are available in high flux densities. Moreover, for most elements the cross section of (n,γ) reactions is relatively high. On the assumption that an activity of 10 Bq allows quantitative determination, the lower limits of determination by (n,γ) reactions at a thermal neutron flux density of $10^{14}\ cm^{-2}\ s^{-1}$ are listed in Table 17.2 for a large number of elements and two irradiation times (1 h and 1 week). Detection limits of the order of 10^{-14} to $^{-10}$ g/g are, in general, not available by other analytical methods.

Li and B are not listed in Table 17.2, because they undergo (n,α) reactions with high yields. Activation of H, Be, C and N is negligible, and these elements are therefore also not listed in the table. Oxygen exhibits only little activation and is found in the last row. If these elements are present in the form of macrocomponents of a sample, for example in water or in biological samples, they are practically not activated and do not interfere. The light elements H, Li, Be, B, C, N and O can be determined by activation with charged particles or with photons. Charged particle or photon activation may also be more favourable for other elements listed in the lower rows of Table 17.2.

The spectrum of neutron energies in nuclear reactors is, in general, relatively broad. Furthermore, it varies with the type of reactor, and in the same reactor with

Table 17.2. Detection limits by neutron activation analysis at a thermal neutron flux density of $10^{14}\,\mathrm{cm^{-2}\,s^{-1}}$ on the assumption that 10 Bq allow quantitative determination.

Detectable in 1 g of sample [g]	Irradiation time 1 h	Irradiation time 1 week
10^{-14}–10^{-13}	Dy[c]	Eu[c], Dy[c]
10^{-13}–10^{-12}	Co, Rh[a][d], Ag[a][d], In[c], Eu[c], Ir	Mn, Co, Rh[a][c], Ag[a][d], In, Sm[c], Ho, Re[c], Ir, Au
10^{-12}–10^{-11}	V, Mn, Se[a], Br[c], I[c], Pr, Er[a], Yb[a], Hf[a], Th[c]	Na, Sc, V, Cu[d], Ga[c], As, Se[a], Br[c], Pd, Sb, I[c], Cs, La, Pr, Er[a], Tm[c], Yb[a], Lu, Hf[a], W, Hg, Th[c]
10^{-11}–10^{-10}	Mg, Al Cl[c], Ar, Cu[c], Ga[d], Nb, Cs, Sm, Ho, Lu, Re, Au, U	Mg, Al, Cl[c], Ar, K[c], Cr[c], Ni[c], Ge[c], Kr, Y[b], Nb, Ru, Gd[c], Tb[c], Tl[c], Os[c], U
10^{-10}–10^{-9}	F[a], Na, Ge[c], As, Kr, Rb[c], Sr, Mo, Ru, Pd, Sb, Te[c], Ba, La, Nd[c], Gd[d], W, Os, Hg, Tl[b]	F[a], P[b], Zn, Rb[c], Sr, Mo, Te[c], Ba, Ce, Nd, Pt, Tl[b]
10^{-9}–10^{-8}	Ne[a], Si[b], K, Sc, Ti, Ni, Y[b], Cd, Sn, Xe, Tb[c], Tm, Ta, Pt	Ne[a], Si[b], Ti, Cd, Sn, Xe, Bi[b]
10^{-8}–10^{-7}	P[b], Cr[c], Zn, Ce	S[b], Ca[b], Fe, Zr
10^{-7}–10^{-6}	S[b], Zr, Pb[b], Bi[b]	Pb[b]
10^{-6}–10^{-5}	O[a], Ca[b]	O[a]

[a] From these elements radionuclides with half-lives between 1 s and 1 min are obtained. Therefore, 100 Bq are assumed necessary for quantitative determination and the elements are relegated to the next group below.

[b] Only β radiation, no γ radiation.

[c][d] If the γ radiation is measured by means of a γ-ray spectrometer, the elements are to be transferred into the next group (c) or the next-but-one group (d) because of the relative abundance of the γ transitions.

the position. As the cross sections of nuclear reactions depend on the energy of the projectiles (excitation functions; section 8.4.), the activity obtained according to eqs. (17.2) or (17.3) varies also with the energy spectrum of the neutrons. Many elements exhibit a high resonance cross section in the range of epithermal neutrons. Therefore, it is necessary to use standards that are irradiated under the same conditions and at the same position as the samples. Some reactors have thermal columns in which only thermal neutrons are available, but at appreciably lower flux densities.

In general, the activity of the radionuclide produced by the nuclear reaction is measured after irradiation, either directly or after chemical separation. However, the prompt γ-ray photons emitted in (n,γ) reactions may also be counted on the site of their production. For that purpose, the samples must be irradiated outside the nuclear reactor or by another neutron source, and the γ-ray photons are recorded by means of a γ-ray spectrometer. The intensity I of the γ rays produced by the nuclear reaction is proportional to the rate of the nuclear reaction (number of transmutations per unit time):

$$I \propto \frac{\mathrm{d}N}{\mathrm{d}t} = \sigma \Phi N_{\mathrm{A}} \tag{17.4}$$

where N_{A} is the number of atoms of the radionuclide undergoing the nuclear reaction.

Activation by high-energy neutrons is of interest if the cross sections of (n,γ) reactions are too low or if the macrocomponents are too highly activated by thermal neutrons. For the purpose of activation analysis, special neutron generators have been developed. In these generators protons or deuterons with energies of the order of 0.1 to 1 MeV are used for the production of neutrons by nuclear reactions. The preferred reactions are

$$t(d, n)\alpha \tag{17.5}$$

and

$$^{9}\mathrm{Be}(d, n)\, ^{10}\mathrm{B} \tag{17.6}$$

The deuterons are accelerated in a van de Graaff generator and hit a tritium or a beryllium target. The neutron yields of these reactions are plotted in Figs. 17.2 and 17.3.

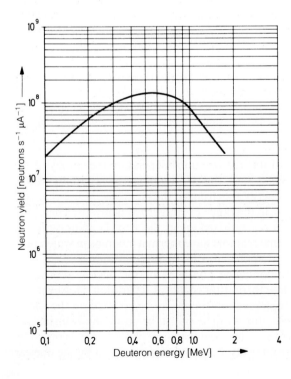

Figure 17.2. Neutron yield of the reaction $t(d, n)\alpha$ as a function of the deuteron energy (target: Ti containing 1 Ci $= 3.7 \cdot 10^{10}$ Bq T).

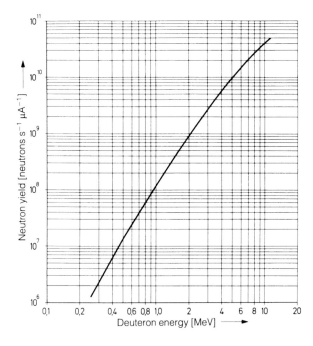

Figure 17.3. Neutron yield of the reaction ^{9}Be(d, n)^{10}B as a function of the deuteron energy (thick Be target).

For application of reaction (17.5), suitable tritium targets have been developed in which T is preferably bound in the form of hydrides such as titanium hydride deposited on copper. The targets must be well cooled to suppress escape of T due to heating by the incident deuterons. Neutron shielding is achieved by a block of paraffin, in which the energy of the neutrons is reduced, and by boron as a neutron absorber.

With tritium targets of 1 Ci ($3.7 \cdot 10^{10}$ Bq) and deuteron fluxes of about 50 mA, neutron yields up to about $5 \cdot 10^{12}$ s^{-1} are obtained. The flux density depends on the distance between the tritium target and the sample. The energy of the neutrons produced by reaction (17.4) is 14 MeV and allows activation by (n,p), (n,γ) or (n,2n) reactions with relatively high yields. Most cross sections of (n,2n) reactions are in the

Table 17.3. Examples of activation by 14 MeV neutrons.

Element determined	Main component of the sample	Nuclear reaction	Detection limit
O	Organic compounds	^{16}O(n,p)^{16}N	$\approx 10 \,\mu$g/g
Si	Oil	^{28}Si(n,p)^{28}Al	$\approx 10 \,\mu$g/g
Ti	Al	^{48}Ti(n,p)^{48}Sc	$\approx 100 \,\mu$g/g
Zn	–	^{68}Zn(n,p)^{68}Cu	$\approx 1 \,\mu$g/g
Al	Si	^{27}Al(n,α)^{24}Na	$\approx 1 \,\mu$g/g
Na	Organic polymers	^{23}Na(n,α)^{20}F	0.3%
K	–	^{41}K(n,α)^{38}Cl	6 μg/g
N	Organic compounds	^{14}N(n,2n)^{13}N	$\approx 100 \,\mu$g/g
F	Organic compounds	^{19}F(n,2n)^{18}F	$\approx 10 \,\mu$g/g
Pb	Petrol	^{208}Pb(n,2n)^{207m}Pb	20 μg/g

range of about 10 to 100 mb, which means that about 10^{-5} to 10^{-4} g of many elements can be determined quantitatively. Examples are given in Table 17.3.

Spontaneously fissioning radionuclides may be applied as neutron sources in those cases in which irradiation in a nuclear reactor is not possible, for example if manganese nodules at the bottom of the sea are to be analysed. For that purpose, ^{252}Cf is a suitable neutron source. It has a half-life of 2.645 y and decays in 96.9% by emission of α particles and in 3.1% by spontaneous fission. It may be installed together with a shielded γ-ray detector in the form of a mobile unit. The neutron production of ^{252}Cf is $2.34 \cdot 10^{12}$ s^{-1} g^{-1}. The neutron flux density is only of the order of 10^9 cm^{-2} s^{-1}, but this is sufficient for applications in which high sensitivity is not needed.

17.4 Activation by Charged Particles

Charged particles must have a minimum energy (threshold energy) to surmount the Coulomb barrier (section 8.3). In general, the excitation functions exhibit maxima in the range of about 0.1 to 1 b. For acceleration of charged particles, van de Graaff generators are often preferred, because the energy of the particles can be kept fairly constant, and because operation of the generator is less costly.

In contrast to neutrons, the penetration depth of charged particles is relatively small, with the result that only the surface layers of thicker samples are activated. Furthermore, the energy of charged particles decreases drastically with the penetration depth, and consequently the cross section varies with the penetration depth. On the other hand, these properties of charged particles offer the possibility of surface analysis.

For example, oxygen can be determined by the following reactions:

$$^{16}O(p,\alpha)\,^{13}N \tag{17.7}$$

$$^{16}O(^3He,p)\,^{18}F \tag{17.8}$$

$$^{16}O(\alpha,d)\,^{18}F \tag{17.9}$$

$$^{16}O(t,n)\,^{18}F \quad (t\text{ produced by }^6Li\,(n,\alpha)t) \tag{17.10}$$

In all cases, the determination of oxygen is limited by the range of the charged particles. For determination by reaction (17.10), a Li compound is mixed with the sample.

The elements Be, B, C, which are not activated by irradiation with thermal neutrons, can also be determined by charged particle activation. Limits of quantitative determination down to the order of about 10^{-9} g/g can be achieved. The same holds for F which gives a nuclide of short half-life (^{20}F, $t_{1/2} = 11$ s).

Heavy ions such as ^{7}Li, ^{10}B or others may also be used for charged particle activation, provided that a suitable heavy-ion accelerator is available. Examples of activation analysis by charged particles are given in Table 17.4.

Table 17.4. Examples of activation by charged particles.

Element determined	Main component of the sample	Nuclear reaction	Projectile energy [MeV]	Detection limit
B	Si	$^{11}B(p,n)^{11}C$	20	$\approx 0.01\,\mu g/g$
Fe	–	$^{56}Fe(p,n)^{56}Co$	12	$6\,\mu g/g$
Cu	–	$^{65}Cu(p,n)^{65}Zn$	12	$3\,\mu g/g$
As	Organic compounds	$^{75}As(p,n)^{75}Se$	12	$3\,\mu g/g$
Mo	–	$^{96}Mo(p,n)^{96}Tc$	12	$2\,\mu g/g$
Pb	–	$^{206}Pb(p,n)^{206}Bi$	12	$11\,\mu g/g$
C	Fe (steel)	$^{12}C(p,\gamma)^{13}N$	0.8	0.04%
F	Si (glass)	$^{19}F(p,\alpha)^{16}O$	1.4	–
B	Si, Ta	$^{10}B(d,n)^{11}C$	6–7	$\approx 0.1\,\mu g/g$
C	Steel	$^{12}C(d,n)^{13}N$	6.7	$\approx 0.1\,\mu g/g$
N	–	$^{14}N(d,n)^{15}O$	>3	$\approx 1\,\mu g/g$
O	–	$^{16}O(d,n)^{17}F$	>3	$\approx 0.01\,\mu g/g$
Si	Al	$^{30}Si(d,p)^{31}Si$	4	0.4%
Ga	Fe	$^{69}Ga(d,p)^{70}Ga$	6.4	$6\,\mu g/g$
Mg	Steel	$^{24}Mg(d,\alpha)^{22}Na$	–	–
S	–	$^{32}S(d,\alpha)^{30}P$	–	$\approx 0.1\,\mu g/g$
Be	–	$^{9}Be(t,p)^{11}Be$	3.5	$1\,\mu g/g$
B	–	$^{10}B(t,2n)^{11}C$	3.5	$0.1\,\mu g/g$
N	–	$^{14}N(t,2n)^{15}O$	3.5	$0.1\,\mu g/g$
O	–	$^{16}O(t,n)^{18}F$	3.5	$0.001\,\mu g/g$
O	Metal surfaces	$^{16}O(t,n)^{18}F$	3	$5\,ng/cm^2$
Mg	–	$^{26}Mg(t,n)^{28}Al$	3.5	$0.02\,\mu g/g$
Si	–	$^{28}Si(t,n)^{30}P$	3.5	$0.01\,\mu g/g$
Fe	Nb, Ta, W	$^{56}Fe(^3He,pn)^{57}Co$	14	$\approx 0.1\,\mu g/g$
Mo	W	$^{95}Mo(^3He.n)^{97}Ru$	14	$\approx 0.1\,\mu g/g$
B	–	$^{10}B(\alpha,n)^{13}N$	>6	$\approx 100\,\mu g/g$
C	–	$^{12}C(\alpha,n)^{15}O$	>10	–
F	–	$^{19}F(\alpha,n)^{22}Na$	>6	–
Al	–	$^{27}Al(\alpha,n)^{30}P$	–	–
O	–	$^{16}O(\alpha,d)^{18}F$	40	$<10\,\mu g/g$
O	–	$^{16}O(\alpha,pn)^{18}F$	40	$<10\,\mu g/g$
Fe	–	$^{56}Fe(\alpha,pn)^{58}Co$	15	$10^{-12}\,g$
C	–	$^{12}C(\alpha,\alpha n)^{11}C$	>10	–
^1H	–	$^1H(^7Li,n)^7Be$	78	$0.1\,\mu g/g$
^1H	–	$^1H(^{10}B,\alpha)^7Be$	60	$0.5\,\mu g/g$
^2H	–	$^2H(^7Li,p)^8Li$	78	$0.1\,\mu g/g$
^2H	–	$^2H(^{11}B,p)^{12}B$	70	$0.1\,\mu g/g$

17.5 Activation by Photons

Photons may induce quite a number of nuclear reactions (section 8.3), and photo-excitation (γ,γ') may also be applied to activation analysis. In general, the photons are obtained in the form of bremsstrahlung: high-energy electrons produced in an electron accelerator hit a target of high atomic number such as tungsten. The maximum energy of the photons is given by the energy of the incident electrons. All kinds of electron accelerators may be applied – van de Graaff generators, betatrons or linear accelerators. Simple electron accelerators of high power such as the "microtrons" are the most suitable.

The most important photon-induced nuclear reactions are (γ,n) and $(\gamma,2n)$ reactions, but (γ,p) reactions may also be applied. The number of possible reactions increases with the energy of the photons. Photons with energies in the range between about 15 to 40 MeV are preferred, and detection limits between about 1 ng and 1 µg are obtained. Photofission of heavy nuclei is achieved with photons of relatively low energy (about 5 to 10 MeV). It leads also to detection limits of about 1 ng to 1 µg.

For photoexcitation (γ,γ') photons with energies between about 1 and 15 MeV are used. The detection limits are in the range of about 0.1 to 10 µg.

Because high-energy photons exhibit only little absorption, reliable results are obtained with compact samples. Activation by photons is favourable if the sample contains macrocomponents with elements of high neutron absorption cross sections, such as Li, B, Cd, In or rare-earth elements. Furthermore, photon activation is applied for determination of elements that cannot be determined by (n,γ) reactions, such as Be, C, N, O, F. These elements can be determined in concentrations down to about 10^{-7} g/g. Other elements such as Si, Zr and Pb can also be determined by (γ,n) reactions. Examples are listed in Table 17.5.

If the concentrations of C, N or O are small (<10 µg/g), chemical separation after irradiation is recommended. For example, by combustion of the sample, ^{11}C and ^{13}N are transformed into $^{11}CO_2$ and $^{13}N_2$, respectively, and sorbed on a molecular sieve for measurement of their activity.

Table 17.5. Examples of activation by γ rays.

Element determined	Main components of the sample	Nuclear reaction	γ energy [MeV]	Detection limit
C	Na, Al, Si, Mo, W	$^{12}C(\gamma,n)^{11}C$	35	0.01–0.1 µg/g
N	Na, Si	$^{14}N(\gamma,n)^{13}N$	35	0.1–1 µg/g
O	Na, Al, Si, Fe, Cu, Nb, Mo, W	$^{16}O(\gamma,n)^{15}O$	35	0.1–1 µg/g
F	Al, Cu, Organic polymers	$^{19}F(\gamma,n)^{18}F$	35	0.01–0.1 µg/g
Cl	Organic polymers	$^{35}Cl(\gamma,n)^{34}Cl$	18	$\approx 0.1\%$
Cu	–	$^{65}Cu(\gamma,n)^{64}Cu$	35	≈ 1 µg/g
As	–	$^{75}As(\gamma,n)^{74}As$	35	≈ 1 µg/g
Cd	–	$^{116}Cd(\gamma,n)^{115}Cd$	35	≈ 1 µg/g
Hg	–	$^{198}Hg(\gamma,n)^{197m}Hg$	35	≈ 1 µg/g
Pb	–	$^{204}Pb(\gamma,n)^{203}Pb$	35	≈ 1 µg/g

17.6 Special Features of Activation Analysis

Due to the high sensitivity, activation analysis is one of the most important methods for determination of microcomponents, in particular trace elements, in materials of high purity (e.g. in semiconductors), in water, in biological samples and in minerals. The main fields of application are:

– geo- and cosmochemistry (terrestrial and lunar samples, meteorites);
– art and archaeology (identification of the origin by the trace element pattern in very small samples);
– environmental samples (atmospheric aerosols, fly ash, water);
– biological samples (blood, organs, body fluids, hair, food).

Another special feature of activation analysis is the fact that, in contrast to other methods, impurities introduced after irradiation in the course of chemical operations by reagents do not affect the results, because these impurities are not activated.

Activation analysis is a blank-free technique. In general, blanks not only determine the limits of detection, but at low concentrations they cause the main problems with respect to accuracy, because the small amounts to be determined have to be conveyed through all the steps of the chemical procedures, from sampling to detection, without introducing systematic errors. These problems are not encountered in activation analysis, because contamination by other radionuclides can, in general, be excluded and losses of the radionuclides to be determined can easily be detected by activity measurements.

For these reasons, activation analysis is preferably applied for certification and calibration purposes in trace element analysis. On the other hand, activation analysis is seldom used as a routine method, because handling of radioactive samples and disposal of the radioctive waste require special precautions.

A great advantage of activation analysis is the possibility of determining a large number of trace elements (up to about 30) simultaneously by γ spectrometry. The γ spectra are preferably recorded by means of Ge(Li) or high-purity Ge detectors in combination with a multichannel analyser (section 7.6). By use of these semiconductor detectors high resolution is obtained, whereas the counting efficiency is relatively low. On the other hand, scintillation detectors such as NaI(Tl) detectors exhibit a relatively high counting efficiency, but low resolution. Radionuclides emitting only β radiation must be measured individually, mostly after preceding chemical separation.

An important aim of activation analysis is high activation of the trace elements to be determined and low activation of the main components. In this respect, the ratios of the cross sections of the elements to be determined (σ_x) and of the main components ($\Sigma\sigma_m$) and the half-lives are important. The higher the ratio $\sigma_x/\Sigma\sigma_m$, the more favourable is the application of activation analysis. Furthermore, if the half-lives of the activation products of the main components are shorter than those of the radionuclides to be measured, the activity of the latter is determined some appropriate time after the end of irradiation. If they are relatively long, the time of irradiation is chosen in such a way that activation of the elements to be determined is high, whereas activation of the main components is low.

For example, activation of the elements H, Be, C, N, O, F, Mg, Al, Si, Ti by thermal neutrons is negligible or low, because the products of (n,γ) reactions are stable (^{2}H, ^{13}C, ^{15}N, ^{17}O, ^{18}O, ^{25}Mg, ^{26}Mg, ^{29}Si, ^{30}Si, ^{49}Ti, ^{50}Ti), short-lived (^{20}F, ^{27}Mg, ^{28}Al, ^{31}Si, ^{51}Ti) or very long-lived (^{10}Be). Samples containing these elements as main components are very well suited to the application of neutron activation analysis (NAA).

In cases in which activation by thermal neutrons causes relatively high activity of the main components, the following measures can be taken:

(a) Variation of the time of irradiation (t_i) and the time of decay after irradiation (t_d). Optimal conditions can be calculated by means of the equation

$$\frac{A_x}{A_m} = \frac{\sigma_x}{\sigma_m} \frac{N_A(x)}{N_A(m)} \frac{(1 - e^{-\lambda_x t_i})e^{-\lambda_x t_d}}{(1 - e^{-\lambda_m t_i})e^{-\lambda_m t_d}} \tag{17.11}$$

where A are the activities, σ the cross sections of the (n,γ) reactions, N the number of atoms and λ the decay constants (subscript x denotes the element to be determined, and subscript m the main component).

(b) Shielding of thermal neutrons with the aim of activating only with high-energy neutrons, for example by wrapping the sample in a cadmium foil.

(c) Choice of other projectiles for activation, for instance activation by 14 MeV neutrons, by charged particles or by γ-ray photons.

Selection of the time of irradiation and the time of measuring is important. If possible, the time of irradiation should correspond to one or several half-lives of the radionuclide to be measured. Long-term irradiation (days or weeks) and short-term irradiation (seconds or minutes) are distinguished. For short-term irradiation a fast transport system is needed, for example a pneumatic dispatch system.

In general, activation analysis relies on the use of standards that are irradiated under the same conditions and in the same position, and are also measured under the same conditions. Monoelement standards contain a known amount of one element. If they are applied to the evaluation of other elements the ratio of the cross sections σ_x/σ_s under the special conditions of irradiation and the ratio H_x/H_s of the relative abundances of the decay processes that are measured must be known (subscript x is for the sample and subscript s for the standard). Knowledge of the ratio σ_x/σ_s may cause problems, because the cross sections may vary drastically with the energy of the projectiles, for instance in the energy range of epithermal neutrons. These problems are not encountered with multielement standards that contain all the elements to be determined. However, the preparation of such multielement standards may be time-consuming.

As already mentioned, two kinds of activation analysis are distinguished:

– direct activity measurement of the samples after activation (instrumental activation analysis, in particular instrumental neutron activation analysis, INAA) and
– chemical separation after irradiation followed by activity measurement of the separated fraction (radiochemical activation analysis, in particular radiochemical neutron activation analysis, RNAA).

Chemical separation is unavoidable if the activities of the individual radionuclides cannot be measured by instrumental methods, as in the case of β radiation. In many cases, α and γ radiation can be measured by means of α and γ spectrometry. However, if the resolution of the spectra is insufficient, chemical separation is also necessary. Isotopic dilution (section 17.7) is often applied, because quantitative separation can be avoided by this method.

17.7 Isotope Dilution Analysis

The principle of isotope dilution analysis (IDA) is illustrated in Fig. 17.4. The sample contains an unknown number N_x of atoms or molecules, and it may also contain an unknown number $^*N_x$ of labelled atoms or molecules of the same kind. Known numbers N_1 and $^*N_1$ are added. These atoms or molecules (subscript 1) must not be identical with the atoms or molecules x, but they must exhibit the same behaviour under the given conditions. After mixing to obtain homogeneous distribution, any fraction is taken and the numbers N_2 and $^*N_2$ are determined in this fraction.

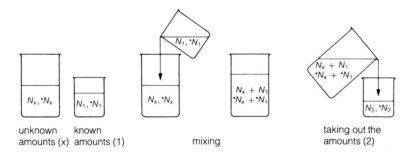

Figure 17.4. Principle of isotope dilution.

Due to homogeneous distribution by mixing, the following relation is valid:

$$\frac{^*N_2}{N_2} = \frac{^*N_x +^* N_1}{N_x + N_1} \tag{17.12}$$

Introducing the specific activities A_s and assuming that $^*N_i \ll N_i$ gives

$$N_x = N_1 \frac{A_s(1) - A_s(2)}{A_s(2) - A_s(x)} \tag{17.13}$$

In the case of determination of inactive atoms or compounds ($^*N_x = 0$), the following relation is valid

$$N_x = N_1 \frac{A_s(1) - A_s(2)}{A_s(2)} \tag{17.14}$$

This equation can also be applied in the form

$$m_x = m_1 \frac{A_s(1) - A_s(2)}{A_s(2)} \tag{17.15}$$

According to eqs. (17.13) to (17.15), the change of specific activity is a measure of the unknown number N_x.

It should be mentioned that the same kind of equations hold if stable isotopes are applied for labelling of the elements or compounds to be determined by isotope dilution analysis. In this application, isotope ratios are taking the place of specific activities and they are preferably measured by mass spectrometry.

The main and unique advantage of this method is the fact that quantitative separation of the element or compound to be determined is not necessary. It is substituted by measuring any fraction. This advantage is most clearly illustrated by an example from biology: if the blood volume of an animal is to be determined, it is obvious that the animal will not survive the extraction of the whole amount of blood to be measured. By application of isotope dilution analysis, a small measured volume V_1 of a solution is injected that contains a measured activity of a radionuclide of low radiotoxicity. Mixing is effected by blood circulation. After some minutes, a small volume of blood is taken and the activity is measured. In analogy to eq. (17.15), the unknown volume is

$$V_x = V_1 \frac{A_1(1) - A_2(2)}{A_s(2)} \tag{17.16}$$

Isotope dilution analysis is a very valuable method for determination of trace elements in all kinds of samples. The radioactive tracers or labelled compounds are added at the beginning of the analysis, and provided that they exhibit the same behaviour as the elements or compounds to be determined, losses in the course of chemical separation procedures are without influence on the results. Furthermore, time-consuming quantitative separation procedures can be substituted by simpler, more qualitative methods.

An important application of isotope dilution in radiochemistry is the determination of a radionuclide by dilution with an inactive nuclide (inactive compound), also called reverse isotope dilution. This application is very valuable if the radionuclide is present in carrier-free form. Again, quantitative separation is avoided; a measured amount m_1 of an inactive isotope of the element to be determined is added and after a non-quantitative separation the amount m_2 is measured. The ratio m_2/m_1 is the yield of the separation procedure and the activity A_x of the carrier-free radionuclide ($N_x = 0$) is obtained from the measured activity A_2:

$$A_x = A_2 \frac{m_1}{m_2} \tag{17.17}$$

Combination of isotope dilution with the principle of substoichiometric analysis offers the possibility of avoiding determination either of the chemical yield of the separation procedure or of the specific activity in the isolated fraction. Two identical aliquots of the radiotracer solution are taken, both containing the tracer with mass

m_0 and activity A_0. One aliquot is added to the solution to be analysed and the other is left as such. Then the same amount m of the substance x to be determined is isolated from both solutions. To ensure this, the concentration of the reagent used for the isolation is adjusted so that it is less than would correspond to quantitative stoichiometric reaction in either solution (substoichiometric principle). The specific activity in the tracer solution is $A_s(0) = A_0/m_0$ and that in the solution to which the tracer solution has been added is $A_s(2) = A_0/(m_0 + m_x)$. The specific activities are not changed by the chemical procedures; in the fractions isolated from the solutions they are $A_s(0) = A_0/m_0 = A_1/m$ and $A_s(2) = A_0/(m_0 + m_x) = A_2/m$. The resulting relation is

$$m_x = m_0 \left(\frac{A_1}{A_2} - 1 \right) \qquad (17.18)$$

where A_1 and A_2 are the activities measured in the isolated fractions.

Eq. (17.18) is similar to eq. (17.15), but the advantage is that the measurement of the relative activities A_1/A_2 is sufficient and the determination of masses is avoided.

Many metals in amounts ranging from micro- to nanograms have been determined by isotope dilution in combination with the substoichiometric principle. Isolation of equal amounts m is usually achieved by solvent extraction of a metal chelate into an organic solvent. It must be certain that the substoichiometric amount of the chelating reagent reacts quantitatively with the metal ions in the concentration range considered.

17.8 Radiometric Methods

Radiometric analysis is also based on the use of radiotracers. However, in contrast to isotope dilution analysis, stoichiometric relations are applied in radiometric methods. The substance to be determined is brought into contact with another substance labelled with a radionuclide or containing a radionuclide. Reaction between these two substances yields a radioactive product that either can be separated and measured or can be measured continuously in the course of the reaction. The activity is proportional to the amount of substance to be determined.

The following applications are distinguished:

– radioreagent methods;
– radiorelease methods;
– isotope exchange methods;
– radiometric titration.

In radioreagent methods, the radioactive product of the reaction between the substance to be determined and a radioactive reagent is separated by various methods, such as precipitation or liquid–liquid extraction. For example, Cl^-, Br^- or I^- in concentrations down to 0.5 µg/l can be determined by addition of an excess of phenylmercury nitrate labelled with ^{203}Hg. The complexes formed with the halide ions are extracted into benzene, whereas the phenylmercury nitrate stays in the aqueous phase. From the difference between the activities in the aqueous phase before and after the reaction, the amount of halide ions is calculated. Traces of Hg^{2+} in water can be determined by shaking with a solution of silver dithizonate in CCl_4 labelled with ^{110m}Ag. Due to the displacement of Ag by Hg, ^{110m}Ag is transferred to the aqueous phase, where it can be measured.

Radiorelease methods are based on the same principle: the substance to be determined is brought into contact with another substance containing a radionuclide reagent, and by their interaction a certain amount of the radionuclide is released and measured. For this method substances loaded with ^{85}Kr (radioactive kryptonates), for example krypton clathrates, may be applied. By reaction with oxygen ^{85}Kr is released and can be measured continuously. Oxygen dissolved in water can be measured by reaction with ^{204}Tl deposited on Cu; ^{204}Tl is oxidized and released into the water where it can be measured. Other oxidizing substances in water such as dichromate can be determined in the same way. Further examples are determination of SO_2 by reaction with IO_3^- labelled with ^{131}I, and determination of active hydrogen in organic substances by reaction with $LiAlH_4$ labelled with T.

The isotope exchange method is based on the exchange between two different forms or compounds of the element M to be determined:

$$MX + M^*Y \rightleftharpoons M^*X + MY \qquad (17.19)$$
$$(1) \quad (2) \qquad (1) \quad (2)$$

The labelled species M^*Y is added and, after equilibration, the specific activity is the same in (1) and (2):

$$\frac{A_1}{m_1} = \frac{A_2}{m_2} \qquad (17.20)$$

After separation of compounds (1) and (2), their activity is measured and the mass m_1 can be calculated from eq. (17.20). Homogeneous as well as heterogeneous exchange reactions may be applied in analytical chemistry. An example is the determination of small amounts of Bi: Bi is selectively extracted by diethyldithiocarbamate in chloroform, a known amount of BiI_4^- labelled with ^{210}Bi is added, after about 30 s BiI_4^- is extracted into water, and the activities in both phases are measured. In the case of heterogeneous exchange reactions, separation of the components is simple.

An advantage of isotope exchange methods is that in special cases individual chemical forms (species) can be determined with high sensitivity. The systems applied have to be selected by the following criteria: knowledge of the species between which exchange occurs, relatively rapid attainment of the exchange equilibrium, and exclusion of side reactions.

In radiometric titration, the radioactivity of one component or in one phase is recorded as a function of addition of titrant. The compound formed is separated by precipitation, extraction or ion exchange in the course of the titration, and the endpoint is determined from the change in the activity in the residual solution. Radiometric titration may be applied in different ways: inactive test solution and active titrant (activity in the solution is low at the beginning and begins to rise at the endpoint); active test solution and inactive titrant (activity in the solution decreases continuously, until the endpoint is reached); both the test solution and the titrant active (activity in the solution decreases until the endpoint is reached and then increases again). Many applications of radiometric titration based on precipitation or complex formation have been described.

17.9 Other Analytical Applications of Radiotracers

Radiotracer techniques have proved to be indispensable with respect to the examination of the individual steps of an analytical procedure, in particular with the aim of revealing the sources of systematic errors. Actually, tracer techniques have contributed essentially to the development of the present state of trace element analysis.

The most significant sources of error in trace element analysis are contamination or losses by adsorption or volatilization. The key property of radiotracers with respect to the investigation of the accuracy of analytical techniques is the emission of easily detectable radiation in any stage of an analytical procedure with extraordinarily high sensitivity.

Mechanisms and yields of analytical procedures such as precipitation or coprecipitation that are essential for their application can be elucidated. Furthermore, general analytical data can be obtained by application of tracer techniques, for example distribution coefficients, stability constants and solubilities.

17.10 Absorption and Scattering of Radiation

Backscattering of β radiation can be taken as the basis for surface analysis. It is due to electron–electron interaction, which is nearly independent of the atomic number Z of the material, and to scattering by atomic nuclei, which increases with Z. Both effects overlap, and the saturation value of backscattering increases approximately with $\sqrt{Z}$. Because the backscattered radiation originates from the layers near the surface, surface analysis is possible. An example is the determination of heavy elements in a solid or liquid matrix of light elements by use of the β^- radiation of ^{90}Sr.

Backscattering of γ rays and X rays depends on the mass per unit area and the effective average atomic number Z. The saturation value of backscattering decreases approximately with this number. For example, the composition of ores can be determined by this method. Elastic scattering of γ radiation ((γ,γ') process) can also be applied for analytical purposes. High selectivity is obtained by resonance absorption, i.e. by application of a radionuclide that decays into a stable ground state of the element to be determined. The γ rays emitted by the (γ,γ') process are measured.

Absorption or moderation of neutrons is used for detection of elements exhibiting high neutron absorption cross sections, such as B or Cd. For this purpose mostly neutron sources, for example ^{252}Cf, or neutron generators are applied (section 17.3). With neutron generators, B can be determined in steel in concentrations down to about 0.001%.

Neutrons give off energy by collision with protons (moderation). This can be applied for determination of hydrogen in samples by measuring the thermal neutron flux density either by means of a detector of by activation of a suitable material such as a gold foil. Examples of application of this technique are determination of the humidity of soil, coke, coal, iron ores or food, and determination of the H : C ratio in organic liquids and in oil. By investigation of boreholes with a neutron source combined with a detector for thermal neutrons, water and oil can be localized.

Boreholes can also be investigated by measuring the prompt γ radiation emitted by (n,γ) reactions. Examples are the investigation of the C : O or the Ca : Si ratio in order to find layers containing oil or coal. Measurement of the backstattering of γ rays allows localization of different layers and gives information about their composition, their density and their ore content.

17.11 Radionuclides as Radiation Sources in X-ray Fluorescence Analysis (XFA)

The application of radionuclides as radiation sources in X-ray fluorescence analysis is illustrated in Fig. 17.5. The X rays or γ rays emitted by a radionuclide are absorbed in the sample and the X rays emitted by the sample are measured by means of a semiconductor in combination with a multichannel analyser. Quantitative evaluation of the spectra is possible by use of suitable standards. In comparison with excitation by means of X-ray tubes, the main advantages of radionuclides are

  monoenergetic radiation,
– the possibility of measuring the K rays of heavy elements by excitation with γ-ray emitters,
– no need for a high-voltage installation.

The detection limits obtained by excitation with radionuclides are of the order of several µg/g. Appreciably higher intensities and lower detection limits are, in general, achieved by X-ray tubes. Monoenergetic radiation may also be obtained with X-ray tubes in combination with a secondary target, and the intensity of the secondary radiation is of the same order as that obtained with radionuclide sources.

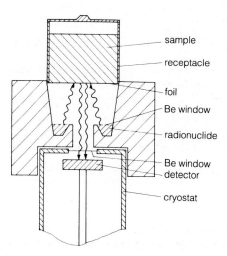

sample

receptacle

foil

Be window

radionuclide

Be window
detector

cryostat

Figure 17.5. Set-up for X-ray fluorescence analysis with radionuclide excitation.

Radionuclides suitable for use as X-ray sources for XFA are listed in Table 17.6. ^{109}Cd is applied most frequently. The use of γ emitters such as ^{57}Co, ^{133}Ba or ^{192}Ir offers the possibility of exciting heavy elements, for example Au, Pb, Th and U, to emit K radiation with high yield. Furthermore, heavy elements can also be measured in thick samples or within tubes or vessels, due to the high energy of their K radiation. These are distinct advantages in comparison with X-ray tubes.

The independence from high-voltage supply and the small amounts of technical equipment required allow construction of mobile units for multielement analysis of mineralogical or geological samples in the field.

Table 17.6. Radionuclides suitable as excitation sources for X-ray fluorescence analysis.

Radionuclide	Half-life	Decay mode	Energy of the emission lines used [keV]
^{55}Fe	2.73 y	ε	5.9 (Mn K)
^{238}Pu	87.74 y	α	12–17 (U L)
^{109}Cd	462.6 d	ε	22.1 (Ag K)
^{125}I	59.41 d	ε	27.4 (Te K); 35.4 (γ)
^{210}Pb	22.3 y	β^-	46.5 (γ)
^{241}Am	432.2 y	α	59.6 (γ)
^{170}Tm	128.6 d	β^-	84.4 (γ)
^{153}Gd	239.47 d	ε	103.2 (γ); 97.4 (γ); 69.7 (γ)
^{57}Co	271.79 d	ε	136 (γ); 122 (γ)

Literature

General

O. Hahn, Applied Radiochemistry, Cornell University Press, Ithaca, NY, **1936**.

A. C. Wahl, N. A. Bonner, Radioactivity Applied to Chemistry, Wiley, New York, **1951**.

M. Haissinsky, Nuclear Chemistry and its Applications, Addison-Wesley, Reading, MA, **1964**.

J. F. Duncan, G. B. Cook, Isotopes in Chemistry, Clarendon Press, Oxford, **1968**.

G. R. Gilmore, Radiochemical Methods of Analysis, in: Radiochemistry, Vol. 1, Specialist Periodical Reports, The Chemical Society, London, **1972**.

J. Tölgyessy, S. Varga, V. Krivan (Vols. I, II), M. Kyrs (Vol. II), J. Krtil (Vol. III), Nuclear Analytical Chemistry, Vols. I, II, III, University Park Press, Baltimore **1971**, **1972**, **1974**.

G. R. Gilmore, G. W. A. Newton, Radioanalytical Chemistry, in: Radiochemistry, Vol. 2, Specialist Periodical Reports, The Chemical Society, London, **1975**.

D. I. Comber, Radiochemical Methods in Analysis, Plenum Press, London, **1975**.

Treatise on Analytical Chemistry, Part I Theory and Practice, 2nd ed., Vol. 14 (Eds. Ph. J. Elving, V. Krivan, I. M. Kolthoff), Section K, Nuclear Activation and Radioisotopic Methods of Analysis, Wiley, New York, **1986**.

W. D. Ehmann, D. E. Vance, Radiochemistry and Nuclear Methods of Analysis, Wiley, New York, **1991**.

More Special

T. Braun, J. Tölgyessy, Radiometric Titrations, Pergamon, Oxford, **1967**.

J. Ruzicka, I. Stary, Substoichiometry in Radiochemical Analysis, Pergamon, Oxford, **1968**.

J. Hoste, J. Op de Beeck, R. Gijbels, F. Adams, P. van den Winkel, D. de Soete, Activation Analysis, Butterworths, London, **1971**.

D. de Soete, R. Gijbels, J. Hoste, Neutron Activation Analysis, Wiley, New York, **1972**.

J. Tölgyessy, T. Braun, M. Kyrs, Isotope Dilution Analysis, Pergamon, Oxford, **1972**.

J. M. Lenihan, S. J. Thomson (Eds.) Advances in Activation Analysis, Academic Press, New York, **1969**, **1972**.

J. A. Cooper, Comparison of Particle and Photon Excited X-Ray Fluorescence Applied to Trace Element Measurements on Environmental Samples, Nucl. Instr. Methods *106*, 525 **(1973)**.

G. Harbottle, Activation Analysis in Archaeology, in: Radiochemistry, Vol. 3, Specialist Periodical Reports, The Chemical Society, London, **1976**.

Proc. Int. Conf. Modern Trends in Activation Analysis, Munich, 1976, J. Radioanal. Chem. *37*, **(1977)**; Toronto, 1981, J. Radioanal. Chem. *69* and *70* **(1982)**.

G. Erdtmann, W. Soyka, The Gamma Rays of the Radionuclides (Ed. K. H. Lieser), Verlag Chemie, Weinheim, **1979**.

N. Pilz, P. Hoffmann, K. H. Lieser, In-Line Determination of Heavy Elements by Gamma-Ray Induced Energy-Dispersive K-Line XRF, J. Radioanal. Chem. *130*, 141 **(1989)**.

S. J. Parry, Activation Spectrometry in Chemical Analysis, Wiley, New York, **1991**.

C. Yonezawa, Prompt γ-Ray Analysis of Elements Using Cold and Thermal Reactor Neutrons, Anal. Sci. *9*, 185 **(1993)**.

18 Radiotracers in Chemistry

18.1 General Aspects

Application of radiotracers in chemistry and in other fields of science is based on two features:

– the high sensitivity of detection of radionuclides, and
– the possibility of labelling of elements or chemical compounds.

The high sensitivity of detection of radionuclides has already been emphasized in section 17.1 with respect to their application in analytical chemistry.

Labelling of elements or compounds by radionuclides allows investigation of the fate of these elements or compounds in the course of a chemical reaction or a transport process. In this way, radiotracer techniques offer unique possibilities for the investigation of reaction mechanisms in homogeneous and heterogeneous systems. These possibilities are illustrated by some typical examples of application in the following sections.

The prerequisite of the application of radiotracer methods is that the tracers show the same behaviour as the atoms or compounds to be investigated. This is, in general, fulfilled, if the chemical forms are identical. However, isotope effects have to be taken into account, if the relative mass differences of the radiotracer and the nuclide to be studied are marked, as in the case of substitution of H by T.

Application of radiotracers is the most important group of indicator methods; the other group based on the same principles comprises labelling of elements or compounds by stable isotopes.

18.2 Chemical Equilibria and Chemical Bonding

The extraordinarily high sensitivity of radiochemical methods makes it possible to measure solubility equilibria of sparingly soluble compounds or distribution equilibria in the range of very low concentrations. This is illustrated for the silver halides AgCl, AgBr and AgI in Fig. 18.1, in which the concentration of Ag in solution is plotted as a function of the concentration of the corresponding halides NaCl, NaBr and NaI. The fall in the curves is due to the solubility product, and the rise is due to complex formation. The minima of the curves, which cannot be found by electrochemical methods, are given by the solubility of the undissociated species.

An example of the investigation of distribution equilibria in the range of very low concentrations is shown in Fig. 18.2. Due to the high sensitivity of radiotracer techniques, distribution equilibria can be measured over wide ranges. Generally, application of radiotracer techniques is favourable in all cases in which very low concentrations are to be measured or in which the distribution coefficients are very high ($K_d > 10^6$).

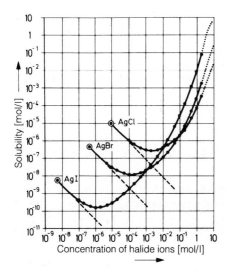

Figure 18.1. Solubility equilibria of silver halides in water ($\odot$) and in sodium halide solutions. (According to K. H. Lieser, Z. Anorg. Allg. Chem., **229**, 97 (1957).)

Various applications of radiotracers aimed at the elucidation of chemical bonding have also been reported. One example is the bonding of lead in Pb_2O_3. To answer the question whether this compound is lead(II) metaplumbate(IV) or whether the lead atoms are equivalent, Pb_2O_3 was precipitated by mixing of solutions of Pb^{2+} labelled with ^{210}Pb and of PbO_3^{2-}; then the precipitate was decomposed by a solution of KOH into Pb^{2+} and PbO_3^{2-}. The activity of ^{210}Pb was only found in the fraction of Pb^{2+}, proving that the Pb atoms in Pb_2O_3 exist in different oxidation states.

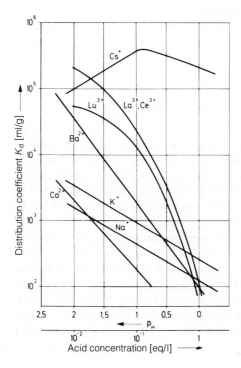

Figure 18.2. Distribution equilibria of various cations on an inorganic ion exchanger. (According to J. Bastian, K. H. Lieser, J. Inorg. Nucl. Chem. **29**, 827 (1967).)

Another example is the bonding of the sulfur atoms in $S_2O_3^{2-}$. This was prepared by heating a solution of SO_3^{2-} with sulfur labelled with ^{35}S, filtered and decomposed by addition of acid. ^{35}S was only found in the sulfur fraction, which is a proof that the sulfur atoms in $S_2O_3^{2-}$ differ in bonding. In both examples, the results are unambiguous, because a mutual change of the oxidation states in Pb_2O_3 and in $S_2O_3^{2-}$ by transfer of oxygen atoms does not occur.

18.3 Reaction Mechanisms in Homogeneous Systems

Radiotracer methods are the methods of choice for investigation of reaction mechanisms. They have also found broad application for determination of kinetic data of chemical reactions such as reaction rates, activation energies and entropies. Only a few examples can be given of the multitude of applications in organic, inorganic and physical chemistry and in biochemistry.

For instance, the Claisen allyl rearrangement has been elucidated by labelling with ^{14}C:

$$O-CH_2-CH={}^{14}CH_2 \qquad OH$$

$$\qquad \longrightarrow \qquad {}^{14}CH_2-CH=CH_2 \qquad (18.1)$$

By investigation of the decomposition products it has been proved that the terminal C atom of the allyl group forms a new bond with the benzene ring.

Measurement of isotope exchange reactions

$$\begin{array}{ccccc} AX + {}^*AY & \rightleftharpoons & {}^*AX + AY & (18.2) \\ (1) \quad (2) & & (1) \quad (2) & \end{array}$$

offers the unique advantage that no chemical change occurs in the system. Only the isotopes A and *A are exchanged between the chemical species AX and AY. Apart from isotope effects, the reaction enthalpy is $\Delta H = 0$. The driving force of the reaction is the reaction entropy ΔS, and the Gibbs free energy of the reaction is $\Delta G \approx - T\Delta S$. The reaction is measurable as long as A and *A are not equally distributed among the two species AX and AY.

If one of the two species AX and AY or both contain several exchangeable atoms, it is important to know whether these atoms are chemically equivalent or not. For example, in the system $AlCl_3/CCl_4$, the Cl atoms in each of the species are equivalent and a simple exchange reaction is observed, characterized by one rate constant. In the system HCl/1-nitro-2,4-dichlorobenzene, however, the Cl atoms in the organic compound are not equivalent; they exchange with different rates. The exchange reaction is more complex and characterized by the overlap of two rate constants. If more than two species that take part in the exchange reaction are present in the system, a corresponding number of exchange reactions has to be taken into account: three exchange reactions in the case of three species and $\frac{1}{2}n(n-1)$ reactions in the case of n species.

The rate of an exchange reaction according to eq. (18.2) is

$$\frac{d^*c_1}{dt} = R(s_2 - s_1) \tag{18.3}$$

where $^*c_1$ is the concentration of the labelled isotopes of the exchanging atoms in the form of species (1), and s_1 and s_2 are the fractions of the labelled isotopes in the species (1) and (2): $s_1 = {}^*c_1/c_1$, $s_2 = {}^*c_2/c_2$; c_1 and c_2 are the concentrations of the exchanging element in the form of the species (1) and (2). R is the reaction rate of the reaction; for instance

$$\left.\begin{array}{l} R = k_1 c_1 \quad \text{(first-order reaction) or} \\ R = k_2 c_1 c_2 \quad \text{(second-order reaction)} \end{array}\right\} \tag{18.4}$$

Eq. (18.3) indicates that the reaction can be measured until the specific activity is equal in both species, as already mentioned above.

By integration of eq. (18.3) the McKay equation is obtained:

$$\ln(1 - \lambda) = -R \frac{c_1 + c_2}{c_1\, c_2} t \tag{18.5}$$

In this equation, λ is the degree of exchange:

$$\begin{aligned} \lambda &= \frac{{}^*c_1 - {}^*c_1(0)}{{}^*c_1(\infty) - {}^*c_1(0)} = \frac{{}^*c_2 - {}^*c_2(0)}{{}^*c_2(\infty) - {}^*c_2(0)} \\ &= \frac{{}^*s_1 - {}^*s_1(0)}{s_1(\infty) - s_1(0)} = \frac{{}^*s_2 - {}^*s_2(0)}{s_2(\infty) - s_2(0)} \end{aligned} \tag{18.6}$$

The indices (0) and (∞) stand for initial and equilibrium values, respectively. Usually, $\ln(1 - \lambda)$ is plotted as a function of time, as illustrated in Fig. 18.3. From the value of $t_{1/2}$ the reaction rates can be calculated, for instance

$$\begin{aligned} k_1 &= \frac{\ln 2}{t_{1/2}} \cdot \frac{c_2}{c_1 + c_2} \quad \text{(first-order reaction)} \\ k_2 &= \frac{\ln 2}{t_{1/2}(c_1 + c_2)} \quad \text{(second-order reaction)} \end{aligned} \tag{18.7}$$

Isotope exchange reactions between organic halides RX and halide ions X^- have been investigated in great detail and valuable information about the mechanism of these reactions has been obtained. The concept of first- and second-order nucleophilic substitution (S_N1 and S_N2 reactions) is based on these investigations. A special example is the reaction

$$\underset{R_3}{\overset{R_1}{\underset{|}{R_2\!-\!C\!-\!X}}} + {}^*X^- \; \rightleftharpoons \; \underset{R_2}{\overset{R_1}{\underset{|}{R_2\!-\!C\!-\!{}^*X}}} + X^- \tag{18.8}$$

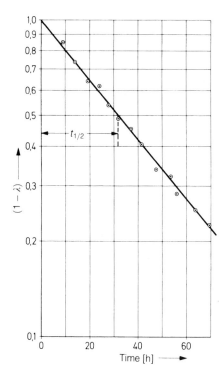

Figure 18.3. Evaluation of an isotope exchange experiment.

If the groups R_1, R_2 and R_3 are different, C is a asymmetric carbon atom. It has been found that the rate of formation of the optical antipode is identical with the exchange rate, which is the proof of the mechanism of second-order nucleophilic substitution (S_N2 reaction).

Radiotracer methods are widely applied in biochemistry to elucidate the mechanisms of reactions in complex systems. Like in other organic reactions, radiotracers make it possible to follow the pathway of atoms or molecules in the course of any kind of chemical transmutation.

In inorganic chemistry valuable information about chemical bonding in complexes has been obtained by investigation of exchange reactions and use of radiotracers. Exchange of ligands L

$$[ML_n] + {}^*L \rightleftharpoons [ML_{n-1}{}^*L] + L \qquad (18.9)$$

and exchange of central atoms M

$$[ML_n] + {}^*M \rightleftharpoons [{}^*ML_n] + M \qquad (18.10)$$

have to be distinguished.

Complexes with the same ligands, but different central atoms, may show very different ligand exchange rates (e.g. fast in the system $[Fe(H_2O)_6]^{3+}/{}^*H_2O$ and slow in the system $[Cr(H_2O)_6]^{3+}/{}^*H_2O$; fast in the system $[(Al(C_2O_4)_3]^{3-}/{}^*C_2O_4{}^{2-}$ and slow in the system $[Fe(C_2O_4)_3]^{3-}/{}^*C_2O_4{}^{2-}$). Accordingly, substitution-labile and substitution-inert complexes are distinguished.

The rate of ligand exchange depends on the number of d electrons in the metal and on the nature of the ligand. However, thermodynamic stability and kinetic stability do not run parallel. In general, high spin octahedral complexes exhibit fast exchange. The same holds for those low-spin complexes in which at least one d orbital is not occupied. In isoelectronic high-spin complexes with the electron configuration sp^3d^2 the lability decreases with increasing atomic number of the central atom, for instance in the order $[AlF_6]^{3-} > [SiF_6]^{2-} > [PF_6]^- > SF_6$. SF_6 is very inert towards substitution.

Qualitative information about the rate of ligand substitution in octahedral complexes is summarized in Table 18.1. Quantitative data are obtained by application of ligand field theory. The rate constants may vary by many orders of magnitude. Thus, for the exchange of water molecules in the first coordination sphere of 3d transition elements the following values have been measured: $Cr^{2+}(d^4)$: $7 \cdot 10^9\,s^{-1}$; $Cr^{3+}(d^3)$: $5 \cdot 10^{-7}\,s^{-1}$; $Mn^{2+}(d^5)$: $3 \cdot 10^7\,s^{-1}$; $Fe^{2+}(d^6)$: $3 \cdot 10^6\,s^{-1}$; $Co^{2+}(d^7)$: $1 \cdot 10^6\,s^{-1}$; $Ni^{2+}(d^8)$: $3 \cdot 10^4\,s^{-1}$.

Table 18.1. Relative rates of ligand exchange with octahedral complexes of transition elements.

Electron configuration	High-spin complexes (weak ligand field)	Low-spin complexes (strong ligand field)
d^0	Fast	Fast
d^1	Fast	Fast
d^2	Fast	Fast
d^3	Always slow	Always slow[a]
d^4	Fast to slow	Relatively slow[a]
d^5	Fast	Relatively show[a]
d^6	Fast	Always slow[a]
d^7	Fast	Fast to slow
d^8	Relatively slow	Relatively slow
d^9	Fast to slow	Fast to slow
d^{10}	Fast	Fast

[a] In the order: d^5, $d^4 > d^3 > d^6$.

The rate of exchange of central atoms increases with decreasing shielding by the ligands and increasing dissociation. Thus, shielding by the ligands leads to an increase in the activation energy of exchange in the following order: linear < planar < tetrahedral < octahedral complexes. In the case of chelate complexes, exchange is appreciably slower if the chelate is formed by only one big molecule.

Another example of the application of radiotracer techniques is the investigation of electron exchange reactions between atoms of the same element, for example

$$Fe(II) + {}^*Fe(III) \rightleftharpoons Fe(III) + {}^*Fe(II) \tag{18.11}$$

These reactions represent a special kind of redox reaction, in which the oxidation states of the same element change reciprocally. They can only be studied by application of labelled atoms or molecules, just like isotope exchange reactions. Investiga-

tion of electron exchange reactions has also contributed essentially to the understanding of bonding in inorganic complexes.

In general, high-spin complexes exhibit slow electron exchange, whereas fast electron exchange is observed with low-spin complexes. Mutual change of the oxidation
states (electron exchange) can occur by transfer of electrons or by transfer of ligands.
Two mechanisms are discussed, the inner sphere and the outer sphere mechanisms.
The transition states of these mechanisms are illustrated in Fig. 18.4. In the outer
sphere mechanism electron exchange proceeds by electron transfer, whereas in the
inner sphere mechanism it can take place by transfer of electrons or by transfer of
ligands. Equal rates of ligand exchange and of electron exchange indicate electron
exchange by transfer of ligands (inner sphere mechanism). This has been found for
complexes of Cr. Fast electron exchange of low-spin complexes is measured for
complexes of Fe(II). It is explained by the fact that the valence electrons are distributed over the whole complex and can therefore be transferred easily (outer sphere
mechanism).

Outer sphere mechanism
(contact of coordination spheres)

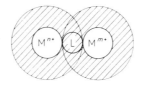

Inner sphere mechanism
(overlap of coordination spheres)

Figure 18.4. Transition
states in electron-exchange
reactions.

Information about details of various chemical reactions can be obtained by application of radiotracers. The photographic process may be mentioned as an example:
if a thick cylindrical pellet of AgI is labelled on one side by isotope exchange with
^{131}I, arranged in such a way that light has no access to the labelled side, and irradiated on the opposite side by visible light, ^{131}I-labelled iodine is immediately liberated. It may be carried away by a stream of an inert gas and measured continuously
in a gas counter. The immediate liberation of I_2 is due to fast transport of defect
electrons in AgI, in agreement with the proposed mechanism of the photographic
process.

18.4 Reaction Mechanisms in Heterogeneous Systems

In heterogeneous systems, radiotracer methods also have a wide range of applications with respect to elucidation of reaction mechanisms as well as to the determination of kinetic data.

In heterogeneous reactions, either the reaction at the phase boundary or the
transport to or from the reaction zone can be rate-determining. In the case of a
reaction between two solid phases, the diffusion of one reactant determines, in general, the rate of the reaction. Diffusion is considered in more detail in section 18.5.

In reactions between a solid and a gas or a solution, three steps are distinguished: transport of the reactant in the gas phase or the solution to the surface of the solid; reaction at the surface of the solid; and transport of the reaction products into the solid, the gas phase or the solution, respectively. Reaction at the solid/gas or solid/liquid interface comprises the following steps: (a) adsorption of the reactant, (b) reaction and (c) desorption of the products. These steps are considered in more detail in the following paragraphs.

(a) The rate R is determined by step (a) (adsorption):

$$R = k_{1a} \cdot \frac{F}{V} \cdot n_2 \qquad (18.12)$$

where k_{1a} is the rate constant, F is the surface area of the solid, n_2 is the mole number of the reactant in the gas or the liquid phase, respectively, and V is the volume of this phase. The rate of the heterogeneous isotope exchange is given by

$$-\frac{d^*n_2}{dt} = k_{1a} \frac{F}{V} n_2 (x_2 - x_1) \qquad (18.13)$$

in which $x_1 = {}^*c_1/c_1 = {}^*n_1/n_1$ and $x_2 = {}^*c_2/c_2 = {}^*n_2/n_2$ are the fractions of the concentrations c or the mole numbers n of the exchangeable particles (atoms, ions or molecules) at the surface of the solid (1) and in the gas or liquid phase (2). By integration and introducing the degree of exchange λ, the following relation is obtained:

$$\ln(1 - \lambda) = -k_{1a} \cdot \frac{F}{V} \cdot \frac{n_2 + n_1}{n_1} t \qquad (18.14)$$

The degree of exchange is

$$\lambda = \frac{{}^*n_2(0) - {}^*n_2}{{}^*n_2(0) - {}^*n_2(\infty)} \qquad (18.15)$$

Eq. (18.14) corresponds to the McKay equation (18.5) for homogeneous exchange reactions (R given by eq. (18.12), $n_1 \simeq c_1$ and and $n_2 \simeq c_2$). k_{1a} is obtained by a plot of $\ln(1 - \lambda)$ as a function of t, as shown in Fig. 18.3:

$$k_{1a} = \frac{\ln 2}{t_{1/2}} \cdot \frac{V}{F} \cdot \frac{n_1}{n_1 + n_2} \qquad (18.16)$$

(b) The rate R is determined by step (b) (reaction at the surface $\simeq$ change of positions):

$$R = k_2 \cdot \frac{1}{V} \cdot n_1 n_2 \qquad (18.17)$$

where n_1 is the mole number of exchangeable particles (atoms, ions or molecules) at the surface. In this case, the rate of heterogeneous isotope exchange is

$$-\frac{d^* n_2}{dt} = k_2 c_2 n_1 (x_2 - x_1) \tag{18.18}$$

Integration and introduction of the degree of exchange gives

$$\ln(1 - \lambda) = -k_2 \cdot \frac{1}{V}(n_2 + n_1)t \tag{18.19}$$

and the reaction constant can be calculated by

$$k_2 = \frac{\ln 2}{t_{1/2}} \cdot \frac{V}{n_1 + n_2} \tag{18.20}$$

(c) The rate R is determined by step (c) (desorption):

$$R = k_{1c} n_1 \tag{18.21}$$

where n_1 is the mole number of exchangeable particles (atoms, ions or molecules) at the surface. The rate of heterogeneous isotope exchange is

$$-\frac{dn_2}{dt} = k_{1c} n_1 (x_2 - x_1) \tag{18.22}$$

Integration and introduction of λ gives

$$\ln(1 - \lambda) = -k_{1c} \frac{n_2 + n_1}{n_2} t \tag{18.23}$$

and the rate constant can be determined by

$$k_{1c} = \frac{\ln 2}{t_{1/2}} \cdot \frac{n_2}{n_2 + n_1} \tag{18.24}$$

If transport in the gas or liquid phase determines the rate of the reaction, the laws of diffusion have to be applied. Two limiting cases are distinguished: the mean radius $\bar{r}$ of the solid particles is much larger or much smaller than the thickness δ of the diffusion layer. The following relations are obtained:

$$R = \frac{FD}{V\delta} c_2 \ (\bar{r} \gg \delta) \quad \text{and} \quad R = \frac{FD}{V\bar{r}} c_2 \ (\bar{r} \ll \delta) \tag{18.25}$$

(D = diffusion coefficient) and

$$\ln(1 - \lambda) = -k_D \frac{n_2 + n_1}{n_1} t \tag{18.26}$$

where

$$k_D = \frac{FD}{V\delta} \ (\bar{r} \gg \delta) \quad \text{or} \quad k_D = \frac{FD}{V\bar{r}} \ (\bar{r} \ll \delta) \tag{18.27}$$

For investigation of the surface reaction, it is, in general, more favourable to label the gas phase or the solution instead of the solid. In measuring the heterogeneous exchange as a function of time, three processes may be observed: adsorption from the gas phase or solution, isotope exchange with particles (atoms, ions or molecules) that are initially present at the surface, and isotope exchange with particles that are entering the surface due to solid diffusion or recrystallization. Adsorption is only observed at low partial pressure in the gas phase or at low concentration in the liquid phase, respectively. It leads to a rapid decrease of activity in the gas phase or solution at the beginning, from which the number of adsorbed molecules or ions can be calculated.

In most cases, curves of the kind shown in Fig. 18.5 are measured. By evaluation of these curves. the kinetics of heterogeneous exchange reactions, the surface area involved in the exchange or the kinetics of diffusion or recrystallization can be determined.

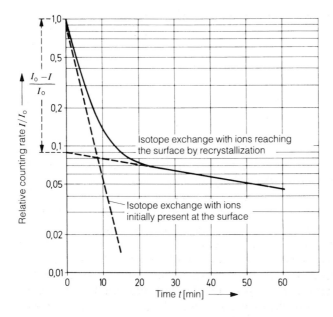

Figure 18.5. Determination of surface area by heterogeneous isotope exchange: activity in solution as a function of time.

An interesting example is the heterogeneous isotope exchange between CO_2 in the gas phase and solid $BaCO_3$:

$$Ba^{14}CO_3 + CO_2 \rightleftharpoons BaCO_3 + {}^{14}CO_2 \tag{18.28}$$

The exchange occurs only in the presence of water vapour, and $^{14}CO_2$ is given off by exchange if $Ba^{14}CO_3$ is stored in humid air. On the other hand, $^{14}CO_2$ produced in the atmosphere under the influence of cosmic radiation, and dissolved in water, exchanges with solid $CaCO_3$, which may cause errors in dating.

The determination of the surface area of ionic crystals by heterogeneous isotope exchange was first shown by Paneth. It is based on the even distribution of labelled ions between the surface and a solution to which these ions have been added:

$$\frac{^*N_1}{N_1} = \frac{^*N_2}{N_2} \tag{18.29}$$

$^*N_1$ and $^*N_2$ are the numbers of labelled ions on the surface and in solution, respectively, and N_1 and N_2 are the total numbers of these ions. The counting rates in solution before equilibration (I_0) and after equilibration (I) are measured:

$$\frac{I_0}{I} = \frac{^*N_1 + {}^*N_2}{^*N_2} = \frac{^*N_1}{^*N_2} + 1 = \frac{N_1}{N_2} + 1 \tag{18.30}$$

and the surface S is calculated by the equation

$$S = \left(\frac{I_0}{I} - 1\right) s_{eq} \, c_2 \, V \, N_{Av} \tag{18.31}$$

where c_2 is the concentration of the exchangeable ions in solution, V is the volume of the solution and N_{Av} is Avogadro's number; s_{eq} is the fraction of the surface that is equivalent to one exchangeable ion and may be calculated approximately from the molar mass M and the density ρ of the ionic compound

$$s_{eq} = \left(\frac{M}{\rho N_{Av}}\right)^{2/3} \tag{18.32}$$

Because of the effect of recrystallization shown in Fig. 18.5, it is recommendable to measure the counting rate in solution as a function of time and to extrapolate the second branch of the curve obtained to $t = 0$.

The correct value of s_{eq} depends on the surface properties, in particular on the presence of an excess of cations or anions. If the surface area S is additionally determined by another method (e.g. the BET method), information is obtained about the surface properties (excess of cations or of anions or equivalent amounts of both).

Heterogeneous isotope exchange has been applied with great success in the investigation of the steps of precipitation reactions and of ageing and recrystallization of precipitates. The mechanism of ripening proposed by Ostwald could be proved by measuring the specific activity of the ions involved in the process of ripening as a function of time; dissolution of small and imperfect particles and simultaneous growth of larger and more perfect crystals (Ostwald ripening) lead to a decrease of specific activity which is a measure of the extent of ripening.

All precipitates show some kind of ageing, which causes a decrease in solubility. Therefore, they are usually filtered after some time of standing or after heating. Sparingly soluble ionic compounds in contact with a solution exhibit continuous recrystallization. Both processes, ageing and recrystallization, can be investigated successfully by application of radiotracers.

Recrystallization and diffusion in solids determine the dynamic behaviour of solid surfaces, including heterogeneous exchange. For example, the exchange of iodine between solid NaI and gaseous CH_3I depends on the mobility of iodide in NaI, which can be studied by labelling of CH_3I with ^{131}I. Heterogeneous reactions in which two solid phases are involved are another field of application of radiotracers.

Reactions at the surface of metals can also be studied by radiotracer methods. Corrosion or adsorption on surface layers may feign isotope exchange. On the other hand, corrosion processes can be elucidated in early stages with high sensitivity. The mechanism of corrosion of brass may be taken as an example: in corrosion of this alloy it is found that the Zn:Cu ratio in the corroding solution is appreciably higher than in the metal. Two mechanisms have been discussed, preferred dissolution of Zn or simultaneous dissolution of Zn and Cu followed by redeposition of Cu. By labelling the solution with ^{65}Zn and ^{64}Cu and measuring the specific activity of both elements in solution as a function of time in the course of corrosion, it was found that preferred dissolution of Zn is the prevailing mechanism. Autoradiography (section 7.12) is another valuable method for the investigation of corrosion processes.

18.5 Diffusion and Transport Processes

The high sensitivity of radiotracer techniques makes these very attractive for determination of diffusion coefficients. Self-diffusion (i.e. diffusion of the intrinsic components of the substance) is of special interest and can only be measured by indicator methods.

Most investigations of self-diffusion have been made with solids. Solution of the diffusion equations requires simple geometric conditions. Generally, samples with plane, cylindrical or spherical surfaces are used. The radiotracer is applied in form of an "infinitely thin" or "infinitely thick" layer. "Infinitely thin" layers may be produced by vapour or electrolytic deposition, chemical reaction or isotope exchange. Layers of "infinite thickness" must have good contact with the sample. At the end of the diffusion experiment, the sample can be sliced into thin discs or dissolved stepwise. After labelling with "infinitely thin" layers, the activity may also be measured from outside before and after the diffusion experiment; in this radiation-absorption method, it is assumed that the tracer spreads into the sample according to the laws of diffusion, and the mean penetration depth and the diffusion coefficient are calculated from the decrease of the counting rate and the absorption coefficient of the radiation in the sample.

Autoradiography provides qualitative information about diffusion. In particular it answers the question of whether diffusion proceeds uniformly in the sample (volume diffusion) or along grain boundaries.

The recoil method was first described by Hevesy. A plane surface of lead is labelled with ^{212}Pb, which decays as follows:

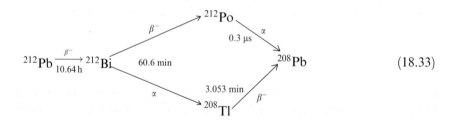

$$(18.33)$$

By α decay of ^{212}Bi, the daughter nuclide ^{208}Tl receives a recoil. Depending on the penetration depth of ^{212}Pb, a greater or smaller fraction of the recoiling ^{208}Tl atoms are sampled on a copper electrode (potential -200 V), and the β^- activity of ^{208}Tl is measured. Application of the recoil method is restricted to α emitters yielding a radioactive daughter nuclide.

The sensitivity of the measurement of diffusion coefficients in solids depends primarily on the method applied. By the radiation-absorption method mentioned above, diffusion coefficients down to about 10^{-10} cm^2 s^{-1} can be measured in the case of β emitters and down to about 10^{-12} cm^2 s^{-1} in the case of α emitters. The recoil method makes it possible to measure diffusion coefficients down to the order of 10^{-19} cm^2 s^{-1}. Relatively low sensitivity $(D \geq 10^{-8}$ cm^2 s$^{-1})$ is achieved with the method of mechanical slicing, whereas diffusion coefficients down to about 10^{-18} cm^2 s^{-1} are obtained with the method of dissolution in steps. Mechanical slicing and dissolution in steps have the advantage that the actual concentration in the solid is obtained.

Examples of the determination of self-diffusion coefficients in solids are the diffusion of hydrogen ions and water molecules (labelled with T and ^{18}O, respectively) in alums, of Cl$^-$ (labelled with ^{36}Cl) in AgCl, and of I$^-$ (labelled with ^{131}I) in AgI. Besides self-diffusion, many other diffusion coefficients of trace elements in metals, oxides, silicates and other substances have been determined by application of radiotracers. Investigation of the migration of trace elements from solutions into glass revealed fast diffusion of relatively small monovalent ions such as Ag$^+$.

In measuring diffusion in liquids or gases, mixing by convection must be excluded. Several experimental arrangements are described, for example use of a diffusion tube, a diaphragm or a capillary. Measurement of the diffusion coefficient of pH-sensitive species as a function of pH allows conclusions with respect to the size of the species.

18.6 Emanation Techniques

Emanation techniques are based on the production of radioactive noble gases by decay of mother nuclides or by nuclear reactions. The emanating power has been defined by Hahn as the fraction of radioactive noble gas escaping from a solid relative to the amount produced in the solid. It depends on the composition of the solid, its lattice structure and its specific surface area. Reactions in the solid have a major influence. Further factors affecting the emanating power are the half-life of the noble gas radionuclide, its recoil energy and the temperature.

The noble gas may escape from the solid by recoil or diffusion. The range R of recoiling atoms produced by α decay is about 100 μm in air and about 0.01 μm in solids. If $R \ll r$, r being the radius of the grains or crystallites of the solid, only a small fraction E_R of the recoiling atoms is able to escape from the solid:

$$E_R = \frac{R}{4}\frac{F}{V} \qquad (18.34)$$

where F is the surface area of a grain or crystallite, respectively, and V its volume. Emanation by diffusion E_D prevails if $R \ll r$ and $r^2\lambda \gg D$ (D = diffusion coefficient):

$$E_D = \sqrt{\frac{D}{\lambda} \frac{F}{V}}$$

(18.35)

The total emanating power E is given by the sum of E_R and E_D.

If the solid consists of an aggregate of grains or crystallites or if it has a porous structure, the fate of the noble gas depends on its interaction with other grains or with the inner surface of the pores, respectively. Noble gas atoms escaping from a certain grain by recoil often hit another grain, enter that grain and leave the solid with some delay. Apart from this effect, adsorption and diffusion play the most important role. In porous solids diffusion is strongly hindered by the presence of water.

In Fig. 18.6 various possibilities of producing radioactive noble gases, suitable for preparation of emanating sources and investigation of emanating power, are listed. The isotopes of radon are applied most frequently. They are all produced by decay of radium isotopes, which are formed by decay of thorium isotopes. Therefore, either the radium isotopes or the thorium isotopes may be incorporated into the solid

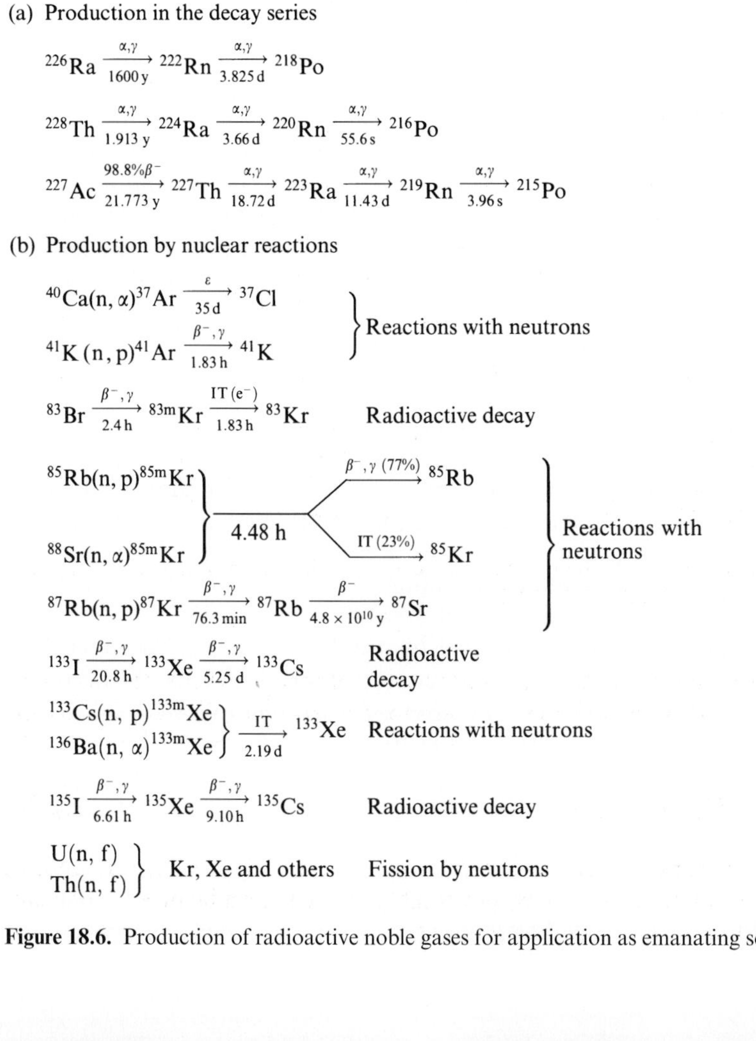

(a) Production in the decay series

$$^{226}\text{Ra} \xrightarrow[1600\,\text{y}]{\alpha,\gamma} {}^{222}\text{Rn} \xrightarrow[3.825\,\text{d}]{\alpha,\gamma} {}^{218}\text{Po}$$

$$^{228}\text{Th} \xrightarrow[1.913\,\text{y}]{\alpha,\gamma} {}^{224}\text{Ra} \xrightarrow[3.66\,\text{d}]{\alpha,\gamma} {}^{220}\text{Rn} \xrightarrow[55.6\,\text{s}]{\alpha,\gamma} {}^{216}\text{Po}$$

$$^{227}\text{Ac} \xrightarrow[21.773\,\text{y}]{98.8\%\beta^-} {}^{227}\text{Th} \xrightarrow[18.72\,\text{d}]{\alpha,\gamma} {}^{223}\text{Ra} \xrightarrow[11.43\,\text{d}]{\alpha,\gamma} {}^{219}\text{Rn} \xrightarrow[3.96\,\text{s}]{\alpha,\gamma} {}^{215}\text{Po}$$

(b) Production by nuclear reactions

$$^{40}\text{Ca(n, }\alpha)^{37}\text{Ar} \xrightarrow[35\,\text{d}]{\varepsilon} {}^{37}\text{Cl}$$

$$^{41}\text{K (n, p)}^{41}\text{Ar} \xrightarrow[1.83\,\text{h}]{\beta^-,\gamma} {}^{41}\text{K}$$

$\left.\right\}$ Reactions with neutrons

$$^{83}\text{Br} \xrightarrow[2.4\,\text{h}]{\beta^-,\gamma} {}^{83\text{m}}\text{Kr} \xrightarrow[1.83\,\text{h}]{\text{IT (e}^-)} {}^{83}\text{Kr}$$ Radioactive decay

$$^{85}\text{Rb(n, p)}^{85\text{m}}\text{Kr}$$
$$^{88}\text{Sr(n, }\alpha)^{85\text{m}}\text{Kr}$$
$\left.\right\}$ 4.48 h
$\nearrow$ $\xrightarrow{\beta^-,\gamma\,(77\%)}$ ${}^{85}\text{Rb}$
$\searrow$ $\xrightarrow{\text{IT (23\%)}}$ ${}^{85}\text{Kr}$

$$^{87}\text{Rb(n, p)}^{87}\text{Kr} \xrightarrow[76.3\,\text{min}]{\beta^-,\gamma} {}^{87}\text{Rb} \xrightarrow[4.8 \times 10^{10}\,\text{y}]{\beta^-} {}^{87}\text{Sr}$$

$\left.\right\}$ Reactions with neutrons

$$^{133}\text{I} \xrightarrow[20.8\,\text{h}]{\beta^-,\gamma} {}^{133}\text{Xe} \xrightarrow[5.25\,\text{d}]{\beta^-,\gamma} {}^{133}\text{Cs}$$ Radioactive decay

$$^{133}\text{Cs(n, p)}^{133\text{m}}\text{Xe}$$
$$^{136}\text{Ba(n, }\alpha)^{133\text{m}}\text{Xe}$$
$\left.\right\}$ $\xrightarrow[2.19\,\text{d}]{\text{IT}}$ ${}^{133}\text{Xe}$ Reactions with neutrons

$$^{135}\text{I} \xrightarrow[6.61\,\text{h}]{\beta^-,\gamma} {}^{135}\text{Xe} \xrightarrow[9.10\,\text{h}]{\beta^-,\gamma} {}^{135}\text{Cs}$$ Radioactive decay

$$\left. \begin{array}{c} \text{U(n, f)} \\ \text{Th(n, f)} \end{array} \right\}$$ Kr, Xe and others   Fission by neutrons

Figure 18.6. Production of radioactive noble gases for application as emanating sources.

whose reactions are to be studied. Other noble gases such as xenon or krypton or their radioactive precursors may be produced by nuclear reactions, as indicated in Fig. 18.6.

The emanating power is determined by measuring either the activity of the noble gas itself or that of its daughter nuclides. In general, the emanating power of salts, glasses or compact oxides is low ($\approx 1\%$), whereas it is high for hydroxides (20 to 100%). It depends strongly on the preparation of the samples.

Samples exhibiting high emanating power (70 to 100%) are prepared for application as emanating sources. Examples are ^{228}Th or ^{226}Ra coprecipitated with thorium hydroxide. ^{222}Rn given off by these emanating sources may be used for chemical or physical investigations with radon. Formerly these sources have also been prepared for application of ^{222}Rn in medicine.

Measurement of the emanating power allows the investigation of transformation, decomposition or other reactions in solids, or of ageing processes in precipitates. In Fig. 18.7 the emanating power of $CaCO_3$ is plotted as a function of the temperature. The transformation of aragonite into calcite at 530 °C and the decomposition of $CaCO_3$ into CaO and CO_2 at 920 °C are clearly discernible. Near 1200 °C the high mobility of the ions in CaO becomes noticeable; it indicates the sintering of CaO.

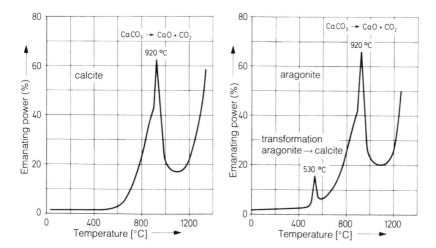

Figure 18.7. Emanating power of calcite and aragonite as a function of temperature. (According to K. E. Zimens, Z. Phys. Chem., B37, 231 (1937).)

Emanation techniques are also used for the study of reactions between two solids, for instance the formation of $PbSiO_3$ (as in the preparation of lead glass) by the reaction

$$PbO + SiO_2 \rightarrow PbSiO_3 \tag{18.36}$$

This reaction also runs parallel to the emanating power. However, quantitative evaluation of the emanating power with respect to the reactions involved is not simple, because of the complexity of the processes.

The ageing of thorium hydroxide and iron hydroxide in the presence of water is obvious from Fig. 18.8. Whereas the emanating power of thorium hydroxide decreases very slowly, that of iron hydroxide falls off relatively fast, indicating faster ageing. Emanation techniques have also been applied to obtain information about surface areas and densities of porous substances.

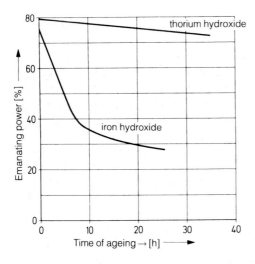

Figure 18.8. Emanating power of thorium hydroxide and iron hydroxide as a function of ageing. (According to O. Hahn, G. Graue: Z. Phys. Chem., Bodenstein Festband 1931, S. 608.)

Literature

General

O. Hahn, Applied Radiochemistry, Cornell University Press, Ithaca, NY, **1936**
A. C. Wahl, N. A. Bonner, Radioactivity Applied to Chemistry, Wiley, New York, **1951**
The Collected Papers of George Hevesy, Pergamon, Oxford, **1962**
M. Haissinsky, Nuclear Chemistry and its Applications, Addison–Wesley, Reading, MA, **1964**
J. F. Duncan, G. B. Cook, Isotopes in Chemistry, Clarendon Press, Oxford, **1968**
E. A. Evans, M. Muramatsu, Radiotracer Techniques and Applications, Marcel Dekker, New York, **1977**
K. H. Lieser, Einführung in die Kernchemie, VCH, Weinheim, **1991**

More Special

International Atomic Energy Agency, Radioisotopes in the Physical Sciences and Industry, IAEA, Vienna, **1962**
International Atomic Energy Agency, Exchange Reactions, IAEA, Vienna, **1965**
H. A. C. McKay, Physicochemical Applications of Radiotracers, in: Principles of Radiochemistry, Butterworths, London, **1971**
F. Basolo, R. G. Pearson, Mechanisms of Inorganic Reactions, A Study of Metal Complexes in Solution, 3rd ed., Wiley, New York, **1973**
E. Buncel, C. C. Lee (Eds.), Isotopes in Organic Chemistry, 2 Vols., Elsevier, Amsterdam, **1975**
P. J. F. Griffiths, G. S. Parks, Diffusion, in: Radiotracer Techniques and Application, Vol. 1 (Eds. E. A. Evans, M. Muramatsu), Marcel Dekker, New York, **1977**
A. Ozaki, Isotopic Studies of Heterogeneous Catalysis, Krdanska–Academic Press, Tokyo, **1977**
S. Hubert, M. Hussonnois, R. Guillaumont, Measurement of Complexing Constants by Radiochemical Methods, in: Structure and Bonding (Ed. J. D. Duntz), Springer, Berlin, **1978**
S. R. Logan, The Kinetics of Isotopic Exchange Reactions, J. Chem. Educ. *67*, 371 **(1990)**

19 Radionuclides in the Life Sciences

19.1 Survey

Application of radionuclides in life sciences is of the greatest importance, and the largest single user of radionuclides is nuclear medicine. Shortly after the discovery of Ra in 1898 by Marie Curie and its subsequent isolation from pitchblende in amounts of 0.1 to 1 g, the finding that this element was useful as a radiation source led to the first application of radionuclides in medicine. In 1921, de Hevesy investigated the metabolism of lead in plants by use of natural radioisotopes of Pb.

The spectrum of radionuclides available for application in the life sciences broadened appreciably with the invention of the cyclotron by Lawrence in 1930 and the possibility of producing radionuclides on a large scale in nuclear reactors in the late 1940s. By application of T and ^{14}C, important biochemical processes, such as photosynthesis in plants, could be elucidated.

Nowadays, nuclear medicine has become an indispensible section of medical science, and the production of radionuclides and labelled compounds for application in nuclear medicine is an important branch of nuclear and radiochemistry. The development of radionuclide generators made short-lived radionuclides available at any time for medical application. New imaging devices, such as single photon emission tomography (SPET) and positron emission tomography (PET) made it possible to study local biochemical reactions and their kinetics in the living human body.

It is an attractive object of research to synthesize labelled compounds that are taking part in specific biochemical processes or able to pass specific barriers in the body, with the aim of detecting malfunctions and of localizing the origin of diseases. Complexes of short-lived no-carrier-added radionuclides and high yields of the syntheses are of special interest. In the case of short-lived radionuclides, such as ^{11}C, the synthesis must be fast and as far as possible automated. Labelled organic molecules can also be used to transport radionuclides to special places in the body for therapeutical application, i.e. as specific internal radiation sources.

The following fields of application of radionuclides in the life sciences can be distinguished:

- ecology (uptake of trace elements and radionuclides from the environment by plants, animals and man);
- analysis (determination of trace elements or compounds in plants, animals and man);
- physiology and metabolism (reactions and biochemical processes of elements and compounds in plants, animals and man);
- diagnosis (identification and localization of diseases);
- therapy (treatment of diseases).

In ecological and metabolic studies and in diagnosis, radioactive tracers are applied (radiotracer techniques). Analytical applications in the life sciences are based either on activation or on tracer techniques, whereas for therapeutical purposes relatively high activities of radionuclides are used.

19.2 Application in Ecological Studies

The uptake of trace metals from the soil by plants and animals can be studied with high sensitivity by radiotracer techniques. In these applications, it is important that the chemical form of the radiotracer is identical with that of the trace element to be studied. For example, in agriculture, the uptake of trace elements necessary for plant growth can be investigated. Essential trace elements, such as Se, are of special interest. By use of radioactively labelled selenium compounds the transfer of this element from soil to plants and animals can be measured. For the investigation of the transfer of radionuclides (radioecology), addition of tracers is, in general, not needed.

In radioecology, transfer factors for relevant radionuclides in various systems (e.g. soil → plant, plant → animal, plant → man, animal → man) have been determined. These transfer factors are used for the assessment of radiation doses received by animals or man due to the presence of natural radionuclides or of radioactive fall-out.

19.3 Radioanalysis in the Life Sciences

Two radioanalytical methods described in chapter 17 are applied preferentially in the life sciences, activation analysis and isotope dilution, the latter mainly in combination with the substoichiometric principle.

Activation analysis can be used for determination of trace elements, in particular heavy metals and essential elements, in various parts or organs, respectively, of plants or animals and man. Making use of the high sensitivity of activation analysis, small samples of the order of several milligrams taken from selected places give information about the concentration of the elements of interest. The results of activation analysis of trace elements also allow conclusions with respect to diseases or malfunctions and are valuable aids to diagnosis. Examples are the determination of Se in man or of trace element concentrations in bones or other parts, with respect to the sufficient supply of essential elements and metabolism. In vivo irradiation has also been proposed.

Isotope dilution in combination with the substoichiometric principle is applied in various ways. The most important examples are radioimmunoassay for protein analysis and DNA analysis. In radioimmunoassay, radionuclides are used as tracers and immunochemical reactions for isolation. Radioimmunoassay was first described in 1959 by Yalow and Berson, and since then has found very broad application in clinical medicine, in particular for the measurement of serum proteins, hormones, enzymes, viruses, bacterial antigens, drugs and other substances in blood, other body fluids and tissues. Only one drop of blood is needed, and the analysis can be performed automatically. Today more than 10^7 immunoassays are made annually in the United States. The most important advantages of the method are the high sensitivity and the high specificity. In favourable cases quantities down to 10^{-13} g can be determined and, in general, only the components of interest are detected. Since measurement of absolute quantities is not necessary, the accuracy is excellent. Difficulties may arise if the immunochemical reactions fail or if they are not selective.

The general procedure of radioimmunoassay is as follows: Two aliquots each containing a known mass m_0 of a labelled protein *P are taken. One aliquot is mixed with a much smaller (substoichiometric) mass of an antibody B which forms the complex *PB. The latter is isolated and its radioactivity A_1 is measured. The other aliquot of the labelled protein *P is mixed with the unknown mass m_x of the protein to be determined and this mixture is also allowed to react with the same amount of antibody B as before. Again, the complex *PB is isolated, and its radioactivity A_2 is measured. The unknown mass m_x is calculated by application of eq. (17.18).

A great number of variations of radioimmunoassay are possible: competitive or non-competitive binding assays and different antibodies may be applied, and various radionuclides may be used for labelling (preferably T, ^{14}C, ^{35}S, ^{32}P, ^{125}I and ^{131}I).

The base sequence of DNA is determined in the following way: The cell walls are broken up by osmosis or other methods and the double-stranded DNA is denaturized to single-stranded pieces, which may be concentrated by centrifugation. By application of different restriction enzymes, the nucleotide chains are sectioned further into different sets of smaller fragments. To these sets labelled compounds are added that attach selectively to the different fragments. The compounds are labelled with radionuclides, such as T, ^{14}C or ^{32}P. On the other hand, the original DNA chain is directly labelled (e.g. by ^{32}P) in a cloning process, and the cloned DNA is also split into fragments by application of restriction enzymes. All samples obtained by treatment with different restriction enzymes are subjected to electrophoresis in a gel, such as agarose or polyacrylamide, by which the fragments are separated according to their migration velocities. By means of autoradiography, characteristic patterns of spots or bands are obtained which give information about the individual from which the DNA was taken.

DNA analysis is of growing importance for various purposes, such as transplantation of organs, detection of genetic diseases, investigation of the evolution of species, or identification of criminals in forensic science.

The substoichiometric principle is also applied for determination of trace elements in biological systems. Some examples are listed in Table 19.1. The detection limits are usually <1 mg/l, in some cases (e.g. $(C_6H_6)_2Hg$) $<10^{-3}$ mg/l.

Table 19.1. Examples of trace element determination in biological samples by isotope dilution in combination with substoichiometric isolation.

Species	Conditions
Ca^{2+}	2-Thenoyltrifluoroacetone in CCl_4
Sr^{2+}	8-Quinolinol in $CHCl_3$
Sn(IV)	N-Benzoyl-N-phenylhydroxylamine in $CHCl_3$
PO_4^{3-}	Extraction of phosphomolybdate formed with molybdate into methyl isobutyl ketone
F^-	$(CH_3)_3SiCl$ in C_6H_6
Ag^+	Dithizone in CCl_4
Au(III)	Cu diethyldithiocarbamate in $CHCl_3$
Cd^{2+}	Dithizone in $CHCl_3$
Hg^{2+}	Thionalide in $CHCl_3$
$(C_6H_5)_2Hg$	Dithizone in $CHCl_3$
Cr(VI)	Diethylammonium diethyldithiocarbamate in C_6H_6
Fe^{3+}	8-Quinolinol in $CHCl_3$
Ni^{2+}	Diacetyldioxime in $CHCl_3$

19.4 Application in Physiological and Metabolic Studies

For investigation of physiological or metabolic processes in plants, animals and man, radiotracer techniques are very useful because of their high sensitivity and the possibility of labelling at certain positions of the molecules. In plant physiology, important biochemical processes can be elucidated by application of radiotracers. An illustrative example is photosynthesis in plants. Plants growing in an atmosphere containing $^{14}CO_2$ synthesize ^{14}C-labelled sugars and cellulose in a sequence of chemical reactions. By measuring the ^{14}C-labelled intermediates and products, it was possible to trace the steps of photosynthesis and to identify the intermediates.

Labelled species taken up by animals are incorporated in various amounts and may be accumulated in certain organs where they undergo chemical reactions, in particular biochemical synthesis and degradation. Finally, products of metabolism are removed from the body. The distribution of the radionuclides in the body gives significant information about normal and abnormal processes. The radionuclides may be measured by direct counting or by autoradiography. Examples are the investigation of metabolism of amino acids, vitamins, drugs or other compounds in animals by means of labelled compounds.

The position of labelling must be chosen in such a way that the products of metabolism can be identified, e.g. by labelling with T or ^{14}C in the side chain of an aromatic compound if the fate of the side chain is of interest, or labelling in the aromatic part of the molecule if this is the main object of investigation. Double labelling, for instance by T and ^{14}C at different positions, is often very helpful.

New drugs developed in laboratories of the pharmaceutical industry must be investigated with respect to their metabolism, before they are admitted for general use. Labelling of the drugs with radionuclides at certain positions of the molecule is the most appropriate method for detailed examination of the metabolism.

The metabolism of trace elements, their possible accumulation in certain organs or parts of the body and their excretion can also be studied with high sensitivity by application of radiotracers.

19.5 Radionuclides Used in Nuclear Medicine

A survey of radionuclides used in nuclear medicine is given in Table 19.2. Many of these radionuclides are also applied in biochemical and agricultural studies with animals and plants.

For diagnostic application in medicine, four aspects are of major importance:

- the radiation exposure of the patients,
- the measurability of the radionuclide from outside,
- the availability of the radionuclide and of suitable labelled compounds, and
- the radionuclide purity.

The radiation exposure of the patients depends on the activity of the radionuclide, the kind of radiation emitted, the half-life of the radionuclide and its residence time in the body. With respect to radiation exposure, α emitters are not suitable for diagnostic application because of the high energy doses transmitted locally to organs or tissues.

However, radiation exposure is low if only γ rays are emitted, as in the case of isomeric transition (IT) or electron capture (ε). Gamma-ray emitters are easily measurable from outside, and γ-ray energies in the range between about 50 and 500 keV are most favourable with respect to penetration through tissues and counting efficiency.

The lower limit of half-lives of radionuclides for diagnostic application is of the order of minutes. It is determined by the time needed for synthesis of suitable compounds and for transport in the body to the place of application. On the other hand, half-lives >1 d are less favourable, because of the longer radiation exposure of the patients and the risk of environmental contamination.

For application of radionuclides with half-lives <10 h, radionuclide generators or suitable accelerators must be available in the hospital or nearby.

Synthesis of labelled compounds requires reliable procedures and experience in radiochemistry. However, special kits are available for preparation of labelled compounds without experience in radiochemistry, and automated procedures have been developed for fast syntheses.

Radionuclides produced in nuclear reactors are of minor interest, because they exhibit β^- decay. Furthermore, reactor-produced radionuclides with half-lives <10 h are hardly applicable, because of the time needed for transport from a nuclear reactor to the site of application.

An example of a reactor-produced radionuclide is ^{131}I, which has been used many years for diagnosis of thyroid function. The γ rays of this radionuclide are easily measurable from outside, but the radiation dose transmitted to the patients by the β^- particles is not negligible, and the half-life (8.02 d) is relatively long. Besides, excretion of ^{131}I may cause contamination problems.

Radionuclide generators are described in section 12.4. They offer the significant advantage that the radionuclides are available at any time for direct application. The

Table 19.2. Radionuclides for application in nuclear medicine.

Radionuclide	Half-life	Decay mode (energy of particles or photons emitted [keV])	Application
Reactor-produced radionuclides			
^{3}H	12.323 y	β^- (18.6)	Biochemistry
^{14}C	5730 y	β^- (156)	Biochemistry
^{24}Na	14.96 h	β^- (1389); γ (1369, 2754)	Biochemistry, circulation
^{32}P	14.26 d	β^- (1710)	Biochemistry (skeleton, bones) Therapy (Leukaemia)
^{35}S	87.5 d	β^- (167)	Biochemistry
^{47}Ca	4.54 d	β^- (1981); γ (1297)	Biochemistry (bones) Diagnosis (bones)
^{59}Fe	44.503 d	β^- (475, 273); γ (1099, 1292)	Diagnosis
^{60}Co	5.272 y	β^- (315); γ (1332, 1173)	Therapy (cancer)
^{131}I	8.02 d	β^- (606); γ (364)	Diagnosis (thyroid function, kidneys) Therapy (hyperthyroidism, thyroid cancer)
^{133}Xe	5.25 d	β^- (346); γ (81)	Diagnosis (lung, brain)
^{137}Cs	30.17 y	β^- (514); γ (662)	Therapy (cancer)
^{186}Re	3.72 d	β^- (1075); ε; γ (137)	Bone cancer
^{198}Au	2.693 d	β^- (961); γ (412)	Therapy (ovarian cancer)
Obtained from radionuclide generators			
^{42}K	12.36 h	β^- (3523, 1970); γ (1525)	Biochemistry (volumetry)
^{68}Ga	1.127 h	β^+ (1830); γ (1077)	Biochemistry (intestine) Diagnosis
^{99m}Tc	6.01 h	IT (141)	Diagnosis (thyroid, heart, lung, liver, kidneys, skeleton, brain etc.)
^{113m}In	1.658 h	IT (392)	Diagnosis
Accelerator-produced short-lived positron emitters			
^{11}C	20.38 min	β^+ (960)	Diagnosis, PET (brain)
^{13}N	9.96 min	β^+ (1190)	Diagnosis, PET
^{15}O	2.03 min	β^+ (1723)	Diagnosis, PET
^{18}F	1.83 h	β^+ (635)	Biochemistry Diagnosis, PET
Other accelerator-produced radionuclides			
^{51}Cr	27.7 d	ε; γ (320)	Biochemistry Diagnosis (kidney function)
^{55}Fe	2.73 y	ε	Biochemistry (red blood cells)
^{57}Co	271.79 d	ε; γ (122, 136)	Biochemistry (liver)
^{58}Co	70.88 d	ε; β^+ (2300); γ (811)	Diagnosis (Vitamin B$_{12}$)
^{75}Se	119.64 d	ε; γ (136, 265)	Biochemistry (kidneys, liver) Diagnosis (methionine)
^{85}Sr	64.9 d	ε; γ (514, 1065)	Biochemistry Diagnosis (skeleton)

Table 19.2. (continued)

Radionuclide	Half-life	Decay mode (energy of particles or photons emitted [keV])	Application
^{111}In	2.807 d	ε; γ (171, 245, 830)	Diagnosis (spinal cord)
^{123}I	13.2 h	ε; γ (159, 1234)	Biochemistry (thyroid) Diagnosis (thyroid, kidney function)
^{125}I	59.41 d	ε; γ (35, 179)	Biochemistry (thyroid) Diagnosis (thrombosis, kidney function)
^{201}Tl	3.046 d	ε; γ (167, 480)	Biochemistry (parathyroid, kidneys) Diagnosis (heart infarct)

^{99}Mo/^{99m}Tc generator is used most frequently, because of the favourable properties of ^{99m}Tc: only γ rays are emitted, the half-life is very suitable for diagnostic application, and the 141 keV rays can be measured from outside with high counting efficiency. Furthermore, the activity of the ground state ^{99}Tc can be neglected, because of its relatively long half-life; the activity ratio of ^{99}Tc formed by decay of ^{99m}Tc to the initial activity of the latter is only $A(^{99}\text{Tc}):A(^{99m}\text{Tc}) \approx 3 \cdot 10^{-9}$. Due to its favourable properties, ^{99m}Tc is the most frequently used radionuclide in nuclear medicine. After elution from the generator as ^{99m}TcO$_4^-$, it is transformed into or attached to suitable compounds for specific application. For this purpose, kits are used that guarantee high yields, good reproducibility and good performance by people not skilled in radiochemistry.

The production of short-lived positron emitters has been described in section 12.2. By interaction with electrons, the positrons are annihilated and two γ-ray photons of 511 keV each are emitted simultaneously in opposite directions. By measuring these photons by means of a suitable arrangement of detectors, exact localization of the radionuclides in the body is possible. This is the basis of positron emission tomography (PET), which has found broad application in nuclear medicine. The most frequently used positron emitters are listed in Table 19.2. They are preferably produced by small cyclotrons in the hospitals or nearby.

Other accelerator-produced radionuclides are also used in nuclear medicine (Table 19.2). One of the most important radionuclides in this group is ^{123}I. This radioisotope of iodine has more favourable properties than ^{131}I: it emits only γ radiation and its relatively short half-life is more appropriate for medical application. Its production is described in section 12.1. Suitable accelerators for the generation of protons of relatively high energy, and transport facilities, are needed.

For therapeutical purposes, natural radionuclides, mainly ^{226}Ra and ^{222}Rn, were the first to be applied as external and internal radiation sources. For example, encapsulated samples of ^{226}Ra have been attached to the skin or introduced into the body, and ^{222}Rn has been recommended for the treatment of the respiratory tract by inhalation in radon galleries or it has been encapsulated in small thin-walled gold tubes and introduced into the body for treatment of cancer.

Various artificial radionuclides were then applied in relatively large amounts for external and internal irradiation, and some of these radionuclides are still used in radiotherapy, either as external radiation sources (e.g. ^{60}Co and ^{137}Cs for treatment of cancer), or for internal application. Examples are ^{131}I in amounts of about 200 to 1000 MBq for treatment of hyperthyroidism or thyroid cancer, ^{32}P for treatment of bone cancer or leukaemia and ^{198}Au for treatment of ovarian cancer.

A fascinating aspect of radiotherapy is the development of monoclonal antibodies that are labelled with α or β emitters and able to seek out particular types of cancer cells, to which they deliver large absorbed doses, whereas neighbouring tissues receive only small doses.

For all internal medical applications, in particular for diagnostic purposes, radionuclide purity is of the highest importance. The absence of long-lived radioactive impurities, in particular α emitters, such as actinides, or high-energy β emitters, such as ^{90}Sr, must be guaranteed. The main problem in checking radionuclide purity of short-lived radionuclides is that many impurities, for example pure β emitters, can only be detected after decay of these short-lived radionuclides or after chemical separation. This makes checking of purity before application difficult, and development of suitable and reliable procedures for control of radionuclide purity is of great importance.

19.6 Single Photon Emission Tomography (SPET)

In computer tomography (CT) with radionuclides, one or several radiation detectors, a computer and a display are used. The detector array is moved in relation to the patient, and the variations in counting rates with the absorbancies of the radiation in the body as a function of the geometry are processed by the software of the computer to give an image on the screen. This procedure is repeated in subsequent sections (slices) of the body, thus providing a three-dimensional picture. The resolution of the scan is of the order of 1 mm. The method is similar to that used in X-ray CT, but in the latter both the radiation source and the detector array can be moved in relation to the patient.

The application of compounds labelled with suitable radionuclides as radiation sources makes it possible to measure the incorporation and discharge of these substances in certain organs of the body, thus providing information about the metabolism and the function of organs of interest. In this way, malfunction and disorder can be detected at very early stages.

Radionuclides applied for SPET should preferably decay by emission of a single γ-ray photon, and the best resolution is obtained at low γ energies.

The γ rays are usually measured by means of NaI(Tl) crystals to obtain high counting efficiencies. Crystals and collimators are combined with a so-called gamma camera which may be used as a stationary detector system to give a single two-dimensional picture. Several cameras at fixed positions or rotating around the patient are used for scanning, in order to obtain a three-dimensional picture and to study the dynamic behaviour of the radionuclides or labelled compounds in certain organs.

Single photon emission tomography (SPET) is primarily used for cardiovascular and brain imaging. For example, brain tumours can be located after intravenous injection of Na^{99m}TcO$_4$, because such tumours exhibit high affinity and slow release of Tc. On the other hand, in the case of brain infarcts the uptake of Tc is low and its release is fast, whereas the release is still faster from healthy parts of the brain.

The radionuclide most frequently used in SPET is ^{99m}Tc, either for static investigations (e.g. secondary spread of malignancy in bones and liver, pulmonary embolism, thyroid function, occult metastases in the brain) or for dynamic investigations (e.g. pulmonary emphysema, renal function, liver function, motion of the cardiac wall, brain drainage, vascular problems). More than 20 different compounds of Tc are commercially available for diagnosis of diseases and misfunctions in bones, thyroid, liver, kidneys, heart and brain.

Less frequently used radionuclides are ^{123}I (e.g. static investigation of thyroid function, dynamic investigation of renal function), ^{133}Xe (dynamic investigation of pulmonary emphysema) and ^{201}Tl (e.g. static investigation of cardiac infarction and ischaemia).

19.7 Positron Emission Tomography (PET)

In positron emission tomography (PET) the two 511 keV γ-ray photons emitted simultaneously in opposite directions are registered by γ-ray detectors, indicating that the positron decay must have occurred somewhere along the line between these two detectors. The same holds for other events of positron decay, and the radionuclide can be localized at the intersection of these lines.

The patients are positioned inside a ring of about 50 to 100 scintillation detectors, and the ring is rotated and moved in a programmed manner. As in SPET, the results are evaluated by computer software to give a three-dimensional picture of the distribution of the radionuclide in the organ of interest. The resolution is also of the order of 1 mm.

Positron emitters most frequently used for positron emission tomography (PET) are listed in Table 19.2. As already mentioned in section 19.5, production of these positron emitters requires the availability of a suitable cyclotron, fast chemical separation techniques and fast syntheses.

PET is primarily used for kinetic investigations in the brain, heart and lungs. For example, ^{11}C-labelled glucose has been applied extensively for the study of brain metabolism. Because glucose is the only energy source used in the brain, the rate of glucose metabolism provides information about brain viability. By application of PET, valuable new information about various forms of mental illness, such as epilepsy, manic depression and dementias, has also been obtained. The development of new radiopharmaceuticals that are able to pass the blood/brain barrier will contribute to better understanding of normal and abnormal functioning of the brain.

Literature

General

G. Hevesy, Radioactive Indicators: Application to Biochemistry, Animal Physiology and Pathology, Interscience, New York, **1948**

W. E. Siri (Ed.), Isotopic Tracers, McGraw-Hill, New York, **1949**

M. D. Kamen, Radioactive Tracers in Biology, Academic Press, New York, **1949**

M. Calvin, C. Heidelberger, J. C. Reid, B. M. Tolbert, P. F. Yankwich, Isotopic Carbon, Wiley, New York, **1949**

G. Wolf, Isotopes in Biology, Academic Press, New York, **1964**

International Atomic Energy Agency, Radiopharmaceuticals and Labelled Compounds, Vol. I, IAEA, Vienna, **1974**

C. H. Wang, D. H. Willis, W. D. Loveland, Radiotracer Methodology in the Biological, Environmental and Physical Sciences, Prentice Hall, Englewood Cliffs, NJ, **1975**

J. H. Lawrence, T. F. Budinger (Eds.), Recent Advances in Nuclear Medicine, Grune and Stratton, New York, **1978**

R. M. Lambrecht, N. A. Morcos (Eds.), Applications of Nuclear and Radiochemistry, Part I, Radiopharmaceutical Chemistry, Pergamon, Oxford, **1982**

Radiochemistry Related to the Life Sciences, Radiochim. Acta *30*, 123 **(1982)** and *34*, 1 **(1983)**

International Atomic Energy Agency, Radiopharmaceuticals and Labelled Compounds, IAEA, Vienna, **1985**

P. P. Van Rijk (Ed.), Nuclear Techniques in Diagnostic Medicine, Martinus Nijhoff, Dordrecht, **1986**

P. H. Cox, S. J. Mather, C. B. Sampson, C. R. Lazarus (Eds.), Progress in Radiopharmacy, Martinus Nijhoff, Dordrecht, **1986**

H. Deckart, P. H. Cox (Eds.), Principles of Radiopharmacology, G. Fischer, Jena, **1987**

A. Theobald (Ed.), Radiopharmaceuticals Using Radioactive Compounds in Pharmaceutics and Medicine, Ellis Horwood, Chichester, **1989**

T. A. Baillie, J. R. Jones (Eds.), Synthesis and Application of Isotopically Labelled Compounds, Elsevier, Amsterdam, **1989**

A. H. W. Nias, An Introduction to Radiobiology, Wiley, New York, **1990**

A. D. Nunn (Ed.), Radiopharmaceuticals: Chemistry and Pharmacology, Marcel Dekker, New York, **1992**

P. A. Schubiger, G. Westera (Eds.), Progress in Radiopharmacy, Kluwer, Dordrecht, **1992**

G. Stöcklin, S. M. Qaim, F. Rösch, The Impact of Radioactivity on Medicine, Radiochim. Acta *70/71*, 249 **(1995)**

More Special

L. E. Feinendegen, Tritium-labelled Molecules in Biology and Medicine, Academic Press, New York, **1967**

J. C. Clark, P. D. Buckingham, Short-lived Radioactive Gases for Clinical Use, Butterworths, London, **1975**

G. E. Abraham, Handbook of Radioimmunoassay, Marcel Dekker, New York, **1977**

S. M. Qaim, Nuclear Data Relevant to Cyclotron Produced Short-lived Medical Radioisotopes, Radiochim. Acta *30*, 147 **(1982)**

W. C. Eckelman (Ed.), Receptor-Binding Radiotracers, Vols. I, II, CRC Press, Boca Raton, FL, **1982**

H. A. O'Brien jr., Radioimmunoimaging and Radioimmunotherapy, Elsevier, Amsterdam, **1983**

J. R. Sharp (Ed.), Guide to Good Pharmaceutical Manufacturing Practice, HMSO, London, **1983**

E. Pochin, Nuclear Radiation; Risks and Benefits, Clarendon Press, Oxford, **1983**

M. F. L'Annunziata, J. O. Legg (Eds.), Isotopes and Radiation in Agricultural Sciences, Academic Press, New York, **1984**

P. F. Sharp, P. P. Dendy, W. I. Keyes, Radionuclide Imaging Technique, Academic Press, New York, **1985**

M. Guillaume, C. Brihaye, Generators of Ultra-short-lived Radionuclides for Routine Clinical Applications, Radiochim. Acta *41*, 119 **(1987)**

C. E. Swenberg, J. J. Conklin, Imaging Techniques in Biology and Medicine, Academic Press, New York, **1988**

S. Webb (Ed.), The Physics of Medical Imaging, Hilger, London, **1988**

H. Hundeshagen (Ed.), Spezielle Syntheseverfahren mit kurzlebigen Radionukliden und Quali-tätskontrolle. Emissions-Computertomographie mit kurzlebigen Zyklotron-produzierten Radio-pharmaka, Springer, Berlin, **1988**

S. C. Srivastava (Ed.), Radiolabeled Monoclonal Antibodies for Imaging and Therapy, Plenum Press, New York, **1988**

R. Bryant, The Pharmaceutical Quality Control Handbook, Aster, Eugene, OR, **1989**

C. Beckers, A. Goffinet, A. Bol (Eds.), Positron Emission Tomography in Clinical Research and Clinical Diagnosis. Tracer Modelling and Radioreceptors, Kluwer, Dordrecht, **1989**

M. R. Kilbourn, Fluorine-18 Labelling of Radiopharmaceuticals, Nuclear Science Series NAS-NS, National Academy Press, Washington, DC, **1990**

The Rules Governing Medical Products in the European Community, Commission of the European Communities, Brussels, **1990**

M. K. Dewanjee, Radioiodination, Theory, Practice and Biomedical Applications, Kluwer, Dor-drecht, **1991**

H. J. Biersack, P. H. Cox (Eds.), Nuclear Medicine in Gastroenterology, Kluwer, Dordrecht, **1991**

D. Kuhl (Ed.), Frontiers in Nuclear Medicine: In Vivo Imaging of Neurotransmitter Functions in Brain, Heart and Tumors, Amer. Coll. Nucl. Phys., Washington, DC, **1991**

R. P. Baum, P. H. Cox, G. Hör, Burragi (Eds.), Clinical Use of Antibodies, Tumors, Infection, Infarction, Rejection, and in the Diagnosis of AIDS, Kluwer, Dordrecht, **1991**

G. Stöcklin, V. W. Pike (Eds.), Radiopharmaceuticals for Positron Emission Tomography, Kluwer, Dordrecht, **1993**

B. M. Mazoyer, W. D. Heuss, D. Comar (Eds.), PET Studies on Aminoacid Metabolism and Pro-tein Synthesis, Kluwer, Dordrecht, **1993**

C. F. Meares, Perspectives in Bioconjugate Chemistry, Amer. Chem. Soc., Washington, DC, **1993**

T. J. McCarthy, S. W. Schwarz, M. J. Welch, Nuclear Medicine and Positron Emission Tomog-raphy, J. Chem. Educ. *71*, 830 **(1994)**

R. E. Weiner, M. L. Thakur, Metallic Radionuclides: Applications in Diagnostic and Therapeutic Nuclear Medicine, Radiochim. Acta *70/71*, 273 **(1995)**

A. G. Jones, Technetium in Nuclear Medicine, Radiochim. Acta *70/71*, 289 **(1995)**

Therapy with heavy ions

G. Kraft, The Radiobiological and Physical Basis for Radiotherapy with Protons and Heavier Ions, Strahlenther. Onkol. *166*, 10 **(1990)**

U. Amaldi, B. Larsson (Eds.), Hadrontherapy in Oncology, Elsevier, Amsterdam, **1994**

Radioanalytical methods

K. Heydorn (Ed.), Neutron Activation Analysis for Clinical Trace Element Research, Vols. I, II, CRC Press, Boca Raton, FL, **1984**

D. M. Wieland, M. C. Tobes, T. J. Mangner, Analytical and Chromatographic Techniques in Radiopharmaceutical Chemistry, Springer, Heidelberg, **1986**

20 Technical and Industrial Applications of Radionuclides and Nuclear Radiation

20.1 Radiotracer Techniques

In the technical sciences and in industry, radiotracer techniques are preferably applied for the investigation of mixing and separation procedures and of transport processes in machines, technical plants and pipelines.

Table 20.1 gives a survey of various radionuclides used as radiotracers (indicators) in industry. T and ^{14}C emit low-energy β^- radiation and therefore cannot be detected from outside. However, samples containing these radionuclides can be measured with high sensitivity by use of liquid scintillation counting, which offers the possibility of applying them over a wide activity range and of measuring them after high dilution. This is illustrated in Fig. 20.1.

Table 20.1. Examples of the application of radiotracers as indicators in technology.

Radionuclide	Half-life	Measured radiation	Application
^{3}H	12.323 y	β^-	In aqueous solutions or organic compounds
^{14}C	5730 y	β^-	In organic compounds
^{24}Na	14.96 h	γ	In aqueous solutions (carbonate); in organic compounds (salicylate or naphthenate)
^{41}Ar	1.83 h	γ	In gases
^{46}Sc	83.82 d	γ	In glasses
^{51}Cr	27.7 d	γ	In aqueous solutions (complexes)
^{82}Br	35.34 h	γ	In aqueous solutions (KBr or HN$_4$Br); in organic compounds (e.g. p-dibromobenzene)
^{85}Kr	10.76 y	γ	In gases
^{125}Xe	16.9 h	γ	In gases
^{133}Xe	5.25 d	γ	In gases
^{140}La	40.272 h	γ	In solids (oxide); in aqueous solution (acetate); in organic compunds (naphthenate)
^{192}Ir	73.83 d	γ	In glasses
^{198}Au	2.6943 d	γ	In organic compounds (colloid)

Radionuclide generators have not found broad application in industry, mainly because of the regulations with respect to handling of radioactive substances and the risk of contamination.

In the metal industry, radionuclides may be used to study diffusion and formation of alloys. Corrosion and wear can also be studied with high sensitivity. An example is the investigation of the wear of piston rings in motors: after activation of the

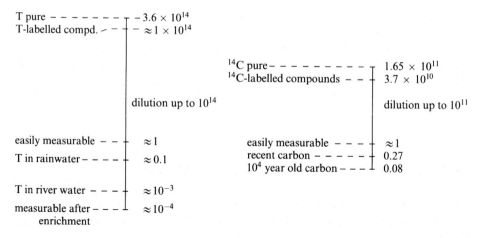

Figure 20.1. Specific activity of T and ^{14}C in Bq per gram of hydrogen and carbon, respectively.

piston rings, the places of greatest wear could be identified at the upper reversal. Application of radiotracers in the investigation of wear offers the advantages of short experiments and relatively exact information. Instead of using activated samples, implants containing ^{85}Kr may be applied. In this case, the liberation of ^{85}Kr is a measure of the wear.

In the water and oil industries, transport processes can be investigated. An example is the study of underground currents by application of tritiated water. In this application, advantage is taken of the high sensitivity of tritium detection (Fig. 20.1; T can be determined in an atomic ratio down to $T:H \approx 10^{-19}$).

For the measurement of the flow of gases and liquids, radionuclides or labelled compounds may be injected and their activity may be measured from a certain distance outside. The activity recorded as a function of time provides information about the flow dynamics.

Material flux and residence time in technical installations, for instance in mixers, can be studied by addition of a radiotracer and taking samples as a function of time. Attainment of constant activity in mixers indicates thorough mixing. In a similar way the residence time in vessels can be determined by adding a radionuclide or a labelled compound at the entrance of the vessel and measuring the activity at the exit as a function of time.

In all these investigations, it is necessary that the behaviour of the radioactive tracer (indicator) and of the substance to be investigated is the same. If the substance to be investigated is chemically changed in the process, it is necessary that the tracer is applied in the same chemical form.

The material balance in technical installations can be studied by single injection of a radiotracer and continuous sampling in all streams, or by continuous addition of the radiotracer up to stationary distribution in all streams and single sampling in these streams. In general, the second method requires application of higher activities, but it gives more exact results with respect to material balance. T and ^{14}C are preferred for these kinds of investigations.

Water-tightness of pipelines may be checked by injecting short-lived γ emitters and measuring possible release from outside. In this way, leaks can be found without digging up the ground. Similarly, leaks in heat exchangers or cooling systems can be investigated. In these applications, the following procedure is frequently applied: after injection of the radionuclide or the labelled compound and homogeneous distribution in the system, the radioactive component is washed out and the outside of the container is checked for radioactivity.

20.2 Absorption and Scattering of Radiation

Absorption or scattering of radioactive radiation is applied in industry for measurement of thickness or for material testing. For example, the production of paper, plastic or metal foils or sheets can be controlled continuously by passing these materials between an encapsulated radionuclide as the radiation source and a detector combined with a ratemeter, as shown in Fig. 20.2. After appropriate calibration, the ratemeter directly indicates the thickness. The radionuclide is chosen in such a way that the radiation emitted is effectively absorbed in the materials to be checked. Thus, the thickness of plastic foils is measured by use of β emitters, whereas ^{137}Cs or other γ emitters are used for measuring the thickness of metal sheets.

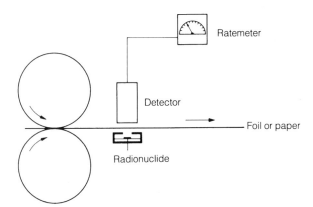

Figure 20.2. Application of radionuclides for thickness measurements.

The thickness of layers of heavy metals, such as Au or Pt, applied on other metals can be determined by measuring the backscattered radiation of β or γ emitters. In this method, use is made of the fact that backscattering increases markedly with the atomic number of the backscattering material (section 6.3). The thickness of very thin layers of heavy metals is measured by use of low-energy β emitters (e.g. ^{14}C), whereas thick layers of heavy metals can be measured by application of high-energy β emitters or of γ emitters.

The thickness of metal coatings can be determined by measuring the characteristic X rays excited by irradiation with γ rays or X rays. In this application, either the characteristic X rays from the layer to be investigated or the absorption in this layer of characteristic X rays emitted from the material underneath can be measured. For example, the thickness of layers of Zn, Al, Sn and Cr on iron or steel has been determined in this way.

Another technical application of γ-ray emitters is the determination of the density of various substances. The basis of this application is the fact that absorption of high-energy γ rays depends mainly on the electron density of the absorber, i.e. approximately on the mass per unit area, and at constant absorber thickness on the density. Density profiles in closed installations, for example in distillation columns, can also be investigated or controlled in this way. Preferred radiation sources are ^{60}Co, ^{137}Cs and ^{241}Am.

Another example of the application of encapsulated sources of radioactive radiation in technical installations is the control of material transport on conveyor belts. The counting rate of the ratemeter is a measure of the thickness or the density, respectively, of the material transported by the belt. By multiplication by the velocity of the conveyor belt, the mass of the transported material is obtained. In the same way, solids transported with gas streams can be determined.

The height of filling in technical installations, e.g. in vessels, containers or tubes, can easily be controlled from outside by means of radionuclides as radiation sources. The radionuclides are selected according to the most appropriate γ energy for the task, in particular the diameters of the equipment and the thickness of the walls. The phase boundary between two liquid phases differing markedly in their properties as neutron moderators can be located from outside by use of a neutron source and a detector for low-energy neutrons.

The size of small particles suspended in a solution can be determined by measuring the scattering of X rays emitted by radionuclides, provided that the atomic numbers of the elements in the particles are high and those in the solvent are low. Under these conditions, Compton scattering depends markedly on the particle size, whereas Rayleigh scattering is only slightly influenced, and the ratio of Compton to Rayleigh scattering is taken as a measure of the particle size. The size of solid particles in a gas can be determined by use of β emitters.

The application of radionuclides as radiation sources for analytical applications has been described in sections 17.10 and 17.11.

20.3 Radiation-induced Reactions

The possibility of initiating chemical reactions by means of the ionizing radiation from radionuclides has been extensively investigated, but it has not found the broad application in industry that had been expected, because of possible radiation hazards.

As radiation sources γ-ray emitters, such as ^{60}Co or ^{137}Cs, and electron accelerators are applied. Nuclear reactors and spent fuel elements have also been discussed as radiation sources.

Chemical reactions induced by ionizing radiation are the field of radiation chemistry. In these reactions, the heat transferred by the radiation to the system is insignificant, as can be illustrated by the following example: The energy of the γ rays emitted by a 1000 Ci (3.7×10^{13} Bq) ^{60}Co source per day is ≈ 1250 kJ, and with this energy ≈ 3.5 l of water could be heated from room temperature to boiling. The interesting feature of the interaction of ionizing radiation with matter is the formation of excited states, ions and radicals, as described in section 6.1.

Radiation chemistry is closely related to photochemistry, which deals with chemical reactions induced by light, in particular ultraviolet light (UV). The energy of light is in the range of $1\,eV$ ($\approx 1240\,nm$) to $10\,eV$ ($\approx 124\,nm$), and the energy of nuclear radiation (γ-ray photons, β and α particles) extends from about $100\,eV$ to about $10\,MeV$. Due to the different energies, one photon of light can usually trigger only one primary reaction (excitation or ionization), whereas one γ-ray photon or one β or α particle induces a great number of primary reactions (section 6.1).

Because of the similarities, photochemistry can be considered to be the low-energy branch of radiation chemistry. In both fields, the primary reactions are followed by secondary and subsequent reactions. Chain reactions are of special interest.

The yield of a photochemical reaction is characterized by the quantum yield, which is the number of molecules formed or decomposed by one photon. In radiation chemistry, the G-value is used as a measure of the chemical yield. It is defined by the number of molecules formed or decomposed per $100\,eV$ energy absorbed in the system. For example, $G(H_2) = 3$ means that 3 molecules of H_2 are formed per $100\,eV$ absorbed and $G(-H_2O) = 11$ means that 11 molecules H_2O are decomposed per $100\,eV$ absorbed.

Some examples of radiation-induced reactions are listed in Table 20.2. Chain reactions exhibiting G-values $> 10^3$ are of greatest interest for technical applications.

Table 20.2. Examples of radiation-induced reactions.

Reactions	G-values
Production of O_3 by irradiation of O_2	6–10
Production of NO_2 by irradiation of N_2/O_2 mixtures (by-products NO and N_2O)	1–7
Production of C_2H_5Br by irradiation of a mixture of C_2H_4 and HBr	$>10^5$
Production of C_2H_5Cl by irradiation of a mixture of C_2H_4 and HCl	$\approx 10^4$
Chlorination (similar to the photochemical reaction by UV), e.g. by irradiation of a mixture of C_6H_6 and Cl_2	10^4–10^5
Oxidation of carbohydrates; production of phenol by irradiation of a mixture of C_6H_6 and O_2	10^4–10^5
Sulfochlorination (similar to the photochemical reaction by UV), e.g. production of sulfonic acid chlorides by irradiation of mixtures of carbohydrates, SO_2 and Cl_2	$\approx 10^7$
Production of alkylsulfonic acids by irradiation of mixtures of carbohydrates, SO_2 and O_2	10^3–10^4

In macromolecular chemistry, ions or radicals formed in primary reactions often trigger chain reactions; this also offers the possibility of improving the properties of products. Radiation may cause interlacing or degradation of macromolecular compounds. Interlacing leads to an increase of hardness, while elasticity and solubility decrease. It predominates in the case of polyethylene, polystyrene and caoutchouc, whereas in other polymers, such as perspex, decomposition prevails. Radiation-induced processes have found interest mainly for the production of polyethylene. Compared with the products obtained by application of high pressure, radiation leads to products of higher density and higher softening temperature. Plants for polyethylene production by application of radiation have been built.

Another application in macromolecular chemistry is radiation-induced graft polymerization, by which favourable properties of two polymers can be combined. In this process, copolymers of A and B are produced by irradiation of the polymer A_n in the presence of the monomer B. Examples are graft polymers of polyethylene and acrylic acid or of polyvinyl chloride and styrene. The properties of textiles (cellulose, wool, natural silk, polyamides, polyesters) can also be modified by graft polymerization, for example for the production of weatherproof products.

Very hard coatings may be obtained by irradiation of polymerized varnishes on wood or metals. Irradiation of wood impregnated with a monomer leads to water-proof products of high stability and hardness. Similarly to these wood–polymer combinations, concrete–polymer combinations can also be produced that exhibit high resistivity in water, particularly in seawater.

The radiation emitted by radionuclides may also be used for sterilization. For example, medical supplies can be sterilized by radiation doses of the order of 10^5 J/kg (10^5 Gy) from ^{60}Co or ^{137}Cs sources of $\approx 10^{11}$ Bq. The safety of food irradiation has been accepted by the IAEA (International Atomic Energy Agency), the FAO (Food and Agricultural Organization) and the WHO (World Health Organization), and food conservation has been used for several decades without any negative health effects. Examples are potatoes, wheat and wheat products, onions, shrimps, fried meat and other agricultural products. The aim is to achieve complete destruction of all bacteria with minimal change in taste due to formation of small amounts of decomposition products. Partial sterilization with lower doses and irradiation at low temperatures cause smaller taste changes.

Treatment of sludge by irradiation with γ rays is applied with the aim of destroying harmful organisms, before the sludge is used as fertilizer.

In all biological applications, it has to be taken into account that the sensitivity to radiation increases markedly with the evolutionary stage of the species.

Radiation doses of ≈ 80 J/kg have been used to sterilize the males of insect species, which are released after sterilization and mate with females. In this way, further reproduction of the species is reduced or prevented. The technique has been applied in the USA, Mexico, Egypt, Libya and other countries to eradicate screw-worm flies and other insect species that are threats to agriculture.

In addition to the applications mentioned so far, the radiation from radionuclides may also be used for ionization and excitation of luminescence. Ionization is desired to remove electric charges (e.g. on analytical balances) or to trigger electric discharge (e.g. in electron valves). ^{204}Tl, for example, has been applied in analytical balances. For excitation of luminescence (e.g. in luminescent substances in watches or in fluorescent screens), ^{228}Ra has been used. However, ^{228}Th causes relatively high activity of the luminescent substances and of watches, and to-day T is preferred for excitation of luminescence, because its low-energy β radiation is absorbed within the watch.

20.4 Energy Production by Nuclear Radiation

Energy production by nuclear fission and by thermonuclear reactions has been discussed in chapter 11. In this section, energy production in radionuclide batteries by the radiation emitted by radionuclides will be considered.

The purpose of radionuclide batteries is to produce energy over longer periods of time without a need for maintenance. An advantage of radionuclide batteries is the relatively high energy output related to mass and volume of the radionuclide. Although the energy of radioactive decay is only of the order of 0.1 to 1% of the energy obtained by nuclear fission, it is attractive for use in maintenance-free energy sources in satellites, remote meteorological stations and oceanography.

Table 20.3. gives a survey of radionuclides applicable in radionuclide batteries. For selection of suitable radionuclides the following criteria are important: The half-life should be long compared with the desired operation time (usually ≥ 10 y), to obtain as far as possible constant power. Furthermore, the power output per mass should be as high as possible. This is achieved if the half-life is not too long ($<10^3$ y) and if the energy of the radiation is high. Alpha emitters have the advantage that the energy of decay is relatively high and that the α particles are effectively absorbed. Radionuclides decaying by subsequent emission of several α particles, such as ^{238}Pu and ^{232}U, are most favourable as energy sources. ^{238}Pu is produced by irradiation of ^{237}Np in nuclear reactors and is used in radionuclide batteries installed in

Table 20.3. Radionuclides for application in radionuclide batteries.

Radionuclide	Half-life [y]	Radiation	Production
^{3}H	12.323	β^-	^{6}Li$(n,\alpha)^3$H
^{14}C	5730	β^-	^{14}N$(n,p)^{14}$C
^{60}Co	5.272	β^-, γ	^{59}Co$(n,\gamma)^{60}$Co
^{63}Ni	100	β^-	^{62}Ni$(n,\gamma)^{63}$Ni
^{85}Kr	10.76	β^-, γ	Fission product
^{90}Sr	28.64	β^-, γ $(^{90}$Y$)$	Fission product
^{106}Ru	1.02	β^-, γ $(^{106}$Rh$)$	Fission product
^{137}Cs	30.17	β^-, γ	Fission product
^{144}Ce	0.78	β^-, γ	Fission product
^{147}Pm	2.62	β^-, γ	Fission product
^{170}Tm	0.35	$\beta^-, (\varepsilon), \gamma, e^-$	^{169}Tm$(n,\gamma)^{170}$Tm
^{171}Tm	1.92	β^-, γ	^{170}Er$(n,\gamma)^{171}$Er $\xrightarrow{\beta^-}$ ^{171}Tm
^{204}Tl	3.78	$\beta^-, (\varepsilon)$	^{203}Tl$(n,\gamma)^{204}$Tl
^{210}Po	0.38	α, γ	Decay product of ^{238}U
^{228}Th	1.913	α, γ	Decay product of ^{232}Th
^{232}U	68.9	$\alpha, (\text{sf}), \gamma$	^{230}Th$(n,\gamma)^{231}$Th $\xrightarrow{\beta^-}$ ^{231}Pa; ^{231}Pa$(n,\gamma)^{232}$Pa $\xrightarrow{\beta^-}$ ^{232}U
^{238}Pu	87.74	$\alpha, (\text{sf}), \gamma$	$\begin{cases} ^{237}\text{Np}(n,\gamma)^{238}\text{Np} \xrightarrow{\beta^-} {}^{238}\text{Pu} \\ \text{Decay product of } ^{242}\text{Cm} \end{cases}$
^{241}Am	432.2	$\alpha, (\text{sf}), \gamma$	^{240}Pu$(n,\gamma)^{241}$Pu $\xrightarrow{\beta^-}$ ^{241}Am
^{242}Cm	0.45	$\alpha, (\text{sf}), \gamma$	^{241}Am$(n,\gamma)^{242}$Am $\xrightarrow{\beta^-}$ ^{242}Cm
^{244}Cm	18.10	$\alpha, (\text{sf}), \gamma$	^{243}Am$(n,\gamma)^{244}$Am $\xrightarrow{\beta^-}$ ^{244}Cm

satellites. ^{232}U can be produced in two stages: neutron irradiation of ^{230}Th (ionium), followed by separation of the ^{231}Pa produced and neutron irradiation of the latter.

The energy of the α or β particles can be converted directly or indirectly to electric energy. Direct conversion is possible by use of charging or contact potentials or by radiophotovoltaic conversion. However, direct conversion is restricted to a power of up to about 10^{-4} W. Indirect conversion is mostly based on the use of the heat generated by absorption of the radiation (thermal conversion). In this application, the encapsulated radiation source serves as the heat source. Thermoelectric conversion operates with thermoelements (e.g. Bi–Te, Pb–Te or Ge–Si, depending on the temperature). The efficiency of thermoelectric conversion is of the order of 5 to 10%. For application in the space, thermoelectric ^{238}Pu-loaded radionuclide batteries of electric power between about 1 W and 1 kW are used. Smaller units of electric power of about 0.1 to 1 mW, also containing ^{238}Pu, were developed as energy sources with a long lifetime for medical application in pacemakers, but these have now been substituted by electrochemical batteries with lifetimes of several years. Prototypes of other radionuclide batteries operating with ^{90}Sr, ^{60}Co, ^{144}Ce, ^{210}Po or ^{244}Cm have also been developed.

The principle of thermionic conversion is that of a diode, in which the cathode (emitter) emits electrons that are collected at the anode (collector). Alloys of W, Re, Mo, Ni or Ta are used as emitters, and the diodes operate at temperatures of about 2200 K. The efficiency varies between about 1 and 10%, depending on the power. In order to take advantage of thermionic conversion, radionuclides of high specific power output (high power per mass unit) are needed, such as ^{238}Pu, ^{232}U, ^{227}Ac, ^{242}Cm. Prototypes of 0.1 to 1 kW have been developed.

In thermophotovoltaic batteries the heat emitted by the radionuclides is converted to electric energy by means of infrared-sensitive photoelements (e.g. Ge diodes), which must be cooled effectively because the efficiency decreases drastically as the temperature rises. With respect to high emitter temperatures, thermophotovoltaic conversion is of interest for power levels between about 10 W and 1 kW, but the efficiency is relatively low (up to about 5%).

Radiophotovoltaic (photoelectric) radionuclide batteries operate in two stages. First the radiation energy is converted to light by means of luminescent substances and then to electric energy by means of photoelements. Because of radiative decomposition of luminescent substances, the number of radionuclides applicable is limited. Alpha emitters are unsuitable, and the most suitable β emitter is ^{147}Pm. The construction of this kind of radionuclide batteries is relatively simple: radionuclide and luminophore are mixed in a ratio of about 1:1 and brought between two photoelements (e.g. Cu–Se or Ag–Si) in form of a thin layer. Efficiencies of the order of 0.1 to 0.5% and powers of the order of 10 µW per cm^2 are obtained. Because of the low efficiency, this type of radionuclide battery has no technical significance.

In contrast to the methods described in the preceding paragraphs, radiophotovoltaic conversion is a direct method. In a semiconductor the incident radiation generates free charge carriers, that are separated in the n,p-barrier layer of the semiconductor. Suitable radiation sources are radionuclides emitting β particles with energies below the limits of radiation damage in the semiconductors. These limits are about 145 keV for Si and about 350 keV for Ge. Therefore only ^{147}Pm, ^{14}C, ^{63}Ni and T are suitable as radiation sources. By use of the combination ^{147}Pm/Si, efficiencies of about 4% are obtained.

Dynamic converters are based on different principles from radionuclide batteries. They contain moving parts and are not maintenance-free. However, relatively high efficiencies of about 20% are obtained with dynamic converters operating with steam engines, Stirling motors or gas turbines. In these types of converters the radiation energy is converted in three stages to electric energy (radiation energy to heat, heat to mechanical energy and mechanical to electric energy).

Literature

General

W. J. Whitehouse, J. L. Putman, Radioactive Isotopes – An Introduction to their Preparation, Measurement and Use, Clarendon Press, Oxford, **1952**

E. Broda, Th. Schönfeld, The Technical Applications of Radioactivity, Vol. 1, Pergamon, Oxford, **1966**

J. A. Heslop, Industrial Applications of Radioisotopes, in: Radiochemistry, Vol. 3, Specialist Periodical Reports, The Chemical Society, London, **1976**

Radiation chemistry

G. J. Dienes, G. H. Vineyard, Radiation Effects in Solids, Interscience, London, **1957**

A. J. Swallow, Radiation Chemistry of Organic Compounds, Pergamon, Oxford, **1960**

A. Charlesby, Atomic Radiation and Polymers, Pergamon, Oxford, **1960**

A. O. Allen, The Radiation Chemistry of Water and Aqueous Solutions, Van Nostrand, London, **1961**

S. C. Lind, Radiation Chemistry of Gases, Reinhold, New York, **1961**

D. W. Billington, J. H. Crawford, Radiation Damage in Solids, Princeton University Press, Princeton, NJ, **1961**

A. Chapiro, Radiation Chemistry of Polymeric Systems, Interscience, London, **1962**

P. J. Dyne, D. R. Smith, J. A. Stone, Radiation Chemistry, Annu. Rev. Phys. Chem. *14*, 313 **(1963)**

R. O. Bolt, J. G. Carroll, Radiation Effects on Organic Materials, Academic Press, New York, **1963**

M. Haissinsky, M. Magat, Radiolytic Yields, Pergamon, Oxford, **1963**

H. A. Schwarz, Radiation Chemistry, Annu. Rev. Phys. Chem. *16*, 347 **(1965)**

B. T. Kelly, Irradiation Damage to Solids, Pergamon, Oxford, **1966**

L. M. Dorfman, R. F. Firestone, Radiation Chemistry, Annu. Rev. Phys. Chem. *18*, 177 **(1967)**

E. J. Hart (Ed.), Radiation Chemistry, Advances in Chemistry Series 81, Amer. Chem. Soc., Washington, DC, **1968**

J. H. O'Donnell, D. F. Sangster, Principles of Radiation Chemistry, Elsevier, Amsterdam, **1970**

A. R. Denaro, G. G. Jayson, Fundamentals of Radiation Chemistry, Butterworths, London, **1972**

G. Hughes, Radiation Chemistry, Oxford Chemistry Series, Clarendon Press, Oxford, **1973**

A. J. Swallow, Radiation Chemistry, Longman, London, **1973**

M. Burton, J. Z. Magee (Eds.), Advances in Radiation Chemistry, Wiley, New York, **1974**

E. R. L. Gaughran, A. J. Goudie (Eds.), Technical Developments and Prospects of Sterilization by Ionizing Radiation, Vols. 1 and 2, Multiscience, Montreal, **1974** and **1978**

H. Drawe, Strahlensynthesen, Chem. Ztg. *100*, 212 **(1976)**

J. Foeldiak (Ed.), Radiation Chemistry of Hydrocarbons, Elsevier, Amsterdam, **1981**

Farhataziz, and M. A. J. Rogers (Eds.), Radiation Chemistry, Principles and Applications, VCH, New York, **1987**

J. Bednar, Theoretical Foundations of Radiation Chemistry, Kluwer, Dordrecht, **1989**

V. K. Milinchuk, V. I. Pupikov (Eds.), Organic Radiation Chemistry Handbook, Ellis Horwood, Chichester, **1989**

J. W. T. Spinks, R. I. Woods, An Introduction to Radiation Chemistry, 3rd ed., Wiley, New York, **1990**

D. W. Clegg, A. A. Collyer (Eds.), Irradiation Effects on Polymers, Elsevier, Amsterdam, **1991**

Other applications

International Atomic Energy Agency, Nuclear Well Logging in Hydrology, Tech. Rep 126, IAEA, Vienna, **1971**

International Atomic Energy Agency, Commercial Portable Gauges for Radiometric Determination of the Density and Moisture Content of Building Materials, Tech. Rep. 130, IAEA, Vienna, **1971**

D. Schalch, A. Scharmann, K.-J. Euler, Radionuklidbatterien, I. Grundlagen und Grenzen, Kerntechnik *17*, 23 **(1975)**; II. Eigenschaften und Anwendungen, Kerntechnik *17*, 57 **(1975)**

R. Spohr, Ion Tracks and Microtechnology, Vieweg, Wiesbaden, **1990**

International Atomic Energy Agency, Isotope Techniques in Water and Resources Development, IAEA, Vienna, **1992**

21 Radionuclides in the Geosphere and the Biosphere

21.1 Sources of Radioactivity

Two groups of radioactivity sources on the earth are to be distinguished, natural and anthropogenic. Natural sources such as ^{40}K, ^{232}Th, ^{235}U and ^{238}U have been produced in the course of nucleogenesis (primordial radionuclides; chapter 15) and have been present on the earth from the beginning. Further sources of natural radioactivity are the cosmogenic radionuclides, such as T and ^{14}C; they are produced continuously by interaction of cosmic rays with the atmosphere.

By mining of ores and minerals, appreciable amounts of natural radionuclides, in particular ^{40}K, ^{232}Th, ^{235}U, ^{238}U and the members of the thorium, uranium and actinium decay series, are brought up to the surface of the earth and contribute to the radioactivity in the environment. In nuclear power stations artificial radionuclides, mainly fission products and transuranium elements, are produced, and great care is usually taken in handling the resulting radioactive waste safely and to localize it to selected places inaccessible to man. The optimal conditions of final storage of high-level waste (HLW) are still a subject of discussion. On the other hand, by application of nuclear weapons, by nuclear weapon tests and nuclear accidents, considerable amounts of fission products and radioelements have been set free and distributed via the atmosphere as radioactive fall-out over large areas, in particular in the northern hemisphere.

Radionuclides of major importance in the geosphere and the biosphere are listed in Table 21.1. Not taken into account are: radionuclides with half-lives $t_{1/2} < 1$ d (in the case of activation products of materials used in nuclear reactors, $t_{1/2} < 1$ y) and with half-lives $t_{1/2} > 10^{11}$ y, radionuclides with fission yields <0.01%, radioisotopes of elements that are not members of the natural decay series,and radionuclides produced solely for medical or technical applications. The radionuclides are arranged according to their position in the Periodic Table of the elements, in order to facilitate the discussion of their chemical behaviour. Radionuclides with half-lives >10 y are underlined, because their behaviour over long periods of time is of special importance.

With respect to radiation doses and possible hazards, the local concentrations and the radiotoxicities of the radionuclides have to be taken into account. Local concentrations of fission products and transuranium elements are high in anthropogenic sources, such as nuclear reactors, reprocessing plants and high-level radioactive waste (HLW). Local concentrations of radionuclides are also high in natural sources, such as uranium or thorium ores. On the other hand, local concentrations are generally low in the case of fall-out (with the exception of the region of the Chernobyl accident) and off-gas and effluents from nuclear installations. They are also low for dispersed natural radionuclides, such as T, ^{14}C and other widespread natural sources containing K, U or Th and daughter nuclides. The radiotoxicities of

T, ^{14}C and K are also low, whereas they are high for many fission products and for the actinides.

Table 21.1. Radionuclides of practical importance in the geosphere and biosphere (radionuclides with lalf-lives >10 y and present in relatively high concentrations are underlined).

Element	Group of the Periodic Table	Radionuclide (half-life)	Source
H		^{3}H (12.323 y)	Cosmic radiation and nuclear fission
C		^{14}C (5730 y)	Cosmic radiation and nuclear fission
K	I	^{40}K (1.28 · 10^9 y)	Potassium salts
Rb		^{87}Rb (4.80 · 10^{10} y)	Rubidium salts; fission
Cs		^{134}Cs (2.06 y); ^{135}Cs (2.0 · 10^6 y); ^{137}Cs (30.17 y)	} Nuclear fission
Sr	II	^{89}Sr (50.5 d); ^{90}Sr (28.64 y)	Nuclear fission
Ba		^{140}Ba (12.75 d)	Nuclear fission
Ra		^{223}Ra (11.43 d); ^{224}Ra (3.66 d); ^{225}Ra (14.8 d); ^{226}Ra (1600 y); ^{228}Ra (5.75 y)	} Ores/minerals
Sn	IV	^{121}Sn (1.125 d); ^{121m}Sn ($\approx$50 y); ^{123}Sn (129.2 d); ^{125}Sn (9.64 d)	} Nuclear fission
Pb		^{210}Pb (22.3 y)	Ores/minerals
Sb	V	^{125}Sb (2.77 y); ^{126}Sb (12.4 d); ^{127}Sb (3.85 d)	} Nuclear fission
Bi		^{210}Bi (5.013 d)	Ores/minerals
Se	VI	^{79}Se (4.8 · 10^5 y)	Nuclear fission
Te		^{127m}Te (109 d); ^{129m}Te (33.6 d); ^{131m}Te (1.25 d) ^{132}Te (3.18 d)	} Nuclear fission
Po		^{210}Po (138.38 d)	Ores/minerals
I	VII	^{129}I (1.57 · 10^7 y); ^{131}I (8.02 d)	Nuclear fission
Kr	0	^{85}Kr (10.76 y)	Nuclear fission
Xe		^{133}Xe (5.25 d); ^{133m}Xe (2.19 d)	Nuclear fission
Rn		^{222}Rn (3.825 d)	Ores/minerals
Ag	IA	^{111}Ag (7.45 d)	Nuclear fission
Cd	IIA	^{109}Cd (1.267 y); ^{113m}Cd (14.6 y); ^{115}Cd (2.224 d); ^{115m}Cd (44.8 d)	} Activation; fission
Y	IIIA	^{90}Y (2.671 d); ^{91}Y (58.5 d)	Nuclear fission
La		^{140}La (1.678 d)	Nuclear fission
Ce		^{141}Ce (32.5 d); ^{143}Ce (1.375 d); ^{144}Ce (284.8 d)	} Nuclear fission
Pr		^{143}Pr (13.57 d)	Nuclear fission
Nd		^{147}Nd (10.98 d)	Nuclear fission
Pm		^{147}Pm (2.623 y); ^{149}Pm (2.212 d); ^{151}Pm (1.183 d)	} Nuclear fission
Sm		^{151}Sm (93 y); ^{153}Sm (1.928 d)	Nuclear fission
Eu		^{155}Eu (4.761 y); ^{156}Eu (15.2 d)	Nuclear fission
Ac		^{225}Ac (10.0 d)	Decay of ^{237}Np
		^{227}Ac (21.773 y)	Ores/minerals

Table 21.1. (continued)

Element	Group of the Periodic Table	Radionuclide (half-life)	Source
Th		^{227}Th (18.72 d); ^{228}Th (1.913 y);	Ores/minerals
		^{229}Th (7880 y);	Decay of ^{237}Np
		^{230}Th (7.54 · 10^4 y);	⎫ Ores/minerals
		^{232}Th (1.405 · 10^{10} y); ^{234}Th (24.10 d)	⎭
Pa		^{231}Pa (3.276 · 10^4 y);	Ores/minerals
		^{233}Pa (27.0 d)	Decay of ^{237}Np
U		^{233}U (1.592 · 10^5 y);	Decay of ^{237}NP
		^{234}U (2.455 · 10^5 y); ^{235}U (7.038 · 10^8 y)	Ores/minerals
		^{236}U (2.342 · 10^7 y); ^{237}U (6.75 d)	Nuclear reactors
		^{238}U (4.468 · 10^9 y)	Ores/minerals
Np		^{236}Np (1.54 · 10^5 y); ^{237}Np (2.144 · 10^6 y)	⎫ Nuclear reactors
		^{238}Np (2.117 d)	⎭
Pu		^{239}Pu (2.411 · 10^4 y); ^{240}Pu (6563 y);	⎫ Nuclear reactors
		^{241}Pu (14.35 y); ^{242}Pu (3.750 · 10^5 y)	⎭
Am		^{241}Am (432.2 y); ^{242m}Am (141 y);	⎫ Nuclear reactors
		^{243}Am (7370 y)	⎭
Cm		^{242}Cm (162.9 d); ^{243}Cm (29.1 d);	⎫
		^{244}Cm (18.10 y); ^{245}Cm (8500 y);	⎬ Nuclear reactors
		^{246}Cm (4730 y); ^{247}Cm (1.56 · 10^7 y);	
		^{248}Cm (3.40 · 10^5 y)	⎭
Bk		^{249}Bk (320 d)	Nuclear reactors
Cf		^{249}Cf (350.6 y); ^{250}Cf (13.08 y);	⎫ Nuclear reactors
		^{251}Cf (898 y); ^{252}Cf (2.645 y)	⎭
Zr	IVA	^{93}Zr (1.5 · 10^6 y); ^{95}Zr (64.02 d)	Fission and activation
Nb	VB	^{94}Nb (2.0 · 10^4 y);	Activation
		^{95}Nb (34.97 d); ^{95m}Nb (3.61 d)	Nuclear fission
Mo	VIA	^{99}Mo (2.75 d)	Nuclear fission
Tc	VIIA	^{99}Tc (2.13 · 10^5 y)	Nuclear fission
Re		^{187}Re (5 · 10^{10} y)	Rhenium compounds
Fe	VIIIA	^{55}Fe (2.73 y)	Activation
Co		^{60}Co (5.272 y)	Activation
Ni		^{59}Ni (7.5 · 10^4 y): ^{63}Ni (100 y)	Activation
Ru		^{103}Ru (39.35 d); ^{106}Ru (1.023 y)	Nuclear fission
Rh		^{105}Rh (1.475 d)	Nuclear fission
Pd		^{107}Pd (6.5 · 10^6 y)	Nuclear fission

21.2 Mobility of Radionuclides in the Geosphere

Gaseous species, aerosols and species dissolved in aquifers are mobile and easily transported with air or water, respectively. Mobility of solid particles, on the other hand, may be caused by dissolution or suspension in water or spreading by wind.

Solubility and leaching of solids depend on the properties of the solid containing the radionuclide and the properties of the solvent, in general water, in which the various components are dissolved. The radionuclide may be present as a micro- or macrocomponent, and dissolution and solubility may vary considerably with the composition, the degree of dispersion and the influence of radiation (radiolytic decomposition). In the case of solid solutions (mixed crystals) the solubility of the components is, in general, different from that of the pure compounds. Eh, pH and complexing agents may be of significant influence.

The inventory of radionuclides on the surface of the earth, including surface waters, is mobile, provided that the species are soluble in water. A great proportion of the relevant radionuclides is of natural origin. ^{40}K is widely distributed in nature and easily soluble in the form of K^+ ions, which are enriched in clay minerals by sorption. Th present as a major or minor component in minerals is immobile, because chemical species of Th(IV) are very sparingly soluble in natural waters. However, some of the decay products of ^{232}Th, such as ^{228}Ra, ^{224}Ra and ^{220}Rn, are mobile. In contrast to Th(IV), U(IV) is oxidized by air to U(VI), which is easily soluble in natural waters containing carbonate or hydrogencarbonate, respectively. The triscarbonato complex $[UO_2(CO_3)_3]^{4-}$) is found in all rivers, lakes and oceans in concentrations of the order of 10^{-6} to 10^{-5} g/l. The daughters of ^{238}U, the long-lived ^{226}Ra and ^{222}Rn are also mobile. ^{226}Ra is found in relatively high concentrations in mineral springs, and ^{222}Rn contributes considerably to the radioactivity in the air. ^{222}Rn and the daughters ^{218}Po, ^{214}Pb, ^{214}Bi, ^{214}Po, ^{210}Pb, ^{210}Bi and ^{210}Po are the major sources of the radiation dose received by man under normal conditions.

In the atmosphere, T and ^{14}C are generated continuously by the impact of cosmic radiation (section 16.2). The natural concentration of T in the air is about $1.8 \cdot 10^{-3}$ Bq m^{-3} and that of ^{14}C is about $5 \cdot 10^{-4}$ Bq m^{-3}.

From near-surface layers, radionuclides are brought to the surface by natural processes and by human activities. ^{222}Rn produced by decay of ^{238}U in uranium ores is able to escape into the air through crevices. Its decay products are found in the air mainly in the form of aerosols. Ions, such as Ra^{2+} and UO_2^{2+}, are leached from ores or minerals by groundwater and may come to the surface. Volcanic activities also lead to the distribution of radionuclides on the surface, where they may be leached out and enter the water cycle.

Mining of uranium ores and of other ores and minerals (e.g. phosphates) brings up appreciable amounts of U, Th and members of their decay series to the surface and initiates mobilization of relevant radionuclides. Large amounts of Rn are released into the air. The residues of mining and processing are stockpiled and slag heaps as well as waste waters contain significant amounts of radionuclides. The isotopes of Ra continue to emit Rn: about 1 GBq of ^{222}Rn is released per ton of ore containing 1% U_3O_8. Ra and its daughters are migrating to natural oil and gas reservoirs and constitute the major radioactive contaminants of crude oil. The global activity of Ra isotopes brought to the surface by oil production is of the order of 10^{13} Bq per year.

Mining of potassium-salt deposits for use of K as a fertilizer brings additional amounts of ^{40}K to the surface of the earth, where it enters the surface waters.

Radionuclides are also liberated by the burning of coal in thermal power stations. Depending on its origin, coal contains various amounts of U and Th, and these as well as the daughters are released by combustion. Volatile species, in particular Rn, are emitted with the waste gas, ^{210}Pb and ^{210}Po are emitted with the fly ash, and the rest, including U and Th, is found in the ash. The global release of Rn is of the order of 10^{14} Bq per year.

Other activities of man have led to the distribution of appreciable amounts of radionuclides in the atmosphere and on the earth's surface. In the first place nuclear explosions and nuclear weapon tests have to be mentioned, by which Pu and fission products have been deposited on the earth, either directly or via the atmosphere in the form of fall-out. The amount of Pu released by nuclear weapon tests between 1958 and 1981 is estimated at 4.2 tons, of which 2.8 tons were dispersed in the atmosphere and 1.4 tons deposited locally. By underground nuclear explosions about 1.5 tons of Pu have been liberated. Radionuclides released into the air are mainly present in the form of aerosols.

Nuclear reactors and reprocessing plants are constructed and operated in such a way that the radioactive inventory is confined to shielded places. Only limited amounts of radionuclides are allowed to enter the environment. The amounts of T and ^{14}C produced in nuclear reactors vary with the reactor type, between about 10^{12} and 10^{13} Bq of T and about 10^{12} Bq of ^{14}C per GW$_e$ per year. Tritium is released as HTO and about one-third of the ^{14}C is in the form of ^{14}CO$_2$. Under normal operating conditions, very small amounts of fission products and radioelements are set free from nuclear reactors and reprocessing plants. In this context, the actinides and long-lived fission products, such as ^{90}Sr, ^{99}Tc, ^{129}I, and ^{137}Cs, are of greatest importance.

Despite the safety regulations, accidents have occurred with nuclear reactors and reprocessing plants, primarily due to mistakes of the operators. By these accidents parts of the radioactive inventory have entered the environment. Mainly gaseous fission products and aerosols have been emitted, but solutions have also been given off. In the Chernobyl accident, gaseous fission products and aerosols were transported through the air over large distances. Even molten particles from the reactor core were carried with the air over distances of several hundred kilometres.

The behaviour of the radionuclides in the environment depends primarily on their chemical and physicochemical form (species). Alkali and alkaline-earth ions, such as ^{137}Cs$^+$ or ^{90}Sr^{2+}, are easily dissolved in water, independently of pH. Their mobility is limited if they are bound in clay minerals or incorporated into ceramics or glass. ^{129}I forms quite mobile species and reacts easily with organic substances. ^{85}Kr and ^{133}Xe stay predominantly in the air. The lanthanides, e.g. ^{144}Ce, ^{147}Pm and ^{151}Sm, are only sparingly soluble in water, because of the hydrolysis of the cations. However, colloids may be formed – either intrinsic colloids, or carrier colloids with natural colloids as the main components. The solubility of the actinides in the oxidation states III and IV is similar to that of the lanthanides, but hydrolysis is more pronounced in the oxidation state IV. On the other hand, the dioxocations MO_2^+ and MO_2^{2+} exhibit relatively high solubility in water in the presence of carbonate or hydrogencarbonate, respectively, as already mentioned for UO_2^{2+}. In general, these species are rather mobile. The mobility of Zr(IV) and Tc(IV) is similar to that of actinides in the oxidation state IV. The influence of the redox potential is very pronounced in the case of Tc: whereas Tc(IV) is not dissolved in water and immobile, Tc(VII) is easily dissolved in the form of TcO_4^-, and very mobile. The oxidation of PuO_2 to PuO_{2+x} in moist air leads to an unexpected solubility of PuO_2 and has a marked influence on the migration behaviour of plutonium.

Large amounts of ores and minerals with considerable contents of U, Th and their daughter nuclides are still buried deep under the earth's surface. They can be considered to be immobile as long as they are not brought up to the surface by geological activities or by man, and as long as contact with water is excluded. The conception for the storage of radioactive waste is to bring it down into layers deeply underground, in order to confine it there safely (section 11.7).

Investigations of the migration behaviour of natural radionuclides in geomedia (natural analogue studies) provide valuable information about the mobility of radionuclides over long periods of time. Examples are the migration of U, Ra and Th in the neighbourhood of natural ore deposits and of nuclides produced by nuclear fission at the natural reactors at Oklo (section 11.8) and the investigation of radioactive disequilibria (section 16.6).

21.3 Reactions of Radionuclides with the Components of Natural Waters

In aqueous solutions, the majority of the radionuclides listed in Table 21.1 are present in cationic forms, for which primarily the following reactions have to be taken into account:

– hydration (formation of aquo complexes),
– hydrolysis (formation of hydroxo complexes),
– condensation (formation of polynuclear hydroxo complexes),
– complexation (formation of various complexes with inorganic or organic ligands),
– formation of radiocolloids (intrinsic or carrier colloids).

Inorganic anions such as Cl^-, CO_3^{2-}, SO_4^{2-} and HPO_4^{2-} and organic compounds containing functional groups compete with the formation of aquo and hydroxo complexes, depending on their chemical properties and their concentrations. Cations of transition elements are known to form relatively strong covalent bonds with ligands containing donor atoms, and chelate complexes exhibit high stability. Formation of radiocolloids has been discussed in section 13.4.

Groundwaters, rivers, lakes and the oceans contain a great variety of substances that may interact with radionuclides. Besides the main component (water), other inorganic compounds have to be considered:

– dissolved gases, such as oxygen and carbon dioxide, which influence the redox potential Eh and the pH;
– salts, such as NaCl, $NaHCO_3$ and others, which affect pH and complexation and are responsible for the ionic strength;
– inorganic colloids, such as polysilicic acid, iron hydroxide or hydrous iron oxide, and finely dispersed clay minerals giving rise to the formation of carrier colloids;
– inorganic suspended matter (coarse particles).

Organic components in natural waters comprise:

– compounds of low molecular mass, e.g. organic acids, amino acids and other metabolites;
– compounds of high molecular mass, such as humic and fulvic acids, and colloids formed by these substances or by other degradation products of organic matter;
– suspended coarse particles of organic matter;
– microorganisms.

The concentrations of these compounds in natural waters vary over a wide range. Coarse particles are observed only in agitated waters and settle down as soon as agitation stops. Even though many of the components listed above may be present in rather low concentrations (microcomponents), their concentration is usually still many orders of magnitude higher than that of the radionuclides in question and therefore cannot be neglected.

The aspect of low concentration is of special importance for short-lived isotopes of radioelements, the concentration of which is, in general, extremely low. In the case of radioisotopes of stable elements, on the other hand, the ubiquitous presence of these elements leads to measurable concentrations of carriers, with the consequence that these radionuclides show the normal chemical behaviour of trace elements.

Due to the large number of components, natural waters are rather complex systems. The relative concentrations of many components, as well as the pH and Eh, are controlled by chemical equilibria. However, there are also components, in particular colloids and microorganisms, for which thermodynamic equilibrium conditions are not applicable. The complexity of the chemistry in natural waters and the non-applicability of thermodynamics are the main reasons for the fact that calculations are very difficult and problematic. The same holds for laboratory experiments with model waters; results obtained with a special kind of water are, in general, not applicable for other natural waters of different origin.

The redox potential Eh has a great influence on the behaviour of radionuclides in geomedia, if different oxidation states have to be taken into account. In this context, the presence of oxygen (aerobic, oxidizing conditions) or of hydrogen sulfide (anaerobic, reducing conditions) in natural waters is significant. H_2S is produced by weathering of sulfidic minerals or by decomposition of organic compounds in the absence of air. It indicates reducing conditions and leads to formation of sparingly soluble sulfides. The redox potential is of special importance for the behaviour of I, U, Np, Pu and Tc. Elemental iodine is volatile and reacts with organic compounds, in contrast to I^- or IO_3^- ions. $U(IV)$ does not form soluble species in natural waters, in contrast to $U(VI)$. But under aerobic conditions $U(IV)$ is oxidized to UO_2^{2+} and dissolved; the latter is reduced again if the water enters a reducing zone and is redeposited as UO_2. The most important feature of Np in water is the great stability range of $Np(V)$ in the form of NpO_2^+; in this respect Np differs markedly from the neighbouring elements U and Pu. Under reducing conditions, $Np(IV)$ is formed that resembles $U(IV)$ and $Pu(IV)$. The chemical form of Tc also depends on the redox potential. Under anaerobic conditions the stable oxidation state is IV, whereas under aerobic conditions Tc is easily oxidized to TcO_4^-.

The pH in natural waters varies between about 6 and 8 (apart from acid rainwater) and influences strongly the chemical behaviour of elements that are sensitive to hydrolysis (elements of groups III, IIIA, IV, IVA, V, VA and VIIIA of the Periodic Table). Relevant radionuclides belong to the lanthanide and actinide groups. The tendency of the actinides to hydrolyse increases in the order $MO_2^+ < M^{3+} < MO_2^{2+} < M^{4+}$. In the presence of complexing agents, formation of other complexes competes with that of hydroxo complexes, as already mentioned. Complexes exhibiting the highest stability under the given conditions are formed preferentially.

Polynuclear hydroxo complexes may be formed by condensation of mononuclear hydroxo complexes of the same kind, provided that the concentration of the latter is high enough. Otherwise, condensation with reactive species of other origins, such as polysilicic acid or hydrous SiO_2, iron(III) hydroxide or hydrous Fe_2O_3, and finely dispersed (colloidal) clay minerals, is preferred. Accordingly, either intrinsic colloids ("Eigenkolloide") or carrier colloids ("Fremdkolloide") may be formed. Due to hydrolysis and interaction of hydroxo complexes with other components, actinides are not found as monomeric species in natural waters in the absence of complexing agents, whereas in the presence of carbonate or hydrogencarbonate, respectively, monomeric carbonato or hydrogencarbonato complexes prevail.

Inorganic salts affect the behaviour of radionuclides in natural waters in various ways. At high salinity (high ionic strength) formation of colloids is hindered and colloids already present are coagulated if salt water enters the system. For instance, precipitation of colloids carried by rivers occurs on a large scale in estuaries. Moreover, dissolved salts influence pH, hydrolysis and complexation. They may act as buffers, e.g. in seawater, where the pH is kept constant at about 8.2 by the presence of $NaHCO_3$. Finally, anions in natural waters form ion pairs and complexes with cationic radionuclides and affect solubility, colloid formation and sorption behaviour. Mobility may be enhanced by complexation; Cl^- ions, for example, are relatively weak complexing agents, but they are able to substitute OH^- ions in hydroxo complexes and to suppress hydrolysis, if they are present in relatively high concentrations.

The logarithms of the stability constants β_1 for the formation of 1:1 complexes of the actinide ions M^{3+}, M^{4+}, MO_2^+ and MO_2^{2+} with various inorganic ligands are plotted in Fig. 21.1. Carbonato complexes of alkaline-earth elements, lanthanides, actinides and other transition elements play an important role in natural waters and may stabilize oxidation states.

Formation of intrinsic colloids in natural waters can be excluded for radioisotopes of elements of groups 0, I and VII, and the probability that they may be formed is small for radioisotopes of elements of other groups as long as the concentration of the elements is low. In general, formation of carrier colloids by interaction of radionuclides with colloids already present in natural waters is most probable. Thus, clay particles have a high affinity for heavy alkali and alkaline-earth ions, which are bound by ion exchange. This leads to the formation of carrier colloids with ^{137}Cs, ^{226}Ra and ^{90}Sr. Formation of radiocolloids with hydrolysing species has already been discussed (section 13.4).

Organic compounds are found primarily in surface waters, but also in groundwaters, if these are or have been in contact with organic substances. Compounds of low molecular mass may be of natural origin (e.g. metabolites, such as organic acids, amines or amino acids) or anthropogenic (e.g. detergents, aromatic sulfonic acids). Many of these organic compounds are strong complexing agents and well soluble in water. They are able to form stable complexes with radioisotopes of elements of groups III, IV, V and in particular transition elements (groups IA to VIIIA).

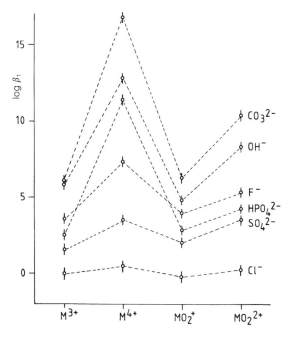

Figure 21.1. Logarithm of the stability constant β_1 of $1:1$ complexes of the actinide ions M^{3+}, M^{4+}, MO_2^+ and MO_2^+ with various inorganic ligands.

The most important representatives of organic compounds of high molecular mass in natural waters are humic and fulvic acids, both degradation products of organic matter. They are polyelectrolytes and contain carboxylic and phenolic hydroxyl groups which make these compounds hydrophilic and enable them to form quite stable complexes. Other compounds of high molecular mass are proteins, lipids and carbohydrates. The concentration of dissolved organic matter in natural waters varies considerably. It may be as low as 0.1 mg/l DOC (dissolved organic carbon) in deep groundwaters, it ranges from 0.5 to 1.2 mg/l in the oceans and it may go up to about 50 mg/l in swamp waters. Relatively high stability constants have been measured for complexes of actinides with humic substances.

Complexation of actinides by organic compounds may also cause an increase in solubility. For example, NpO_2^+ ions exhibit strong complexation by organic complexing agents, even in high salt concentrations. The influence of hydrolysis increases in the order $MO_2^+ < M^{3+} < MO_2^{2+} < M^{4+}$. Whether the formation of hydroxo complexes or of other complexes prevails depends on the stability constants of the complexes, the concentration of the complexing agents and the pH.

With increasing mass of the organic compounds, colloidal properties may prevail and a pronounced difference between non-colloidal and colloidal species no longer exists. Radiocolloids containing organic substances may be formed by sorption or ion exchange of radionuclides on macromolecular organic compounds or by sorption of organic complexes of the radionuclides on inorganic colloids. In all cases, the surface properties determine the behaviour (section 13.4).

Microorganisms may also influence the fate of radionuclides in natural waters, in particular in surface waters and near-surface groundwaters. Depending on the metabolism of the microorganisms and their preference for certain elements, they incorporate radionuclides which then migrate with the microorganisms. With uptake by microorganisms, the radionuclides may enter the food chain.

The migration behaviour of radionuclides may also be affected by precipitation and coprecipitation. In the range of low concentrations, coprecipitation is generally the most important process. Radionuclides may be coprecipitated by isomorphous substitution or by adsorption (section 13.3). Precipitation reactions often observed in natural waters are the precipitation of $CaCO_3$, caused by a shift of the equilibrium between $Ca(HCO_3)_2$ and $CaCO_3$ due to escape of CO_2, and precipitation of iron(III) hydroxide due to oxidation of iron(II), dissolved as $Fe(HCO_3)_2$, by access of air. Depending on the conditions and the properties of the compounds, precipitation may stop at an intermediate stage with formation of a colloid (e.g. formation of a sol of colloidal iron(III) hydroxide in water). Coprecipitation of trace elements with iron(III) hydroxide is used as an effective procedure in the preparation of drinking water. Generally, coprecipitation of microcomponents is to be expected if these would also precipitate under the given conditions, provided they were present in higher concentrations. However, actinides(IV) and actinides(III) are also coprecipitated with $BaSO_4$ and $SrSO_4$ by formation of anomalous solid solutions. Coprecipitation may influence considerably the mobility of radionuclides in natural waters, in particular that of lanthanides and actinides.

21.4 Interactions of Radionuclides with Solid Components of the Geosphere

Mobility and transport of radionuclides in the geosphere are influenced markedly by their interaction with solids. Migration is retarded, or even stopped, if the interaction is strong, in particular if the radionuclides are incorporated into the solids. Sorption of radionuclides on solids has been investigated extensively for materials in the neighbourhood of planned high-level waste repositories.

Various kinds of interaction have to be taken into account:

– fixation by predominantly ionic bonds (ion exchange);
– sorption by mainly covalent bonds (chemisorption);
– sorption by weak (van der Waals) bonds (physisorption);
– interaction at the outer surface of solids;
– interaction at inner surfaces of porous substances;
– ion exchange at positions within the solids (e.g. clay minerals);
– incorporation into solids by formation of solid solutions.

Reactions of hydroxo complexes or anionic forms of radionuclides with hydroxyl groups at the surface of solids ($\geqslant SiOH, > AlOH$ or $> FeOH$) are frequently observed, for example

$$\geqslant SiOH \, (s) + HO-M^{(n-1)+} \, (aq) \; \rightleftharpoons \; \geqslant Si-O-M^{(n-1)+} \, (s) + H_2O \qquad (21.1)$$

(additional water molecules are omitted). In this way, the complex of the metal M in solution is converted into a complex at the surface of the sorbent ("surface complex") with a partly changed coordination sphere. The bonds formed are of a predominantly covalent nature. This type of reaction has been described in the older literature by the term "hydrolytic adsorption". It is very common and has been applied in analytical chemistry for selective separations.

Exchange of non-hydrolysed cationic species of radionuclides at the outer surface or within the layer structure of solids plays an important role in the case of $^{137}Cs^+$, $^{90}Sr^{2+}$ and $^{226}Ra^{2+}$. It is very effective if clay minerals are present that exhibit high exchange capacity. The predominant type of interaction is ion exchange. Exchange of cationic or anionic forms of radionuclides on the surface of ionic compounds such as $CaCO_3$ or $BaSO_4$ may also contribute to sorption, in particular if the radionuclides are incorporated in the course of recrystallization.

Adsorption of complexes of radionuclides with inorganic or organic ligands (in particular complexes with humic substances) and of colloidal species of radionuclides may also markedly influence the migration behaviour. The predominant kind of interaction is physical adsorption.

The solids in the geosphere are of very different kind and composition. With respect to the interaction with radioactive species, the surface properties are of special interest. Keeping this aspect in mind, the following main components of the geosphere can be distinguished:

- consolidated rocks (magmatic rocks, such as basalt, granite, feldspar, quartz, olivine, plagioclases and pyroxenes, and sedimentary rocks, e.g. sandstone, limestone or dolomite);
- unconsolidated rocks (more or less loose-packed, consisting mainly of glacial deposits of gravel, sand and clay);
- sediments in rivers, lakes and oceans;
- soils (mainly sand, clay, humus with plant residues, small animals and plenty of microorganisms).

Besides the main components, many other minerals have to be taken into account:

- oxides and hydroxides (e.g. hydrargillite, diaspore, corundum, spinels, haematite, magnetite, perovskites);
- halides (e.g. rock salt, cryolite, carnallite);
- sulfides (e.g. pyrite);
- sulfates (e.g. gypsum, anhydrite, alum);
- phosphates (e.g. apatites, monazites – some phosphates contain exchangeable protons);
- carbon and carbonaceous material.

Of special significance with respect to their properties as sorbents are the clay minerals (e.g. kaolinite, montmorillonite, vermiculite, illite, chlorite), mainly due to their high exchange capacity.

The surfaces of silicate rocks are altering in the course of time. Cations such as Na^+, K^+, Mg^{2+} and Ca^{2+}, that are integral components of the silicates, are leached and hydroxyl groups are formed at the surface which may add or give off protons, depending on the pH. The resulting sorption sites are of a different nature and quality. $\geqslant SiOH$, $>AlOH$, and $>FeOH$ groups are found most frequently. In neutral media radionuclides present as hydroxo complexes or divalent anions are sorbed preferentially. Most investigations with consolidated rocks have been made with granites, because they are possible host rocks for high-level waste repositories.

With sedimentary rocks radionuclides may also interact in different ways. They may be sorbed on sandstone, limestone and other sedimentary rocks. Heterogeneous exchange may contribute to the sorption, e.g. exchange of $^{14}CO_3^{2-}$ (aq) for $^{12}CO_3^{2-}$ (s) or exchange of $^{90}Sr^{2+}$ (aq), $^{210}Pb^{2+}$ (aq) or $^{226}Ra^{2+}$ (aq) for Ca^{2+} (s) at the surface of calcite, limestone or dolomite. In addition, incorporation into the inner parts of the crystals due to recrystallization and formation of solid solutions may take place. Interaction of radionuclides with consolidated volcanic tuffs has been investigated intensively, because the deposits in the Yucca Mountains, USA, have been found to be suitable as repositories for high-level waste.

The loose-packed material of unconsolidated rocks consists mainly of sand and clay. Clay minerals are the most important components, because of their high sorption capacity and their selectivity for heavy alkali and alkaline-earth ions. Compared with clay minerals, sand is a rather poor sorbent, although hydrated silica also exhibits exchange properties due to the presence of $\geqslant SiOH$ groups. The exchange capacity of clay minerals with a layer structure (e.g. montmorillonite, vermiculite) goes up to about 1 to 2 meq/g. Clay minerals with high charge densities, like illite and the micas, exhibit a marked preference for monovalent cations ($Cs^+ > Rb^+ > K^+$), which are bound more or less irreversibly. Divalent cations ($Ra^{2+} \geqslant Ba^{2+} > Sr^{2+}$) are also firmly bound. Mica-type clay minerals play an important role in nature, because they are present in all fertile soils. ^{137}Cs, ^{90}Sr and ^{226}Ra are fixed very strongly in positions between the layers. Hydroxo complexes of lanthanides and actinides are mainly bound at the outer surfaces by interaction with $\geqslant SiOH$ or $> AlOH$ groups. In Germany layers of unconsolidated rocks above salt domes proposed for the storage of high-level waste have been investigated in great detail with respect to the migration behaviour of radionuclides. In other countries, clay is discussed as a host for high-level waste repositories, because of the favourable sorption properties of clay minerals.

Compounds present in relatively small amounts are often decisive for the sorption behaviour of natural solids. Examples are small amounts of clay in association with large amounts of sand with respect to sorption of ^{137}Cs, and small amounts of carbonaceous material with respect to sorption of ^{129}I.

Heterogeneous exchange of radionuclides on carbonates has already been mentioned. Exchange on other sparingly soluble minerals (e.g. halides, sulfates, phosphates) may lead to rather selective separation of radionuclides. Following the exchange at the surface, ions may be incorporated into the solids in the course of recrystallization, which is a very slow but continuous process. Anomalous solid solutions with radioactive ions of different charges may also be formed.

Aquatic sediments are formed in all surface waters by the settling of coarse and fine inorganic and organic particles. They are present in rivers, in lakes and in the oceans, and radionuclides deposited on the surface of the earth will sooner or later come into contact with these sediments. They may enter the sediments by sorption of molecularly-dispersed species (ions, molecules), by precipitation or coprecipitation, by coagulation of colloids (in particular carrier colloids) followed by sedimentation of the particles formed, or by sedimentation of coarse particles (suspended matter). By desorption, the radionuclides may be remobilized and released again into the water.

The main components of these sediments are similar to those in the sediments formed at earlier times: sand and clay minerals. However, river and lake sediments contain also relatively large amounts of organic material and microorganisms. Appreciable fractions of radionuclides present in rivers and lakes are sorbed in sediments, but usually it is difficult to discriminate between the influences of the various processes taking place and to correlate the fixation in the sediments with certain components. As far as the inorganic components are concerned, clay minerals play the most important role.

With respect to components and chemistry, soils are even more complex than river or lake sediments. On the other hand, large areas of the continents are covered with soils of various compositions, and therefore interest in the behaviour of radionuclides in soils is justified. Above that, radionuclides are easily transferred from soils to plants and animals, and in this way they enter the biosphere and the food chain.

The main components of soils are sand, clay and humus. Whereas interaction between radionuclides and sand is rather weak, as in the case of sediments, sorption by clay minerals and reactions with the organic compounds in humus are most important for the migration behaviour of radionuclides. $^{137}Cs^+$ ions are quite strongly bound in clay particles, as already mentioned. $^{90}Sr^{2+}$, $^{226}Ra^{2+}$ and $^{210}Pb^{2+}$ are also retained by clay particles or bound on chalky soil via precipitation or ion exchange. $^{129}I^-$ ions are oxidized, and I_2 reacts easily with organic compounds which take part in the metabolism of microorganisms. Lanthanides and actinides are either present as hydroxo complexes or as organic complexes. These species are rather firmly bound to the components of soils, but organic compounds may also stay in solution, possibly in the form of colloids. $^{99}TcO_4^-$ reacts with proteins and may thus be incorporated into organic matter.

Besides the composition of the soils, other factors have a major influence on the migration of radionuclides: rainfall, the thickness of the soil layers, their permeability to water, and the nature of the layers underneath. For example, ^{137}Cs is washed down quickly through layers of sand, but it will stay in layers of clay. ^{239}Pu is sorbed by clay more strongly than by sand and may stay in soils for rather long times, if it is not dissolved by complexation or displaced by other compounds.

The large number of measurements in various systems lead to the result that the interaction of radionuclides with solids is rather complex and depends on many parameters: the species of the radionuclides in the solution, their properties and their dispersion, the components of the solid, the surface area of the particles, the nature of the sorption sites, the presence of organic substances and of microorganisms, and the interference or competition of other species. Therefore, an investigation of the behaviour of the radionuclides in the specific system of interest is unavoidable, if reliable information about their migration behaviour is required.

21.5 Radionuclides in the Biosphere

Radionuclides in the atmosphere, in surface waters and on the surface of the earth have immediate access to the biosphere. From surface waters as well as from near-surface groundwaters, from soils and from the air, radionuclides may be taken up by microorganisms, plants, fish and other animals, thus also entering the various pathways of the food chain. The fractions of radionuclides transferred to living things depend on the chemical form (species) of the radionuclide considered and the metabolism of the microorganisms, plants and animals, and may differ by several orders of magnitude. The study of the behaviour of radionuclides in ecosystems comprising air, water, soil, microorganisms, plants, animals and man is the field of radioecology. Various ecosystems, such as aquatic, agricultural, forest and alpine, are distinguished.

The pathways of radionuclides in ecosystems are illustrated schematically in Fig. 21.2. Plants may take up radionuclides from the air by deposition on the leaves, or from the soil by the roots with water and minerals. In this step, the species of the radionuclides and their solubility are most important. Microorganisms incorporate radionuclides present in water or in the soil. Animals and man may be contaminated by radionuclides from the air (gases, aerosols, dust) by inhalation or deposition on the skin, or by uptake of water and via the food chain by digestion. Animals may also be contaminated by ingestion of contaminated soil particles.

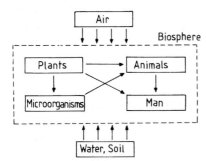

Figure 21.2. Pathways of radionuclides in ecosystems.

The metabolism of radionuclides in plants, animals and man can be compared with that of trace elements. Radioisotopes of essential trace elements, such as T, ^{14}C, ^{55}Fe, ^{54}Mn, ^{60}Co and ^{65}Zn, show the same behaviour as the stable isotopes of these elements. Other radionuclides behave in a similar way to other elements (e.g. $^{90}Sr \approx Ca$, $^{137}Cs \approx K$), and radioisotopes of heavy elements are comparable with other heavy elements.

Metabolism in man, animals and plants comprises the following steps:

- resorption in the lung or in the gastrointestinal tract (man and animals) or by the leaves or roots (plants);
- distribution in the body or in the plant;
- retention in the body or in the plant, often in special organs or places.

Resorption depends on the chemical properties and the chemical form (species) of the element. Biological inert and bioavailable species are distinguished. For example, mono- and divalent cations are easily resorbed, whereas elements of higher valency are, in general, not able to pass the intestinal walls or the membranes, respectively, and to enter the body fluids or the plants. The presence of other substances may diminish the resorption of radionuclides. For example, resorption of ^{137}Cs is reduced if cows take up soil particles together with grass, because this radionuclide is bound quite firmly on the clay particles in soil. The resorption factor f_R is given by

$$f_R = \frac{A}{A_0} \tag{21.2}$$

where A_0 is the activity taken up and A the activity resorbed.

In most cases, the concentration of radionuclides in animals and man decreases with excretion and decay. In the ideal case, decrease of activity A due to excretion can be described by an exponential law

$$A = A_0 e^{-\lambda_b t} = A_0 \left(\frac{1}{2}\right)^{t/t_{1/2}(b)} \tag{21.3}$$

where $\lambda_b A$ is the rate of excretion and $t_{1/2}(b)$ is the biological (ecological) half-life of the radionuclide in the animal or in man. In general, the decay, given by the physical decay constant λ_p, must also be considered, and the effective rate constant of decrease of activity in the body or in the plant is

$$\lambda_{eff} = \lambda_b + \lambda_p \tag{21.4}$$

Accordingly, the time dependence of the activity in the body or the plant is

$$A = A_0 e^{-\lambda_{eff} t} \tag{21.5}$$

Effective, biological (ecological) and physical half-lives ($t_{1/2}(eff)$, $t_{1/2}(b)$ and $t_{1/2}(p)$, respectively) may also be distinguished, where

$$t_{1/2}(eff) = \frac{\ln 2}{\lambda_{eff}} = \frac{t_{1/2}(b)\, t_{1/2}(p)}{t_{1/2}(b) + t_{1/2}(p)} \tag{21.6}$$

In the case of short-lived radionuclides, λ_{eff} is mainly determined by the physical decay constant, and in the case of long-lived radionuclides mainly by the rate constant of excretion.

In general, radionuclides are often not distributed evenly in the body or in plants, but enriched in certain organs or parts, as illustrated in Fig. 21.3. For example, in man and animals Sr and Pu are enriched in bones, and I in the thyroid gland. Considering a single uptake of a radionuclide in a certain organ or part X of a body or plant, respectively, the activity in that organ or part after the time t is

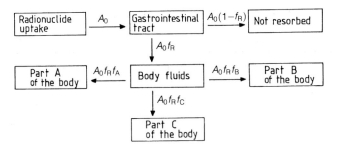

Figure 21.3. Distribution of radionuclides in an animal. f_A, f_B and f_C are the fractions transferred to different parts of the body.

$$A_X = A_0 f_R f_X e^{-\lambda_{eff} t} \tag{21.7}$$

Continuous uptake of ag radionuclide at a certain rate R leads to an increase in the activity in the body, organ, plant or part of the plant which is given by

$$A = \frac{R}{\lambda_{eff}} (1 - e^{-\lambda_{eff} t}) \tag{21.8}$$

In animals and man, the activity A often approaches a saturation value in which the rates of decrease in the body (by excretion and decay) and of ingestion are the same

$$A = \frac{R}{\lambda_{eff}} \tag{21.9}$$

R is usually measured in Bq d^{-1} and λ_{eff} in d^{-1}.

Resorption and metabolism in plants vary with the composition and pH of the soils and with the seasons; in animals and man they vary with age, sex, health, diet and other factors. In cases in which radionuclides are resorbed selectively and accumulated in certain organs of animals or man or in certain parts of plants or in which the rate of excretion in animals or man is small compared with the rate of sorption, these radionuclides will be enriched in the body or in the plant (bioaccumulation), with the result that the activity may reach high values.

In radioecology, the transfer of radionuclides from water, soil or food to animals, man or plants is described in various ways. In aquatic ecosystems, concentration factors CF are determined, given by the concentration c_i (2) of a certain radionuclide i in microorganisms or animals in relation to the concentration c_i (1) of that radionuclide in water at the same time: CF $= c_i(2)/c_i(1)$. Some typical concentration factors for ^{137}Cs and ^{90}Sr measured in freshwater and marine ecosystems are listed in Table 21.2. The influence of the competition of K and Na with ^{137}Cs, and of Ca with ^{90}Sr, in marine ecosystems is obvious.

The transfer from soil to vegetation is described by transfer factors TF, given by the activity in Bq per kg of dry plants, divided by the activity in Bq per m^2 of the soil (m^2/kg). In agricultural ecosystems the transfer of radionuclides from grass to agricultural products, such as meat or milk, is characterized by various factors, which are listed in Table 21.3.

Table 21.2. Typical concentraton factors in freshwater and marine ecosystems.

Radionuclide	Ecosystem	Molluscs	Crustaceans	Fish muscle
^{137}Cs	Freshwater	600	4000	3000
	Marine	8	23	15
^{90}Sr	Freshwater	600	200	200
	Marine	1	3	0.1

Table 21.3. Factors used to characterize the transfer of radionuclides to meat (or milk).

Factor	Definition	Unit
Concentration factor (CF)	Activity [Bq] per kg of meat (per l of milk) divided by the activity [Bq] per kg of grass	kg/kg (l/kg)
Transfer factor T (also called transfer coefficient)	Activity [Bq] per kg of meat (per l of milk) divided by the activity [Bq] taken up with the food per day	d/kg (d/l)
Transfer factor (TF) (also called aggregated transfer factor T_{ag})	Activity [Bq] per kg of meat (per l of milk) divided by the activity [Bq] per m^2 of the soil	m^2/kg (m^2/l)

Concentration factors, transfer factors and transfer coefficients depend on the radionuclide considered, its chemical form (species), the conditions in the soil, and the kind of vegetation and animal; they may vary appreciably with the seasons.

The concentration of many elements, such as C, K or Ca, in animals or plants is often regulated in such a way that it remains constant or nearly constant. Because radioisotopes (subscript 1) and stable isotopes (subscript s) of these elements exhibit the same behaviour, their ratios in the substance taken up and in the plants or animals are the same:

$$\frac{c_i(2)}{c_s(2)} = \frac{c_i(1)}{c_s(1)}$$

This relationship is called the specific activity model. Its main advantage is its independence of the kind of animal or plant. The model can also be applied if the radionuclides show a similar behaviour to stable isotopes of other elements, for example ^{90}Sr^{2+} and Ca^{2+}. Different behaviour can be taken into account by a correction factor (called the observed ratio OR):

$$c_i(2) = \frac{c_i(1)}{c_s(1)} c_s(2) \cdot \text{OR} \tag{21.11}$$

For radionuclides for which values of OR are not known, the specific activity model is not applicable.

Natural radionuclides are present in all plants and animals and in man. The activity of ^{40}K is 31 Bq per g of K, and the average activities in meat and in milk are about 120 Bq/kg and 50 Bq/l, respectively. The transfer of U and Th to plants and animals is very small due to the low solubility and the low resorption of these elements. Their activities in milk are of the order of 10^{-4} Bq/l and in meat and fish of the order of $5 \cdot 10^{-3}$ Bq/kg. Ra has better access to the food chain and, due to its similarity to Ca, Ra is enriched in bones, where it is found in amounts of the order of 10^{-12} g/g. The activity of ^{226}Ra in other parts of animals and man is about 10^{-2} Bq/kg. ^{210}Pb and ^{210}Po, decay products of ^{222}Rn, are present in aerosols and deposited with precipitations on plants. Their uptake from the air is much higher than that by the roots. They enter the food chain and are found in concentrations of 1 to 10 Bq/kg in meat. In reindeer livers values >100 Bq/kg have been measured.

The activity of cosmogenic T in ecosystems is negligible, because of the dilution of T in the water cycle and its short residence time in all living things. The specific activity of cosmogenic ^{14}C in the biosphere is 0.23 Bq/g of carbon. As the average carbon content of organisms is about 23% (w/w), the average activity of ^{14}C in organisms is about 50 Bq/kg.

Artificial radionuclides released by nuclear explosions, weapon tests and accidents have been deposited from the air as fall-out on soil and vegetation. In 1963 values of up to 0.8 Bq/l ^{90}Sr and up to 1.2 Bq/l ^{137}Cs were measured in precipitations in central Europe. In 1964, the concentration of ^{137}Cs in beef reached values of about 36 Bq/kg. Consequently, the concentration of ^{137}Cs in man went up to about 11 Bq/kg.

Much higher activities of radionuclides were found after the reactor accident at Chernobyl, for example in Bavaria up to $\approx 10^4$ Bq ^{131}I and up to $\approx 3 \cdot 10^3$ Bq ^{137}Cs per kg of grass. In the following days, the activity of these radionuclides in milk increased quickly to values of ≈ 400 Bq ^{131}I and ≈ 60 Bq ^{137}Cs per litre. The local activities varied considerably, depending on the local deposition with the precipitations. Whereas the activity of ^{131}I in the biosphere decreased quite rapidly with the half-life of this radionuclide, that of ^{137}Cs decreased slowly and varied with the agricultural activities. For example, by ploughing, radionuclides may be brought from the surface deeper layers and vice versa.

After the Chernobyl accident, the behaviour of ^{137}Cs and ^{90}Sr in ecosystems has been investigated in detail in many countries. The resorption factor for ^{137}Cs in cows is $f_R \approx 0.6$, and its biological half-life in animals varies between about 40 and 150 d, increasing with the size of the animal. ^{137}Cs is quite evenly distributed in the body and shows some similarity to K. Transfer factors TF (soil–grass) in the range between 0.0005 and 0.1 m^2/kg, concentration factors CF between 0.3 and 0.7 and aggregated transfer factors T_{ag} (soil–meat) between 0.0001 and 0.1 m^2/kg have been measured. The aggregated soil–milk transfer factors varied between 0.0001 and 0.005 m^2/l. High transfer factors were observed for moose (TF $= 0.01$–0.3 m^2/kg) and in particular for fungi (TF $= 1$–2 m^2/kg). Consequently, aggregated transfer factors were high for roe deer and reindeer ($T_{ag} = 0.03$–0.2 m^2/kg). Typical transfer coefficients are $T \approx 0.06$ d/kg for beef, $T \approx 0.3$ d/kg for pork, $T \approx 0.6$ d/kg for lamb and $T \approx 0.01$ d/kg for milk. Very little or no decrease in the activity of ^{137}Cs was found in many plants, which means that in these plants the ecological half-life is given mainly by the physical half-life.

For ^{90}Sr the resorption factor in cows is $f_R = 0.05$ to 0.4 (in young animals $f_R = 0.2$ to 1.0). About 90% of the resorbed ^{90}Sr is deposited in the bones. The biological half-life in domestic animals is of the order of 100–500 d for the skeleton and of the order of 20–100 d for the rest of the body. In man the biological half-life of ^{90}Sr is in the range 200–600 d. Aggregated transfer factors (soil–meat) vary between 0.01 and 0.06 m^2/kg, and typical transfer coefficients (food–product) are $T \approx$ 0.0003 d/kg for beef, $T \approx 0.002$ d/kg for pork, and $T \approx 0.002$ d/kg for milk.

In ecosystems, Pu is present mainly in the form of sparingly soluble Pu(IV) dioxide or hydroxide and is therefore rather immobile. It stays mainly in the upper layers of the soil and its uptake by roots is very small (soil–plant transfer coefficients <0.001 d/kg). However, plants may be contaminated with Pu by deposition from the air. Resorption factors in the gastrointestinal tract of animals are also very small ($f_R < 10^{-4}$). On the other hand, up to 5% of inhaled Pu is found in blood and up to 15% in the lymph glands. About 80% of resorbed Pu is deposited in bones, the rest in kidneys and liver. Biological half-lives reported in the literature vary between 500 and 1000 d for the lymph glands and between 1 and 100 y for the skeleton.

Literature

General

United Nations, The Environmental Impacts of Production and Use of Energy, UNEP, Energy Report Series 2-79, Nairobi, **1979**

E. J. Hall, Radiation and Life, 2nd ed., Pergamon, Oxford, **1984**

Chemistry and Migration Behaviour of Actinides and Fission Products in the Geosphere, Radiochim. Acta *44/45* (**1988**); *52/53* (**1990**); *58/59* (**1992**); *66/67* (**1994**); *74* (**1996**)

Proc. Conf. Scientific Basis for Nuclear Waste Management, Materials Research Society, North-Holland, Amsterdam, I (**1978**)–XII (**1989**) and subsequent conferences

G. Pfennig, H. Klewe-Nebenius, W. Seelmann-Eggebert, Chart of the Nuclides (Karlsruher Nuklidkarte), 6th ed., revised, Forschungszentrum Karlsruhe, **1998**

E. Browne, R. B. Firestone (Ed. V. S. Shirley), Table of Radioactive Isotopes, Wiley, New York, **1986**

Radionuclides in the Environment

C. E. Junge, Air Chemistry and Radioactivity, Academic Press, New York, **1963**

D. Lal, H. A. Suess, The Radioactivity of the Atmosphere and the Hydrosphere, Annu. Rev. Nucl. Sci. *18*, 407 (**1968**)

Radioactivity in the Marine Environment, National Academy of Sciences, Washington, DC, **1971**

Long-term Worldwide Effects of Nuclear-Weapons Detonations, National Academy of Sciences, Washington, DC, **1975**

International Atomic Energy Agency, The Oklo Phenomenon, IAEA, Vienna, **1975**

International Atomic Energy Agency, Transuranium Nuclides in the Environment, IAEA, Vienna, **1976**

P. Fritz, and J. Ch. Fontes (Eds.), Handbook on Environmental Isotope Chemistry, Vol. 1, The Terrestrial Environment, Elsevier, Amsterdam, **1980**

International Atomic Energy Agency, Nuclear Power, the Environment and Man, IAEA, Vienna, **1982**

International Atomic Energy Agency, Environmental Migration of Long-lived Radionuclides, IAEA, Vienna, **1982**

L. Stein, The Chemistry of Radon, Radiochim. Acta *32*, 163 (**1983**)

R. L. Kathren, Radioactivity in the Environment; Sources, Distribution and Surveillance, Harwood, Chur, **1984**

D. G. Brookins, Geochemical Aspects of Radioactive Waste Disposal, Springer, Berlin, **1984**

G. S. Barney, J. D. Navratil, W. W. Schulz (Eds.), Geochemical Behaviour of Radioactive Waste, Amer. Chem. Soc. Symp. Series, Amer. Chem. Soc., Washington, DC, **1984**

E. M. Durrance, Radioactivity in Geology, Principles and Applications, Ellis Horwood, Chichester, **1986**

J. Tadmor, Radioactivity from Coal-fired Power Plants, J. Environ. Radioact. *4*, 177 **(1986)**

Chemical Phenomena Associated with Radioactivity Releases during Severe Nuclear Plant Accidents, Amer. Chem. Soc. Symp. Series, Amer. Chem. Soc., Washington, DC, **1986**

International Atomic Energy Agency, Summary Report on the Post-accident Review Meeting on the Chernobyl Accident, International Nuclear Safety Advisory Group (INSAG), Safety Series No. 75-INSAG-1, IAEA, Vienna, **1986**

G. Desmet, C. Myttenaere (Eds.), Technetium in the Environment, Elsevier, Amsterdam, **1986**

M. Eisenbud, Environmental Radioactivity from Natural, Industrial and Military Sources, 3rd ed., Academic Press, New York, **1987**

International Atomic Energy, The Environmental Behaviour of Radium, IAEA, Vienna, **1990**

G. Desmet, P. Nassimbeni, M. Belli (Eds.), Transfer of Radionuclides in Natural and Semi-Natural Environments, Elsevier, Amsterdam, **1990**

International Atomic Energy Agency, Proc. Int. Symp. Environmental Contamination Following a Major Accident, IAEA, Vienna, **1990**

A. C. Chamberlain, Radioactive Aerosols, Cambridge Environmental Chemistry Series 3, Cambridge University Press, Cambridge, **1991**

M. Ivanovich, R. S. Harmon, Uranium Series Disequilibrium: Applications to Earth, Marine, and Environmental Sciences, 2nd ed., Clarendon Press, Oxford, **1992**

International Atomic Energy Agency, The Radiobiological Impact of Hot Beta-particles from the Chernobyl Fallout: Risk Assessment, IAEA, Vienna, **1992**

K. H. Lieser, Technetium in the Nuclear Fuel Cycle, in Medicine and in the Environment, Radiochim. Acta *63*, 5 **(1993)**

F. J. Sandalls, M. G. Segal and N. Victorova, Hot Particles from Chernobyl: A Review, J. Environ. Radioactivity *18*, 5 **(1993)**

P. Strand, E. Holm (Eds.), Environmental Radioactivity in the Arctic and Antarctic, Proc. Int. Conf. Kirkenes, August *1993*, Osterås, Norway

A. M. Ure, C. M. Davidson (Eds.), Chemical Speciation in the Environment, Chapman and Hall, London, **1994**

H. W. Gäggeler, Radioactivity in the Atmosphere, Radiochim. Acta *70/71*, 345 **(1995)**

K. H. Lieser, Radionuclides in the Geosphere: Sources, Mobility, Reactions in Natural Waters and Interactions with Solids, Radiochim. Acta *70/71*, 355 **(1995)**

R. J. Silva, H. Nitsche, Actinide Environmental Chemistry, Radiochim. Acta *70/71*, 377 **(1995)**

Chemistry and Speciation

M. Haissinsky, Les Radiocolloides, Hermann, Paris, **1934**

R. A. Bulman, J. R. Cooper (Eds.), Speciation of Fission and Activation Products in the Environment, Elsevier, Amsterdam, **1985**

C. Keller, The Chemistry of the Transuranium Elements (Ed.: K. H. Lieser), Verlag Chemie, Weinheim, **1971**

J. M. Cleveland, The Chemistry of Plutonium, Gordon and Breach, New York, **1971**

W. Stumm, P. A. Brauner, Chemical Speciation, in: Chemical Oceanography, 2nd ed. (Eds. J. P. Riley, G. Skirow), Academic Press, New York, **1975**

W. C. Hanson (Ed.), Transuranic Elements in the Environment, US Department of Energy, Technical Information Center, Washington, DC, **1980**

T. F. Gesell, W. M. Loder (Eds.), Natural Radiation Environment III, NTIS, Springfield, **1980**

International Atomic Energy Agency, Transuranic Speciation in Aquatic Environments, IAEA, Vienna, **1981**

N. M. Edelstein (Ed.), Actinides in Perspective, Pergamon, Oxford, **1982**

W. T. Carnall, G. R. Choppin (Eds.), Plutonium Chemistry, Amer. Chem. Soc. Symp. Series 216, Amer. Chem. Soc., Washington, DC, **1983**

R. Stumpe, J. I. Kim, W. Schrepp, H. Walther, Speciation of Actinide Ions in Aqueous Solution by Laser-induced Pulsed Spectrophotoacoustic Spectroscopy, Appl. Phys. B *34*, 203 **(1984)**

T. W. Newton, J. C. Sullivan, Actinide Carbonate Complexes in Aqueous Solution, in: Handbook on the Physics and Chemistry of the Actinides, Vol. 3 (Eds. A. J. Freeman, C. Keller), Elsevier, Amsterdam, **1985**

G. R. Choppin, B. Allard, Complexes of Actinides with Naturally Occurring Organic Compounds, in: Handbook on the Physics and Chemistry of the Actinides, Vol. 3 (Eds. A. J. Freeman, C. Keller), Elsevier, Amsterdam, **1985**

J. J. Katz, G. T. Seaborg, L. R. Morss (Eds.), The Chemistry of Actinide Elements. 2nd ed., Chapman and Hall, London, **1986**

M. Bernhard, F. E. Brinckman, P. J. Sadler (Eds.), The Importance of Chemical Speciation in Environmental Processes, Springer, Berlin, **1986**

J. I. Kim, Chemical Behaviour of Transuranic Elements in Natural Aquatic Systems, in: Handbook on the Physics and Chemistry of Actinides, Vol. 4 (Eds. A. J. Freeman, C. Keller), Elsevier, Amsterdam, **1986**

L. Landner (Ed.), Speciation of Metals in Water, Sediment and Soil Systems, Springer, Berlin, **1987**

J. R. Kramer, H. E. Allen (Eds.), Metal Speciation: Theory, Analysis and Application, Lewis, Chelsea, MI, **1988**

G. E. Batley (Ed.), Trace Element Speciation, Analytical Problems and Methods, CRC Press, Boca Raton, FL, **1989**

J. A. C. Broekart, S. Gücer, F. Adams (Eds.), Metal Speciation in the Environment, Springer, Berlin, **1990**

J. I. Kim, R. Klenze, Laser-induced Photoacoustic Spectroscopy for the Speciation of Transuranic Elements in Natural Aquatic Systems, Top. Curr. Chem. *157*, 129 **(1990)**

K. H. Lieser, A. Ament, R. Hill, R. N. Singh, U. Stingl, B. Thybusch, Colloids in Groundwater and Their Influence on Migration of Trace Elements and Radionuclides, Radiochim. Acta *49*, 83 **(1990)**

S. Ahrland, Hydrolysis of the Actinide Ions, in: Handbook on the Physics and Chemistry of the Actinides, Vol. 9 (Eds. A. Freeman, C. Keller), Elsevier, Amsterdam, **1991**

International Atomic Energy Agency, Reports of the VAMP (Validation of Environmental Model Predictions) Terrestrial Working Group, IAEA Techn.Doc. 647, Vienna, **1992**, **1995**; Reports of the Urban Working Group, IAEA Techn.Doc. 760, Vienna, **1994**, **1995**

M. Dozol, R. Hagemann, Radionuclide Migration in Groundwaters: Review of the Behaviour of Actinides, Pure Appl. Chem. *65*, 1081 **(1993)**

Trace Elements in Water

U. Förstner, G. Müller, Schwermetalle in Flüssen und Seen, Springer, Berlin, **1974**

C. F. Baes, jr., R. Mesmer, The Hydrolysis of Cations, Wiley, New York, **1976**

R. M. Smith, A. E. Martell, Critical Stability Constants, Plenum Press, New York, **1976**

P. Beneš, V. Majer, Trace Chemistry in Aqueous Solutions, Elsevier, Amsterdam, **1980**

M. C. Kavanough, J. O. Leckie, Particulates in Water, Amer. Chem. Soc. Adv. Chem. Series *189*, Amer. Chem. Soc. Washington, DC, **1980**

W. Stumm, J. J. Morgan, Aquatic Chemistry, 2nd ed., Wiley, New York, **1981**

U. Förstner, G. T. W. Wittmann, Metal Pollution in the Aquatic Environment, 2nd ed., Springer, Berlin, **1981**

E. K. Duursma, H. J. Dawson, Marine Organic Chemistry, Elsevier, Amsterdam, **1981**

E. Högfeld, Stability Constants of Metal-ion Complexes, Pergamon, Oxford, **1982**

F. M. N. Morel, Principles of Aquatic Chemistry, Wiley, New York, **1983**

C. S. Wong, E. Boyle, K. W. Bruland, J. D. Burton, E. D. Goldberg (Eds.), Trace Metals in Sea Water, Plenum Press, New York, **1983**

G. G. Leppard (Ed.), Trace Element Speciation in Surface Waters and its Geological Implications, Plenum Press, New York, **1983**

W. Salomons, U. Förstner, Metals in the Hydrocycle, Springer, Berlin, **1984**

J. Buffle, Complexation Reactions in Aquatic Systems: An Analytical Approach, Ellis Horwood, Chichester, **1988**

G. Bidoglio, W. Stumm (Eds.), Chemistry of Aquatic Systems: Local and Global Perspectives, Kluwer, Dordrecht, **1994**

Interaction with Solids

C. B. Amphlett, Inorganic Ion Exchangers, Elsevier, Amsterdam, **1964**

S. J. Gregg, K. S. W. Sing, Adsorption, Surface Area and Porosity, Academic Press, New York, **1964**

International Atomic Energy Agency, Exchange Reactions, Proceedings, IAEA, Vienna, **1965**

R. F. Gould, Adsorption from Aqueous Solutions, Amer. Chem. Soc. Adv. Chem. Series No. 79, Amer. Chem. Soc., Washington, DC, **1968**

K. Hauffe, S. R. Morrison, Adsorption, Walter de Gruyter, Berlin, **1974**

K. H. Lieser, Sorption Mechanisms, in: Sorption and Filtration for Gas and Water Purification (Ed. M. Bonnevie-Svendson), NATO Advanced Study Institute Series, Noordhoff-Leyden, **1975**

H. L. Bohn, B. L. McNeal, G. A. O'Connor, Soil Chemistry, Wiley, New York, **1979**

W. L. Lindsay, Chemical Equilibria in Soils, Wiley, New York, **1979**

B. E. Davies (Ed.), Applied Soil Trace Elements, Wiley, New York, **1980**

M. A. Anderson, A. J. Rubin (Eds.), Adsorption of Inorganics at Solid–Liquid Interfaces, Ann Arbor Science, Ann Arbor, MI, **1981**

A. Clearfield, Inorganic Ion Exchange Materials, CRC Press, Boca Raton, FL, **1982**

G. Sposito, The Surface Chemistry of Soils, Oxford University Press, Oxford, **1984**

T. H. Sibley, C. Myttenaere (Eds.), Application of Distribution Coefficients in Radiological Assessment Models, Elsevier, Amsterdam, **1985**

W. Stumm, Aquatic Surface Chemistry, Wiley, New York, **1987**

M. F. Hochella, jr., A. F. White (Eds.), Mineral–Water Interface Geochemistry, Reviews in Mineralogy Vol. 23, Mineralogical Society of America, Washington, DC, **1990**

D. A. Dzombak, F. F. M. Morel, Surface Complexation Modelling: Hydrous Ferric Oxide, Wiley, New York, **1990**

A. Weiss, E. Sextl, Clay Minerals as Ion Exchangers, in: Ion Exchangers (Ed. K. Dorfner), Walter de Gruyter, Berlin, **1991**

K. H. Lieser, Non-Siliceous Inorganic Ion Exchangers, in: Ion Exchangers (Ed. K. Dorfner), Walter de Gruyter, Berlin, **1991**

Radioecology

V. Schultz, A. W. Klement, jr. (Eds.), Radioecology of Aquatic Organisms, North-Holland, Amsterdam, **1966**

R. S. Russell (Ed.), Radioactivity and Human Diet, Pergamon, Oxford, **1966**

B. Åberg, F. P. Hungate (Eds.), Radioecological Concentration Processes, Pergamon, Oxford, **1967**

R. J. Garner, Transfer of Radioactive Materials from the Terrestrial Environment to Animals and Man, CRC Press, Boca Raton, FL, **1972**

The Metabolism of Compounds of Plutonium and other Actinides, ICRP Publ. 19, Pergamon, Oxford, **1972**

D. Greenberg, Radioecology, Wiley, New York, **1973**

International Atomic Energy Agency, Effects of Ionizing Radiation on Aquatic Organisms and Ecosystems, IAEA Techn. Rep. Ser. 172, Vienna, **1976**

F. W. Whicker, V. Schultz, Radioecology, Nuclear Energy and the Environment, Vols. I and II, CRC Press, Boca Raton, FL, **1982**

International Atomic Energy Agency, Sediment K_d's and Concentration Factors in the Marine Environment, Tech. Rep. 247, IAEA, Vienna, **1985**

S. Nair et al., Nuclear Power and Terrestrial Environment: The Transport of Radioactivity through Foodchains to Man, CEGB Research **1986**, No. 19

K Haberer, Umweltradioaktivität und Trinkwasserversorgung, Oldenbourg, Munich, **1989**

Commission of European Communities, Proc. Int. Symp. Radioecology, Znojmo, Brussels, **1992**

International Atomic Energy Agency, Handbook of Parameter Values for the Prediction of Radionuclide Transfer in Temperate Environments, Tech. Rep. 364, IAEA, Vienna, **1994**

H. Dahlgaard (Ed.), Nordic Radioecology. The Transfer of Radionuclides through Nordic Ecosystems to Man, Studies in Environmental Science 62, Elsevier, Amsterdam, **1994**

22 Dosimetry and Radiation Protection

22.1 Dosimetry

The units used in radiation dosimetry are summarized in Table 22.1. The energy dose and the ion dose are also used in radiation chemistry, whereas the equivalent dose is only applied in radiation biology and in the field of radiation protection.

Table 22.1. Radiation doses and dose rates.

Dose	Symbol	Unit	Abbreviation	SI unit
Energy dose	D	gray	Gy	$1\,\mathrm{Gy} = 1\,\mathrm{J\,kg^{-1}}$
(formerly:		rad	rd	$1\,\mathrm{rd} = 0.01\,\mathrm{J\,kg^{-1}})$
Ion dose	J	roentgen	R	$1\,\mathrm{R} = 2.580 \cdot 10^{-4}\,\mathrm{C\,kg^{-1}}$
Equivalent dose	H	sievert	Sv	$1\,\mathrm{Sv} \triangleq 1\,\mathrm{J\,kg^{-1}}$
(formerly:		rem	rem	$1\,\mathrm{rem} \triangleq 0.01\,\mathrm{J\,kg^{-1}})$

Dose rate	Unit	SI unit
Energy dose rate dD/dt	Gy/s (Gy/h, ...)	$1\,\mathrm{Gy/s} = 1\,\mathrm{J\,kg^{-1}\,s^{-1}}$
(formerly:	rd/s (rd/h, ...)	$1\,\mathrm{rd/s} = 0.01\,\mathrm{J\,kg^{-1}\,s^{-1}})$
Ion dose rate dJ/dt	R/s (R/h, ...)	$1\,\mathrm{R/s} = 2.580 \cdot 10^{-4}\,\mathrm{C\,kg^{-1}\,s^{-1}}$
Equivalent dose rate dH/dt	Sv/s (Sv/h, ...)	$1\,\mathrm{Sv/s} \triangleq 1\,\mathrm{J\,kg^{-1}\,s^{-1}}$
(formerly:	rem/s (rem/h, ...)	$1\,\mathrm{rem/s} \triangleq 0.01\,\mathrm{J\,kg^{-1}\,s^{-1}})$

The energy dose D is the energy dE transmitted by ionizing radiation to the mass dm of density ρ in the volume dV:

$$D = \frac{dE}{dm} = \frac{dE}{\rho\,dV} \tag{22.1}$$

In contrast to the ion dose, the energy dose is independent of the nature of the absorbing substance. The integral energy dose is

$$E = \int D\,dm \quad \text{(SI unit 1 J)} \tag{22.2}$$

Because direct determination of the energy dose D is difficult, the ion dose J is measured in radiation dosimeters. This is the charge dQ of ions of one sign generated by the ionizing radiation in the volume dV of air containing the mass dm:

$$J = \frac{dQ}{dm} = \frac{dQ}{\rho\,dV} \tag{22.3}$$

The unit of the ion dose is the roentgen (R). It is used in radiology and defined as the dose of X or γ rays that produces in air under normal conditions of temperature and pressure ions and electrons of one electrostatic unit each. From this definition it follows that $1\,R = 2.580 \cdot 10^{-4}\,C/kg$ (1 C (coulomb) = 1 As). Because 34 eV are needed to produce one ion pair (ion + electron) in air, 1 R is equivalent to an energy absorption of $0.877 \cdot 10^{-2}\,J$ per kg air. At the same ion dose, the energy absorption in various substances can be rather different. However, for substances of interest in radiation protection (aqueous solutions and tissue) the energy absorption is similar to that in air. For example, an ion dose of 1 R of X rays or γ rays in the energy range between 0.2 and 3.0 MeV causes an energy absorption of $0.97 \cdot 10^{-2}\,J/kg$ in water, of $0.93 \cdot 10^{-2}\,J/kg$ in soft tissue and of $0.93 \cdot 10^{-2}\,J/kg$ in bone. Therefore, in the praxis of radiation protection the following approximative relation is valid:

$$1\,R \approx 10^{-2}\,Gy = 10^{-2}\,J/kg \tag{22.4}$$

In order to take into account the biological effects of different kinds of radiation, radiation weighting factors w_R were introduced by the International Commission on Radiological Protection (ICRP) in 1990 (Table 22.2). The weighting factor w_R indicates the ratio of the degree of a certain biological effect caused by the radiation considered, to that caused by X rays or γ rays at the same energy absorption. It is laid down on the basis of the experience gained in radiation biology and radiology.

Table 22.2. Radiation weighting factors w_R (ICRP 1990).

LET in water [eV/nm]	Weighting factor w_R	Type of radiation
0.2–35	1	Photons (X and γ rays)
0.2–1.1	1	Electrons and positrons >5 keV
≈ 20	5	Slow neutrons <10 keV
≈ 50	20	Intermediate neutrons 0.1–2 MeV
	10	Fast neutrons 2–20 MeV
	5	Protons >2 MeV
≈ 130	20	α particles
	20	High-energy ions

The equivalent dose H is measured in sievert (Sv) and defined as

$$H = w_R \cdot D \tag{22.5}$$

where D is the absorbed energy dose, measured in Gy. Considering the effects of various radiations R on a special tissue T, the equivalent dose H_T received by this tissue is

$$H_T = \Sigma w_R \cdot D_{T,R} \tag{22.6}$$

$D_{T,R}$ is the absorbed dose averaged over the tissue T due to the radiation R. The sum is taken over all radiations R. In the case of low linear energy transfer (LET; section

6.1), w_R is equal to 1 and $H_T = D_{T,R}$. Instead of the weighting factor w_R, the terms quality factor Q and factor of relative biological effectiveness f_{RBE}, respectively, were used before 1990. An earlier, but similar, concept was the rem (radiation equivalent man, 1 rem $\simeq 10^{-2}$ Sv).

The advantages of the equivalent dose are that the biological effectiveness is directly taken into account and that equivalent doses received from different radiation sources can be added. However, the dimension J/kg is only correct if $w_R = 1$.

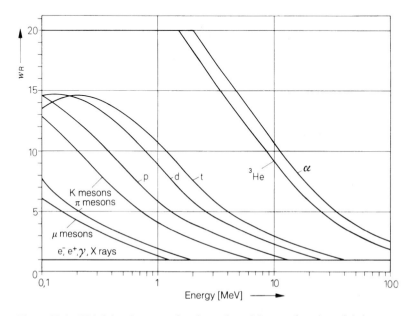

Figure 22.1. Weighting factor w_R for charged particles as a function of their energy.

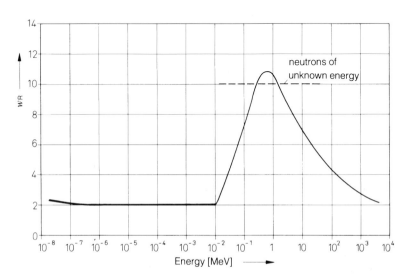

Figure 22.2. Weighting factor w_R for neutrons as a function of their energy.

Weighting factors are assessed on the basis of the LET value. The influences of spatial ionization density and of temporal distribution of ionization have to be taken into account separately. Weighting factors for various particles and for γ-ray photons are plotted in Fig. 22.1 and those for neutrons in Fig. 22.2 as a function of the energy.

22.2 External Radiation Sources

Sensitive parts of the body with respect to external radiation sources are the haematogenous organs, the gonads and the eyes. Less sensitive parts are the arms and hands, the legs and feet, the head (except the eyes) and the neck. The ion dose rate transmitted by a point-like γ-radiation source of activity A at the distance r is

$$\frac{\mathrm{d}J}{\mathrm{d}t} = k_\gamma \frac{A}{r^2} \tag{22.7}$$

The dose rate constant k_γ for γ radiation depends on the energy of the γ rays and on the decay scheme of the radionuclide. Values of k_γ for various radionuclides are listed in Table 22.3. For rough estimation, it is useful to know that a point-like radiation source of 1 GBq emitting γ rays with energies of ≈ 1 MeV transmits an ion dose rate of about 0.03 R h^{-1} at a distance of 1 m.

Table 22.3. Dose rate constants k_γ for various radionuclides (The frequency of the γ transitions is taken into account).

Radionuclide	k_γ [mGy m^2 h^{-1} GBq^{-1}]	Radionuclide	k_γ [mGy m^2 h^{-1} GBq^{-1}]
^{22}Na	0.32	^{82}Br	0.40
^{24}Na	0.49	^{85}Kr	0.0003
^{42}K	0.038	^{99m}Tc	0.016
^{51}Cr	0.0049	^{110m}Ag	0.40
^{52}Mn	0.48	^{124}Sb	0.24
^{54}Mn	0.13	^{123}I	0.019
^{56}Mn	0.24	^{131}I	0.057
^{59}Fe	0.17	^{132}I	0.31
^{58}Co	0.15	^{137}Cs + ^{137m}Ba	0.086
^{60}Co	0.35	^{140}Ba	0.032
^{64}Cu	0.032	^{144}Ce	0.0065
^{65}Zn	0.081	^{182}Ta	0.18
^{68}Ga	0.15	^{192}Ir	0.14
^{76}As	0.068	^{198}Au	0.062

The ion dose rate transmitted by a point-like β emitter can be calculated by an equation similar to eq. (22.7),

$$\frac{\mathrm{d}D}{\mathrm{d}t} = k_\beta(r)\frac{A}{r^2} \tag{22.8}$$

In contrast to k_γ, however, $k_\beta(r)$ depends strongly on the distance, because of the much stronger absorption of β particles in air. $k_\beta(r)$ is higher than k_γ by about two orders of magnitude. Below $r \approx 0.3\ R_{max}$ ($R_{max} =$ maximum range), a rough estimate is $k_\beta \approx 1\ \text{R m}^2\ \text{h}^{-1}\ \text{GBq}^{-1}$.

The influence of the distance r indicates the importance of remote handling of higher activities.

22.3 Internal Radiation Sources

Internal radiation sources are always more dangerous than external ones, because shielding is impossible, and incorporated radionuclides may be enriched in certain organs or parts of the body and affect them over long periods of time.

With respect to radiotoxicity, possible storage in the body and half-lives of radionuclides are therefore most important. Removal from the body is characterized by the biological half-life, and the effective half-life $t_{1/2}(\text{eff})$ is given by the relation (section 21.5):

$$\frac{1}{t_{1/2}(\text{eff})} = \frac{1}{t_{1/2}(\text{b})} + \frac{1}{t_{1/2}(\text{p})} \tag{22.9}$$

where $t_{1/2}(\text{b})$ and $t_{1/2}(\text{p})$ are the biological and physical half-lives, respectively. Physical half-lives and effective half-lives of some radionuclides in the human body are listed in Table 22.4. The latter depend largely on the part of the body considered.

Radiotoxicity depends on the radiation emitted by the radionuclide considered, the mode of intake (e.g. by air, water or food), the size of the ingested or inhaled particles, their chemical properties (e.g. solubility), metabolic affinity, enrichment, effective half-life and ecological conditions. The radiotoxicity of some radionuclides is listed in Table 22.5. The limits of free handling of radionuclides and the acceptable limits of radionuclides in air, water and food are also laid down on the basis of their radiotoxicity.

Most of the properties mentioned in the previous paragraph are taken into account in the ALI and DAC concept (ALI = Annual Limits of Intake; DAC = Derived Air Concentration, based on the ALI value). 1 ALI corresponds to an annual committed equivalent dose of 50 mSv.

Table 22.4. Physical half-lives $t_{1/2}(p)$ and effective half-lives $t_{1/2}(eff)$ of radionuclides in the human body (ICRP 1993).

Radionuclide	$t_{1/2}(p)$	Part of the body considered	$t_{1/2}(eff)$
T	12.323 y	Body tissue	12 d
^{14}C	5730 y	Fat	12 d
^{24}Na	14.96 h	Gastrointestinal tract	0.17 d
^{32}P	14.26 d	Bone	14 d
^{35}S	87.5 d	Testis	76 d
^{42}K	12.36 h	Gastrointestinal tract	0.04 d
^{51}Cr	27.7 d	Gastrointestinal tract	0.75 d
^{55}Fe	2.73 y	Spleen	390 d
^{59}Fe	44.5 d	Gastrointestinal tract	0.75 d
^{60}Co	5.272 y	Gastrointestinal tract	0.75 d
^{64}Cu	12.7 h	Gastrointestinal tract	0.75 d
^{65}Zn	244.3 d	Total	190 d
^{90}Sr	28.64 y	Bone	16 y
^{95}Zr	64.0 d	Bone surface	0.75 d
^{99}Tc	$2.1 \cdot 10^5$ y	Gastrointestinal tract	0.75 d
^{106}Ru	373.6 d	Gastrointestinal tract	0.75 d
^{129}I	$1.57 \cdot 10^7$ y	Thyroid	140 d
^{131}I	8.02 d	Thyroid	7.6 d
^{137}Cs	30.17 y	Total	70 d
^{140}Ba	12.75 d	Gastrointestinal tract	0.75 d
^{144}Ce	284.8 d	Gastrointestinal tract	0.75 d
^{198}Au	2.6943 d	Gastrointestinal tract	0.75 d
^{210}Po	138.38 d	Spleen	42 d
^{222}Rn	3.825 d	Lung	3.8 d
^{226}Ra	1600 y	Bone	44 y
^{232}Th	$1.405 \cdot 10^{10}$ y	Bone	200 y
^{233}U	$1.592 \cdot 10^5$ y	Bone, lung	300 d
^{238}U	$4.468 \cdot 10^9$ y	Lung, kidney	15 d
^{238}Pu	87.74 y	Bone	64 y
^{239}Pu	$2.411 \cdot 10^4$ y	Bone	200 y
^{241}Am	432.2 y	Kidney	64 y

Table 22.5. Radiotoxicity of radionuclides and radioelements.

Radiotoxicity	Radionuclides and radioelements
Group I: very high	^{90}Sr, Ra, Pa, Pu
Group II: high	^{45}Ca, ^{55}Fe, ^{91}Y, ^{144}Ce, ^{147}Pm, ^{210}Bi, Po
Group III: medium	^{3}H, ^{14}C, ^{22}Na, ^{32}P, ^{35}S, ^{36}Cl, ^{54}Mn, ^{59}Fe, ^{60}Co, ^{89}Sr, ^{95}Nb, ^{103}Ru, ^{106}Ru, ^{127}Te, ^{129}Te, ^{137}Cs, ^{140}Ba, ^{140}La, ^{141}Ce, ^{143}Pr, ^{147}Nd, ^{198}Au, ^{199}Au, ^{203}Hg, ^{205}Hg
Group IV: low	^{24}Na, ^{42}K, ^{64}Cu, ^{52}Mn, ^{76}As, ^{77}As, ^{85}Kr, ^{197}Hg

22.4 Radiation Effects in Cells

With respect to radiation protection, the relation between radiation dose and damage, in particular the effects of radiation in cells, is of the greatest importance

Cells contain about 70% water, and the radiation is largely absorbed by interaction with the water molecules and formation of ions, free radicals and excited molecules. The ions may react at ionizable positions of the DNA (e.g. phosphate groups). Radicals, such as $\cdot OH$ and $\cdot H$, and oxidizing products, such as H_2O_2, may be added at unsaturated bonds or they may break the bonds between two helices. Excited molecules may transfer the excitation energy to the DNA and also cause breaks. A great number of different products of DNA damage have been identified.

On the other hand, living cells contain natural radical scavengers, and as long as these are present in excess of the radiolysis products, they are able to protect the DNA. However, when the concentration of radiolysis products exceeds that of the scavengers, radiation damage is to be expected. This leads to the concept of a natural threshold for radiation damage which should at least be applicable for low LET values and low radiation intensities, e.g. for low local concentrations of ions and radicals. The scavenging capacity may vary with the age and the physical conditions of the individuals considered.

Furthermore, the cells are protected by various repair mechanisms with the aim of restoring damages. Details of these mechanisms are not known, but most single-strand breaks are correctly repaired. Sometimes, however, repair fails (e.g. by replacement of a lost section by a wrong base pair), which may lead to somatic effects, such as cancer or inheritable DNA defects. Results obtained with cells of living organisms and samples studied in the laboratory indicate that repair mechanisms are more effective in living organisms, i.e. in the presence of surrounding cells and body fluids.

Gamma rays exhibit low LET values and cause only few ionizations in a DNA segment several nanometres in length, corresponding to an energy transmission of up to about 100 eV. If this energy is distributed over a large number of bonds, these will not receive enough energy to break, and the DNA segment will not be changed. However, formation of clusters of ions by low-LET radiation will increase the risk of damage. This risk is appreciably higher in the case of the high LET values of α rays, e.g. of high ionization density. At low LET values, the ratio of double-strand breaks to single-strand breaks is about 3 : 100, whereas at high LET values this ratio is much higher. Repair of double-strand breaks is more difficult and takes more time (on average several hours) than that of single-strand breaks (on average ≈ 10 min). Therefore, the probability of repair errors causing permanent damage or mutation is much higher in the case of double-strand breaks. Consequently, chromosome aberrations are only observed after double-strand breaks.

In the nucleus of a single cell, a γ-ray photon produces up to about 1500 ionizations (on average ≈ 70), up to 20 breaks of single-strand DNA (on average ≈ 1) and only up to a few breaks of double-strand DNA (on average ≈ 0.04). For an α-particle crossing a nucleus the corresponding values are: up to 10^5 ionizations (on average 23 000), up to 400 single-strand breaks (on average 200), and up to 100 double-strand breaks (on average 35).

With respect to the damage, the dose rate (i.e. the time during which a certain dose is transmitted to the body) is of great importance. This is illustrated by the following example: A low-LET dose of 3 Gy produces ≈ 3000 single-strand breaks and ≈ 100 double-strand breaks in every cell of a human body. If this dose is transmitted within a short time of several minutes, the damage in the cells cannot be repaired and may result in death. However, if the same dose is spread over a period of about a week, only aberrations in the chromosomes are observed.

With respect to radiation effects, two types of cells are distinguished, those involved in the function of organs (e.g. in bone marrow, liver and the nervous system) and those associated with reproduction (gonads). In the first group, radiation damage may give rise to somatic effects (e.g. cell death or cancer) and in the second group to genetic effects (e.g. effects transferred to future generations).

22.5 Radiation Effects in Man, Animals and Plants

The effects of radiation on microorganisms, plants and animals differ appreciably. From Table 22.6, it is evident that organisms at a low stage of evolution exhibit much higher radiation resistance than those at a higher evolutionary stage.

Table 22.6. Effect of γ irradiation on enzymes, microorganisms, plants, animals and man.

Organism	Dose of inactivation (D_i) or 50% lethal dose within 30 days ($D_{50/30}$)
Enzymes	$D_i > 20\,000\,\text{Gy}$
Viruses	$D_i = 300–5000\,\text{Gy}$
Bacteria	$D_i = 20–1000\,\text{Gy}$
Flowers	$D_i > 10\,\text{Gy/d}$ $\left.\right\}$ during the season of growing
Trees	$D_i > 1\,\text{Gy/d}$
Amoeba	$D_{50/30} \approx 1000\,\text{Gy}$
Drosophila	$D_{50/30} \geqslant 600\,\text{Gy}$
Shellfish	$D_{50/30} \approx 200\,\text{Gy}$
Goldfish	$D_{50/30} \approx 20\,\text{Gy}$
Rabbit	$D_{50/30} \approx 8\,\text{Gy}$
Monkey	$D_{50/30} \approx 6\,\text{Gy}$
Dog	$D_{50/30} \approx 4\,\text{Gy}$
Man	$D_{50/30} \approx 4\,\text{Gy}$

Two kinds of radiation exposure are distinguished, stochastic exposure (the effects are distributed statistically over a large population), and deterministic exposure (the effects are inevitable or intended, as in the case of deliberate irradiation, e.g. in radiotherapy).

The survivors of the bombs dropped over Japan in 1945 are the most important source of information about human whole-body irradiation. The health of about 76 000 persons who had been exposed to doses of up to 7–8 Gy of instantaneous neutron and γ radiation and that of their children has been investigated in detail for

50 years. The frequency of leukaemia observed as a function of dose is plotted in Fig. 22.3. Following large-dose exposure, leukaemia is first observed after a latent period of about 2–3 y, reaches a peak frequency after about 6–8 y and almost disappears after about 25 y. The large statistical variations are obvious from Fig. 22.3. The higher frequency of leukaemia after the Hiroshima explosion is explained by the higher neutron-flux density of this nuclear explosion. With respect to the influence of γ irradiation, the results obtained from Nagasaki are more meaningful.

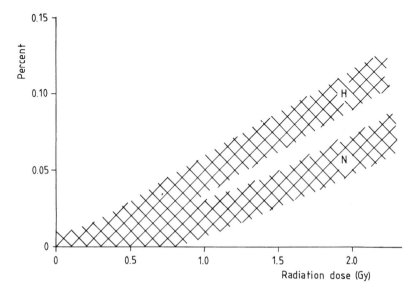

Figure 22.3. Frequencies of leukaemia observed in Hiroshima (H) and Nagasaki (N): percentage of persons as a function of the radiation dose.

For other types of cancer, the excess malignancies are relatively low. They appear usually after a latent period of about 10 y, and their frequency increases with time.

Deterministic irradiation is applied in nuclear medicine for therapeutic purposes (chapter 19). For tumour treatment, large doses are delivered to selected tissues. Gamma rays emitted by ^{60}Co (up to $2 \cdot 10^{14}$ Bq) or ^{137}Cs (up to $2 \cdot 10^{13}$ Bq) are preferred for irradiation of deeply located organs. The doses transmitted to malignant tumours vary between about 10 and 100 Sv, and the individual responses of patients to certain radiation doses may vary appreciably.

In order to take into account the radiation sensitivity of different tissues, tissue weighting factors w_T are introduced, and the effective equivalent dose H_E received by the tissue E is defined by

$$H_E = \Sigma w_T \cdot H_T \tag{22.10}$$

where H_T is given by eq. (22.6). H_E is also measured in Sv. For low-LET whole-body irradiation Σw_T is equal to 1 and $H_E = H_T$. The weighting factors w_T are used in nuclear medicine (therapeutic applications of radionuclides).

The possibility of genetic effects has received special attention since radiation-induced mutations were observed for drosophila in 1927. However, the vast majority of changes in the DNA are recessive. In mammals, no radiation-induced hereditary effects have been observed. Irradiation of the genitals of mice with $\approx 2\,Sv$ for 19 generations did not lead to observable genetic changes, and careful investigations of the 75 000 children born to parents who were exposed to the irradiation released by the nuclear explosions in 1945 did not show any increase in the frequency of cancer or hereditary diseases.

Information about radiation effects due to ingestion or inhalation of radioactive substances is mainly obtained from the following groups:

- workers in uranium mines (inhalation of Rn and ingestion of U and its daughters);
- people living in areas of high Rn concentrations (inhalation of Rn);
- people who were painting luminous dials in Europe before 1930 (ingestion of Ra by licking the brushes);
- patients who received therapeutic treatment with ^{131}I or other radionuclides.

Therapeutic treatment of hyperthyroidism with ^{131}I led to large local doses of several hundred Sv of β radiation, but with respect to leukaemia no significant difference was observed in comparison with untreated people.

Investigation of tumour frequencies in animals after incorporation of ^{90}Sr or ^{226}Ra, or irradiation with X rays or electrons, indicates a threshold dose of $\leq 5\,Gy$, below which no effects are observed. The effects of large single doses on man are summarized in Table 22.7.

Large irradiation doses applied to plants cause mutations which either improve the properties of the species or produce disadvantageous effects. Although irradiation of plant seeds results in a ratio of only about 1 : 1000 of advantageous to

Table 22.7. Effects of accidental radiation exposure on man (approximate values).

Dose	Effects
Whole-body irradiation	
0.25 Sv	No clinically recognizable damage
0.25 Sv	Decrease of white blood cells
0.5 Sv	Increasing destruction of the leukocyte-forming organs (causing decreasing resistance to infections)
1 Sv	Marked changes in the blood picture (decrease of leukocytes and neutrophils)
2 Sv	Nausea and other symptoms
5 Sv	Damage to the gastrointestinal tract causing bleeding and $\approx 50\%$ death
10 Sv	Destruction of the neurological system and $\approx 100\%$ mortality within 24 h
Irradiation of the hands	
2 Gy	No proven effects
4 Gy	Erythema, skin scaling
6 Gy	Skin reddening, pigmentation
8.5 Gy	Irreversible degeneration of the skin
50 Gy	Development of non-healing skin cancer (amputation necessary)

harmful effects, new plant variations have been obtained by selection and cultivation of the few plants exhibiting improved properties. For example, most of the grain grown today in Scandinavian countries consists of radiation-produced species showing greater resistance to the cold weather conditions.

In the range of low radiation doses, reliable conclusions with respect to radiation effects on man are difficult, because malignancies or mortalities caused by radiation have to be calculated from the difference between those observed under the influence of radiation and those without radiation, and are therefore very uncertain. For example, any significant increase in cancer or malignant mortality of the population in the USA, UK, Canada, France, Sweden, Finland, China and other countries due to the natural background in the range of dose rates between about 1 and 5 mSv/y has not been found. Furthermore, except for uranium miners, a statistically proved relation between the frequency of lung cancer and the concentration of Rn in the air could not be established, although such a relation is assumed to exist.

Because of the uncertainties caused by the impossibility of obtaining statistically significant data, the construction of dose–effect curves is difficult in the range of small doses <0.1 Sv or small dose rates <0.5 Sv/y. This is also evident from Fig. 22.3, in which the frequency of leukaemia in the populations of Hiroshima and Nagasaki after receiving relatively large doses is plotted as a function of the dose. In the range of doses <0.2 Sv the statistical errors are appreciably greater than the effects.

Several assumptions are made for the range of small doses:

– a linear increase of the effects with the dose (ICRP recommendation);
– a threshold at about 50 mSv, below which there is no increase of cancer or other radiation-induced diseases (this assumption is favoured by many radiologists);
– a quadratic linear model assuming an initial increase with the square of the dose in the range of small doses, followed by a linear increase.

22.6 Non-occupational Radiation Exposure

Average equivalent dose rates received from natural radiation sources are listed in Table 22.8. The values vary appreciably with the environmental conditions. The influence of cosmic radiation increases markedly with the height above sea level, and terrestrial radiation depends strongly on the local and the living conditions.

The largest contribution to the everyday radiation dose of the population comes from the concentration of Rn and its daughter products in the air, as can be seen from Table 22.8. The concentration of ^{222}Rn is relatively high in regions of high uranium concentration in the ground and in poorly ventilated housing built of materials containing small amounts of U, Ra or Th. Accordingly, the dose rate varies considerably (e.g. between ca. 0.3 and 100 mSv/y in the UK, on average 1.2 mSv), but epidemiological investigations have failed to exhibit any statistically significant correlation between these different radiation doses and lung cancer. Only at larger doses is such a correlation evident, for example for uranium miners working from 1875 to 1912 in the Erzgebirge: 25 to 50% of these miners died from lung cancer caused mainly by the decay products of ^{222}Rn present in the form of aerosols in the air.

Table 22.8. Average radiation exposure by natural radiation sources.

Kind of exposure	Equivalent dose [mSv/y]		
	Whole body and gonads	Bone	Lung
External radiation sources			
Cosmic radiation[a]	0.35	0.35	0.35
(sea level, 50° north)			
Terrestrial radiation[b]	0.49	0.49	0.49
(K; U, Th and decay products)			
Internal radiation sources			
Uptake by ingestion			
T	<0.00002	–	–
^{14}C	0.016	0.016	0.016
^{40}K	0.19	0.11	0.15
^{87}Rb	0.003	–	–
^{210}Po	–	0.14	–
^{220}Rn + ^{222}Rn	0.02	0.02	0.02
^{226}Ra + ^{228}Ra	0.03	0.72	0.05
^{238}U	0.0008	–	–
Uptake by inhalation			
^{220}Rn	–	–	1.75[c]
^{222}Rn	–	–	1.30[c]
Sum	≈1.10	≈1.85	≈4.10

[a] On the ground, locally up to ≈2 Sv/y. Intensity of cosmic radiation increases by a factor of ≈1.6 per 1000 m above sea level.

[b] Locally up to ≈4.3 Sv/y. On average, in the open air ≈25% less than in buildings. Minimum values × 1/10, maximum values × 10 of the values listed.

[c] Values for brick buildings and 3.5-fold exchange of air per hour. In concrete buildings without exchange of air the values are higher by a factor of 4–7.

Some caves in granitic rocks containing relatively high concentrations of Rn are used for radiotherapy of the respiratory tract. At activity levels of Rn in domestic air <400 Bq/m^3, carcinogenic effects are not proven, but action is proposed for dwellings containing more than 200–600 Bq Rn per m^3 air.

Average equivalent dose rates due to artificial radiation sources are listed in Table 22.9. These dose rates originate from application of X rays and radionuclides for diagnostic and therapeutic purposes, from various radiation sources applied in daily life and from radioactive fall-out.

Table 22.9. Radiation exposure by artificial radiation sources.

Kind of exposure	Single radiation dose		Average radiation dose per capitem of the population
	Local dose	Whole body and gonad dose	
Medical applications			
External radiation sources (medical application of X rays or radionuclides)			
Diagnosis	1–10 mSv	0.1–1 mSv	≈0.5 mSv/y
Therapy	Up to 50 Sv	50 mSv	≈0.01 mSv/y
Internal radiation sources (medical application of radionuclides such as ^{99m}Tc)			
Diagnosis	1–1000 mSv	0.1–10 mSv	≈0.02 mSv/y
Therapy	10 Sv	50 mSv	<0.01 mSv/y
Other radiations			
Technical applications of radionuclides and ionizing radiation	–	–	<0.02 mSv/y
Luminous dials	–	0.3 mSv/y	<0.01 mSv/y
Occupational radiation exposure	500 mSv/y	50 mSv/y	<0.001 mSv/y
Radioactive fall-out	(a)	(a)	<0.1 mSv/y
Nuclear installations	(a)	(a)	<0.01 mSv/y
Sum			0.6–0.7 mSv/y

(a) By severe accidents, such as the Chernobyl accident, doses up to several sieverts have been transmitted.

22.7 Safety Recommendations

The assumption by ICRP of a linear relation between radiation effects and dose (section 22.5) implies the highest degree of safety. It may overestimate the risk, but for safety reasons its application is recommended. In the linear ICRP approach, the risk L_c of cancer is assumed to be given by

$$L_c = 0.05\, H_E \tag{22.11}$$

In this equation, the risk of cancer is assumed to be 100% for an effective equivalent dose of 20 Sv. The equation is based on the probabilities listed in Table 22.10. In order to take into account that a dose delivered with a relatively low dose rate over a longer period of time has an appreciably smaller effect than a single dose, a dose reduction factor of 2 is recommended for smaller dose rates. However, in the report of the United Nations Scientific Committee on the Effects of Atomic Radiation

Table 22.10. Probability coefficients assessed for stochastic effects in detriment (ICRP 1990)

Population sector	Probability [% per Sv]		
	Fatal cancer	Non-fatal cancer	Severe hereditory effects
Adult workers	4	0.8	0.8
Whole population	5	1.0	1.2

(UNSCEAR 1993) it is noted that a reduction factor between 2 and 10 is more appropriate.

Another dose concept is that of the collective dose. It is based on the assumption that cancer is induced by a stochastic single process, independently of the dose rate and the dose fractionation, and it implies that the detriment is the same whether one person receives 20 Sv or 20 000 persons receive 1 mSv each. In both cases the collective dose is 20 man sieverts (man Sv) and there should be a 100% probability of one disease of cancer. The collective dose concept is often applied to assess the effect of the natural radiation background. At an average level of 3 mSv/y, 0.015% of the population should die each year due to natural radiation. As the frequency of deaths by cancer is about 0.2% per year, it is not possible to confirm the collective dose concept by epidemiological investigations.

In the committed dose concept, also introduced by ICRP, the total dose contribution to the population over all future years due to a specific release or exposure is considered. The committed dose is defined as the time integral per capitem dose rate between the time of release and infinite time, and is measured in Sv

$$H_{\text{comm}}(t) = \int_t^\infty \frac{\mathrm{d}H}{\mathrm{d}t} \mathrm{d}t \tag{22.12}$$

Similarly, a collective committed dose is obtained by integrating the collective dose.

Dose limits recommended by ICRP (1990) are listed in Table 22.11. With respect to possible genetic damage, an upper limit of 50 mSv in 30 y should not be exceeded. The maximum permissible single dose without consideration of genetic damage is 0.25 Sv. At this dose clinical injuries are not observed (Table 22.7). To take into account the different sensitivity of various organs and parts of the human body, respectively, coefficients and weighting factors have been published 1991 by ICRP.

Table 22.11. Recommended dose limits (ICRP 1990).

Dose	Occupational	Public
Effective dose	20 mSv/y[a]	1 mSv/y[b]
Equivalent dose		
In the lens of the eye	150 mSv	15 mSv
In the skin	500 mSv	50 mSv
In the hands and feet	500 mSv	50 mSv

[a] Averaged over 5 y, but in one year 50 mSv should not be exceeded.
[b] May be exceeded as long as the 5-year dose does not exceed 5 mSv.

22.8 Safety Regulations

Special regulations have been established for persons working professionally with X rays or with radioactive substances in radiology, nuclear medicine, in chemical or physical laboratories, technical installations, at accelerators, nuclear reactors, or in reprocessing and other technical plants. With respect to possible damage, in particular genetic damage, dose and dose rate limits are laid down.

Rooms are classified according to the activities permitted, e.g. controlled areas and supervised areas, α laboratories and β laboratories.

In rooms in which high activities are handled, the stay should be as short as possible and the operations in these rooms should be restricted to a minimum.

Regular measurements of the dose rates are obligatory in all rooms in which radioactive substances are handled, and the doses received by the persons working in these rooms have to be noted down. Furthermore, the rooms have to be checked regularly with respect to contamination. Finally, all persons working in these rooms have to be supervised medically.

The radiation doses received by persons working in radiochemical laboratories vary appreciably with the conditions. For example, continuous working with 1 GBq of a high-energy γ emitter (1 MeV) without shielding at a distance of 40 cm during 8 h transmits a dose of ≈ 1 mSv. If the radiation is shielded by 5 cm lead, the dose is only ≈ 0.01 mSv and can be neglected. Shielding of β radiation is much simpler, because it is largely absorbed in the samples and in the walls of the equipment used. However, the bremsstrahlung has to be taken into account, particularly in the case of high β activities. Because the intensity of bremsstrahlung increases with the energy of the β radiation and with the atomic number of the absorber, materials containing elements of low atomic numbers, such as perspex, are most suitable absorbers. Generally, a shield of 1 cm thickness is sufficient for quantitative absorption of β radiation. Shielding of α radiation is unproblematic, because it is absorbed in several cm of air.

Practical experience in handling of radioactive substances and safety regulations have led to general rules for radiochemical laboratories concerning the distinction between laboratories, measuring rooms and storage rooms, the equipment used in the laboratories, the handling of radionuclides, radioactive waste, used air and waste water, and the supervision of the persons working in the laboratories.

With respect to practical work with radioactive substances, the risk of a hazardous effect should be as low as possible, and <0.1%. The following measures are recommended to keep radiation exposure to a minimum:

– an adaquate distance to the radioactive substance;
– use of shielding, particularly in handling of high activities;
– control of working place, equipment, hands and clothing for radioactivity.

Safety regulations and rules for operation of radiochemical laboratories depend strongly on the activities to be handled and on the kind of radiation emitted. In accordance with the classification by the International Atomic Energy Agency (IAEA) it is useful to distinguish between laboratories for low activities (type C), medium activities (type B) and high activities (type A). This classification is given in Table 22.12. Laboratories for handling of higher activities are often classified

Table 22.12. Classification of laboratories for open handling of radioactive substances (IAEA recommendations).
Factor 100 for storage in closed ventilated containers
Factor 10 for simple wet chemical procedures
Factor 1 for usual wet chemical procedures
Factor 0.1 for wet chemical procedures with the danger of spilling, and simple dry procedures
Factor 0.01 for dry procedures with the danger of production of dust

Radiotoxicity of the radionuclide	Type of laboratory		
	Type C (simple equipment)	Type B (better equipment)	Type A (hot laboratory)
Very high	<10 μCi	10 μCi–10 mCi	>10 mCi
High	<100 μCi	100 μCi–100 mCi	>100 mCi
Medium	<1 mCi	1 mCi–1 Ci	>1 Ci
Low	<10 mCi	10 mCi–10 Ci	>10 Ci

$(1 \, Ci = 3.7 \cdot 10^{10} \, Bq.)$

according to the kind of radiation emitted by the radionuclides (α laboratories, β laboratories, γ laboratories).

Type C laboratories can be fitted out in normal chemical laboratories. The floors, walls and benches should be free of grooves, and the ventilation should be good. It is recommended that all operations with radionuclides be carried out in tubs and that suitable containers be provided for solid and liquid waste. Because of the risk of incorporation, all mouth operations (e.g. pipetting with the mouth) are strongly forbidden. For wiping of pipettes and other equipment, paper tissue is used. Monitors must be available at the working place, to detect radioactive substances on the equipment and to check hands and working clothes for radioactivity.

Type B laboratories require sufficient equipment for radiation protection, such as tongs for remote handling, lead walls constructed of lead bricks and lead-glass windows for handling γ emitters, and perspex shielding for handling β emitters. The use of gloves is recommended. Dosimeters for measurement of dose rates must be available. For handling of α emitters, glove boxes are used that are equipped for the special chemical operations and may be combined in glove-box lines. Ventilation must be efficient, and people as well as laboratories must be supervised with care. Treatment of waste water and of solid and liquid radioactive waste needs special attention.

Radioactive contamination of the laboratories must be strictly avoided, because reliable activity measurements would otherwise become impossible and low activities could not be measured any more. Therefore, careful handling of radioactive substances and regular activity checking in the laboratories is indispensible. Personal danger due to radioactive contamination is, in general, only to be feared in the case of high contamination of the laboratories and great carelessness.

Handling of materials of high activity in type A laboratories requires special installations, in particular hot cells equipped with instruments for remote handling.

All operations with radionuclides need careful preparation, and the time needed for that preparation increases with the activity to be handled. All operations should be as simple and as safe as possible, which implies the use of small numbers of receptacles. Boiling of solutions containing radioactive substances is to be avoided, because of the risk of spraying. Closed receptacles are preferable whenever possible. Often preliminary experiments with inactive substances are useful, in order to find the optimal conditions.

Cleaning of receptacles and other equipment is recommended immediately after the end of a chemical operation, because radionuclides may enter the surface layers of glass and other materials by diffusion, with the result that cleaning becomes very difficult. For the purpose of decontamination, solutions containing carriers are often useful. All the parts used, the working place, the hands and clothes must finally be checked carefully for radioactivity.

For the supervision of the persons working in radiochemical laboratories, pocket dosimeters (generally ionization dosimeters) and film dosimeters are used. The lower detection limits of these dosimeters vary between about 1 and 40 mR. Furthermore, hand–foot monitors are installed near the exit of the laboratories, by which external contamination can be detected. In the case of suspected internal contamination, the person is checked by means of a whole-body counter which allows detection of γ-ray emitters with high sensitivity. The presence of natural ^{40}K contributes essentially to the γ activity in the body. Besides that, about 100 Bq of ^{137}Cs can be determined quantitatively.

If a radioactive substance is taken up by ingestion or inhalation, attempts are to be made to remove it as fast as possible, before it may be incorporated into tissue where it may exhibit a relatively long biological half-life. For this purpose, complexing agents forming stable complexes with the radionuclides may be applied.

Monitors in the laboratory are used for control of the working place and the equipment, as well as for control of hands and working clothes. Counters with large sensitive areas (generally flow counters operating with methane) allow more exact control of the working place and the floor for contamination. Wiping tests are also very useful for contamination control. Humid filter papers or other absorbent materials used for wiping benches or floors are measured for α or β activity by means of a suitable counter. In this way, the radionuclide causing the contamination can also be identified.

Radioactive waste solutions can be handled in different ways: the solutions may be collected separately, according to the radionuclides, in labelled bottles. This makes sorting easy and separate processing possible. Alternatively, all radioactive waste solutions may be collected in tanks and processed together.

Waste water from radiochemical laboratories which may contain radioactive substances, e.g. by mistake, must be checked for radioactivity. If the condition is such that the activity should not be higher than the acceptable limit for drinking water (e.g. 1 Bq/l), very sensitive methods are necessary to detect these low concentrations. For comparison, 1 Bq/l is the activity of 20 mg of natural K per litre, river water contains about 1 Bq/l, water from natural springs up to several kBq/l, and rainwater sampled after test explosions of nuclear weapons also contained up to 1 kBq/l. The safest method for determination of concentrations of the order of 1 Bq/l is evaporation of about 1 l and measurement of the residue by means of a large-area flow counter. This method makes it possible to detect several mBq of α or β emitters.

In the off-gas system of radiochemical laboratories, filters are installed which retain aerosols and vapours. These filters have to be checked regularly for radioactivity and to be exchanged from time to time. Automated supervision of the off-gas is also possible. In this case, the activity is recorded continuously.

Combustible and non-combustible solid radioactive waste is often collected separately. Waste containing only short-lived radionuclides is usually stored until the activity has vanished. Solid waste containing long-lived radionuclides is collected in polyethylene bags, which are closed by welding, filled in drums and taken to a central collecting station for radioactive waste.

22.9 Monitoring of the Environment

In many countries the radioactivity in the environment is continuously measured by means of monitoring stations, in particular at and in the neighbourhood of nuclear power stations and other nuclear facilities. Monitors are installed at elevated positions or on the ground, to measure the radioactivity in the air and on the surface of the earth, respectively. Furthermore, samples of rainwater and river water are taken at certain intervals.

For automated routine measurements usually the γ activity is determined by means of Geiger–Muller or scintillation counters. Additional measurements are made by filtering off aerosols from the air and by taking samples of water, soil or plants. The samples may be dried or processed by chemical methods for more detailed investigations. The α, β or γ activities on the aerosol filters and in the processed samples are measured by means of proportional counters or α and γ spectrometers, by which the radionuclides can be identified.

The results of routine measurements at various stations are evaluated and can be fed into a computerized network in which all the data are collected. By suitable programs such as IMIS (integrated measuring and information system for the surveillance of environmental radioactivity) the data can be evaluated further to give, at any time, an up-to-date overview of the radioactivity in various regions of the country.

The principle objective of radiation protection and monitoring is the achievement of appropriate safety conditions with respect to human exposure. Consequently, radionuclides in those parts of the environment which are of immediate influence on human life (e.g. air, foodstuffs or surroundings) are taken into account by safety regulations and monitored. However, the effects of radionuclides in other parts of the environment, particularly in other living things, have not been investigated in detail and the radioactivity in these parts is not kept under surveillance, even though radioactive fall-out has widely been distributed in the natural environment and radioactive waste has been disposed of in remote areas, above all into the deep sea. Therefore, radiological protection and monitoring of the whole natural environment has found increasing interest in recent years.

Literature

Dosimetry

A. H. Snell (Ed.), Nuclear Instruments and their Uses, Wiley, New York, **1962**
F. H. Attix, W. C. Roesch, E. Tochilin (Eds.), Radiation Dosimetry, 2nd ed., Vols. I, II, III and Supplement I, Academic Press, New York, **1966–1972**

J. R. Cameron, N. Suntharalingham, G. N. Kenney, Thermoluminescent Dosimetry, University of Wisconsin Press, Wisconsin, **1968**

N. E. Holm, R. J. Berry (Eds.), Manual on Radiation Dosimetry, Marcel Dekker, New York, **1970**

United Nations, Radiation, Doses, Effects, Risks, United Nations Environment Programme (UNEP), Nairobi, Kenya, **1985**

J. R. Greening, Fundamentals of Radiation Dosimetry, Hilger, London, **1989**

Health Effects and Radiation Protection

B. T. Price, C. C. Horton, K. T. Spinney, Radiation Shielding, Pergamon, Oxford, **1957**

J. C. Rockley, An Introduction to Industrial Radiology, Butterworths, London, **1964**

K. Z. Morgan, J. E. Turner (Eds.), Principles of Radiation Protection, Wiley, New York, **1967**

W. S. Snyder et al.(Eds.), Radiation Protection, Parts 1 and 2, Pergamon, Oxford, **1968**

H. Cember, Introduction to Health Physics, Pergamon, Oxford, **1969**

Basic Radiation Protection Criteria, National Council on Radiation Protection, Report 39, NCRP Publications, Washington, DC, **1971**

H. Kiefer, R. Maushart, Radiation Protection Measurement, Pergamon, Oxford, **1972**

Recommendation of the International Commission on Radiological Protection (ICRP), ICRP Publ. No. 26, Pergamon, Oxford, **1977**

A. Martin, S. A. Harbisson, An Introduction to Radiation Protection, 2nd ed., Chapman and Hall, London, **1979**

International Atomic Energy Agency, Nuclear Power, the Environment and Man, IAEA, Vienna, **1982**

United Nations, Ionizing Radiation: Sources and Biological Effects, United Nations, New York, **1982**

H. Aurand, I. Gans, H. Rühle, Radioökologie und Strahlenschutz, Erich Schmidt Verlag, Stuttgart, **1982**

E. Pochin, Nuclear Radiation; Risks and Benefits, Clarendon Press, Oxford, **1983**

O. F. Nygaard, M. G. Simic (Eds.), Radioprotectors and Anticarcinogens, Academic Press, New York, **1983**

E. J. Hall, Radiation and Life, 2nd ed., Pergamon, Oxford, **1984**

C. von Sonntag, The Chemical Basis of Radiation Biology, Taylor and Francis, Philadelphia, **1987**

National Council of Radiation Protection, Report No. 94, 95, NCRP Publications, Washington, DC, **1987**

G. Kraft, Radiobiological Effects of very Heavy Ions: Inactivation, Induction of Chromosome Aberrations and Strand Breaks, Nucl. Sci. Appl. *3*, 1 **(1987)**

M. E. Burns (Ed.), Low-Level Radioactive Waste Regulation; Science, Politics and Fear, Lewis, Chelsea, MI, **1988**

International Commission on Radiological Protection, Recommendations, Pub. 60, Annals of the ICRP, *20*(4), Pergamon Press, Oxford **1990**

International Commission on Radiological Protection, Risks Associated with Ionizing Radiation, Annals of ICRP, *22*(1), **1991**

National Council on Radiation Protection and Measurements, Evaluation of Risk Estimates for Radiation Protection Purposes, Report *115*, NCRP, Bethesda, MD, **1993**

Protection against Radon-222 at Home and at Work, ICRP Publ. No. 65, Pergamon, Oxford, **1993**

A. Kaul, W. Kraus, A. Schmitt-Hannig, Exposure of the Public from Man-made and Natural Sources of Radiation, Kerntechnik *59*, 98 **(1994)**

A. Kaul, Present and Future Tasks of the International Commission on Radiological Protection, Kerntechnik *59*, 238 **(1994)**

European Commission, Proc. Int. Symp. Remediation and Restoration of Radioactive-contaminated Sites in Europe, Antwerp, Oct. **1993**, EC Doc. XI-5027/94, **1994**

W. Nimmo-Scott, D. J. Golding (Eds.), Proc. 17th IRPA Regional Congress on Radiological Protection, Portsmouth, June **1994**, Nuclear Technol. Publ., Ashford, Kent, UK, **1994**

International Atomic Energy Agency, Advances in Reliability Analysis and Probabilistic Safety Assessment for Nuclear Power Reactors, IAEA Tec. Doc. No. 737, Vienna, **1994**

International Atomic Energy Agency, Assessing the Radiological Impact of Past Nuclear Activities and Events, IAEA Tec. Doc. No. 755, Vienna, **1994**

G. van Kaick, A. Karaoglou, A. M. Kellerer (Eds.), Health Effects of Internally Deposited Radionuclides: Emphasis on Radium and Thorium, EUR 15877 EN, World Scientific, Singapore, **1995**
Standards of Protection against Radiation, Title 10, Code of Federal Regulations, Part 20 (published annually)
International Radiation Protection Association (IRPA), Proceedings of the 1996 International Congress on Radiation Protection, Vol. *1–4*, Vienna, **1996**
International Atomic Energy, One Decade after Chernobyl: Summing up the Consequences of the Accident, IAEA, Vienna **1997**

Monitoring

N. A. Chieco, D. C. Bogan (Eds.), Environmental Measurements Laboratory (EML), Procedures Manual (HASL-300), 27th ed., Vol. 1, US Department of Energy, New York, **1992**
Reaktorsicherheit, Meßanleitungen für die Überwachung der Radioaktivität in der Umwelt und zur Erfassung radioaktiver Emissionen und kerntechnischer Anlagen, Gustav Fischer, Stuttgart, **1992**
A. Bayer, H. Leeb, W. Weiss, Measurement, Assessment and Evaluation within an Integrated Measurement and Information System for Surveillance and Environmental Radioactivity (IMIS); in: Proc. Topical Meeting on Environmental Transport and Dosimetry, Charleston, SC, Sept. **1993**, p. 108.
W. Weiss, H. Leeb, IMIS – The German Integrated Radioactivity Information and Decision Support System, Rad. Prot. Dosimetry *50*, 163 **(1993)**
R. J. Pentreath, A System for Radiological Protection of the Environment: Some Initial Thoughts and Ideas, J. Radiol. Prot. *19*, 117 **(1999)**

Appendix

Glossary

Activation: Production of radionuclides by nuclear reactions

Artificial elements: Man-made elements produced by nuclear reactions

Autoradiography: Picture produced by nuclear radiation in photographic films or plates

Becquerel: Unit of (radio)activity (abbr. Bq; $1 \, Bq = 1 \, s^{-1}$)

Carriers: Isotopic or non-isotopic atoms or molecules exhibiting the same or very similar chemical properties to the radioactive tracer present in the system

Counting rate: Result of an activity measurement by means of a detector and counter

Cross section: Measure of the probability of a nuclear reaction $[cm^2, m^2 \text{ or barn(b)}]$

Curie: Former unit of radioactivity (abbr. Ci; defined as activity of $1 \, g \, ^{226}Ra$; $1 \, Ci = 3.7 \cdot 10^{10} \, s^{-1}$)

Decay: See radioactive decay

Decay constant: Measure of the probability of the decay of radioactive atoms $[s^{-1}]$

Decay series: Sequence of successive decay processes

Disintegration: Synonym of (radioactive) decay

Disintegration rate: Disintegrations per unit time $[s^{-1}]$

Elementary particles: Great number of particles (including protons, neutrons, electrons and many others) that have been considered to be the most simple forms of matter (many are very short-lived)

Emanation: Gaseous products of radioactive decay, in particular radon

Excitation functions: Cross sections of nuclear reactions as a function of the energy of the projectiles

Fundamental particles: Three groups of particles (quarks, leptons and mesons) not exhibiting inner structure, regarded as the fundamental constituents of all matter (many of them are not able to exist in the free state)

Half-life: Time in which the activity decreases to one-half of its initial value

Heavy ions: Ions heavier than $^4He^{2+}$, in particular ions of intermediate or high mass numbers A

Hot atoms: High-energetic atoms produced by radioactive decay or nuclear reactions

Hot-atom chemistry: Chemical effects of nuclear transformations (radioactive decay or nuclear reactions)

Internal conversion: Emission of an electron instead of a γ-ray photon

Isobars: Atoms with the same mass number $A = Z + N$

Isotones: Atoms with the same number N of neutrons, but different atomic number Z

Isotopes (Greek: at the same place): Atoms with the same atomic number Z containing different numbers N of neutrons

Isotope effects: Influence of different mass numbers A of atoms of the same atomic number Z on kinetics, chemical equilibria and spectroscopic properties

Isotope exchange: Chemical reactions in which isotopes are exchanged between different chemical species

Labelled compounds: Compounds in which one or several atoms are substituted by radioactive or stable isotopes

Mass number: Sum of the number of protons (equal to the atomic number Z) and the number N of neutrons, $A = Z + N$

Nuclear isomers: Atoms with the same atomic number Z and the same mass number A in different states of excitation, the higher states being metastable with respect to the ground state

Nuclear radiation: Radiation emitted by nuclei, in particular α particles, electrons (β^- particles), positrons (β^+ particles) and photons (γ rays)

Nucleons: Protons and neutrons

Nuclides: Any kind of atoms that may differ in atomic number Z and/or neutron number N from other atoms

Radiation chemistry: Radiation-induced chemical reactions, either by nuclear radiation, by X rays or by UV or visible light (photochemistry)

Radioactive contamination: Contamination with radioactive matter

Radioactive decay: Change of unstable atomic nuclei into other stable or unstable nuclei, associated with emission of nuclear radiation

Radioactive equilibria: Definite ratios between the activities of mother and daughter nuclides, given by their decay constants

Radioactive fall-out: Radionuclides, in particular fission products and actinides, deposited from the air

Radioactivity (activity): Property of matter exhibiting (radioactive) decay or isomeric transition of atomic nuclei and emission of nuclear radiation [$Bq = s^{-1}$]

Radioanalysis: Analysis by means of radioactive atoms (radionuclides)

Radiocolloids: Colloids (i.e. matter in the colloidal state) consisting of the radioactive matter considered (intrinsic colloids) or containing microamounts of radioactive matter (carrier colloids)

Radioelements: Elements existing only in the form of radionuclides but not in a stable form

Radionuclides: Any kind of unstable (radioactive) atoms that may differ in atomic number Z and/or neutron number N from other atoms

Radionuclide generators: Set-up to separate repeatedly short-lived daughter nuclides from longer-lived mother nuclides by chemical methods

Radiotoxicity: Toxicity of radionuclides

Radiotracers (also called radioactive indicators): Traces of radionuclides (or labelled compounds) added to follow (trace) the fate of elements or compounds

Specific activity: (Radio)activity per mass unit [Bq/g or Bq/mol]

Transition, in particular γ transition: Change of the excited state of nuclei to the ground state or a lower excited state

Transmutation: Change of atomic nuclei into other nuclei by radioactive decay or nuclear reactions (see also radioactive decay)

Fundamental Constants

(In parentheses: uncertainties in the last digits)
Velocity of light in vacuo: $c = 2.997924580(12) \cdot 10^{-8} \, \mathrm{m \, s^{-1}}$
Avogadro number: $N_{Av} = 6.022045(31) \cdot 10^{23} \, \mathrm{mol^{-1}}$
Planck constant: $h = 6.626176(36) \cdot 10^{-34} \, \mathrm{J \, s}$
Boltzmann constant: $k = 1.380662(44) \cdot 10^{-23} \, \mathrm{J \, K^{-1}}$
Gas constant: $R = 8.314409(45) \, \mathrm{J \, K^{-1} \, mol^{-1}}$
Faraday constant: $F = 9.648456(27) \cdot 10^4 \, \mathrm{C \, mol^{-1}}$
Electron charge: $e = 1.602191(7) \cdot 10^{-19} \, \mathrm{A \, s}$
Atomic mass unit (amu): $u = 1.6605655(86) \cdot 10^{-27} \, \mathrm{kg}$
Electron rest mass: $m_e = 5.4858026(21) \cdot 10^{-4} \, \mathrm{u}$
Hydrogen atom rest mass: $m_H = 1.007825037(10) \, \mathrm{u}$
Neutron rest mass: $m_n = 1.008665012(37) \, \mathrm{u}$
Bohr magneton: $\mu_B = 9.274078(36) \cdot 10^{-24} \, \mathrm{J \, T^{-1}}$
Nuclear magneton: $\mu_N = 5.050824(20) \cdot 10^{-27} \, \mathrm{J \, T^{-1}}$

Conversion Factors

Energy: $1 \, \mathrm{eV} = 1.6021892(46) \cdot 10^{-19} \, \mathrm{J}$
$\qquad\qquad = 1.6021892(46) \cdot 10^{-12} \, \mathrm{erg}$
$\qquad\qquad = 3.829324(28) \cdot 10^{-20} \, \mathrm{cal}$
$\qquad 1 \, \mathrm{J} = 1 \, \mathrm{kg \, m^2 \, s^{-2}} = 10^7 \, \mathrm{erg}$
Energy equivalent of $1 \, \mathrm{u}$: $931.5016(26) \, \mathrm{MeV}$
Photon wavelength associated with $1 \, \mathrm{eV}$: $1.239852(3) \cdot 10^{-6} \, \mathrm{m}$
Power (watt): $1 \, \mathrm{W} = 1 \, \mathrm{J \, s^{-1}}$
Temperature corresponding to $1 \, \mathrm{eV}$: $1.160450(36) \cdot 10^4 \, \mathrm{K}$
Pressure (pascal): $1 \, \mathrm{Pa} = 1 \, \mathrm{kg \, m^{-1} \, s^{-2}}$
$\quad 1 \, \mathrm{atm}$ (standard atmosphere) $= 1.01325 \, \mathrm{bar} = 1.01325 \cdot 10^5 \, \mathrm{Pa}$
Electric charge (coulomb): $1 \, \mathrm{C} = 1 \, \mathrm{A \, s}$
Magnetic flux density (tesla): $1 \, \mathrm{T} = 1 \, \mathrm{V \, s \, m^{-2}}$

Relevant Journals

Radiochimica Acta, R. Oldenbourg Verlag, Munich, Germany
Journal of Radioanalytical and Nuclear Chemistry, Elsevier Sequoia, Lausanne, and
    Akadémiai, Budapest
Applied Radiation and Isotopes, Pergamon Press, Oxford
Radioactivity and Radiochemistry, Caretaker Communications, Atlanta, GA
Soviet Radiochemistry, Consultant Bureau, New York and London
Nuclear Technology, American Nuclear Society, Chicago
Radioactive Waste Management and the Nuclear Fuel Cycle, Harwood, Yverdon
Nuclear Engineering International, Quadrant Services, Haywards Heath, UK
Kerntechnik, Thiemig Verlag, Munich

Journal of Labelled Compounds and Radiopharmaceuticals, London
The Journal of Radiation Physics and Chemistry, Pergamon Press, Oxford (regularly contains survey papers on radiation chemistry)
Journal of Environmental Radioactivity, Pergamon Press, Oxford
Health Physics, Williams and Wilkins, Baltimore
International Atomic Energy Agency Bulletin – and other publications, IAEA, Vienna

Name Index

Subject Index